Y0-BGD-372

# NITROGEN METABOLISM IN MAN

*International Symposium organised by The Rank Prize Funds and held at Kingston, Jamaica, on 21–25 November, 1980*

# NITROGEN METABOLISM IN MAN

Edited by

## J. C. WATERLOW and J. M. L. STEPHEN

*Department of Human Nutrition,*
*London School of Hygiene and Tropical Medicine, London, UK*

## APPLIED SCIENCE PUBLISHERS
LONDON and NEW JERSEY

APPLIED SCIENCE PUBLISHERS LTD
Ripple Road, Barking, Essex, England

APPLIED SCIENCE PUBLISHERS INC.
Englewood, New Jersey 07631, USA

**British Library Cataloguing in Publication Data**

Nitrogen metabolism in man.
1. Metabolism—Congresses
2. Nitrogen metabolism—Congresses
I. Waterlow, J.C.     II. Stephen, J.M.L.
612'.398     QP171

ISBN 0-85334-991-6

WITH 108 TABLES AND 82 ILLUSTRATIONS

© APPLIED SCIENCE PUBLISHERS LTD 1981

Printed in Great Britain by Galliard (Printers) Ltd, Great Yarmouth

# Foreword

The Rank Prize Funds were created by the late Lord Rank shortly before his death to encourage further progress in the acquisition of knowledge in the fields of human nutrition and crop husbandry on the one hand, and opto-electronics on the other hand. One of the stimuli directed towards attaining these objectives is the award of prizes for significant advances, and another is the holding of international symposia by which a selected group of professionals is secluded by residence for a few days in conditions conducive to promoting good fellowship and free scientific discussion.

Human nutrition is not and never has been one of the more glamorous research subjects. The centres of activity, therefore, are harder to find and those so employed tend to have a high degree of dedication. One such centre is the Tropical Metabolism Research Unit of the University of the West Indies situated on the University campus at Mona, on the outskirts of Kingston, Jamaica. This Unit was established in 1954 by the British Medical Research Council with the objective of acquiring a better understanding of the metabolic changes which occur in malnourished children. The first and founding Director was Professor John C. Waterlow. It was inevitable that the principal impact of the work of the Unit would be in Jamaica, but to ensure a more regional approach, there was established, in 1956, an Advisory Committee for Medical Research in the British Caribbean. In line with the various political changes that subsequently occurred, this Committee was superseded by the Commonwealth Caribbean Medical Research Council.

For the last ten years, the Unit has formally been part of the University of the West Indies with the withdrawal of the UK Medical Research Council and with Professor Waterlow then being replaced as Director by Professor David Picou who in turn was succeeded by Dr Alan Jackson. In view of the excellent work carried out by the Unit, it seemed fitting that note should be taken of the 25th anniversary of its foundation.

The Trustees of the Rank Prize Funds willingly responded to the recommendation of the Funds' Advisory Committee on Nutrition and Crop Husbandry that this anniversary should be marked by the holding of a scientific symposium on the subject of nitrogen metabolism in man. This took place in Jamaica from 21st to 25th November, 1980. This volume records the proceedings of the symposium. The initiative for this meeting was taken by Professor John C. Waterlow, who has been responsible for starting useful and important work on protein malnutrition and protein metabolism. He has trained and influenced a considerable number of people in the Caribbean, and his activities and those of his colleagues have been important for the development of medical research in that part of the world.

This was the fourth symposium organized by the Advisory Committee and the high standard set on previous occasions was indeed enhanced by the quality of the contributions and the discussions during this meeting. One can feel confident that by this means the concern felt by Lord Rank for improving the well-being of mankind is receiving support of a constructive nature.

Sir William Henderson
*Chairman, Advisory Committee on Nutrition*
    *and Crop Husbandry*
*Rank Prize Funds*
*London, UK*

# Introduction

A symposium on nitrogen metabolism seemed timely at the beginning of the 1980s because in the last few years this subject has been developing rapidly and will undoubtedly continue to expand. One main idea running through these discussions is that although a great deal is known about the biochemical pathways of amino acid and protein metabolism, our knowledge of the rates at which the various reactions and interconversions occur in the living subject is still very inadequate. We have a reasonably good map, but we do not know much about the flow of traffic along the different routes. A further problem is how the rates of flow are organized and controlled. Many of the papers in this book touch on that subject. Perhaps it will merit another symposium in a few years' time.

A second theme of the meeting is the subject of protein turnover in the whole animal and in man. It is a curious fact that after the pioneer work with $^{15}$N-labelled amino acids in the period from 1947 to 1960, the stimulus for which can be traced back to Schoenheimer, interest lapsed and for nearly ten years virtually no papers on this subject appeared. A new impetus was provided by the clinical problem of protein malnutrition in children, and the hypothesis that the machinery for synthesizing protein might be impaired, perhaps irreversibly. The work that was undertaken at the Tropical Metabolism Research Unit of the University of the West Indies to test this idea was tentative and unsophisticated by modern standards. Nevertheless, it aroused some interest, and a growing number of scientists are now studying the turnover of protein in the whole body from different points of view and with a range of new methods.

At this stage in the development of the subject it was important to have an opportunity to exchange information, to discuss critically the underlying theories and assumptions, and to examine the new possibilities being opened up by the advances in instrumentation. I speak for all the participants in thanking the Rank Prize Funds for making such discussions possible. Those of us who have worked in the TMRU in the past, or are working there now, appreciate very much the decision to hold the meeting in Jamaica, to coincide with the Unit's 25th anniversary. Indeed, Drs Jackson and Golden provided the initial ideas for the planning of the symposium. Its success owes much to the thorough and careful organization of Dr M. H. N. Golden.

J. C. WATERLOW

# Contents

## Part III: Turnover Methods

## Part IV: Turnover in Various States

# PART I

# Free Amino Acids and their Metabolism

# 1

# Free Amino Acid Pools and their Regulation

J. C. Waterlow and E. B. Fern

*Department of Human Nutrition,*
*London School of Hygiene and Tropical Medicine, London, UK*

Since free amino acids are the 'units of currency' of protein metabolism, they provide a logical starting point for the discussions in this symposium. Ten years ago Munro (1970) put forward five propositions about free amino acid pools, all of which have stood the test of time. These are:

1.  Free amino acids represent a very small proportion of the total amounts of amino acids in the body. Munro calculated that in the rat free amino acids account for only $0.5\%$ of the total.
2.  Most free amino-N comes from the four non-essential amino acids—alanine, glutamic acid, glutamine and glycine.
3.  Half or more of the free pool of each amino acid is present in skeletal muscle.
4.  The concentrations of amino acids, especially of the non-essentials, are higher in cells than in plasma.
5.  The relative amounts of the different free amino acids bear little relation to the average composition of tissue proteins, the amino acid pattern of dietary protein or the known requirements for essential amino acids.

It is quite difficult to attach precise numbers to these propositions because of wide variations in the results obtained in the same tissue by different authors. For example, if we take the eight essential amino acids in the free pool of muscle, the proportions found by four different authors range from 5 to 25 % of total essentials for histidine and from 1.5 to 19 % for leucine (Waterlow *et al.*, 1978). The variation is even greater for the non-essentials, particularly alanine and glutamine, as one might expect from their active role in the transport of amino-N (Felig, 1973). Therefore in what follows I shall confine myself mainly to the essential amino acids (EAA), considering first patterns and then some quantitative aspects.

## Patterns of Free Essential Amino Acids

In spite of the variability mentioned above, it seems possible to define certain consistencies of pattern. In doing this, it is simpler to look at rank orders than actual concentrations.

Table 1 shows the rank order of the concentrations of EAA in plasma in rat and man. There is a fairly close correspondence, which is statistically significant. The total concentration of these eight EAA is about three times as great in rat as in man (Table 2). One may speculate whether this is related to the more rapid rate of protein turnover in the rat. The EAAs are divided here into four groups. The large share taken by lysine + threonine in the rat is remarkable. As we shall see, this pattern is quite consistent.

Tables 1 and 2 also show a comparison between rat and man of the EAA concentrations in muscle. In what follows most emphasis will be placed on muscle, because, as Munro pointed out, it contains the largest proportion of the body's free pool. As with plasma, the total concentration in man is

TABLE 1

RANK ORDER OF CONCENTRATIONS OF FREE ESSENTIAL AMINO ACIDS IN PLASMA AND MUSCLE OF RAT AND MAN

|  | Plasma | | Muscle | |
|---|---|---|---|---|
|  | Rat[a] | Man[b] | Rat[c] | Man[d] |
| Threonine | 1 | 3 | 2 | 2 |
| Lysine | 2 | 2 | 1 | 1 |
| Valine | 3 | 1 | 4 | 4 |
| Leucine | 4 | 4 | 5 | 5 |
| Isoleucine | 5 | 6 | 7 | 6 ⎫ |
| Methionine | 6 | 8 | 6 | 6 ⎭ |
| Histidine | 7 | 5 | 3 | 3 |
| Phenylalanine | 8 | 7 | 8 | 8 |

[a] Lunn *et al.* (1976).
[b] Pozefsky *et al.* (1969).
[c] Data of Fern (unpublished).
[d] Bergström *et al.* (1974).
For plasma, Spearman rank correlation coefficient = 0.79; $0.02 > P > 0.01$.
For muscle, order is identical.

TABLE 2

TOTAL AND RELATIVE AMOUNTS OF FREE ESSENTIAL AMINO ACIDS IN
PLASMA ($\mu$MOL/ML) AND MUSCLE ($\mu$MOL/ML INTRACELLULAR WATER)

| | Plasma | | Muscle | |
| | Rat[a] | Man[b] | Rat[c] | Man[d] |
|---|---|---|---|---|
| Total essentials | 2·25 | 0·87 | 11·55 | 3·25 |
| % of total: | | | | |
|   Lysine + threonine | 63 | 35 | 68 | 67 |
|   BCAA | 26 | 48 | 14 | 16 |
|   Histidine | 3·6 | 9 | 13 | 11·5 |
|   Methionine + phenylalanine | 7·3 | 8 | 5 | 5·5 |

[a] From Lunn *et al.* (1976).
[b] From Pozefsky *et al.* (1969).
[c] Data of Fern (unpublished), recalculated on the assumption that extracellular water of muscle = 0·2 ml/g and intracellular water = 0·6 ml/g.
[d] From Bergstrom *et al.* (1974).

considerably less than half that in the rat. However, in muscle the rank order in the two species is identical and the relative amounts of the different amino acids are almost the same.

## Comparisons Between Tissues

The experiments of Lunn *et al.* (1976) provide values for free amino acid concentrations in plasma, liver and muscle of normal rats. Their findings are shown in Table 3. The correspondence in rank order between the tissues,

TABLE 3

RANK ORDER OF CONCENTRATIONS OF FREE ESSENTIAL AMINO ACIDS
IN PLASMA, LIVER AND MUSCLE OF RAT (LUNN *et al.*, 1976)

| | Plasma | Liver | Muscle |
|---|---|---|---|
| Threonine | 1 | 1 | 1 |
| Lysine | 2 | 2 | 3 |
| Valine | 3 | 4 | 4 |
| Leucine | 4 | 6 | 6 |
| Isoleucine | 5 | 7 | 7 |
| Methionine | 6 | 5 | 5 |
| Histidine | 7 | 3 | 2 |
| Phenylalanine | 8 | 8 | 8 |
| Total ($\mu$mol/ml or $\mu$mol/g) | 2·25 | 8·36 | 8·26 |
| Agreement of rank order significant ($P < 0·01$) | | | |

TABLE 4

RANK ORDER OF CONCENTRATIONS AND TOTAL AND RELATIVE
AMOUNTS OF FREE ESSENTIAL AMINO ACIDS IN LIVER OF RATS ON
NORMAL AND LOW-PROTEIN DIETS (LUNN *et al.*, 1976)

|  | Normal | Low protein |
|---|---|---|
| Threonine | 1 | 1 |
| Lysine | 2 | 2 |
| Histidine | 3 | 3 |
| Valine | 4 | 5 |
| Methionine | 5 | 4 |
| Leucine | 6 | 6 |
| Isoleucine | 7 | 7 |
| Phenylalanine | 8 | 8 |
| Total: ($\mu$mol/g) | | |
| essentials (E) | 8·4 | 5·0 |
| non-essentials (NE) | 17·1 | 21·0 |
| Ratio: NE/E | 2·03 | 4·2 |
| % of total essentials: | | |
| Lysine + threonine | 67 | 71 |
| BCAA | 14 | 9 |
| Histidine | 11 | 11 |
| Methionine + phenylalanine | 7·5 | 9 |

although not exact, is statistically highly significant (Friedmann's test). All these results show the same general pattern: threonine and lysine come at the top, followed by the branched chain amino acids (BCAAs)—valine, leucine and isoleucine, almost invariably in that order. At the end come methionine and phenylalanine. Histidine is by far the most variable.

**Different Nutritional States**
The pattern seems to be maintained even under nutritional stress. Some experiments of Lunn *et al.* (1976) compared free amino acid concentrations in plasma, liver and muscle in control and protein-deficient rats. Table 4, from their data, shows the rank order in liver: it remains the same in the protein-deficient animals. The table also shows the relative and absolute amounts of EAA in the liver on these regimes. It illustrates some points already well known—the fall in total essentials compared with a rise in non-essentials, and the relative decrease in the BCAA.

In a study by Fern (unpublished) free amino acid concentrations were measured in two muscles from six groups of rats: weight controls, age controls, rats starved for 2 days, rats on 3·5% protein or a protein-free diet

TABLE 5

RANK ORDER OF CONCENTRATIONS AND TOTAL AND RELATIVE AMOUNTS OF
FREE ESSENTIAL AMINO ACIDS IN MUSCLE FROM RATS UNDER DIFFERENT
DIETARY CONDITIONS (FERN (UNPUBLISHED))

| | *Control* | *Fasted* | *Protein-free* |
|---|---|---|---|
| Lysine | 1 | 1 | 1 |
| Histidine | 2 | 3 | 2 |
| Threonine | 3 | 2 | 3 |
| Valine | 4 | 4 | 4 |
| Methionine | 5 | 7 | 8 |
| Leucine | 6 | 5 | 5 |
| Isoleucine | 7 | 6 | 6 |
| Phenylalanine | 8 | 8 | 7 |
| Total essentials ($\mu$mol/g) | 3·99 | 3·55 | 2·11 |
| % of total: | | | |
|     Lysine + threonine | 68 | 62 | 49 |
|     BCAA | 12 | 17 | 9 |
|     Histidine | 15 | 15 | 40 |
|     Methionine + phenylalanine | 5 | 5 | 2·5 |

for 2 weeks, and rats refed after the 3·5 % protein diet. Some of the results are shown in Table 5. Even in these rather extreme dietary conditions the pattern, as shown by the rank order, remains very constant. The total concentration of EAA falls on the protein-free diet, and again the well known decrease, both relative and absolute, in the concentrations of the BCAA, phenylalanine and methionine is seen. What is less well recognized is the very large relative increase in histidine, mainly at the expense of threonine. The data of Lunn *et al.* (1976) on muscle show a similar rise in histidine concentration in protein-deficient rats. It does not occur in starvation. One may speculate whether the two basic amino acids, lysine and histidine, are playing a role as intracellular cations.

**Concentration Ratio**
Table 6, from Fern's data, shows the distribution ratios, in intracellular compared to extracellular water, of the EAA in muscle of normal and protein-deficient rats. The amino acids with the largest pool sizes—histidine, lysine and threonine—have the highest ratios. These large concentration differences are maintained by active transport. The BCAA have ratios of 1·5–2·5. On the protein-free diet there is a small fall in the ratio for most amino acids, except for lysine, which is reduced by 50 %.

The non-essentials in general have much higher concentration ratios,

TABLE 6

DISTRIBUTION RATIOS[a] OF ESSENTIAL AMINO ACIDS IN MUSCLE OF
NORMAL AND PROTEIN-DEFICIENT RATS (FERN (UNPUBLISHED))

|  | *Normal* (N) | *Protein-deficient* (PD) | *PD/N* (%) |
|---|---|---|---|
| Histidine | 8·3 | 7·9 | 95 |
| Isoleucine | 2·5 | 2·1 | 81 |
| Leucine | 1·6 | 1·3 | 79 |
| Lysine | 4·7 | 2·4 | 52 |
| Methionine | 3·3 | — | — |
| Phenylalanine | 1·8 | 1·3 | 74 |
| Threonine | 7·3 | 6·5 | 88 |
| Valine | 1·8 | 1·35 | 75 |

[a] Distribution ratio = intracellular/extracellular concentration.

ranging in our material from 9 for alanine to 50 for aspartic acid. Presumably in some cases the reason is that the amino acids are formed from Krebs cycle intermediates in the cell.

## Distribution of Free and Protein-bound Amino Acids

As Munro pointed out, the pattern of free amino acids bears no relation to that in protein. Table 7 shows the rank order of free and protein-bound EAA in rat muscle. There is absolutely no correlation.

Table 8 shows the ratios of the concentrations of protein-bound to free amino acids in muscle. The ratios for the EAA vary over a 20-fold range. This difference in composition between free and protein-bound pools has implications for the measurement of protein synthesis. It means that one cannot treat the free pool as a whole as if it were the precursor pool. The amino acids used for protein synthesis represent a selection of those in the free pool.

How are these differences in concentration between free and bound pools maintained? We have tried to show that although the free EAA pool is small, the pattern of its composition is relatively constant. Therefore, although the values in Table 8 may not be numerically very accurate, these huge differences in the ratio of protein-bound to free amino acids cannot be attributed to chance.

In general, the pattern of amino acids coming in from the food is broadly similar to that in the tissue proteins, and hence similar to the pattern of the

TABLE 7

RANK ORDER AND RELATIVE CONCENTRATIONS OF FREE
AND PROTEIN-BOUND ESSENTIAL AMINO ACIDS IN RAT
MUSCLE (FERN (UNPUBLISHED))

|  | Rank order | |
|  | Free | Bound |
| --- | --- | --- |
| Lysine | 1 | 4 |
| Threonine | 2 | 3 |
| Histidine | 3 | 8 |
| Valine | 4 | 2 |
| Leucine | 5 | 1 |
| Methionine | 6 | 7 |
| Isoleucine | 7 | 5 |
| Phenylalanine | 8 | 6 |
| % of total: | | |
| Lysine + threonine | 68 | 25 |
| BCAA | 13·5 | 52 |
| Histidine | 13·5 | 5·5 |
| Methionine + phenylalanine | 5 | 17·5 |

No significant correlation in rank order.

amino acid mixture being synthesized to protein or derived from it by degradation. In this situation, the only way in which a different pattern can arise in the free amino acid pool is by differences in the relative rates of disposal by oxidation. Krebs (1972) has pointed out that because of the values of their Michaelis constants, the activity of the oxidative enzymes is sensitive to substrate concentration. Presumably then, if a substrate has a high steady state concentration, e.g. lysine and threonine, this means that

TABLE 8

RELATIVE CONCENTRATIONS (PROTEIN-BOUND:FREE) OF AMINO
ACIDS IN RAT MUSCLE (FERN (UNPUBLISHED))

| Essentials | | Semi- and non-essentials | |
| --- | --- | --- | --- |
| Histidine | 67 | Arginine | 269 |
| Isoleucine | 306 | Tyrosine | 266 |
| Leucine | 556 | Alanine | 40 |
| Lysine | 31 | Aspartic acid (+amide) | 97 |
| Methionine | 225 | Glutamic acid (+amide) | 15 |
| Phenylalanine | 646 | Glycine | 60 |
| Threonine | 31 | Serine | 38 |
| Valine | 272 | | |

TABLE 9

REPRESENTATIVE VALUES OF MICHAELIS CONSTANTS OF
SOME AMINO ACID DEGRADING ENZYMES IN RAT LIVER
(KREBS, 1972)

|                                   | $K_m$ (mM) |
| --------------------------------- | ---------- |
| Lysine-1-oxoglutarate reductase   | 18         |
| Serine (threonine) dehydratase    | 52         |
| Phenylalanine hydroxylase         | 0·66       |
| Tryptophan pyrrolase              | 0·15       |

the activity of its oxidative system must be relatively low, and vice versa. The under-representation of the BCAA in the free pool could be attributed to a high activity of their oxidative enzymes. Perhaps this idea is borne out by the values for $K_m$ shown in Table 9. It should also be noted that lysine and threonine are the two essential amino acids which apparently do not undergo transamination (Weissman and Schoenheimer, 1941; Elliott and Neuberger, 1950).

Perhaps these speculations may help to explain how this strange pattern of the free EAA pool is maintained. Having been trained as a physiologist I always like to ask 'Why?', but that is an even more difficult question to answer.

## Dynamic Behaviour of the Free Amino Acid Pool

**Turnover Rate**

Munro (1970) calculated that in the rat free amino acids accounted for 0·5 % of total amino acids in the whole body. Our calculations would put the figure at more like 1 %, but the difference is unimportant. From known values for whole body protein flux in the rat, say 60 g/kg/day (Waterlow *et al.*, 1978), it can easily be calculated that the free pool as a whole is being renewed about once every hour. It would be better to call it a pipe, through which there is a rapid flow, than a pool according to the conventional model.

This calculation is based on the contributions to the flux from protein breakdown and from the food, and ignores the unknown amount of recycling of amino-N in newly synthesized non-essential amino acids. It may be more illuminating to consider the essential amino acids alone. Table 10 shows such a calculation for the turnover rate in the whole body of the EAA taken as a group. Although the free pool is relatively smaller in man than in the rat, it still turns over much more slowly.

TABLE 10
TURNOVER RATE OF TOTAL FREE ESSENTIAL AMINO ACID POOL
IN RAT AND MAN

| | Rat | Man |
|---|---|---|
| Total free EAA pool (mmol/kg) | $4^a$ | $2 \cdot 4^b$ |
| Flux (mmol/kg/day) | $200^c$ | $12^d$ |
| Turnover rate (%/day) | 50 | 5 |
| Half-life (min) | 20 | 200 |

[a] Data for rat muscle from Fern (unpublished).
[b] Data for human muscle from Bergström et al. (1974).
[c] EAAs taken as 40% of total protein, flux as 60g protein/kg/day.
[d] EAAs taken as 40% of total protein, flux as 4g protein/kg/day.

This overall picture conceals differences between individual amino acids. In man, using the data of Bergström et al. (1974) for the free amino acid concentrations in muscle, we can calculate that the half-life of free leucine would be about 45 min and that of lysine about 10 h. This latter figure agrees with our finding many years ago, that in man with constant infusion of $^{14}$C-lysine it took well over a day to reach plateau specific activity (Waterlow, 1967). The overall picture also conceals differences between tissues. Table 11 shows some calculated half-lives for free leucine and lysine in liver and muscle of the rat.

**Uptake from the Free Amino Acid Pool**
It is perhaps of interest to compare the supply of amino acids brought to the tissues by the blood with the amounts needed to maintain protein synthesis. If cardiac output in adult man is assumed to be 5 litres/min, and only free

TABLE 11
ESTIMATED HALF-LIVES OF FREE LEUCINE AND LYSINE IN
LIVER AND MUSCLE OF THE RAT

| | Half-life (min) | | Turnover rate (%/day) |
|---|---|---|---|
| | Leucine | Lysine | |
| Liver | 6·5 | 50 | 60 |
| Muscle | 10·5 | 290 | 10 |

Pool sizes from Lunn et al. (1976).

amino acids in plasma are taken into account, ignoring those in red cells, then the total amount of EAA circulated to the tissues is of the order of 100 mmol/kg/day. This is some eight times the rate at which these amino acids are taken up into protein (Table 10). There may, however, be situations where there is less margin of safety. The uptake of leucine into muscle is used as an example. In man the blood flow to the total muscle mass at rest has been estimated at 1 litre/min (Samson Wright, 1971). If blood contains 0·12 mmol leucine/litre, if leucine constitutes 8 % of muscle protein, and if the protein is turning over at 2%/day (Halliday and McKeran, 1975), then it is found that the daily leucine supply in the blood (about 170 mmol) is only about 2·5 times the amount needed to maintain protein synthesis in muscle (about 70 mmol/day). It is, perhaps, worth considering whether under any conditions, e.g. low protein intake, the actual supply of amino acids to a tissue might limit the rate of protein synthesis.

It is a matter of dispute whether the precursor pool for protein synthesis is inside or outside the cell, or even in the cell membrane. If there is a barrier between the blood and the site of synthesis, then active transport across that barrier could in theory be a step which limits the rate of protein synthesis. This subject will be dealt with in more detail by Pratt (Paper 2), but one relevant point will be dealt with here.

Table 12 shows estimates of the total entry rate of EAA into the

TABLE 12

UPTAKE INTO PROTEIN COMPARED WITH RATE OF ENTRY OF ESSENTIAL AMINO ACIDS IN RAT MUSCLE

|  | *Total entry* ($E$) *(from plasma + protein breakdown)* (*nmol/g/min*) | *Uptake into protein* ($S$) (*nmol/g/min*) | $S/E$ |
|---|---|---|---|
| Histidine | 3·3 | 1·8 | 0·55 |
| Isoleucine | 7·4 | 3·5 | 0·47 |
| Leucine | 16·0 | 7·4 | 0·46 |
| Lysine | 16·9 | 3·9 | 0·23 |
| Methionine | 4·2 | 2·5 | 0·60 |
| Phenylalanine | 9·9 | 3·1 | 0·31 |
| Threonine | 18·9 | 4·1 | 0·22 |
| Valine | 12·9 | 5·6 | 0·43 |

Entry rates from plasma from Baños *et al.* (1973).
Protein breakdown taken as 7%/day.
Protein synthesis taken as 10%/day.

intracellular pool of rat muscle, calculated as the sum of values obtained by Baños *et al.* (1973) for the rate of uptake from plasma, plus the rate at which these amino acids are being discharged into the pool by protein breakdown. Synthesis is assumed to be greater than breakdown in growing rats. The ratio of synthesis rate to total entry rate suggests that for some amino acids, e.g. the BCAA and methionine, there is not a very great margin, and if there were any large increase in the net rate of synthesis, the entry rate might become limiting.

However, calculations of this kind, based on data from different sources, are never very satisfactory. The same sort of information can be obtained by comparison of intracellular and extracellular specific activities at plateau during infusion of a labelled amino acid (Aub and Waterlow, 1970). Results of such experiments suggest that in rat muscle some 20–25% of the methionine entering is taken up into protein (Millward, unpublished data). For liver the corresponding value is about 50%.

In ending I would like to emphasize the point which seems to me to stand out when we look at the free pool of essential amino acids. I find it most remarkable that, in spite of a very rapid turnover, it maintains a relatively constant pattern, which is quite different from the pattern of amino acids entering it.

## Discussion

*Munro (Tufts University, Boston, USA):* At different levels of lysine intake it has been shown by Pawlak and Pion (1968) and others that the amount of lysine in muscle increases much more steeply than in blood. O'Dell and Savage (1966) have demonstrated that the potassium content in the hen declines as the free lysine in muscle increases so that your speculation on the exchange of cations is probably substantiated by these data.

*Waterlow:* The concentration of lysine in any tissue can indeed go up or down without necessarily altering the rank order. On the other hand, concentrations expressed as percentages are very misleading because the change in concentration of one amino acid, such as histidine, can upset all the others. I am not really suggesting that the amounts remain constant, but it is this relative constancy of pattern which is extraordinary. I should like to throw out the challenge: why is there so much threonine and lysine in the free pool, and also very often so much histidine, when there is only about 2% of histidine in protein?

*Reeds* (*Rowett Research Institute, Aberdeen, UK*): Substantial amounts of histidine exist as carnosine, a dipeptide with $\beta$-alanine. 1-methyl and 3-methyl histidine also form dipeptides with $\beta$-alanine and the physiological function of these dipeptides in muscle is not known.

*Munro:* Another example is tryptophan, which is unique among free amino acids in that 80 % of 'free' tryptophan is bound to albumin, which leads to unusually high concentrations of tryptophan in plasma due to the reservoir of loosely bound tryptophan.

*Walser* (*Johns Hopkins University, Baltimore, USA*): I would like to pursue the point Waterlow has raised as to how amino acid concentrations tend to vary in parallel. We have also examined this question in a preliminary report (Walser *et al.*, 1979) and have suggested that amino acid concentrations expressed as fractions of the total circulating amino acids may be more meaningful and more useful than when they are considered in absolute terms. In many physiological as well as pathological states, it seems that fractional concentrations vary less than absolute concentrations. Do you think it is possible that what is regulated is the concentration of individual amino acids as fractions of the total? Or is this exactly the point that you were making?

*Waterlow:* What I am trying to say is that even when the total of essential amino acids is more or less halved, as in protein deficiency, the fractional concentrations seem relatively unchanged. I say relatively because of course it is well known that the branched chain amino acids go down a bit more than the others. I would agree with you, that this is a more fruitful approach.

*Harper* (*University of Wisconsin-Madison, USA*): If the $K_m$'s of degradative enzymes differ greatly, one would expect rates of degradation to change as the amino acid concentrations go up or down, so that the relative proportions of the amino acids might change.

*Waterlow:* But am I wrong in thinking that the steady state concentration of an amino acid whose degrading enzyme has a high $K_m$, for example lysine, will be greater than that of an amino acid whose enzyme has a low $K_m$, as in Table 9 of our paper?

*Harper:* No, you are right; but if concentrations fall quite low as, say, in protein deficiency, you would expect that the rate of removal by enzymes with a high $K_m$ would be relatively low, whereas the rate of removal of phenylalanine or tryptophan whose degradative enzymes have very low

$K_m$'s, would be unchanged. Thus you might get a disproportion in the amino acid concentrations.

*Waterlow:* To some extent this does indeed happen. According to my argument, the enzyme $K_m$ must be lowest for those amino acids which have relatively low concentrations in plasma, that is, the branched chain amino acids as well as phenylalanine and methionine. In protein deficiency the levels of the branched chains go much lower than those of lysine and threonine.

*Lund (Radcliffe Infirmary, Oxford, UK):* One cannot always extrapolate from $K_m$'s obtained *in vitro* to the situation in the intact cell, especially for mitochondrial enzymes which may be membrane-bound, and may show totally different properties when separated from the membrane or when the membrane configuration is changed on swelling and shrinking.

*Bier (Washington University School of Medicine, St. Louis, USA):* I would like you to comment on what seems to be an inconsistency with regard to lysine. If the fractional turnover rates of the amino acids are very fast, how then does the lysine in plasma take a day or more to reach an isotopic plateau? We have data which show that lysine flux is of similar magnitude to that measured for other amino acids such as leucine. Secondly, it seems to me that there are also serious problems in interpreting any study with a labelled amino acid which takes a day or more to reach plateau. The subject must either be fed or fasted. In the former case flux will be changing consequent to food intake and in the latter case it will change because of the metabolic adaptation to fasting. Thus, it is inconceivable that a true 'steady state' will be reached under such circumstances.

*Waterlow:* I am not defending the use of lysine for flux measurements. When we did our work 15 years ago (Waterlow, 1967), it was the only amino acid we could handle because we did not have an amino acid analyser and we had to use an amino acid which could be measured enzymatically by decarboxylation. At that time also we had no idea of the differences in behaviour of amino acids such as lysine and leucine. However, I do not think the time taken to reach plateau is inconsistent with the pool size of lysine, which is undoubtedly very big. In our data, both in rat and in man, the time taken to reach plateau with lysine is long. If you have different data for lysine, showing that it behaves like leucine, that is new to me and I cannot explain it.

*Bier:* However, if the rate constant for plasma lysine removal is very fast, as you and others have found, then one should reach plateau far sooner than in a day and a half.

*Waterlow:* But the half-life for lysine in muscle was estimated to be about 200 min in the rat, whereas that of leucine was about 50 min. The times will be even longer in man. There is a big difference between these two amino acids, which results from their different pool sizes (Waterlow *et al.*, 1978).

*Matthews (Washington University School of Medicine, St. Louis, USA):* I may have missed this, but when you were doing your calculations for the rate of supply of amino acids to the muscle did you take into consideration A–V differences, or are these just total amino acid concentrations times blood flow?

*Waterlow:* These were total concentrations times blood flow, in order to show that much more was being supplied than is actually being taken out of the blood.

*Matthews:* When you consider the forelimb or leg balance studies of Felig and colleagues (e.g. Pozefsky *et al.*, 1969; Felig and Wahren, 1971), are the A–V differences that were seen across the forearm appropriate for the uptake that you would expect?

*Waterlow:* That is a very interesting question. I tried to make such a comparison but there are difficulties: first, I do not know how much muscle there is in Felig's forearm, and secondly, the A–V differences do not exactly reflect the uptake of individual amino acids, because we know that the branched chain amino acids, after being taken up into muscle, will come out again as something quite different, such as alanine or keto acids. So I could not make this calculation from these data. All I could say in a general way is that the A–V differences seem to show that about 10 % of what is supplied to the muscle is extracted; that would fit with the figure I gave of about eight times as much being supplied as is needed.

*Felig (Yale University School of Medicine, USA):* I would agree wholeheartedly with Waterlow that the net output of amino acids from the leg or forearm of intact humans cannot be equated with net exchange across muscle tissue. Particularly in the case of leg studies involving arterial–femoral venous differences, the net exchange reflects balance across adipose tissue and skin as well as muscle. Another point: when we look at overall changes in metabolic clearance of amino acids, it is becoming apparent that some of the changes we observe in such conditions as starvation may reflect clearance in tissues which we often did not consider as being major sites of amino acid turnover, specifically adipose tissue. For example, the elevation in plasma levels of branched chain amino acids with starvation cannot be explained either by an increased outflow from liver or from muscle (see also p. 52). When one looks at the degradative enzymes there appears to be an increase in branched chain amino acid oxidation in muscle, yet an accumulation of these amino acids in plasma. On the other hand, when one looks at Goodman's data from adipose tissue, oxidation of branched chain amino acids falls there, which might explain what is happening (Frick and Goodman, 1979). Fluxes from the periphery to the liver thus represent a number of tissues which may reflect varying changes in amino acid degradation versus uptake.

# References

AUB, M. R. and WATERLOW, J. C. (1970). Analysis of a five-compartment system with continuous infusion and its application to the study of amino acid turnover, *J. Theor. Biol.*, **26**, 243–50.

BAÑOS, G., DANIEL, P. M., MOORHOUSE, S. R. and PRATT, O. E. (1973). The movement of amino acids between blood and skeletal muscle in the rat, *J. Physiol.*, **235**, 459–75.

BERGSTRÖM, J., FÜRST, P., NORÉE, L.-O. and VINNARS, E. (1974). Intracellular free amino acid concentration in human muscle tissue, *J. Appl. Physiol.*, **36**, 693–7.

ELLIOTT, D. F. and NEUBERGER, A. (1950). The irreversibility of the deamination of threonine in the rabbit and rat, *Biochem. J.*, **46**, 207–10.

FELIG, P. (1973). The glucose-alanine cycle, *Metabolism*, **22**, 179–207.

FELIG, P. and WAHREN, J. (1971). Amino acid metabolism in exercising man, *J. Clin. Invest.*, **50**, 2703–14.

FRICK, G. P. and GOODMAN, H. M. (1979). Physiological regulation of leucine metabolism in rat adipose tissue, *Fed. Proc.*, **38**, 947.

HALLIDAY, D. and MCKERAN, R. O. (1975). Measurement of muscle protein synthetic rate from serial muscle biopsies and total body protein turnover in man by continuous intravenous infusion of L[$\alpha$-$^{15}$N]lysine, *Clin. Sci. Mol. Med.*, **49**, 581–90.

KREBS, H. A. (1972). Some aspects of the regulation of fuel supply in omnivorous animals, in *Advances in Enzyme Regulation*, (Ed. G. Weber), Vol. 10, Pergamon Press, New York, pp. 397–420.

LUNN, P. G., WHITEHEAD, R. G. and BAKER, B. A. (1976). The relative effects of a low-protein–high-carbohydrate diet on the free amino acid composition of liver and muscle, *Br. J. Nutr.*, **36**, 219–30.

MUNRO, H. N. (1970). Free amino acid pools and their regulation, in *Mammalian Protein Metabolism*, (Ed. H. N. Munro), Vol. IV, Academic Press, New York, Ch. 34.

O'DELL, B. L. and SAVAGE, J. E. (1966). Arginine–lysine antagonism in the chick and its relationship to dietary cations, *J. Nutr.*, **90**, 364–70.

PAWLAK, M. and PION, R. (1968). Influence de la supplémentation des protéines de blé par des doses croissants de lysine sur la teneur en l'acides aminés libres du sang et du muscle du rat en croissance, *Ann. Biol. anim. Bioch. Biophys.*, **8**, 517–30.

POZEFSKY, T., FELIG, P., TOBIN, J. D., SOELDNER, S. and CAHILL, G. F. (1969). Amino acid balance across tissues of the forearm in post-absorptive man. Effects of insulin at two dose levels, *J. Clin. Invest.*, **48**, 2273–81.

WALSER, M., HAMMOND, V. and MEARNS, D. (1979). Reinterpretation of plasma amino acid concentrations expressed as a function of total amino acids, *Clin. Res.*, **27**, 519A.

WATERLOW, J. C. (1967). Lysine turnover in man measured by intravenous infusion of L[U-$^{14}$C]lysine, *Clin. Sci.*, **33**, 507–15.

WATERLOW, J. C., GARLICK, P. J. and MILLWARD, D. J. (1978). *Protein Turnover in Mammalian Tissues and in the Whole Body*, Elsevier North-Holland, Amsterdam.

WEISSMAN, N. and SCHOENHEIMER, R. (1941). The relative stability of L(+)-lysine in rats studied with deuterium and heavy nitrogen, *J. Biol. Chem.*, **140**, 779–95.

WRIGHT, SAMSON (1971). *Samson Wright's Applied Physiology*, 12th edition, (Revised by C. A. Keele and E. Neil), Oxford University Press, London.

# 2

# The Entry of Amino Acids into Cells

O. E. PRATT

*Institute of Psychiatry, London, UK*

Amino acids are needed continually by all the cells of the body (with the possible exception of mature red blood cells) for the synthesis of protein. In addition, amino acids are used by certain cells for a number of other purposes, e.g. to synthesize neurotransmitters or to act themselves as neurotransmitters, to provide material from which substances such as ketone bodies and glucose (which are needed for energy-yielding metabolism) can be made, and to act as sources of special chemical groups, such as methyl groups, which are required for a variety of synthetic purposes. About 20 amino acids are needed to make protein and about a dozen more are used for other purposes.

In considering how some 30 amino acids pass from the circulation into cells we are concerned with substances which differ considerably in their physical and chemical properties. Their acid–base characteristics and lipid solubilities are of particular importance. Most amino acids are electrically neutral, having no charge at the pH of body fluids, but the diacidic amino acids have a negative, and the dibasic, a positive charge. Lipid solubilities range from very low (e.g. arginine) to moderate (e.g. tryptophan). The great differences in the ways in which the individual amino acids reach the cells from the circulating blood are explained in part by these variations in their properties.

Although the majority of amino acids needed by the body are derived from the digestion of proteins in the gut, some non-essential amino acids are made in the cells, especially those of the muscles, liver, kidney and brain and these may be released into the circulation. Some amino acid movement arises from recycling, that is, part of the protein formed in the cells is broken down and the amino acids released become available for re-use (Waterlow *et al.*, 1978). Generally this recycling takes place locally within the cells, but

## TABLE 1

THE PASSAGE OF NUTRIENTS FROM THE CIRCULATING BLOOD INTO THE CELLS OF VARIOUS ORGANS (THE SITES OF THE MAJOR BARRIERS ARE GIVEN, AS ARE ESTIMATES OF THE BLOOD FLOW, THE FLUXES ACROSS THE BARRIERS AND THE NET MOVEMENT FROM BLOOD TO CELLS. FIGURES RELATE TO HUMAN MATERIAL WHERE DATA ARE AVAILABLE. ALL FLUXES ARE IN UNITS OF $CM^3/G/SEC \times 10^3$)

| Source | Destination | First barrier | Intervening space and % of tissue | Second barrier | Blood flow | First barrier | | Second barrier | | Flux from blood to cells | Net movement from blood to cells |
|---|---|---|---|---|---|---|---|---|---|---|---|
| | | | | | | Due to diffusion | Carrier-mediated | Due to diffusion | Carrier-mediated | | |
| blood | muscle cell | typical capillary wall | extracellular extravascular 25% | surface membranes of cells | 1·8 to 13[a] | 0·8 to 1·4[e] | ? | 0·064 | 0·7 to 1·3[b] | 0·48 to 2·7[c] | −2·3 to 1·1[d] |
| blood | cardiac cell | typical capillary wall | extravascular extracellular, 13 to 25% | surface membranes of cells | 12·8 ± 0·5[e] | 5·4 ± 0·4[e] | ? | 0·1[f] | ? | 0·7 to 4·1[f] | ? |
| blood | cerebral cell | capillaries with tight junctions | extravascular extracellular 17 to 20% | surface membranes of cells | 12·0 ± 0·6[g] | 0·005 to 0·083[h] | 0·2 to 1·8[i] | 0·02 to 0·4[j] | 3 to 15[k] | 0·2 to 1·8[i] | 0·1 to 0·6[h] |
| blood | liver cell | fenestrated capillaries | extracellular extravascular | surface membranes of cells | 19 ± 8[l] | 16 to 30[m] | ? | 0·75[n] | 16 to 18[p] | 2 to 11[q] | −4 to 8[r] |
| blood | cell of nephron | fenestrated capillaries | 6·8 to 0·4% glomerular filtrate | surface membranes of cells | 20 ± 5[s] | *6·9 ± 1·1[x] | ? | 6·8[t] | 16·9[u] | — | 6·6 to 6·9[u] |
| maternal blood | fetal blood | relatively impermeable capillaries | extravascular extracellular | fetal–placental interface | 12 ± 1·3[v] | ? | 4 to 16[w] | ? | 4 to 16[w] | ? | 0·3 to 2·3[x] |
| blood | lumen† of gut | fenestrated capillaries | extravascular extracellular | brush-border of intestinal cells | 20 ± 3[y] | 8 to 15 | ? | 0·3[z] | 3 to 7[z] | ? | 2 to 8[z] |

* Filtration

† Movement in reverse direction, i.e. from lumen of gut to blood.

Figures in bold type represent energy-dependent transport.

[a] Honig et al., 1980; [b] Akedo and Christensen, 1962; Riggs and McKirahan, 1973; [c] Baños et al., 1973b; [d] Daniel et al., 1977a,b,c; [e] Haunso et al., 1980; [f] Baños et al., 1978; [g] Gjedde et al., 1980; [h] Pratt, 1980b; [i] Baños et al., 1973a, 1975; Pardridge and Oldendorf, 1975; Pratt, 1976; [j] Blasberg and Lajtha, 1965; Pratt, 1980b; [k] Neidle et al., 1973; [l] Ossenberg et al., 1974; [m] Crone, 1963; [n] Le Cam et al., 1979; [p] Joseph et al., 1978; [q] Pardridge and Jefferson, 1975; Pardridge, 1977; [r] Aikawa et al., 1973; Yamamoto et al., 1974; [s] Smith, 1951; [t] Lingard et al., 1973, 1978; [u] Stanbury et al., 1978; [v] Laga et al., 1972; McFadyen, 1979; [w] Schneider et al., 1979; Yudilevich and Eaton, 1980; Yudilevich et al., 1980; [x] Hill and Young, 1973; Lemons et al., 1976; Young, 1979; [y] Hafström et al., 1979; [z] Nathans et al., 1960; Robinson and Alvardo, 1979.

this is only partly true of the brain. A considerable proportion of the protein which is made in the cell bodies of neurones passes down the axonal processes to be broken down peripherally. If the amino acids so formed happen to be released a considerable distance from the body of the cell and cannot be re-used locally, they enter the blood. The brain has to replace them by amino acids from the systemic circulation (Pratt, 1979*b*).

## The Barriers Between the Circulating Blood and Somatic Cells

As a general rule amino acids in the circulating blood have to cross two barriers before they can enter the interior of the cells of the body. The two barriers are commonly situated; (a) in the walls of the blood capillaries, and (b) in the surface membranes of the various somatic cells. They are separated by the extravascular, extracellular tissue space. The relative importance of the two barriers and the size of the intervening space vary very considerably in different tissues (Table 1). In the central nervous system the permeability of the capillaries is very low and this first barrier is the most important. At the other extreme are tissues like the liver where the capillary permeability is very high (Palade *et al.*, 1979). Thus, in these latter organs only the second barrier is of importance in regulating the supply of amino acids to the cells.

## Methods for Studying Amino Acid Transport

The rate of movement of an amino acid into a tissue or the flux, $J$, can be studied in a number of different ways. Thus, the uptake of a radioactive tracer by a tissue may be measured during a constant input, that is, while a steady level of the tracer is maintained in the arterial blood over a known period of time by means of a specially programmed intravenous injection (Daniel *et al.*, 1975); or by the single injection method in which a bolus of tracer is injected into an artery supplying the organ and the extraction of the tracer by the tissue compared with that of a highly diffusible reference marker (Oldendorf, 1971); or by the dilution of an indicator (Crone, 1963; Yudilevich *et al.*, 1972).

Kinetic analysis of the data obtained by these methods gives valuable information about the mechanism of transport, e.g. whether or not it is carrier-mediated, as well as the relative importance of blood flow or of the barriers in limiting the flux from the blood into the cells. Table 1 shows the specific fluxes, $J/c$, for separate stages of movement into various tissues, where $c$ is the concentration of the amino acid in blood plasma. Various expressions are used to describe the rates at which solutes move. Thus, in different contexts specific flux is equivalent to a permeability-surface product, to a clearance, to a capillary diffusion capacity, or, for convective transport, to a rate of blood flow.

In Table 1 specific fluxes have been calculated, using data from various sources, as follows:

1.  For transport in the blood stream: $J/c = F$, where $F$ is the flow of blood plasma in unit time to unit weight of tissue.
2.  For passive diffusion as a first order rate constant: $J/c = P_d A$, where $P_d$ is the permeability of the membrane to amino acids and $A$ is its area per unit weight of tissue.
3.  For carrier-mediated transport, which is subject to saturation by excess substrate, and also to competitive inhibition:

$$\frac{J}{c} = \frac{J_{max}}{c + K_m \left[ 1 + \sum_{n=1}^{n=n} \left( \frac{c_i}{K_i} \right)_n \right]} \qquad (1)$$

$$= \frac{J_{max}}{c + K_a}$$

where $J_{max}$ is the maximum carrier-mediated flux, when $c$ is large enough to saturate the carrier, and $K_m$ is a constant inversely proportional to the affinity of the amino acid for the carrier. Amino acids clearly enter tissue cells mainly by carrier-mediated transport processes (Table 1).

The term

$$\sum_{n=1}^{n=n} \left( \frac{c_i}{K_i} \right)_n$$

in eqn (1) represents the inhibitory effects of a number of other amino acids which are each present in a concentration, $c_i$, with an affinity constant, $K_i$. The effect of these inhibitory amino acids is to raise the apparent value of $K_m$ to $K_a$ (Pratt, 1979a). In Table 1 it is assumed that $c$ is small compared with $K_a$, so that $J/c = J_{max}/K_a$ approximately. However, the specific flux due to any of the carrier-mediated processes will be reduced below the values shown in Table 1 if $c$ is not small compared with its $K_a$. A number of different types of transport carriers for amino acids have been characterized and their distribution in different tissue sites is shown in Table 2. Passive diffusion across cell membranes can be measured, but the results are not reliable if there is tissue damage. In such cases permeability-surface products were calculated using permeabilities derived from measurements made *in vivo* on the cerebral capillary barrier (Pratt, 1980$b$).

Influx into a tissue, $J_{in}$, is usually offset in part by efflux, $J_{out}$. The net movement into the tissue, $J_{net}$, is given by the difference:

$$J_{net} = J_{in} - J_{out}$$

Thus, the transport of an amino acid into and out of a cell largely determines the availability of that amino acid for cell metabolism. This is true even if the amino acid is made within the cell (Christensen, 1977).

The other approach to amino acid transport is to measure these net movements. To do this the amino acid content of venous blood draining from the organ is compared with that of arterial blood. From these data the extraction ratio, $E$, can be calculated by:

$$E = \frac{c_a - c_v}{c_a}$$

where $c_a$ and $c_v$ are the concentrations of the amino acid in the arterial and venous blood respectively. Efflux is usually rather difficult to measure but it can be assessed indirectly if the blood flow through an organ and the extractions of the amino acid are known. Thus

$$J_{net} = EFc_a$$

where $F$ is the blood plasma flow and the efflux is given by:

$$J_{out} = J_{in} - EFc_a$$

The margin by which the influx exceeds the net uptake of an amino acid by any tissue provides a measure of safety in the supply of the amino acid to the organ. If for any reason (including reduced blood flow) the influx falls or the rate of use of an amino acid rises so that this safety margin disappears, then the cells will not receive sufficient amino acid due to a restricted rate of movement of the amino acid across one or other of the two barriers. This state of affairs can usually be best detected by measuring either the net uptake or the rate of utilization of the amino acid and comparing it with the influx, that is, by showing that the efflux has fallen to a low value.

TABLE 2

THE MORE IMPORTANT TRANSPORT CARRIERS FOR AMINO ACIDS AND WHERE THEY APPEAR TO OPERATE

| Transport[a] carrier | Muscle cells | Brain capillaries | Brain cells | Liver cells | Kidney tubules | Placenta | Intestinal epithelium | Main amino acids transported |
|---|---|---|---|---|---|---|---|---|
| L | + | + | + | + | + | + | + | Large neutral amino acids |
| A | + | (+)[b] | + | + | + | | + | Alanine and small neutral amino acids |
| B | + | + | + | + | + | + | + | Basic amino acids |
| N | | | + | + | | + | | Amides and histidine |
| ASC | + | | + | + | + | + | | Cysteine, alanine, threonine, serine |

[a] For references see Table 1. For carrier types see Christensen (1977), Sershen and Lajtha (1979), Kilberg et al. (1980).
[b] One-way-only, brain to blood (Betz and Goldstein, 1978; Christensen, 1979).

In this situation the extraction ratio may not rise very much, but, if the overall flux is limited by movement across the first barrier, the concentration of the amino acid will fall to a low level in the space between the barriers. If the supply is inadequate because not enough amino acid is being brought to the organ by the blood, then the extraction ratio will rise sharply to nearly 100%.

An unsolved problem is the extent to which not only the amino acids in the plasma, but also those within the red blood cells, are extracted as the blood passes through the tissues. Work *in vitro* shows that the time constants for the movement of amino acids across the erythrocyte membranes are minutes or hours (Winter and Christensen, 1964; Young and Ellory, 1979), but Felig *et al.* (1973) report that the extraction of amino acids from the red cells is sometimes large enough to increase the flux of particular amino acids by up to 35–40%. All calculations for Table 1 have been made on the assumption that amino acids can only slowly escape from red blood cells but if this is not so, some of the fluxes may be appreciably higher, especially for glutamine, alanine and glycine. These amino acids may pass directly from erythrocytes to tissue cells by cell membrane to cell membrane contact (Felig *et al.*, 1973).

## Transport into Specific Organs or Tissue

### Skeletal Muscle

According to the degree of activity, the blood flow varies to an extent not seen in other tissues, over a range of some 7 to 1 (Table 1). Some amino acids (e.g. phenylalanine and tyrosine) can move faster into the tissue than the rate at which they are supplied to it in the plasma when the blood flow is at its lowest, suggesting that blood flow may sometimes limit the overall flux. Another unusual characteristic of the muscle is the wide range of net flux of gluconeogenic amino acids, especially alanine and glutamine, which can move either into or out of the tissue, since the muscles act as a metabolic organ helping to regulate the levels of these substances in the blood (Daniel *et al.*, 1977a,b).

Nothing is known about the amino acid level in the extracellular tissue spaces of either skeletal or cardiac muscle. The levels of at least some amino acids are likely to fall below those in the blood plasma since many amino acids (e.g. lysine and threonine) are taken into the cells by energy-dependent transport against a concentration gradient (Baños *et al.* 1973b).

Movement of any particular amino acids *into* the muscle cells may be limited according to circumstances by blood supply, diffusion across the capillary wall or transport across the cell surface membrane. It is not clear what controls the rate at which amino acids pass *out of* the muscle cells back into the blood. Probably one transport system moves amino acids into the cells and another moves them out. It is mainly the aromatic, branched chain and other essential amino acids which move inwards and the gluconeogenic amino acids, especially alanine and glutamine, which move outwards, but threonine and serine are common to both groups.

**Cardiac Muscle**

Blood flow is remarkably constant and is.comparable with the high flows through skeletal muscle during exercise. The rich supply of fairly permeable capillaries and the rhythmical pressure changes within the tissue ensure that equilibrium between the amino acids in the blood and those in the extracellular space is reached rapidly. For at least 10 amino acids the overall flux into the heart shows the characteristics of a carrier-mediated process and nine amino acids enter the cells against an adverse concentration gradient by energy-dependent transport (Baños *et al.*, 1978). It seems likely that it is the carrier-mediated transport across the second barrier, the surface membranes of the cardiac cells, which limits the overall flux from the blood into the cells.

**Cerebral Tissue**

In this case it is the first barrier which is most important. Only a few small lipid molecules such as oxygen, alcohol and volatile anaesthetics, pass freely through the walls of the cerebral capillaries. Amino acids, in common with other substances which have oil:water partition coefficients of less than about 0·02, diffuse only very slowly across this barrier (Table 1). Those amino acids which the brain needs have to cross this first barrier almost entirely by carrier-mediated transport (Baños, *et al.* 1973a, 1975). This barrier is the most effective of the two, not only because of its low permeability to diffusion, but also because the carrier-mediated specific fluxes across this barrier are less than those across the surface membranes of brain cells. Not only is the surface area of the capillaries less than one fifth of the cells in the brain, but there are probably no more than two important carrier systems for amino acids (Baños *et al.* 1973a, 1974; Pratt, 1976; Christensen, 1979), whereas a much wider variety of carrier systems (Table 2) take the amino acids into the cells (Christensen and Handlogten, 1978; Sershen and Lajtha, 1979). Some of these different carriers are probably associated with uptake by special cell types, e.g. astrocytes. The faster specific flux at the second barrier will remove amino acids from the extracellular space of the brain as fast as they enter it by crossing the capillary barrier, thus reducing the concentrations in the tissue spaces considerably below the blood plasma levels. Confirmation that this happens is provided by the low levels of amino acids in the cerebrospinal fluid (assumed to equilibrate with the tissue spaces), usually no more than 10–40% of those in either the blood plasma or inside the brain cells (Bradbury, 1979).

The direction of net movement of amino acids (except glutamine) is from the blood to the brain and it seems clear that this movement needs energy, which is provided by the carrier-mediated systems at the second barrier. Thus movement across the capillary wall barrier is always down a concentration gradient, that is, it is carrier-facilitated diffusion. The net movement of glutamine is out of the brain and provides a means of removing surplus amino groups from the cells (Pratt, 1976). It is not clear how glutamine is transported from the brain to blood, although it may well leave the cells by diffusion since the intracellular level is more than ten times that in the tissue space and may cross the capillary barrier by a one-way-only system A carrier (Table 2).

## Liver

Blood flow is rapid and the fenestrated capillaries are exceptionally permeable to amino acids. Amino acids are also transported rapidly by carrier-mediated processes in both directions across the cell membranes of the hepatocytes and the flux of amino acids into the liver is probably regulated by a complex interaction between different transport carriers (Table 2). For some of the gluconeogenic amino acids, e.g. alanine, proline and threonine, net movement into the liver may roughly equal the unidirectional flux (Pardridge, 1977), that is, there is no efflux. Thus the rate at which the liver can make glucose from these amino acids during fasting is limited by their flux from the blood to the liver cells, which, in turn, depends upon their blood concentration. This is affected to a large extent by the rate at which these amino acids are released by the muscles. On the other hand the uptake by the liver of the branched chain amino acids, isoleucine, leucine and valine (used only for protein synthesis) represents only a small fraction of the flux. Amino acids are sometimes released from the liver, as from muscle, that is, the net flux can be negative. It is believed that many neutral amino acids enter liver cells by one carrier system and leave by another (Table 2).

**Kidney**

Movement across the first barrier is by convective flow, that is, by filtration through the glomerulus. The compartment between the two barriers is not the tissue space but the glomerular filtrate within the convoluted tubule, with one of more further but less important barriers between that and the venous blood. Comparison of the specific fluxes in Table 1 shows that the rate of carrier-mediated transport across the second barrier is high enough to ensure that reabsorption from the glomerular filtrate, especially of nutritionally essential amino acids, is substantially complete. Thus, these amino acids are more or less completely taken back from the glomerular filtrate before it has completed its passage through the convoluted tubule, so minimizing urinary loss. If the carrier-mediated systems on the luminal surfaces of the tubule cells fail for any reason reabsorption will only be delayed. As the water is reabsorbed from the tubular fluid the concentrations of the amino acids in it will rise well above the levels in the blood and the rate of passive diffusion across the wall of the tubules will be fast enough to ensure that most of the amino acids are reabsorbed before the fluid leaves the nephron.

**Placenta**

The two major barriers appear to be between the maternal blood and the placenta and between the placenta and the fetal blood. How amino acids move through these barriers is far from clear. It is certain that there is a continuous net movement of nutritionally important amino acids (not including glutamate) from the mother to the fetus and that the levels of most amino acids are rather higher in the fetal blood than in the maternal blood. Still higher levels of amino acids, of the order of 3 to 20 times those in fetal or maternal blood, are found in placental tissue (Hill and Young, 1973). These amino acids must cross the first barrier from the maternal blood by energy-dependent, carrier-mediated processes. Since similar carrier-mediated processes take amino acids from the fetal blood back into the placenta it is not clear why there should be a net movement into the fetal blood. A possible explanation may be that the placental–fetal barrier is readily permeable to passive diffusion down the steep concentration gradient from placenta to fetus.

**Intestine**

Here the net movement of amino acids is in the reverse direction to that in all the previous examples. It seems clear that movement across the brush-border of the epithelial cells of the gut will limit the flux from the lumen of the gut to the blood. Efflux of amino acids back into the lumen of the gut seems to be unimportant. Either movement across the brush-border is unidirectional and energy-dependent or the high blood flow washes away amino acids from the mucosa before efflux can occur.

**Other Tissues**
Other sites in which carrier-mediated amino acid transport systems have been studied include the salivary gland (Bustamente *et al.*, 1981) and the mammary gland (Daniel *et al.*, 1968).

# Disorders of Amino Acid Transport

The transport of amino acids into cells may be interfered with in one of two ways. The carrier-mediated system may be abnormal, usually as a result of an inborn error, or a competitive inhibitor may impede transport. Four abnormalities of the transport systems are known, affecting transport in the brush-border of the intestine or in the kidney tubules (see Stanbury *et al.*, 1978). Cystinuria is caused by a defect in the transport system for the dibasic amino acids which disturbs cystine metabolism. In Hartnup disease the defect lies in the transport of most neutral amino acids (Milne *et al.*, 1961). In familial aminoglycinuria the transport of glycine, proline and hydroxyproline is defective, as probably is that of methionine (Jepson *et al.*, 1958; Hooft *et al.*, 1965; Jepson, 1978). These metabolic defects have less damaging effects than might be expected because the absorption of peptides is not affected and these can then be broken down to provide the amino acids. Failure to reabsorb glucose leaves a considerable osmotic load in the renal tubule, whereas only a small osmotic load arises from failure to reabsorb amino acids and they are readily reabsorbed by passive diffusion after the water is taken back. The inborn errors of transport reported so far seem to affect mainly transport in the kidney and the intestine.

A striking example of how a normal transport system may be prevented from working properly is provided by the severe effects of maternal phenylketonuria upon the development of the otherwise normal fetus. Excessive amounts of phenylalanine cross the placenta to the fetal circulation and cause the partial exclusion from the fetal brain of various other amino acids by competing for transport across the blood–brain barrier (Pratt, 1980*b*).

## Acknowledgements

I am grateful to Professor Peter Daniel for many helpful discussions. My own work reported above has been assisted by grants from the Wellcome Trust and the National Fund for Research into Crippling Diseases.

## Discussion

*Munro (Tufts University, Boston, USA):* How are your interpretations affected by the anatomical microfeatures of the liver where there are no traditional blood vessels but sinusoids which are lined with Kupffer cells? Do the amino acids have to go through these cells to reach the hepatocytes or can they by-pass them?

*Pratt:* It seems likely that amino acids do not have to go through the cells lining the sinusoids because Goresky (1964) has shown that small diffusible molecules rapidly enter the extracellular space of the liver at a rate mainly limited by the blood flow. Decreases in liver blood flow reduce hepatic clearance rates (Lautt, 1977).

*Harper (University of Wisconsin-Madison, USA):* How significant is competition among amino acids for uptake into brain or other tissues in the regulation of the concentrations in the tissue? Many transport studies have been done with a single amino acid, often in the absence of some other amino acids. The rates of uptake when a complete amino acid mixture is included in the medium in studies *in vitro* are much lower than they are if you are looking at uptake of one individual amino acid into a tissue.

*Pratt:* This is a very important field and there are three areas of interest. One is the inborn errors of metabolism, which we shall be hearing more about from Nyhan (Paper 6). In untreated phenylketonurics, not only are the phenylalanine levels high in the blood, but as a result levels of other amino acids such as methionine, tyrosine and histidine may be quite low. Unless man is totally different from other mammalian species, the movement of these amino acids across the blood–brain barrier must be very severely reduced in untreated phenylketonuria. This is one possible explanation for the anomalies of brain development. Similar mechanisms may operate in the other inborn errors of amino acid metabolism. Because phenylalanine has quite a low Michaelis constant for transport, it is dangerous to have it around in large amounts.

Another field is inhibition of tryptophan transport into the brain. Wurtman's group in America (Fernstrom, 1979; Wurtman, 1979) and Curzon's group in Britain (Curzon, 1979) have shown that changes in the supply of tryptophan to the brain are reflected in the metabolism of the neurotransmitter serotonin. These changes may be caused by giving insulin or by a diet which causes insulin secretion. Tryptophan transport is complicated, as we heard in the previous discussion (p. 12), by the binding of tryptophan to plasma albumin. What happens is disputed, but many people, myself included, believe that a large proportion of the tryptophan that is bound to albumin can be stripped off as the blood goes through the cerebral capillaries. This happens if the transport carriers on the walls of the endothelial cells have a higher affinity for tryptophan than does the albumin. However, the amount stripped off may vary with circumstances, and with the proportion of inhibitors and other components in the blood which compete for the tryptophan-binding sites on the albumin (Daniel *et al.*, 1981).

Another problem, which is an even more difficult one, is severe hepatic failure in which one gets bizarre patterns of amino acids in the blood, (e.g. very high levels of amino acids such as methionine and aromatic amino acids and reduced levels of branched chain amino acids) and at the same time damage to the blood–brain barrier. On the one hand transport inhibition may be depriving the brain of some essential amino acids; on the other hand breakdown of the blood–brain barrier occurs and allows very large amounts of amino acids to pass into the brain and cause trouble there. One or other of these factors, or toxic by-products of amino acid metabolism, seem likely to account for hepatic coma or encephalopathy in liver failure.

*Harper:* Our impression is that these competitive relationships influence amino acid uptake by other organs and tissues much less, presumably because they do not have a physical barrier comparable to the blood–brain barrier?

*Pratt:* Yes, it is not only that the carrier-mediated limiting process is on the capillary wall in the blood–brain barrier, but also that the high affinities predispose to competition. Also in the capillary barrier in the brain only two carrier systems are important, the neutral amino acid system (L-system) and the system for basic amino acids; whereas on brain cells there are ten or more transport systems, and on the cells of cardiac or skeletal muscle there are a number of different systems. In fact the N-system, transporting especially glutamine, has been studied mainly on hepatocytes.

Nevertheless, as we have shown (Baños *et al.*, 1978), inhibition may sometimes deprive the heart cells of particular amino acids and this may be of some clinical significance.

*Munro:* That implies, does it not, that the most significant and selective barrier is at the capillary level and consequently that a number of *in vitro* experiments on separated cells or tissue slices may be giving erroneous deductions, particularly in the case of the brain.

*Pratt:* Yes, the results of work on transport into brain cells are meaningless unless they are related to what happens at the brain–capillary barrier.

*Garlick (London School of Hygiene and Tropical Medicine, UK):* Does the counter-transport exhibited by some transport systems (e.g. the L-system) have any significance in the competition for uptake between different amino acids? If the competing amino acid were to be taken up into the cell first, might it not then stimulate uptake of other amino acids by counter-transport?

*Pratt:* In our short-term experiments the inhibitor is in the blood stream only, but counter transport is likely to be important especially in tissues or organs other than the brain. The transport mediators in the brain capillary are not sodium-dependent, not affected by insulin and not energy-dependent—they are simply carriers that facilitate diffusion. But elsewhere such factors complicate transport processes. Two-way exchanges by separate but often interacting transport processes may well help to explain the remarkable shifts in net amino acid movement into or out of liver or muscle (see Table 1).

*Golden (University of the West Indies, Kingston, Jamaica):* The intracellular/extracellular specific activity ratios during infusion of a radioactive amino acid are different in different tissues. They reflect the resultant effects of protein synthesis rates and amino acid transport rates. Do the numbers generated for protein synthesis rates, the intracellular/extracellular specific activity ratios and the transport rates match in the different tissues?

*Pratt:* I think the answer is probably no for the reasons that Waterlow has already referred to|(p. 10).

*Garlick:* In liver the specific activity during constant infusion of a labelled amino acid is indeed much lower than in most other tissues, but this results from uptake of unlabelled dietary amino acids into the portal vein. Our

observations show that typically portal vein specific activity is about 30 %
lower than in systemic venous blood (Fern and Garlick, 1976).

*Waterlow* (*London School of Hygiene and Tropical Medicine, UK*): I was
not quite clear from what Golden said whether he was talking about
extracellular and intracellular differences or differences between different
amino acids. I am a firm believer in the idea that the plateau specific
activities reached with an infusion can give us values for rates of entry, but
we do not have enough comparisons of different amino acids in terms of
their intracellular and extracellular plateaux. Nor do we know if the
behaviour of these different amino acids fits with the other parameters that
Pratt has described.

*Pratt:* I did not have time to deal with the very big differences in transport
rates between different amino acids. It is clear that the essential amino acids
form one pattern and the non-essentials another, but glutamine is very often
the odd one out. Its net movement across the blood–brain barrier is not
inwards like most amino acids but outwards, for glutamine excretion seems
to be the main, if not the only, way in which the brain gets rid of nitrogenous
waste material. How it gets out, whether it moves down a concentration
gradient or is transported, is not clear.

*Bier* (*Washington University School of Medicine, St. Louis, USA*): My
question is a conceptual one which goes back to an earlier question from
Harper concerning the relative quantitative importance of the various
competitive inhibiting processes under usual physiological circumstances.
After a meal, most plasma amino acids rise and yet each enters the
appropriate cells in the appropriate amounts since this is a function of
eating in the first place. Does this allow us to conclude that competitive
inhibition is not terribly important within the range of amino acid
concentrations commonly encountered in man?

*Pratt:* I think this is a very interesting point. In the brain amino acid levels
and protein content change very little in quite severe metabolic disturbance,
for example, in long fasts. But how this is managed is not yet fully clear.
Again, if raised levels of amino acids are maintained in the blood
experimentally, over a period of time, the amino acids do get into the brain.
What the effect of this is, is not clear. On the other hand severe competition
can almost entirely exclude amino acids from the brain and this must cause
trouble. As Rose *et al.* (1948) showed many years ago, feeding a diet to rats
lacking one particular amino acid is one of the most metabolically
disturbing things you can do. Regulation of brain amino acid levels must

depend to a large extent upon reasonable regulation of blood levels. Even in severe fasting they do not vary over a wide range. The brain somehow manages to obtain the amino acids it needs provided the blood levels are within certain limits and the proportions reasonable.

*Young (Massachusetts Institute of Technology, USA):* There are some interesting associations among specific amino acids with respect to plasma amino acid concentrations. For example, a low level of dietary lysine brings about a specific and marked increase in plasma threonine (Zimmerman and Scott, 1967). A low intake of leucine results in a marked increase in plasma valine (Hambraeus *et al.*, 1976) (see also p. 108). I wonder whether, in the light of your discussion, you could offer any insight as to the possible mechanisms behind these specific associations.

*Pratt:* These changes are likely to be due to effects not only of transport competition, but also of disturbances in the metabolism of these amino acids in tissues including some which we do not normally think of as sites of amino acid metabolism. Some indication of what may happen is given by the changes that occur in maternal phenylketonuria. In a mother who is a phenylketonuric with a baby in her uterus, things go wrong at a number of different stages. There is a high concentration of phenylalanine in the digestive juices, movement of other amino acids across the gut wall is impeded, movement across the placental barrier goes wrong, and renal reabsorption may be affected. Inhibition of transport occurs most readily when the inhibitory amino acid is one with a high affinity for the carriers and is present in a concentration high enough to contribute a big affinity term. The effect of inhibition is seen in the term $1 + (c_i/K_i)_1 + (c_i/K_i)_2$ etc. in the denominator of the modified Michaelis equation (see p. 20). If the concentration of the inhibitor is high and its $K_i$ small, the factor added will be large.

*Harper:* I wonder if there is not a simpler explanation for some of the observations which Young mentioned. If the level of leucine, for example, is low in the blood as a result of low intake, protein synthesis will be depressed and therefore there will be more isoleucine and valine circulating.

*Young:* I do not think that explanation is satisfactory because that mechanism would have to apply to all the essential amino acids and this does not appear to be the case.

*Millward (London School of Hygiene and Tropical Medicine, UK):* There is another pathway by which amino acids can enter cells and that is through the uptake of proteins into cells by heterophagy and their subsequent breakdown. In those cells where this occurs, such as the Kupffer cells in the liver, this may well be the major route of entry of amino acids.

## References

AIKAWA, T., MATSUTAKA, H., YAMAMOTO, H., OKUDA, T., ISHIKAWA, E., KAWANO, T. and MATSUMURA, E. (1973). Gluconeogenesis and amino acid metabolism. II. Interorganal relations and roles of glutamine and alanine in the amino acid metabolism of fasted rats, *J. Biochem. (Tokyo)*, **74**, 1003–17.

AKEDO, H. and CHRISTENSEN, H. N. (1962). Nature of insulin action on amino acid uptake by the isolated diaphragm, *J. Biol. Chem.*, **237**, 118–22.

BAÑOS, G., DANIEL, P. M., MOORHOUSE, S. R. and PRATT, O. E. (1973a). The influx of amino acids into the brain of the rat *in vivo*: the essential compared with some non-essential amino acids, *Proc. Roy. Soc. Lond. B.*, **183**, 59–70.

BAÑOS, G., DANIEL, P. M., MOORHOUSE, S. R. and PRATT, O. E. (1973b). The movement of amino acids between blood and skeletal muscle in the rat, *J. Physiol.*, **235**, 459–75.

BAÑOS, G., DANIEL, P. M. and PRATT, O. E. (1974). Saturation of a shared mechanism which transports L-arginine and L-lysine into the brain of the living rat, *J. Physiol.*, **236**, 29–41.

BAÑOS, G., DANIEL, P. M., MOORHOUSE, S. R. and PRATT, O. E. (1975). The requirements of the brain for some amino acids, *J. Physiol.*, **246**, 539–48.

BAÑOS, G., DANIEL, P. M., MOORHOUSE, S. R., PRATT, O. E. and WILSON, P. A. (1978). The influx of amino acids into the heart of the rat, *J. Physiol.*, **280**, 471–86.

BETZ, A. L. and GOLDSTEIN, G. W. (1978). Polarity of the blood–brain barrier: neutral amino acid transport into isolated brain capillaries, *Science*, **202**, 225–7.

BLASBERG, R. and LAJTHA, A. (1965). Substrate specificity of steady-state amino acid transport in mouse brain slices, *Arch. Biochem.*, **112**, 361–77.

BRADBURY, M. (1979). *The Concept of a Blood–Brain Barrier*, Wiley, New York.

BUSTAMENTE, J. C., MANN, G. E. and YUDILEVICH, D. L. (1981). Specificity of neutral amino acid uptake at the basolateral side of the epithelium in the cat salivary gland *in situ*, *J. Physiol.*, **313**, 65–80.

CHRISTENSEN, H. N. (1977). Implications of the cellular transport step for amino acid metabolism, *Nutr. Rev.*, **35**, 129–33.

CHRISTENSEN, H. N. (1979). Developments in amino acid transport, illustrated for the blood–brain barrier, *Biochem. Pharmacol.*, **28**, 1989–92.

CHRISTENSEN, H. N. and HANDLOGTEN, M. E. (1978). Cellular uptake of lithium via amino acid transport system. A. *Biochim. Biophys. Acta*, **512**, 598–602.

CRONE, C. (1963). The permeability of capillaries in various organs as determined by use of the 'indicator diffusion' method, *Acta Physiol. Scand.*, **58**, 292–305.

CURZON, G. (1979). Relationships between plasma, CSF and brain tryptophan, *J. Neural Transm. Suppl.*, **15**, 81–92.

DANIEL, P. M., MOORHOUSE, S. R. and PRATT, O. E. (1968). The effect on lactation of feeding large quantities of single amino acids, *J. Physiol.*, **196**, 106–8P.

DANIEL, P. M., DONALDSON, J. and PRATT, O. E. (1975). A method for injecting substances into the circulation to reach rapidly and to maintain a steady level, *Med. Biol. Engineer*, **13**, 214–27.

DANIEL, P. M., PRATT, O. E. and SPARGO, E. (1977a). The mechanism by which glucagon induces the release of amino acids from muscle and its relevance to fasting, *Proc. Roy. Soc., London., B.*, **196**, 347–65.

DANIEL, P. M., PRATT, O. E. and SPARGO, E. (1977b). The metabolic homoeostatic role of muscle and its function as a store of protein, *Lancet*, **ii**, 446–8.

DANIEL, P. M., PRATT, O. E., SPARGO, E. and TAYLOR, D. E. M. 1977c). Effect of bilateral hind-limb ischaemia upon the patterns of free amino acids in the circulation in the rat, *J. Physiol.*, **272**, 103–4P.

DANIEL, P. M., LOVE, E. R., MOORHOUSE, S. R. and PRATT, O. E. (1981). The effect of insulin upon the influx of tryptophan into the brain of the rabbit, *J. Physiol.*, **312**, 551–62.

FELIG, P., WAHREN, J. and RÄF, L. (1973). Evidence of inter-organ amino acid transport by blood cells in humans, *Proc. Nat. Acad. Sci. USA*, **70**, 1775–9.

FERN, E. B. and GARLICK, P. J. (1976). Compartmentation of albumin and ferritin synthesis in rat liver *in vivo*, *Biochem. J.*, **156**, 189–92.

FERNSTROM, J. D. (1979). Diet-induced changes in plasma amino acid pattern: effects on the brain uptake of large neutral amino acids, and on brain serotonin synthesis, *J. Neural Transm. Suppl.*, **15**, 55–67.

GJEDDE, A., HANSEN, A. J. and SIEMKOWICZ, E. (1980). Rapid simultaneous determination of regional blood flow and blood–brain glucose transfer in brain of rat, *Acta Physiol. Scand.*, **108**, 321–30.

GORESKY, C. A. (1964). Initial distribution and rate of uptake of sulfobromophthalein in the liver, *Am. J. Physiol.*, **207**, 13–26.

HAFSTRÖM, I., PERSSON, B. and SUNDQUIST, K. (1979). Measurements of cardiac output and organ blood flow in rats using $^{99}Tc^m$ labelled microspheres, *Acta. Physiol. Scand.*, **106**, 123–8.

HAMBRAEUS, L., BILMAZES, C., DIPPEL, C., SCRIMSHAW, N. S. and YOUNG, V. R. (1976). Regulatory role of dietary leucine on plasma branched-chain amino acid levels in young men, *J. Nutr.*, **106**, 230–40.

HAUNSØ, S., PAASKE, W. P., SEJRSEN, P. and AMTORP, O. (1980). Capillary permeability in canine myocardium as determined by bolus injection, residue detection, *Acta Physiol. Scand.*, **108**, 389–97.

HILL, P. M. M. and YOUNG, M. (1973). Net placental transfer of free amino acids against varying concentrations, *J. Physiol.*, **235**, 409–22.

HONIG, C. R., ODOROFF, C. L. and FRIERSON, J. L. (1980). Capillary recruitment in exercise: rate, extent, uniformity, and relation to blood flow, *Am. J. Physiol.*, **238**, H31–42.

HOOFT, C., TIMMERMANS, J., SNOECK, J., AUTENER, T., OYAERT, W. and VAN DEN HENDE, C. (1965). Methionine malabsorption syndrome, *Ann. Pediat. Basel*, **205**, 73–104.

JEPSON, J. B. (1978). Hartnup's disease, in *Metabolic Basis of Inherited Disease* 4th Ed. (Ed. J. B. Stanbury, J. B. Wyngaarden and D. S. Fredrickson) McGraw-Hill, New York and London.

JEPSON, J. B., SMITH, A. J. and STRANG, L. B. (1958). An inborn error of metabolism with urinary excretion of hydroxyacids, ketoacids and aminoacids, *Lancet*, **ii**, 1334–5.

JOSEPH, S. K., BRADFORD, N. M. and McGIVAN, J. D. (1978). Characteristics of the transport of alanine, serine and glutamine across the plasma membrane of isolated rat liver cells, *Biochem. J.*, 1976, 827–36.

KILBERG, M. S., HANDLOGTEN, M. E. and CHRISTENSEN, H. N. (1980). Characteristics of an amino acid transport system in rat liver for glutamine, asparagine, histidine, and closely related analogues. *J. Biol. Chem.*, **255**, 4011–19.

LAGA, E. M., DRISCOLL, S. G. and MUNRO, H. N. (1972). Comparison of placentas from two socio-economic groups. I. Morphometry, *Pediatrics*, **50**, 24–32.

LAUTT, W. W. (1977). The hepatic artery: subservient to hepatic metabolism or guardian of normal hepatic clearance rates of humoral substances, *Gen. Pharmacol.*, **8**, 73–8.

LE CAM, A., REY, J. F., FEHLMANN, P. K. and FREYCHET, P. (1979). Amino acid transport in isolated hepatocytes after partial hepatectomy in the rat, *Am. J. Physiol.*, **236**, E594–E602.

LEMONS, J. A., ADCOCK, E. W. III, JONES, M. D. JR, NAUGHTON, M. A., MESCHIA, G. and BATTAGLIA, F. C. (1976). Umbilical uptake of amino acids in the unstressed lamb, *J. Clin. Invest.*, **58**, 1428–34.

LINGARD, J., RUMRICH, G. and YOUNG, J. A. (1973). Kinetics of L-histidine transport in the proximal convolution of the rat nephron studied using stationary microperfusion technique, *Pflügers Arch.*, **342**, 13–28.

LINGARD, J. M., COOK, D. I. and YOUNG, J. A. (1978). A mathematical analysis of the role of passive diffusion in the renal reabsorption of amino acids and other organic compounds under free flow conditions, *Aust. J. Exp. Biol. Med. Sci.*, **56**, 395–408.

McFADYEN, I. R. (1979). Maternal blood flow to the uterus, in *Placental Transfer* (Ed. G. Chamberlain and A. Wilkinson) Pitman Medical, Tunbridge Wells.

MILNE, M. D., ASATOOR, A. and LOUGHBRIDGE, L. W. (1961). Hartnup disease and cystinuria, *Lancet*, **i**, 51–2.

NATHANS, D., TAPLEY, D. F. and ROSS, J. E. (1960). Intestinal transport of amino acids studied *in vitro* with L-$^{131}$I monoiodotyrosine, *Biochim. Biophys. Act.*, **41**, 271–82.

NEIDLE, A., KANDERA, J. and LAJTHA, A. (1973). The uptake of amino acids by the intact olfactory bulb of the mouse: a comparison with tissue slice preparations, *J. Neurochem.*, **20**, 1181–93.

OLDENDORF, W. H. (1971). Brain uptake of radiolabeled amino acids, amines and hexoses after arterial injection, *Am. J. Physiol.*, **221**, 1629–39.

OSSENBERG, F. W., DENIS, P. and BENHAMOU, J. P. (1974). Hepatic blood flow in the rat: effect of portacaval shunt, *J. Appl. Physiol.*, **37**, 806–8.

PALADE, G. E., SIMIONESEN, M. and SIMIONESEN, N. (1979). Structural aspects of the permeability of the microvascular endothelium, *Acta. Physiol. Scand. Suppl.*, **463**, 11–32.

PARDRIDGE, W. M. (1977). Unidirectional influx of glutamine and other neutral amino acids into liver of fed and fasted rats *in vivo*, *Am. J. Physiol.*, **232**, E492–6.

PARDRIDGE, W. M. and JEFFERSON, L. S. (1975). Liver uptake of amino acids and carbohydrates during a single circulatory passage, *Am. J. Physiol.*, **228**, 1155–61.

PARDRIDGE, W. M. and OLDENDORF, W. H. (1975). Kinetic analysis of blood–brain barrier transport of amino acids, *Biochim. Biophys. Acta*, **401**, 128–36.

PRATT, O. E. (1976). Transport of metabolizable substances into the living brain, in *Transport Phenomena in the Nervous System: Physiological and Pathological Aspects*, (Ed. G. Levi, L. Battistin and A. Lajtha) Plenum Press, New York, pp. 55–75.

PRATT, O. E. (1979a). Kinetics of tryptophan transport across the blood–brain barrier, *J. Neurol. Trans., Suppl.*, **15**, 29–42.

PRATT, O. E. (1979b). Adequate nutrition of the developing brain, in *Advances in Perinatal Neurology*, Vol. 1 (Ed. R. Korobkin and C. Guilleminault), Spectrum, New York and London, pp. 21–55.

PRATT, O. E. (1980a). The transport of nutrients into the brain: the effect of alcohol on their supply and utilisation, in *Addiction and Brain Damage* (Ed. D. Richter) Croom Helm, London, pp. 94–128.

PRATT, O. E. (1980b). A new approach to the treatment of phenylketonuria, *J. Ment. Def. Res.*, **24**, 203–17.

RIGGS, T. R. and MCKIRAHAN, J. J. (1973). Action of insulin on transport of L-alanine into rat diaphragm *in vitro*, *J. Biol. Chem.*, **248**, 6450–5.

ROBINSON, J. W. L. and ALVARADO, F. (1979). Interactions between tryptophan, phenylalanine and sugar transport in the small intestinal mucosa, *J. Neural Trans., Suppl.*, **15**, 125–37.

ROSE, W. C., OESTERLING, M. J. and WOMACK, M. (1948). Comparative growth on diets containing ten and nineteen amino acids with further observations upon the role of glutamic and aspartic acids, *J. Biol. Chem.*, **176**, 753–62.

SCHNEIDER, H., MÖHLEN, K. -H. and DANCIS, J. (1979). Transfer of amino acids across the *in vitro* perfused human placenta, *Pediat. Res.*, **13**, 236–40.

SEJRSEN, P. (1979). Capillary permeability measured by bolus injection, residue and venous detection, *Acta Physiol. Scand.*, **105**, 73–92.

SERSHEN, H. and LAJTHA, A. (1979). Inhibition pattern by analogs indicates the presence of ten or more transport systems for amino acids in brain cells, *J. Neurochem.*, **32**, 719–26.

SMITH, H. W. (1951). *The Kidney*, Oxford University Press, New York.

STANBURY, J. B., WYNGAARDEN, J. B. and FREDRICKSON, D. S. (1978). *The Metabolic Basis of Inherited Disease*, 4th Ed., McGraw-Hill, New York and London.

WATERLOW, J. C., GARLICK, P. J. and MILLWARD, D. J. (1978). *Protein Turnover in Mammalian Tissues and in the Wole Body*, Elsevier North-Holland, Amsterdam.

WINTER, C. G. and CHRISTENSEN, H. N. (1964). Migration of amino acids across the membrane of the human erythrocyte, *J. Biol. Chem.*, **239**, 872–8.

WURTMAN, R. J. (1979). When—and why—should nutritional state control neurotransmitter synthesis? *J. Neural Transm. Suppl.*, **15**, 69–79.

YAMAMOTO, H., AIKAWA, T., MATSUTAKA, H., OKUDA, T. and ISHIKAWA, E. (1974). Interorganal relationships of amino acid metabolism in fed rats, *Am. J. Physiol.*, **226**, 1428–33.

YOUNG, J. D. and ELLORY, J. C. (1979). Transport of tryptophan and other amino acids by mammalian erythrocytes, *J. Neural Trans., Suppl.*, **15**, 139–51.

YOUNG, M. (1979). Transfer of amino acids, in *Placental Transfer* (Ed. G. Chamberlain and A. Wilkinson) Pitman Medical, Tunbridge Wells.

YUDILEVICH, D. L. and EATON, B. M. (1980). Amino acid carriers at maternal and fetal surfaces of the placenta by single circulation paired-tracer dilution. Kinetics of phenylalanine transport, *Biochim. Biophys. Acta*, **596**, 315–9.

YUDILEVICH, D. L., DE ROSA, N. and SEPULVEDA, F. V. (1972). Facilitated transport of amino acids through the blood–brain barrier of the dog studied in a single capillary circulation, *Brain Res.*, **44**, 569–78.

YUDILEVICH, D. L., EATON, B. M. and MANN, G. E. (1980). Carriers and receptors at the maternal and fetal sides of the placenta studied by a single circulation paired-tracer dilution technique, in *Placental Transfer: Methods and Interpretations* (Ed. M. Young, L. G. Longo and G. Talegdy) Saunders, New York.

ZIMMERMAN, R. A. and SCOTT, H. M. (1967). Plasma amino acid pattern of chicks in relation to length of feeding period, *J. Nutr.*, **91**, 503–6.

# 3

# Adaptive Enzyme Changes

JOAN M. L. STEPHEN

*Formerly, External Staff, Medical Research Council, UK*

At an early stage in our work on protein turnover, we were struck by the remarkable capacity of the organism to economize amino acids. On a normal protein intake about 20 % of the amino-N flux is oxidized to urea, whereas on a low protein intake this proportion falls to about 5 %. Thus on a low intake an amino acid is somehow diverted away from oxidation and into a pathway which leads to synthesis. It seemed likely that the mechanism of this diversion depended at least in part, on adaptive enzyme changes.

As is well known, many of the transaminating and deaminating enzymes which are concerned with amino acid degradation vary in activity according to nutritional conditions. Table 1 shows the extent of this variation as summarized by Krebs (1972).

Schimke (1962) demonstrated that the urea cycle enzymes also adapt to changes in the level of protein intake. Das and Waterlow (1974) followed up this work, and concluded that the different urea cycle enzymes adapt at the same rate, and that in the rat the adaptation takes about 30 hours. This fits in with the time it takes for nitrogen balance to reach a new equilibrium after a change in the dietary protein intake.

These results relate to the enzymes of the degradative pathway, and in all cases the activity of the enzymes is greater on a high protein intake than on a low one.

At the time we were working on protein-energy malnutrition in Jamaica, the only work on adaptive changes in enzymes of the synthetic pathway was that of Spadoni and her colleagues in Rome (Gaetani *et al.*, 1964). They showed that in the liver the activity of the group of enzymes then called amino acid synthetases but now called t-RNA amino acyl transferases

39

TABLE 1

SOME AMINO ACID DEGRADING ENZYMES IN LIVER WHICH
VARY ACCORDING TO NUTRITIONAL CONDITIONS (MODIFIED
FROM KREBS, 1972)

| *Enzyme* | *Extent of variation* |
|---|---|
| Aminotransferases | |
| Tyrosine | 50-fold |
| Ornithine | 20 |
| Leucine-1-oxoglutarate | 5 |
| Glutamate-pyruvate | 15 |
| Glutamate-oxaloacetate | 1·5 |
| Others | |
| Threonine dehydratase | 300 |
| Arginase | 3 |

varied in the opposite direction in response to changes in the protein intake, the activity being doubled in animals on a protein-free diet.

It was thought important to find out whether adaptive changes of the same kind occur in man in response to different nutritional conditions. Table 2 shows results obtained in liver biopsy specimens taken from children when they were malnourished and after they had recovered (Stephen and Waterlow, 1968). The changes were in precisely the direction expected from the animal work. These measurements of enzyme activity were combined with measurement of the fat content of the liver. At that time fatty infiltration of the liver was thought to be one of the main causes of

TABLE 2

ACTIVITY OF TWO ENZYMES IN BIOPSY SPECIMENS OF LIVER

| | *Amino acid-activating enzymes* ($\mu mol$ P exchanged/ mg protein /h) | *Argininosuccinase* ($\mu mol$ urea/mg protein /h) |
|---|---|---|
| Number of children | 18 | 11 |
| Biopsy: | | |
| Initial          I | 1·44 | 1·06 |
| Intermediate   II | 1·04 | 1·31 |
| Recovered      III | 0·915 | 1·46 |
| Differences between biopsies: | | |
| I and II | $0·05 > P > 0·02$ | Not significant |
| I and III | $0·01 > P > 0·002$ | $P = 0·02$ |

death in our malnourished children and serial biopsies of the liver to control the fat content were thought to be justified.

These results related to liver enzymes, and the question then arose as to whether similar adaptive changes occur in muscle. The amino acid synthetase activity in muscle of rats on a low protein diet was measured and it was found that the activity did not increase in muscle as a result of depletion, but that there was a significant increase when the animals were refed (Stephen, 1968); this agreed with the findings of the Italian workers.

It has been suggested that in both animals and man on low protein intakes the branched chain amino acids become limiting, (e.g. Grimble and Whitehead, 1971) and these amino acids appear to be oxidized mainly, if not entirely in muscle. This question was studied by Reeds (1974) who found that in protein-deficient rats there was a reduced output of $^{14}CO_2$ from 1-$^{14}$C-valine *in vivo*, and a reduced oxidation *in vitro*.

These findings suggest that muscle enzymes can adapt as well as liver enzymes. If the result can be extrapolated to man it could be very important, since in man $60\%$ of the branched chain keto acid dehydrogenase activity of the whole body is said to be in skeletal muscle, compared with only $10\%$ in the rat (Khatra *et al.*, 1977). The difficulties of studies in man are obvious, though perhaps some useful information could be obtained from leucocytes.

All these results are rather old, but, as far as we can find, not much more work has been done on this subject in recent years. Many immunochemical studies have shown that these changes in enzyme activity in general represent changes in the amount of enzyme protein. Therefore this subject is relevant not only to the mechanism of regulation of protein metabolism in the body, but also to the means by which amounts and turnover rates of individual proteins are controlled.

It seems to me that the outstanding questions which remain to be answered are:

How far do the conventional measurements of enzyme activity *in vitro* reflect events *in vivo*? For example, in the urea cycle arginino-succinate synthetase is supposed to be the rate-limiting enzyme. In the experiments of Das and Waterlow (1974) the total liver activity of this enzyme in rats on a low-protein diet corresponded to the production of 5·5 mg N/h: the actual excretion was 3·5 mg N/h. The activity of arginase was 400 times as great.

How is it that the activity of these enzymes varies in a coordinated way—not only the urea cycle enzymes themselves, but also those which have a key note in feeding nitrogen into the cycle—alanine and aspartate aminotransferases and glutamic dehydrogenase?

In some cases it has been shown that changes in the amount of an enzyme are mediated through changes in the availability of m-RNA (Porter, 1978). Even with this advance in understanding, how can one visualize that a single stimulus produces changes in opposite directions in the synthesis and degradation of different enzyme proteins?

## Discussion

*Munro* (*Tufts University, Boston, USA*): This is a useful reminder of an area which needs further exploration. Your final questions are well justified. In the cell, t-RNA is usually fully charged in the normally nourished animal, which suggests that the changes in enzyme activity may have some meaning other than producing lower levels of charging. At the moment the status of t-RNA charging is very complicated because of the presence of multi-enzyme complexes—they apparently function in this form rather than as free enzymes. It may thus be, as you say, that the status of the real enzyme working in its environment differs from that of the enzyme that one is able to measure *in vitro*.

*Cohen* (*University of Wisconsin-Madison, USA*): I think these questions will come up again in relation to the enzymes of urea biosynthesis (p. 215).

*Swick* (*University of Wisconsin-Madison, USA*): We have shown in the pig that by increasing the protein content of the diet the transaminases are induced at the same rate and to the same degree as they are in the rat (Swick and Benevenga, 1973). So I would assume that this sort of induction takes place in man as well.

*Waterlow* (*London School of Hygiene and Tropical Medicine, UK*): I should like to ask Cohen whether he can give us some explanation of how, as he showed also in the tadpole, these enzymes vary in a co-ordinated way.

*Cohen:* We do have some information that bears on this, which suggests that for the two mitochondrial enzymes involved in urea biosynthesis, under certain nutritional conditions, the rate of synthesis is regulated by the messenger RNA level. The large unanswered question which I think will emerge from this discussion is the basis for selective control of the rate of degradation. This is the bottleneck which is urgently in need of more study.

# References

DAS, T. K. and WATERLOW, J. C. (1974). The rate of adaptation of urea cycle enzymes, aminotransferases and glutamic dehydrogenase to changes in dietary protein intake, *Br. J. Nutr.*, **32**, 353–73.

GAETANI, S., PAOLUCCI, A. M., SPADONI, M. A. and TOMASSI, G. (1964). Activity of amino acid-activating enzymes in tissues from protein-depleted rats, *J. Nutr.*, **84**, 173–8.

GRIMBLE, R. F. and WHITEHEAD, R. G. (1971). The effect of an oral glucose load on serum free amino acid concentrations in children before and after treatment for kwashiorkor, *Br. J. Nutr.*, **25**, 253–8.

KHATRA, B. S., CHAWLA, R. J., SEWELL, C. W. and RUDMAN, D. (1977). Distribution of branched chain α-keto acid dehydrogenases in primate tissues, *J. Clin. Invest.*, **59**, 558–64.

KREBS, H. A. (1972). Some aspects of the regulation of fuel supply in omnivorous animals, in *Advances in Enzyme Regulation*, Vol. 10, (Ed. G. Weber), Pergamon Press, New York, pp. 397–420.

PORTER, J. W. (1978). Regulation of fatty acid and glycerolipid metabolism, in *Enzyme Adaptation to Nutritional and Hormonal Changes*, FEBS 11th Meeting, Copenhagen, Vol. 46, Symposium A5, Pergamon Press, Oxford, pp. 41–52.

REEDS, P. J. (1974). The catabolism of valine in the malnourished rat. Studies *in vivo* and *in vitro* with different labelled forms of valine, *Br. J. Nutr.*, **31**, 259–70.

SCHIMKE, R. T. (1962). Adaptive characteristics of urea cycle enzymes in the rat, *J. Biol. Chem.*, **237**, 459–68.

STEPHEN, J. M. L. (1968). Adaptive enzyme changes in liver and muscle of rats during protein depletion and refeeding, *Br. J. Nutr.*, **22**, 153–63.

STEPHEN, J. M. L. and WATERLOW, J. C. (1968). Effect of malnutrition on activity of two enzymes concerned with amino acid metabolism in human liver, *Lancet*, **i**, 118–19.

SWICK, R. W. and BENEVENGA, N. J. (1973). Rapid turnover of liver proteins in a large animal, *Fed. Proc.*, **32**, 507A.

# 4

# Inter-Organ Amino Acid Exchange

P. FELIG

*Department of Internal Medicine,*
*Yale University School of Medicine,*
*New Haven, Connecticut, USA*

## Introduction

The major reservoir of body nitrogen is in the form of amino acids in body proteins. Quantitatively, muscle constitutes the largest reservoir of peptide-bound amino acids. Much smaller amounts of amino acids are present in the free form within cells and in extracellular compartments such as plasma. Despite the relatively small circulating pool of amino acids, studies on net flux of amino acids between various tissues provide insights into the factors regulating overall body nitrogen metabolism. In this paper we will review data on inter-organ amino acid exchange in intact human subjects based on measurements of arterial-venous differences and blood flow. It should be emphasized that such measurements determine *net* exchange of substrates and do not permit conclusions regarding the relative contributions of changes in rates of uptake versus output in determining the overall effect on amino acid exchange of a given physiological or pathological condition.

## The Post-absorptive State

After an overnight 10–14 h fast, there is a net outflow of amino acids from muscle tissue. As reflected by measurements across the deep tissues of the forearm (London *et al.*, 1965; Pozefsky *et al.*, 1969; Felig *et al.*, 1970), or across the leg (arterial-femoral venous differences) (Felig and Wahren, 1971; Felig *et al.*, 1973*b*), there is a net release of virtually all amino acids (Felig, 1975). The pattern of this release is distinctive in that alanine and glutamine are the major amino acids released by peripheral tissues, each accounting for 30–40 % of total α-amino N release (Fig. 1).

45

*P. Felig*

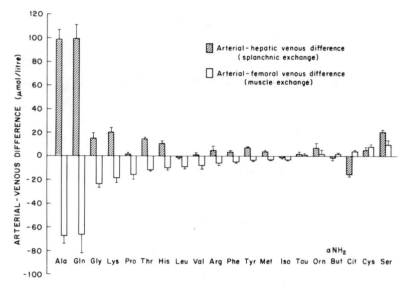

FIG. 1.   Splanchnic and leg exchange of amino acids in normal humans after an overnight fast. The data are presented on arterial-hepatic and arterial-femoral venous differences. From Felig (1975).

Complementing this release of amino acids, a net uptake of amino acids is observed across the splanchnic bed (Felig and Wahren, 1971; Felig *et al.*, 1973*b*; Felig *et al.*, 1969). The pattern of amino acid uptake is again distinctive with alanine and glutamine predominating (Felig and Wahren, 1971; Felig *et al.*, 1973*b*; Felig *et al.*, 1969) (Fig. 1). Since the splanchnic bed includes both hepatic and extrahepatic (intestinal) tissue, it is of particular interest to determine the relative contributions of these organs to total splanchnic amino acid uptake. Studies involving measurements of arterial-portal venous differences obtained at the time of elective cholecystectomy have revealed that the gut is releasing alanine and consuming glutamine (Felig *et al.*, 1973*b*). As in the case of muscle, alanine accounts for 35–40 % of total amino acid output from intestinal tissue (Felig *et al.*, 1973*b*). The rate of glutamine uptake by the gut is such as to account for more than 50 % of the net splanchnic uptake of this amino acid (Felig *et al.*, 1973*b*).

More recently, the metabolic fate of glutamine extracted by gut tissue was examined *in vivo* by means of arterial-venous differences and isotopic tracer studies by Windmueller and Spaeth (1978). In perfused segments of rat jejunum, over 50 % of the extracted glutamine was converted to $CO_2$, accounting for 35 % of total $CO_2$ production by intestine. In contrast, only

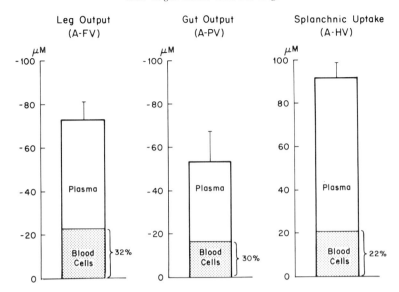

FIG. 2.  The proportion of net exchange of alanine across the leg, gut and splanchnic tissues contributed by plasma and by blood cells. A – FV = arterial-femoral venous difference; A – PV = arterial-portal venous differences; A – HV = arterial-hepatic venous differences. Based on the data of Felig *et al.* (1973).

4 % of glutamine carbon was converted to alanine; over 80 % of the alanine carbon released by intestine was derived from glucose. These data thus indicate that glutamine is an important oxidative fuel for intestine. The findings also demonstrate that no more than 30 % of the carbon skeletons of glutamine taken up by the gut are released and made available to the liver as gluconeogenic substrate in the form of lactate, alanine or other amino acids (Windmueller and Spaeth, 1978).

In addition to muscle, liver and gut, a fourth major organ involved in amino acid exchange is the kidney. The kidney is an important site of uptake of glutamine which provides the nitrogen for urinary ammonia excretion (Owen and Robinson, 1963). The kidney also releases serine and small amounts of alanine (Owen and Robinson, 1963). The production of serine by the kidney is of particular interest since this amino acid is normally being taken up by muscle as well as liver tissue (Fig. 1).

Finally, limited observations on arterial-jugular venous differences have also shown a net uptake of most plasma amino acids by the brain, notably valine and proline (Felig *et al.*, 1973a). However, the pattern of amino acid exchange in various areas of the brain may be quite distinctive and cannot be gleaned from arterial-jugular venous measurements.

A major question raised by Elwyn in studies on protein-fed dogs concerns the importance of blood cells, particularly red blood cells, in the exchange of amino acids among various organs (Elwyn et al., 1972). In confirmation of Elwyn's observations, studies in man demonstrated that blood cells contribute 20–30 % of total alanine flux from muscle and gut to liver (Felig et al., 1973b) (Fig. 2). In the case of glutamine, plasma measurements overestimate total splanchnic glutamine uptake, suggesting a net release of glutamine from plasma to blood cells within the splanchnic bed (Felig et al., 1973b).

From a quantitative standpoint, it is of interest to compare the rates of observed net flux of alanine and glutamine from muscle to the splanchnic bed with the known rates of urea production in 24-h fasted man. As pointed out by Lund elsewhere in this symposium (Paper 12), assuming the rates of alanine and glutamine flux observed in the post-absorptive state (Felig et al., 1973b) were to continue for 24 h, one can estimate a total net release of these amino acids of approximately 180 mmol/day which would provide 10–11 g urea-N per day. The latter figure corresponds quite well with urinary urea-N losses observed in 24-h fasted man (Cahill et al., 1966).

## The Glucose–Alanine Cycle

The unique role of alanine in inter-organ amino acid exchange in the post-absorptive state, its rapid conversion by liver to glucose (Mallette et al., 1969), and the stimulatory effects of exercise on alanine production (see below), led to the formulation of the glucose–alanine cycle (Felig et al., 1970; Mallette et al., 1969; Felig, 1973). In this cycle, which is analogous to the Cori cycle for lactate, glucose taken up by muscle is converted to pyruvate which in turn is transaminated to form alanine. The alanine is then released by muscle and taken up by liver where it is reconverted to glucose (Fig. 3).

Two questions which immediately arise regarding the alanine cycle are: (1) of the total alanine released by muscle what proportion represents release of preformed alanine from muscle proteins and what proportion constitutes newly synthesized alanine? and (2) what are the sources of the nitrogen and pyruvate which are utilized for alanine synthesis? Specifically, is the pyruvate generated for alanine synthesis derived from glucose or is it an end product in the degradation of other amino acids?

LIVER        BLOOD        MUSCLE

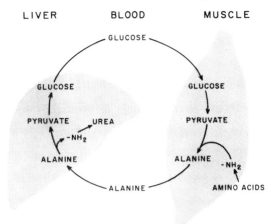

FIG. 3.   The glucose–alanine cycle.

The studies on human forearm and leg permit an estimation of the proportion of total alanine release which is attributable to *de novo* synthesis. Such estimations are based on the observation that lysine and alanine are present in muscle proteins in comparable amounts (Kominz *et al.*, 1954), and the assumption that lysine does not undergo transamination. Lysine release from muscle thus serves as an index of 'protein-derived' alanine. The remainder of the alanine released thus represents 'synthesized' alanine. By these calculations, we have estimated that 66–72 % of the total alanine released from muscle tissue is synthesized *in situ* (Table 1) (Felig, 1973).

With respect to the source of the amino groups for alanine production, *in vitro* studies have provided strong evidence for an important role for the branched chain amino acids, leucine, isoleucine and valine (Buse *et al.*, 1972; Odessey *et al.*, 1974; Fulks *et al.*, 1975). In support of the latter observations are studies in intact man demonstrating that after protein feeding, the branched chain amino acids escape hepatic uptake and are taken up by muscle in amounts exceeding that of other amino acids (Wahren *et al.*, 1976) (see below).

Studies *in vitro* designed to determine the source of the carbon skeletons for alanine production in skeletal muscle have yielded somewhat conflicting results. The studies of Goldberg and colleagues in diaphragm, soleus muscle and extensor digitorum longus have provided strong evidence that approximately 60 % of the carbon skeletons in alanine are derived from

TABLE 1

ESTIMATED PROPORTION OF MUSCLE ALANINE RELEASE
WHICH IS SYNTHESIZED *IN SITU*[a]

*Leg*
   A − FV, Alanine[b] = − 68 μmol/litre
   A − FV, Lysine = − 19 μmol/litre
   A − FV, 'Synthesized alanine'[c] = − 49 μmol/litre
   % Alanine synthesized = 72%

*Forearm muscle*
   A − DV, Alanine[d] = − 111 μmol/litre
   A − DV, Lysine = − 37 μmol/litre
   A − DV, 'Synthesized alanine'[c] = − 74 μmol/litre
   % Alanine synthesized = 67%

[a] Based on the data of Felig *et al.* (1970) and Felig and Wahren (1971).
[b] A − FV: Arterial-femoral venous difference.
[c] A − V: 'Synthesized alanine': (A − V) Alanine−(A − V) Lysine. See text for details.
[d] A − DV: Arterial-deep venous difference.

glucose (Fulks *et al.*, 1975; Chang and Goldberg, 1978*a*). The latter figure fits well with the estimated rates of net alanine synthesis determined by forearm and leg balance studies in intact humans (Table 1) and suggests that virtually all of the newly synthesized alanine in muscle tissue is formed via the glucose–alanine cycle. In contrast to the work of Goldberg, other studies have provided evidence that the carbon skeletons of alanine may be derived from other amino acids which are catabolized to citric acid cycle intermediates and then undergo decarboxylation reactions (Snell and Duff, 1977; Lee and Davis, 1979). While muscle clearly does possess the enzymatic capacity for such reactions (Lee and Davis, 1979), their quantitative importance in alanine synthesis has not been established. Thus the quantitative isotopic data which are available (Fulks *et al.*, 1975; Chang and Goldberg, 1978*a*) strongly favour the conclusion that most of the carbon skeletons for the synthesis of alanine by muscle tissue are derived from glucose.

## *Effect of Exercise*

One of the earliest physiological conditions to be examined with respect to its effects on inter-organ amino acid exchange was the effect of muscular exercise (Felig and Wahren, 1971). In healthy subjects performing leg exercises at various intensities, an increase in arterial alanine was observed which correlated directly with the rise in plasma pyruvate (Felig *et al.*, 1970). Furthermore, net release of alanine from the exercising extremity was stimulated in proportion to the intensity of the exercise performed. The changes in alanine were particularly noteworthy because exercise failed to raise the plasma levels of other amino acids nor did it stimulate the net release of other amino acids from muscle tissue.

The relationship of alanine production to glucose utilization by muscle is particularly evident in exercise. A direct linear relationship between arterial alanine and pyruvate concentrations is observed in exercising subjects (Felig *et al.*, 1970). Secondly, in subjects with McArdle's syndrome in whom pyruvate production is markedly limited because of lack of myophosphorylase, a corresponding decline in alanine production is observed (Wahren *et al.*, 1973). Thirdly, in patients with a myopathy characterized by excessive production and accumulation of pyruvate and lactate there is an accompanying exaggeration of alanine production (Wahren *et al.*, 1979).

A major problem concerning augmented formation and release of alanine by muscle has been the traditional view that exercise fails to stimulate net protein breakdown as reflected by urine nitrogen excretion (Young *et al.*, 1966). On the other hand, more recent studies have shown a net increase in urea production during exercise (Lemon and Mullin, 1980). Furthermore, studies employing stable isotopes have shown an increased formation of $CO_2$ from leucine during exercise (Wolfe *et al.*, 1981). Noteworthy in this regard is the observation that prolonged starvation is associated with a net flux of branched chain amino acids from the splanchnic bed to exercising muscles (Ahlborg *et al.*, 1974). Thus, increased oxidation of branched chain amino acids may provide the nitrogen for augmented alanine synthesis in exercise.

## *Starvation*

When food is withheld beyond 12–14 h, changes are observed in amino acid flux between muscle and liver which depend upon the duration of the fast. With relatively short term fasts (2–3 days), there is an increased release of alanine from muscle (Pozefsky *et al.*, 1976) as well as increased uptake by the splanchnic bed (Felig *et al.*, 1969). The augmented hepatic uptake of alanine in short term fasting is due in part to increased fractional extraction

*P. Felig*

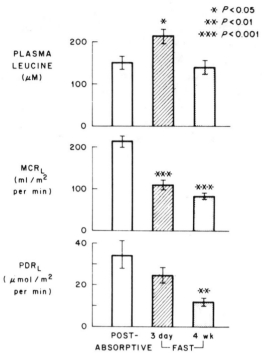

FIG. 4.   The effects of starvation on the plasma concentration, metabolic clearance rate (MCR) and the plasma delivery rate (PDR) of leucine. From Sherwin (1978).

of this amino acid by the liver (Felig *et al.*, 1969). In contrast, after prolonged fasting (3–6 weeks), there is a marked reduction in muscle output as well as splanchnic uptake of alanine (Felig *et al.*, 1970; Felig *et al.*, 1969). The hyperketonaemia which occurs in prolonged fasting has been postulated to be responsible in part for the fall in muscle alanine output (Sherwin *et al.*, 1975; Palaiologos and Felig, 1976). However more recent studies have cast doubt on this hypothesis (Féry and Balasse, 1980).

The role of organ exchange in determining the metabolism of branched chain amino acids in starvation is less clear cut than in the case of alanine. Early in starvation (2–7 days), an increase in plasma leucine, isoleucine and valine is observed (Felig *et al.*, 1969). As starvation extends beyond one week there is a progressive decline in the plasma levels of each of these amino acids (Felig *et al.*, 1969). The initial rise in circulating branched chain amino acids occurs in the absence of increased release of these amino acids from either muscle (Pozefsky *et al.*, 1976) or liver tissue (Felig *et al.*, 1969).

A progressive decrease in the metabolic clearance rate of leucine has, however, been observed throughout starvation (Fig. 4) (Sherwin, 1978). The initial rise in levels of branched chain amino acids in plasma is thus a consequence of decreased clearance in association with unchanged rates of net release from muscle tissue. The later decline in plasma levels reflects the fact that net release from muscle tissue (Felig *et al.*, 1970) falls to an even greater extent than the overall rate of removal from plasma (Sherwin, 1978).

It should be noted that in contrast to the evidence of decreased removal of leucine from plasma *in vivo*, studies *in vitro* have shown an increase in leucine oxidation by muscle tissue from fasted rats (Goldberg and Odessey, 1972). Perhaps this seeming discrepancy can be resolved by the observation that oxidation of branched chain amino acids by adipose tissue is reduced in starvation (Frick and Goodman, 1979). Although fat tissue has generally not been considered to be of overriding importance in the metabolism of branched chain amino acids, it may in fact be responsible for the reduced utilization of these amino acids observed during starvation.

An unanticipated finding in the studies involving leucine infusions in prolonged starvation was the demonstration by Sherwin that such infusions result in a rise in plasma glucose (Sherwin, 1978). Subsequent studies *in vitro* have shown that leucine inhibits glucose utilization by brain slices (Palaiologos *et al.*, 1979) and incubated diaphragm (Chang and Goldberg, 1978*b*) from fasted rats. Thus, if plasma leucine levels become sufficiently elevated during starvation, they may to some extent be similar to ketones in replacing glucose as an oxidative fuel for brain as well as other tissues.

## Protein Feeding

In as much as the fasting state is characterized by a net release of amino acids from muscle tissue, the maintenance of body muscle mass necessitates a net repletion of muscle amino acids after feeding. The pattern of this net amino acid flux from the splanchnic bed to muscle tissue after the ingestion of a protein meal is quite distinctive (Wahren *et al.*, 1976).

After protein feeding there is a large amino acid release from the splanchnic bed predominantly involving the branched chain amino acids (Wahren *et al.*, 1976). The latter account for more than half of total splanchnic amino acid output, despite the fact that they contribute only 20 % of the amino acid residues in the ingested protein. Furthermore, large increments (100–200 %) are seen in the arterial concentrations of the branched chain amino acids while other amino acids show substantially

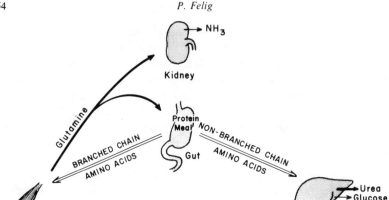

FIG. 5.    The effect of protein feeding on inter-organ amino acid exchange. Based on the data
of Wahren *et al.* (1976).

smaller (5–50%) increments. Most importantly, the peripheral (muscle) exchange of most amino acids reverts from a net output to a net uptake after protein feeding. As in the case of the splanchnic exchange, the branched chain amino acids account for more than half of the total amino acid uptake. Interestingly, the net flux of alanine and glutamine from muscle tissue to the splanchnic bed which occurs in the post-absorptive state persists after the ingestion of a protein meal (Wahren *et al.*, 1976).

More recent studies in dogs have provided some clarification of the importance of hepatic and extrahepatic tissues in the metabolism of the constituent amino acids in a protein meal. In unanaesthetized dogs with chronically implanted catheters in the portal vein, hepatic vein and an artery, Barrett *et al.* (1981) have examined amino acid flux after oral administration of an amino acid mixture. Approximately 80–90% of the administered load of branched chain amino acids appears in the portal vein. Of that which appears in the portal vein, 15–20% is extracted by the liver and the remainder (80–85%) escapes hepatic uptake and is made available for peripheral (muscle) tissues. In the case of the non-essential glycogenic amino acids, particularly alanine and glycine, there is a net production of these amino acids by the gut, so that the amounts appearing in the portal vein exceed those ingested. Furthermore, there is virtually complete net removal by the liver of the ingested and gut-produced alanine and glycine.

These observations in humans and dogs thus indicate that after a protein meal there is a clear distinction between the branched chain amino acids and all other amino acids with respect to uptake by the liver as well as muscle (Fig. 5). Of the absorbed branched chain amino acids, only 15–20 % are taken up by the liver. The large proportion of ingested branched chain amino acids escape hepatic uptake and are taken up by muscle tissue. The branched chain amino acids thus constitute the principal source of nitrogen repletion after protein feeding. In contrast, all other amino acids are largely taken up by the liver and are made available to muscle in much smaller amounts. Furthermore, net muscle release of alanine and glutamine persists after a protein meal.

## Diabetes

Insulin has long been recognized as having a primary role in the regulation of amino acid metabolism. In accordance with this concept, abnormalities in the hepatic and muscle exchange of amino acids have been observed in diabetic patients in the fasted state as well as after protein feeding.

In the fasted state, a decrease in plasma alanine is observed in the diabetic in association with an increase in net uptake of this amino acid by the splanchnic bed (Wahren *et al.*, 1972). The augmented splanchnic uptake of alanine is a consequence of increased fractional extraction rather than increased substrate delivery (Wahren *et al.*, 1972). Thus, in the absence of adequate insulin, the liver increases its uptake of alanine, presumably as a consequence of stimulation of hepatic gluconeogenesis.

An elevation in plasma concentrations of branched chain amino acids is also observed in the diabetic (Wahren *et al.*, 1972). Studies *in vitro* have demonstrated an increase in the rate of oxidation of these amino acids by muscle tissue obtained from diabetic rats (Buse *et al.*, 1976). Thus, as in the case of starvation, increased circulating levels of branched chain amino acids in the diabetic (in the post-absorptive state) may reflect decreased utilization of these amino acids in an extramuscular site, notably adipose tissue.

Exaggerated elevations of the branched chain amino acids are also observed in the diabetic after the feeding of a protein meal (Wahren *et al.*, 1976). In such patients net escape of branched chain amino acids from the splanchnic bed is comparable to that in controls. However, the uptake of these amino acids by peripheral (muscle) tissue is reduced. These observations suggest that the rise in plasma insulin which normally accompanies protein feeding may be a major regulatory factor in the repletion of nitrogen by muscle tissue. In the absence of adequate insulin this anabolic response is reduced.

## Discussion

*Harper* (*University of Wisconsin-Madison, USA*): In relation to the output of branched chain amino acids from the splanchnic area, I am always concerned when more comes out than goes in. You are measuring the concentration differences between arterial and hepatic venous blood, without taking account of the concentrations in the portal vein. Could this lead to an overestimate of the output?

*Alleyne* (*University of West Indies, Kingston, Jamaica*): On a similar point, the alanine taken up by the liver is a combination of hepatic arterial alanine and portal vein alanine. You have assumed that it is the hepatic arterial alanine that is the main determinant of uptake by the liver. Could it be that the portal vein alanine derived from glutamine is equally important?

*Felig:* I am sure it is important, but we do not have data on portal alanine concentrations in man in conditions such as feeding and fasting. When we do get this information, some of our conclusions may have to be modified. Recently, however, we have been making some studies in the triply catheterized, unanaesthetized dog, and we can calculate the hepatic and extra-hepatic balances after a protein meal. About 85 % of the branched chain amino acids in the meal enter the portal blood. The remaining 15 % is either not absorbed or is metabolized in the gut. Of that which is absorbed, 25–30 % is taken up by the liver and the remainder goes to the peripheral tissues.

*James* (*Dunn Clinical Nutrition Centre, Cambridge, UK*): You have shown a loss of a whole range of different amino acids from muscle during fasting, but most of the output from the splanchnic area after a meal seems to be in the form of branched chain amino acids. Surely muscle has to get the other amino acids back somehow?

*Felig:* In the first 30–60 min there is an uptake of other amino acids as well.

*James:* Is the uptake by muscle equivalent to the amount that you calculate to be coming out from the periphery during fasting?

*Felig:* It is probably less during the later part of the period. However, you have to remember that these are studies in the post-absorptive state.

*Garlick* (*London School of Hygiene and Tropical Medicine, UK*): An explanation for the continued production of glutamine by muscle during feeding is the high rate of amino acid oxidation at this time. In measurements with leucine in man, we find that the rate of oxidation is $2\frac{1}{2}$ times as high in the fed as in the fasted state (Garlick *et al.*, 1980).

*Reeds (Rowett Research Institute, Aberdeen, UK):* Dr Felig, you made the point that one cannot interpret limb or forearm studies only in terms of skeletal muscle, and you raised the subject of adipose tissue. I suspect that for protein metabolism a much more important component is skin and bone. Lobley and I (unpublished, but see Lobley *et al.*, 1980) and, independently, Preedy and Garlick (1981), have shown very high rates of protein synthesis in skin and bone, an observation made originally by Waterlow and Stephen (1966). A substantial proportion of this probably involves collagen. I am not sure how this affects the metabolism, particularly of glutamate, across the forearm or leg.

*Felig:* Your point is well taken. There is evidence that skin may be an important glycolytic tissue and hence a major site of lactate production. One could then envisage a substantial outflow of alanine in association with a rapid rate of protein turnover. However, with the techniques available to us we cannot make that kind of partitioning of what is happening in the forearm. We have made some studies of the concentration differences between arterial and superficial venous blood. The A-V differences are smaller (i.e. there is less output of amino acids) than those found with the deep tissues, but there are differences in blood flow as well.

*Rennie (University College Hospital Medical School, London, UK):* Do you have any information about the lung—a tissue which is not inconsiderable from a metabolic point of view?

*Felig:* We have enough difficulty in studying changes in the splanchnic bed in intact subjects, let alone the pulmonary bed!

*Alleyne:* You showed that there is an increased uptake of glutamine in red blood cells in transit to the gut. This is a remarkable phenomenon, and I wonder whether you have any explanation for it.

*Felig:* I have no idea what is determining uptake into red cells or efflux from them in blood going through the splanchnic bed.

*Harper:* The effect of exercise is very interesting. I get the impression that the increased output of alanine during exercise is a function of pyruvate accumulation. What happens if you exercise a starved individual? Does the alanine output go up? If it does, one might think of this as a protective mechanism against acidosis; a reduced output of lactate into the blood would have survival value, particularly in people who are likely to be exercising when they are very hungry.

*Felig:* I do not know what happens to alanine under those conditions. However, Phinney *et al.* (1980), in a study on obese subjects who were put on a so-called protein-modified fast, showed that at 3 days of fasting exercise performance was markedly diminished, but with a longer fast it improved. At this stage the subjects are replacing glucose as a major fuel with the ketones which become available.

*Alleyne:* The output from resting muscle is alanine plus glutamine. You showed that in exercise it is the output of alanine which increases. What determines the balance between transamination of pyruvate and amidation of glutamate?

*Felig:* I suppose that it is substrate-dependent, and related to the increased availability of pyruvate.

*Waterlow* (*London School of Hygiene and Tropical Medicine, UK*): What about the big increase in ammonia output from the purine nucleotide cycle during exercise? Is that got rid of as alanine?

*Felig:* There may be some question about the quantitative significance of ammonia release through the purine nucleotide cycle. From our standpoint the key issue is that exercise does increase the production of ammonia from muscle. Reasoning teleologically, we and others before us have suggested that alanine may represent a non-toxic alternative to the release of free ammonia.

*Young* (*Massachusetts Institute of Technology, USA*): The increased output of alanine during exercise, in conjunction with the glucose–alanine cycle, suggests that exercise should bring about an increased loss of body nitrogen, and hence increase the total protein requirement. Do you have nitrogen balance data for your exercising subjects?

*Felig:* We do not. Some years ago Munro called attention to studies on an expedition in the Swiss Alps, which suggested that there is no increased loss of nitrogen with exercise. However, more recent work from a number of laboratories suggests that there may well be an increase in protein breakdown. One may speculate whether protein in muscle might not be analogous to glycogen. We know that, with exercise, glycogen is depleted, but that repeated exercise stimulates an increase in glycogen stores during resting periods. Is muscle hypertrophy with repeated exercise a comparable phenomenon? During the period of exercise, which even in a marathon is only a small proportion of the 24-h period, there may be an increase in protein breakdown, followed by a period of anabolism, perhaps depending on the type of exercise.

*Munro (Tufts University, Boston, USA):* The event to which you referred is the expedition of Fick and Wislicenus on the Faulhorn on 30 August 1865. They performed the analyses of their urines in a hotel on the top of the mountain.

*Millward (London School of Hygiene and Tropical Medicine, UK):* In answer to Young's question, there is one study in which subjects exercised regularly over long periods of time (Gontzea *et al.*, 1975). In the first week or so there was a decrease in nitrogen balance, but after a further week the subjects adapted and moved back into balance. We have also examined the effect of acute exercise on nitrogen balance (see p. 509).

## *References*

AHLBORG, G., FELIG, P., HAGENFELDT, L., HENDLER, R. and WAHREN, J. (1974). Substrate turnover during prolonged exercise in man: splanchnic and leg metabolism of glucose, free fatty acids and amino acids, *J. Clin. Invest.*, **53**, 1080–90.

BARRETT, E., FERRANNINI, E., GUSBERG, R., FELIG, P. and DEFRONZO, R. (1981). Amino acid metabolism by liver and intestinal tissues following oral and i.v. amino acids, *Clin. Research*, in press (abstract).

BUSE, M. G., BIGGERS, J. F., FRIDERICI, K. H. and BUSE, J. F. (1972). Oxidation of branched chain amino acids by isolated hearts and diaphragm of the rat. The effect of fatty acids, glucose and pyruvate, *J. Biol. Chem.*, **247**, 8085–96.

BUSE, M. G., HERLONG, H. F. and WEIGARD, D. A. (1976). The effect of diabetes, insulin and the redox potential on leucine metabolism by isolated rat hemidiaphragm, *Endocrinology*, **98**, 1166–75.

CAHILL, G. F., JR., HERRERA, M. G., MORGAN, A. P., SOELDNER, J. S., STEINKE, J., LEVY, P. L., REICHARD, G. A., JR. and KIPNIS, D. M. (1966). Hormone–fuel interrelationships during fasting, *J. Clin. Invest.*, **45**, 1751–69.

CHANG, T. W. and GOLDBERG, A. L. (1978a). The origin of alanine produced in skeletal muscle, *J. Biol. Chem.*, **253**, 3677–84.

CHANG, T. W. and GOLDBERG, A. L. (1978b). Leucine inhibits oxidation of glucose and pyruvate in skeletal muscle during fasting, *J. Biol. Chem.*, **253**, 3696–701.

ELWYN, D. H., LAUNDIER, W. J., PARIKH, H. C. and WIRE, E. M., JR. (1972). Roles of plasma and erythrocytes in inter-organ transport of amino acids in dogs, *Am. J. Physiol.*, **222**, 1333–42.

FELIG, P. (1973). The glucose-alanine cycle, *Metabolism*, **22**, 179–207.

FELIG, P. (1975). Amino acid metabolism in man, *Annual Rev. Biochem.*, **44**, 933–55.

FELIG, P., OWEN, O. E., WAHREN, J. and CAHILL, G. F., JR. (1969). Amino acid metabolism during prolonged starvation, *J. Clin. Invest.*, **48**, 584–94.

FELIG, P., POZEFSKY, T., MARLISS, E. and CAHILL, G. F., JR. (1970). Alanine: Key role in gluconeogenesis, *Science*, **167**, 1003–4.

FELIG, P. and WAHREN, J. (1971). Amino acid metabolism in exercising man, *J. Clin. Invest.*, **50**, 2703–14.

FELIG, P., WAHREN, J. and AHLBORG, G. (1973a). Evidence for the uptake of amino acids by the human brain, *Proc. Soc. Exp. Biol. Med.*, **142**, 230–1.

FELIG, P., WAHREN, J. and RÄF, L. (1973b). Evidence of inter-organ amino acid transport by blood cells in man, *Proc. Nat. Acad. Sci. USA*, **70**, 1775–9.

FÉRY, F. and BALASSE, E. O. (1980). Differential effects of sodium acetoacetate and acetoacetic acid infusions on alanine and glutamine metabolism in man, *J. Clin. Invest.* **66**, 323–31.

FICK, A. and WISLICENUS, J. (1865). Cited by Munro, H. N. in *Mammalian Protein Metabolism* (Eds. H. N. Munro and J. B. Allison), Vol. 1, pp. 1–28, Academic Press, New York, (1964).

FRICK, G. P. and GOODMAN, H. M. (1979). Physiological regulation of leucine metabolism in rat adipose tissue, *Fed. Proc.*, **38**, 947.

FULKS, R. M., LI, J. B. and GOLDBERG, A. L. (1975). Effects of insulin glucose, and amino acids on protein turnover in rat diaphragm, *J. Biol. Chem.*, **250**, 290–8.

GARLICK, P. J., CLUGSTON, G. A., SWICK, R. W. and WATERLOW, J. C. (1980). Diurnal pattern of protein and energy metabolism in man, *Am. J. clin. Nutr.*, **33**, 1983–6.

GOLDBERG, A. L. and ODESSEY, R. (1972). Oxidation of amino acids by diaphragms from fed and fasted rats, *Am. J. Physiol.*, **223**, 1384–91.

GONTZEA, I., SUTZESCU, R. and DUMITRACHE, S. (1975). The influence of adaptation to physical effort on nitrogen balance in man, *Nut. Rep. Intl.*, **11**, 231–6.

KOMINZ, D. R., HOUGH, A., SYMOND, P. and LAKI, K. (1954). The amino acid composition of actin, myosin, tropomyosin, and the meromysines, *Arch. Biochem. Biophys.*, **50**, 148.

LEE, S. H. and DAVIS, E. J. (1979). Carboxylation and decarboxylation reactions. Anaplerotic flux and removal of citrate cycle intermediates in skeletal muscle, *J. Biol. Chem.*, **254**, 420–30.

LEMON, P. W. R. and MULLIN, J. P. (1980). Effect of initial muscle glycogen levels on protein catabolism during exercise, *J. Appl. Physiol.*, **48**, 624–9.

LOBLEY, G. E., MILNE, V., LORIE, J. M., REEDS, P. J. and PENNIE, K. (1980). Whole body and tissue protein synthesis in cattle, *Br. J. Nutr.*, 43, 491–501.

LONDON, D. R., FOLEY, T. H. and WEBB, C. G. (1965). Evidence for the release of amino acids from the resting human forearm, *Nature*, **208**, 588–9.

MALLETTE, L. E., EXTON, J. H. and PARK, C. R. (1969). Control of gluconeogenesis from amino acids in the perfused rat liver, *J. Biol. Chem.*, **244**, 5713–23.

ODESSEY, R. E., KHAIRALLAH, E. A. and GOLDBERG, A. L. (1974). Origin and possible significance of alanine production by skeletal muscle, *J. Biol. Chem.*, **249**, 7623–9.

OWEN, E. E. and ROBINSON, R. R. (1963). Amino acid extraction and ammonia metabolism in the human kidney during the prolonged administration of ammonium chloride., *J. Clin. Invest.*, **42**, 263–76.

PALAIOLOGOS, G. and FELIG, P. (1976). Effects of ketone bodies on amino acid metabolism in isolated rat diaphragm, *Biochem. J.*, **154**, 709–16.

PALAIOLOGOS, G., KOIVISTO, V. A. and FELIG, P. (1979). Interaction of leucine, glucose and ketone metabolism in rat brain in vitro, *J. Neurochem.*, **32**, 67–72.

PHINNEY, S. D., HORTON, E. S., SIMS, E. A. H., HANSON, J. S., DANFORTH, E. JR. and LA GRANGE, B. M. (1980). Capacity for moderate exercise in obese subjects after adaptation to a hypocaloric, ketogenic diet, *J. Clin. Invest.*, **66**, 1152–61.

POZEFSKY, T., FELIG, P. TOBIN, J., SOOELDNER, J. S. and CAHILL, G. F., JR. (1969). Amino acid balance across the tissues of the forearm in postabsorptive man: effects of insulin at two dose levels, *J. Clin. Invest.*, **48**, 2273–82.

POZEFSKY, T., TANCREDI, R. G., MOXLEY, R. T., DUPRE, J. and TOBIN, J. D. (1976). Effects of brief starvation on muscle amino acid metabolism in non-obese man, *J. Clin. Invest.*, **57**, 444–9.

PREEDY, V. R. and GARLICK, P. J. (1981). Rates of protein synthesis in skin and bone and their importance in the assessment of protein degradation in the perfused rat hemicorpus, *Biochem. J.*, **194**, 373–6.

SHERWIN, R. S. (1978). Effect of starvation on the turnover and metabolic response to leucine, *J. Clin. Invest.*, 61, 1471–81.

SHERWIN, R., HENDLER, R. G. and FELIG, P. (1975). Effect of ketone infusions on amino acid and nitrogen metabolism in man, *J. Clin. Invest.*, **55**, 1382–90.

SNELL, K. and DUFF, D. A. (1977). The release of alanine by rat diaphragm muscle *in vitro*, *Biochem. J.*, **162**, 399–403.

WAHREN, J., FELIG, P., CERASI, E. and LUFT, R. (1972). Splanchnic and peripheral glucose and amino acid metabolism in diabetes mellitus, *J. Clin. Invest.*, **51**, 1870–8.

WAHREN, J., FELIG, P. and HAGENFELDT, J. (1976). Effect of protein ingestion on splanchnic and leg metabolism in normal man and in patients with diabetes mellitus, *J. Clin. Invest.*, **57**, 987–99.

WAHREN, J., FELIG, P., HAVEL, R. J., JORFELDT, L., PERNOW, B. and SALTIN, B. (1973). Amino acid metabolism in McArdle's syndrome, *New Engl. J. Med.*, **288**, 774–7.

WAHREN, J., LINDERHOLDM, H. and FELIG, P. (1979). Amino acid metabolism in patients with a hereditary myopathy and paroxysmal myoglobinuria, *Acta Med. Scand.*, **206**, 309–14.

WATERLOW, J. C. and STEPHEN, J. M. L. (1966). Adaptation of the rat to a low-protein diet: the effect of a reduced protein intake on the pattern of incorporation of L-[$^{14}$C]lysine, *Br. J. Nutr.*, **20**, 461–84.

WINDMUELLER, N. G. and SPAETH, A. E. (1978). Identification of ketone bodies and glutamine as the major respiratory fuels in vivo for post-absorptive rat small intestine, *J. Biol. Chem.*, **253**, 69–76.

WOLFE, R. R., GOODENOUGH, R., ROYLE, G., WOLFE, M. and NADEL, E. (1981). Leucine oxidation during exercise in humans, *Fed. Proc.*, **40**, 900.

YOUNG, D. R., PELLIGRA, R. and ADACKI, R. R. (1966). Serum glucose and free fatty acids in man during prolonged exercise, *J. Appl. Physiol.*, **21**, 1047–52.

# 5

# Protein Utilization and Glucose Production in Fasting Malnourished and Hypoglycaemic Children

D. S. KERR

*Rainbow Babies' and Children's Hospital,*
*Case Western Reserve University, Cleveland, Ohio, USA*

The classical studies of Benedict demonstrated that after prolonged starvation in man, nitrogen excretion, reflecting net utilization of protein, is minimized (Benedict, 1915). The investigations of Cahill and co-workers related these observations to utilization of amino acids for gluconeogenesis and emphasized the importance of fat in substituting for glucose as an energy source, thereby sparing protein (Owen *et al.*, 1969). Alanine was found to be the major gluconeogenic amino acid extracted by the liver in the starved state, accounting for approximately 10 % of total gluconeogenesis (Felig *et al.*, 1969). Shortly thereafter it was found that children with fasting 'ketotic' hypoglycaemia have low plasma levels of alanine; it was proposed that this reflected a deficiency of substrate for gluconeogenesis (Pagliara *et al.*, 1972). According to the concepts of the glucose–alanine cycle, alanine carbon could be derived either from glucose or from gluconeogenic amino acids (Felig, 1973). It is still not clear what adaptive mechanisms link glucose homoeostasis and protein conservation and how this might be deranged in an abnormal state such as childhood hypoglycaemia.

Severely malnourished infants have markedly decreased body fat protein, as confirmed by tissue analyses conducted by Garrow, Alleyne and co-workers (Garrow *et al.*, 1965; Alleyne and Scullard, 1969). In our studies we considered how the substrate deficiency of severe infant undernutrition would affect fasting energy balance and glucose production (Kerr *et al.*, 1978*a* and *b*). Similar studies were later carried out in older children with fasting hypoglycaemia to determine the relationships between plasma alanine, net protein catabolism, and glucose production (Kerr *et al.*, 1981; Kerr and Picou, 1981). To monitor the metabolic transitions during the fasting state in children, nitrogen excretion was determined as a measure of net protein catabolism, oxygen consumption and carbon dioxide

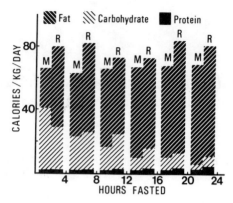

FIG. 1.   Sources of energy in fasting malnourished and recovered infants as determined from indirect calorimetry (average results from five infants). M, malnourished; R, recovered.

production were determined as measures of glycogen and fat oxidation, and glucose production was determined by constant infusion of U-$^{13}$C-glucose.

Fasting metabolism was compared in five infants before and after recovery from malnutrition. To avoid differences due to the immediately preceding diet, studies were carried out after 3 days of a diet which provided minimal maintenance energy and protein requirements. After food was withdrawn, total energy utilization remained relatively constant (Fig. 1). Oxidation of carbohydrate was replaced by that of fat which provided more than 90% of total energy within 24 h of fasting. Utilization of protein remained relatively small, but increased as a result of fasting much more when the children were recovered than when they were malnourished. Total nitrogen excretion approximately doubled in the recovered children, but increased very little in the malnourished state (Fig. 2). The predominant form of urinary nitrogen was urea. Plasma urea reflected the same difference. By the time glycogen oxidation was nearly complete, the availability of potentially gluconeogenic amino acids was twice as great in the recovered infants.

Plasma glucose decreased in both groups as glycogen was depleted but was not lower in the malnourished group. Total glucose flux, measured by dilution of constantly infused U-$^{13}$C-glucose in the steady state following completion of glycogen oxidation, was very similar in the two groups, approximately 3 mg/kg/min (Fig. 3). The calculated availability of various potential sources of glucose was estimated simultaneously. These data are superimposed on the estimation of total flux. It is not known if these potential substrates were in fact converted to glucose. Since continued

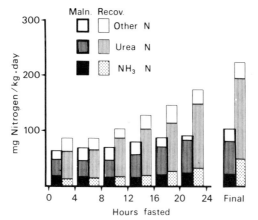

FIG. 2. Urinary nitrogen excretion in fasting malnourished and recovered infants (reproduced from *Metabolism* with permission of the publisher; Kerr *et al.*, 1978*a*).

oxidation of glycogen was very small, most of the glucose production must have been due to gluconeogenesis. Glycerol could have provided about 20 % of the total glucose. In the malnourished infants gluconeogenic amino acids could not have provided more than 10 % of total glucose, whereas in the recovered state amino acids could have accounted for as much as 20 %. The major fraction of glucose production is unaccounted for by any of these sources and must be due to recycling of glucose carbon via lactate, pyruvate, and alanine. It can be calculated that because of dilution

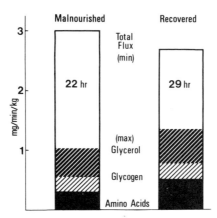

FIG. 3. Comparison of total fasting plasma glucose flux and possible sources of glucose. Glucose flux was determined by dilution of constantly infused U-[13]C-glucose. The possible sources of glucose were calculated from the indirect calorimetry data (Fig. 1).

by fatty acid carbon at the oxalacetate level, very little $^{13}$C from glucose would be expected to recycle under these conditions (Kerr *et al.*, 1978*b*).

There is no definite explanation of why nitrogen excretion remains so much less in the malnourished state. Total muscle mass, reflected by creatinine excretion and total body potassium, is proportionately reduced, as is protein turnover (Waterlow *et al.*, 1977). These findings indicate that the response of malnourished infants to brief fasting is similar to that

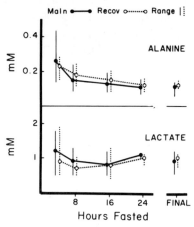

FIG. 4. Plasma levels of alanine and lactate in fasting malnourished and recovered infants (reproduced from *Metabolism* with permission of the publisher; Kerr *et al.*, 1978*a*).

observed after prolonged starvation in adults (Benedict, 1915; Owen *et al.*, 1969) but accelerated. Protein losses are minimized without interfering with glucose homoeostasis since fat becomes the predominant energy source.

No significant difference in plasma alanine was observed between the malnourished and recovered infants (Fig. 4), in spite of the large difference in nitrogen excretion. While urea production was increasing in the recovered infants plasma alanine decreased, keeping a positive correlation with plasma glucose. To test to what extent availability of substrate was a limiting factor in gluconeogenesis, excess alanine was infused at the end of the fasting period (Fig. 5). The relatively high rate of infusion (4 mg/kg/min) resulted in a comparable large increase in plasma alanine in both groups. There was an accompanying increase in plasma urea and plasma glucose. The increase in glucose was the result of a combination of small increases in production and decreases in utilization. No difference was found between the two groups. This suggests that availability of alanine as a

Malnourished ●——● Recovered o·······o

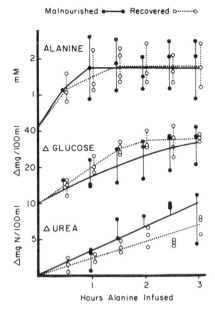

FIG. 5. Response to infusion of alanine (4 mg/kg/min) in malnourished and recovered infants (reproduced from *Metabolism* with permission of the publisher; Kerr *et al.*, 1978*b*).

substrate for gluconeogenesis was not a major rate-limiting factor and did not distinguish the malnourished group.

Turning now to similar studies in children with fasting hypoglycaemia, it should be noted that while the majority of malnourished children do not become hypoglycaemic when fasted and adapt remarkably well to their emaciated state, there is a subgroup of malnourished children who do become hypoglycaemic (Kerr *et al.*, 1973). We were intrigued to find that most of these individuals showed a persistent susceptibility to fasting hypoglycaemia after recovery. The fortuitous observation of this phenomenon in one of a pair of identical twins led to an investigation of fasting hypoglycaemia in three sets of identical twins who were not malnourished. In each pair one twin, the one who had been smaller at birth, had a definite history of recurrent symptomatic fasting hypoglycaemia of the so-called 'ketotic' variety, whereas the second twin did not and served as a control (Kerr *et al.*, 1981; Kerr and Picou, 1981).

Pagliara and co-workers had described hypoalaninaemia in children with 'ketotic' hypoglycaemia and showed that plasma glucose increased after infusion of alanine (Pagliara *et al.*, 1972). In the present three sets of

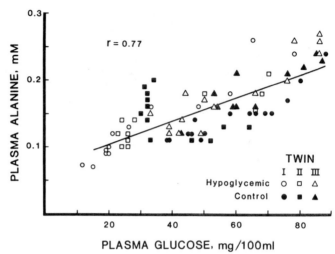

FIG. 6. Relationship of plasma glucose and alanine in hypoglycaemic and control twins (reproduced from *Metabolism* with permission of the publisher; Kerr and Picou, 1981).

twins, plasma alanine was indeed less in the affected subjects during hypoglycaemia. However, in both groups there was a similar positive correlation to plasma glucose, as has been found in the malnourished infants (Fig. 6). After fasting until becoming hypoglycaemic, urea production was greater in the affected twins, while glucose production, estimated by infusion of U-$^{13}$C-glucose, was very similar in the two groups (Table 1). In this case there was significant residual oxidation of glycogen in

TABLE 1

UREA AND GLUCOSE PRODUCTION DURING FASTING HYPOGLYCAEMIA IN AFFECTED TWINS (H) AND CONTROL (C)

| Twin pair | Urea production[a] (mg urea N/kg/min) | | Glucose production[b] (mg glucose/kg/min) | |
|---|---|---|---|---|
| | Twin H | Twin C | Twin H | Twin C |
| I | 0·31 | 0·15 | 3·0 | 3·1 |
| II | 0·28 | 0·14 | 2·3 | 2·6 |
| III | 0·32 | 0·16 | 2·5 | 2·9 |
| Mean | 0·30 | 0·15 | 2·6 | 2·9 |

[a] Urea production measured as urinary excretion plus change of body pool.
[b] Glucose production measured by dilution of constantly infused U-$^{13}$C-glucose.

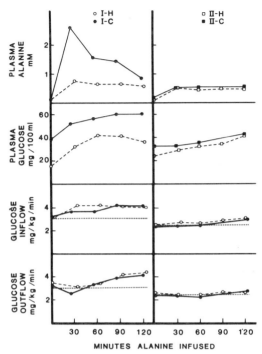

FIG. 7.  Response to alanine infusion in two pairs of hypoglycaemic and control twins. The rate of infusion was 4 mg/kg/min in Twins I and 1 mg/kg/min in Twins II (reproduced from *Metabolism* with permission of the publisher; Kerr and Picou, 1981).

the control twins; the difference, due to gluconeogenesis, was not less in the affected twins. Therefore, low plasma alanine correlated with low glucose but not with decreased urea or glucose production.

In two of these twin pairs glucose production was measured while infusing alanine (Fig. 7). At a high rate of alanine infusion (4 mg/kg/min) there was a substantial increase in plasma glucose, but this also occurred in the control and did not eliminate the difference between the two. A smaller rate of alanine infusion in the second pair (1 mg/kg/min) increased plasma alanine above the fasting range without a difference in the response of the two subjects. Plasma glucose production increased very little, again confirming that lack of alanine was not a major rate-limiting factor in gluconeogenesis.

Additional data obtained from these studies in the twins showed that carbohydrate oxidation was inappropriately increased during the period of rapidly decreasing plasma glucose, before becoming symptomatic. Plasma β-hydroxybutyrate and free fatty acids were paradoxically lower than in the controls relative to plasma glucose. Urinary epinephrine did not increase normally during fasting hypoglycaemia. Furthermore, during simulation of acute hypoglycaemia by infusion of 2-deoxyglucose, plasma glucose, free fatty acids, and epinephrine failed to increase as they did in the controls. These observations led to the conclusion that hypoglycaemia was due to epinephrine deficiency and that low alanine was secondary to hypoglycaemia (Kerr *et al.*, 1981).

Subsequent observations in thirty-three other children with hypoglycaemia have confirmed that hypoalaninaemia is not a specific finding, being associated with hypoglycaemia in a variety of disorders including hyperinsulinism, deficiencies of epinephrine, cortisol and growth hormone, and certain types of glycogen storage disease. Indeed the only abnormalities of the glucose–alanine relationship which have been described in childhood hypoglycaemia are in defects of gluconeogenesis, in which alanine is increased (Felig, 1973).

In normal children infusion of 2-deoxyglucose (50 mg/kg over 30 min) results in a prompt increase of plasma epinephrine, glucose, and lactate without an increase in glucose production. Under these conditions lactate

TABLE 2

CHANGES OF PLASMA LACTATE AND ALANINE FOLLOWING ADMINISTRATION OF 2-DEOXYGLUCOSE TO CONTROL AND EPINEPHRINE-DEFICIENT CHILDREN

|  | Control subjects (n = 5) | Epinephrine deficient (n = 4) |
|---|---|---|
| Δ Lactate (mM) | $+0.64 \pm 0.17$ | $-0.23 \pm 0.08*$ |
| Δ Alanine (mM) | $-0.06 \pm 0.03$ | $-0.03 \pm 0.01$ |

Results expressed as mean $\pm$ SEM.
* $p < 0.01$ (controls versus epinephrine deficient).

increases while alanine decreases (Table 2). However, in children who are unable to produce epinephrine these changes do not occur. This is consistent with the finding in adults that infusion of epinephrine increases glucose and lactate but not alanine (Rizza *et al.*, 1979). Incubation of skeletal muscle with epinephrine results in increased lactate and decreased alanine efflux (Garber *et al.*, 1976). Under these conditions, where glycogenolysis is increased and pyruvate oxidation probably decreased, availability of amino-N for transamination may be rate-limiting and may account for the difference between lactate and alanine. If, on the other hand, pyruvate dehydrogenase is activated by infusion of dichloroacetate, both lactate and alanine sharply decrease (Stacpoole *et al.*, 1978).

In summary, these observations have indicated that normal rates of fasting gluconeogenesis can be maintained in spite of severe depletion of body protein. This requires reutilization rather than oxidation of pyruvate. The concentration of plasma alanine correlates with plasma glucose but not with urea formation or glucose production. Low plasma alanine is not a specific phenomenon and is not evidence for decreased availability of substrate for gluconeogenesis. Increased pyruvate oxidation in the fasting state, associated with decreased glucose and alanine, requires increased utilization of protein to replace this irreversible loss.

## Acknowledgements

Co-investigators in these studies included Michael Stevens, Hazel Robinson, David Picou, Oliver Brooke, Inger Hansen, and Marilyn Levy. Support was provided by the Medical Research Council, the Helen Hay Whitney Foundation, the Diabetes Association of Cleveland, and the National Institutes of Health (HD 11089). Most of the work was conducted at the Tropical Metabolism Research Unit with the invaluable assistance of the nursing and technical staffs.

## Discussion

*Millward* (*London School of Hygiene and Tropical Medicine, UK*)*:* Where do you think the increased recycling of glucose is occurring? More specifically, is it possible that the brain may also have been participating in glycolysis and therefore increasing the overall recycling in these malnourished children?

*Kerr:* I don't think it is known exactly to what extent pyruvate dehydrogenase is inactivated in various tissues or parts of the body in the fasting state or under conditions of increased epinephrine.

*Coore* (*University of the West Indies, Kingston, Jamaica*)*:* In brain as in other tissues, pyruvate dehydrogenase is controlled by a cycle of phosphorylation which inactivates and dephosphorylation which activates the enzyme. However, brain pyruvate dehydrogenase is not inactivated by starvation in contrast to the enzyme in other tissues. The measurement is a little unsatisfactory in the brain which cannot be freeze-clamped, but in other tissues even without freeze-clamping one can detect a fall in the activity of pyruvate dehydrogenase after starvation when compared to the fed state.

*Bier* (*Washington University School of Medicine, St. Louis, USA*)*:* The question that got me into the stable isotope field was that of the role of substrate deficiency in the pathogenesis of low blood glucose in children with 'ketotic' hypoglycaemia. Because substrate levels alone couldn't answer this question and since radioisotopes were contra-indicated in infants, we undertook to develop stable isotope methods to measure alanine flux in such children. Since then, we and others have come to appreciate the caveats associated with attempting to quantify alanine transport with tracers. For example, there are the problems of rapid distribution of label into other related 3-carbon substrates and the need to sample blood from the arterial mixing pool. Nevertheless, we measured venous deuterium-labelled alanine flux in several children with ketotic hypoglycaemia. Since it is extraordinarily difficult to interpret precursor–product relationships in the non-steady state, we couldn't carry out these studies while the child was becoming hypoglycaemic. Thus, we studied these subjects in the 8–12 h post-absorptive state, reasoning that if faulty alanine delivery was responsible for hypoglycaemia, alanine flux would be decreased in the post-absorptive state prior to the development of low blood sugar. We were not able to demonstrate this decrease. In my laboratory, Dr Teresa Frazer carried out complementary studies on the newborn during the first 8 h of life (Frazer *et al.*, 1981). Several people have proposed that certain neonates become hypoglycaemic because their liver can't convert alanine to glucose. The evidence for this is that their plasma alanine levels are raised. We infused a variety of infants with L-[2,3-$^{13}$C]-alanine and measured the entry of $^{13}$C into blood glucose. Interestingly, several infants with the high alanine levels (in the 600 $\mu$M range) had substantial $^{13}$C labelling of blood glucose. Thus, one must be extremely cautious about extrapolating static measurements to estimates of substrate transport.

*James* (*Dunn Clinical Nutrition Centre, Cambridge, UK*)*:* Could I just take up this question of the epinephrine metabolism: are you implying that the release of epinephrine normally regulates substrate flow? If so, are you suggesting that normally there is an increase in epinephrine secretion during starvation? Young and Landsberg (1978) suggested a reduction of sympathetic activity during fasting.

*Kerr:* Not in the adult. But in children between about 2 to 8 years, the ages when ketotic hypoglycaemia is prevalent, it has been demonstrated several times that epinephrine secretion definitely increases as a result of fasting.

*Felig (Yale University School of Medicine, USA):* Dr Kerr, you found that the rate of glucose oxidation in hypoglycaemic children was increased. How would you relate that to the changes in epinephrine? A number of us have wondered whether hypoalaninaemia was an effect rather than a cause of ketotic hypoglycaemia. Is there something about these children which makes them require more glucose? Are these the smarter kids—or are they more nervous or what is it that correlates with glucose uptake?

*Kerr:* I think there is some relationship between lack of epinephrine and increased glucose uptake. After giving 2-deoxyglucose to normal children there is an increase in plasma glucose without an increase in glucose production. Furthermore, there is a decrease in the disappearance rate of intravenously injected glucose. However, in epinephrine-deficient subjects, giving 2-deoxyglucose has no effect on the plasma concentration of glucose or on the disappearance rate of IV glucose. So I think that fits with the hypothesis that increased glucose utilization is a major problem.

*Felig:* Yes, but I think that here one has to keep in mind the effects of epinephrine in interfering with glucose utilization. I am not aware how one could invoke changes in epinephrine as the basis for increased glucose turnover when much of that increase is due to glucose being taken up by the brain. May I put one other question? In the malnourished children that you showed initially, the glucose turnover rates were in the range 2–3 mg/kg/min, both before and after they were fed. Wouldn't that be much lower than the rate which you normally find in infants as compared to adults? In infants there is a much higher rate of turnover because of the size of the brain relative to that of the body. Is the rate of glucose turnover in the malnourished children reduced both before and after refeeding and if so what are they using as a brain substrate?

*Kerr:* There is a substantial rise in plasma hydroxybutyrate. The levels are not quite as high in malnourished children as in normal children, although from the calorimetric data which I showed you they are oxidizing fat. Bier showed that with di-deuterated glucose the rates of glucose utilization are more like 4–6 mg/kg/min after an overnight fast (Bier *et al.*, 1977). I think the difference from our data is due to the length of fasting. In the earlier stages of fasting the availability of glycogen would make a much greater

74        *D. S. Kerr*

contribution to glucose production; our results were obtained after a longer
time when glycogen was no longer a significant energy source.

*Pratt* (*Institute of Psychiatry, London, UK*): With regard to brain
metabolism in fasting, we and other groups have observed that during
metabolism tracer glucose undergoes extensive exchange with non-essential
amino acids via Krebs cycle intermediates. This process of exchange is one
of the few things in the brain which, as we have shown, is stimulated by
insulin. In fasting, it will therefore be greatly reduced. This may be part of
the general adaptation of the body's amino acid metabolism to fasting.

# References

ALLEYNE, G. A. O. and SCULLARD, G. H. (1969). Alterations in carbohydrate
metabolism in Jamaican children with severe malnutrition, *Clin. Sci.*, **37**,
631–42.
BENEDICT, F. G. (1915). A study of prolonged fasting, Carnegie Institute,
Washington. Publication No. 203.
BIER, D. M., LEAKE, R. D. and HAYMOND, M. W. (1977). Measurement of 'true'
glucose production rates in infancy and childhood with 6,6 dideuteroglucose,
*Diabetes*, **26**, 1016–23.
FELIG, P., OWEN, O. E., WAHREN, J. and CAHILL, G. F. JR. (1969). Amino acid
metabolism during prolonged starvation, *J. Clin. Invest.*, **48**, 584–94.
FELIG, P. (1973). The glucose–alanine cycle, *Metabolism*, **22**, 179–207.
FRAZER, T. E., KARL, J. E., HILLMAN, L. S. and BIER, D. M. (1981). *Amer. J. Physiol.*
(In press).
GARROW, J. S., FLETCHER, K. and HALLIDAY, D. (1965). Body composition in
severe infantile malnutrition, *J. Clin. Invest.*, **44**, 417–25.
GARBER, A. J., KARL, I. E. and KIPNIS, D. M. (1976). Alanine and glutamine
synthesis and release from skeletal muscle, *J. Biol. Chem.* **251**, 851–7.
KERR, D. S., STEVENS, M. C. G., ROBINSON, H. M. and PICOU, D. I. M. (1973).
Hypoglycemia and the regulation of fasting glucose metabolism in
malnutrition, in *Endocrine Aspects of Malnutrition*, (Eds Gardner, L. I. and
Amacher, P.), Santa Ynez, California, The Kroc Foundation, pp. 313–41.
KERR, D. S., STEVENS, M. C. G. and ROBINSON, H. M. (1978a). Fasting metabolism
in infants. I. Effect of severe undernutrition on energy and protein utilization,
*Metabolism*, **27**, 411–32.
KERR, D. S., STEVENS, M. C. G. and PICOU, D. I. M. (1978b). Fasting metabolism in
infants. II. The effect of severe undernutrition and infusion of alanine on
glucose production estimated with U-$^{13}$C-glucose. *Metabolism*, **27**, 831–48.
KERR, D. S., BROOKE, O. G. and ROBINSON, H. M. (1981). Fasting energy utilization
in the smaller of twins with epinephrine deficient hypoglycemia, *Metabolism*,
**30**, 6–17.

KERR, D. S. and PICOU, D. I. M. (1981). Fasting glucose production in the smaller of twins with epinephrine deficient hypoglycemia, *Metabolism*, **30**, 18–25.

OWEN, O. E., FELIG, P., MORGAN, A. P., WAHREN, J. and CAHILL, G. F. JR. (1969). Liver and kidney metabolism during prolonged starvation, *J. Clin. Invest.*, **48**, 574–83.

PAGLIARA, A. S., KARL, I. E., DEVIVO, D. C., FEIGIN, R. D. and KIPNIS, D. M. (1972). Hypoalaninemia: A concomitant of ketotic hypoglycemia, *J. Clin. Invest.*, **51**, 1440–9.

RIZZA, R., HAYMOND, M., CRYER, P. and GERICH, J. (1979). Differential effects of epinephrine on glucose production and disposal in man, *Am. J. Physiol.*, **237**, E356–62.

STACPOOLE, P. W., MOORE, G. W. and KORNHAUSER, D. M. (1978). Metabolic effects of dichloroacetate in patients with diabetes mellitus and hyperlipoproteinemia, *New Engl. J. Med.*, **298**, 526–30.

WATERLOW, J. C., GOLDEN, M. and PICOU, D. (1977). The measurement of rates of protein turnover, synthesis, and breakdown in man and the effects of nutritional status and surgical injury, *Am. J. Clin. Nutr.*, **30**, 1333–9.

YOUNG, J. B. and LANDSBERG, L. (1978). Fasting, feeding and regulation of the sympathetic nervous system, *New Engl. J. Med.*, **298**, 1295–301.

# 6

# Information on Nitrogen Metabolism to be Learned from Inborn Errors of Metabolism

W. L. NYHAN, CAROLYN BAY and D. KELTS

*Department of Pediatrics, University of California,
San Diego, California, USA*

## Introduction

The study of heritable disorders of human metabolism has been a rich source of fundamental principles in biochemistry, as well as in genetics and in cell biology. It was out of the study of alkaptonuria and other inborn errors of metabolism that Garrod (1909), who coined the expression, first developed the current concept of gene action which determines the structure of enzyme protein. More recently the relevance of the metabolism of purines to the development of immunocompetence has been recognized through the discovery of adenosine deaminase deficiency (Giblett *et al.*, 1972). The dihydroxyphenylalanine pathway to the neurotransmitter dopamine was first recognized as a component of human biochemistry in the study of the first patient with tyrosinosis (Medes, 1932). In studies of patients with propionic acidaemia and methylmalonic acidaemia (Ando *et al.*, 1972a) we have obtained evidence that valine, like isoleucine, threonine and methionine, is first converted to propionyl CoA before being converted to methylmalonyl CoA, whereas the pathway as given in most textbooks of biochemistry shows valine, unlike the others, being converted directly to methylmalonyl CoA.

It is clear that over the years it has been possible to learn quite a lot that is relevant to nitrogen metabolism in man from the inborn errors of metabolism. In order to focus on a single theme, the concept has been developed that the patient with an inborn error of metabolism may provide an exquisite approach to determining the true requirements for essential amino acids in infancy and childhood.

## Requirements for Essential Amino Acids

The accepted values for the requirements of essential amino acids were determined in a series of studies by Holt and Snyderman on normal infants (Snyderman, 1974). The requirements given for isoleucine, leucine, methionine, threonine and valine are 100, 45, 87 and 105 mg/kg, respectively. The ranges found in the infants studied were 80–100, 33–45, 45–87 and 85–105 mg/kg, respectively. It is important to recognize that for each amino acid there was a substantial range and that the accepted requirements are not mean values but rather the maximum among the infants studied. In general, the data indicated that the actual requirement was in all cases less than the value currently being employed as the requirement. In order to develop these numbers, Snyderman and her colleagues fed normal infants synthetic diets in which nitrogen was provided entirely by mixtures of 18 L-amino acids in the proportions in which they occur in human milk. Then they varied the intake of an amino acid under study without altering the other amino acid components. The diet was kept iso-nitrogenous at all times by the replacement of the reduced amino acid with glycine. Nitrogen balances, concentrations of serum proteins, and body weight were measured. Comparisons were made between a study period when the intake of a single amino acid was reduced and a control period during which time the infant was receiving a balanced amino acid intake.

For example, in determining the requirement for tryptophan, complete deletion of tryptophan reduced nitrogen retention but did not abolish it over a 5-day period. However, there was no weight gain. Both the rate of increase in weight and the balance of nitrogen returned to the normal level when the infant received either 16·5 or 22 mg/kg of tryptophan, but the rate of gain decreased when the infant received a tryptophan intake of 13·5 mg/kg. Thus, it was concluded that this particular baby required 16·5 mg/kg of tryptophan. Similar observations were made with each of the amino acids. These studies established the fact that histidine is an essential amino acid in this age group and also that a skin rash is associated with histidine deficiency. Nitrogen retention was adequate even with no histidine intake, but the infant did not gain weight on a histidine-free diet. In general, weight gain was a more sensitive index of amino acid deficiency than nitrogen balance.

It is important to remember that the data generally accepted as the basis for requirements depend on no more finely tuned measures of nutritional control than the maintenance of a gain in weight similar to that obtained when the amino acids were supplied in the proportion in which they occur in casein. In determining requirements it would be of considerable interest to employ some test for excessive intake. Intuitively it would appear that in that way we might be more likely to obtain data for the minimal daily requirements of the amino acids. The patient with an inborn error of the metabolism of an amino acid should provide an ideal model for such an assessment. As long as the amounts of the relevant amino acid provided are those required for growth, they will be used for the synthesis of proteins; the amino acid will not accumulate, and unusual metabolites which characterize the disease will not be found in body fluids.

## Information from the Treatment of Phenylketonuria

The treatment of phenylketonuria (PKU) is an important example of nutritional therapy in children with inborn errors of metabolism. It established the framework for nutritional intervention in children with metabolic diseases. A number of lessons have been learned from the treatment of these patients. Systematic studies relevant to phenylalanine requirements have not been carried out in patients with PKU, but they are essentially all treated with diets restricted in phenylalanine, and the aim of therapy is to minimize the accumulation of phenylalanine and to avoid the formation of metabolites such as phenylpyruvic acid.

It was learned early on that in the treatment of PKU it is important to adhere to sound nutritional principles. It has been clearly demonstrated that the patient with an inborn error in the metabolism of an individual essential amino acid does not require less of that amino acid than does a normal individual. They both require the same amount for growth and the maintenance of body proteins. While the normal infant can catabolize the large excesses that most infants ingest, the patient with PKU accumulates phenylalanine when he ingests an excess. Phenylalanine deficiency has been documented in patients with PKU. In general the symptoms are those of deficiency of any amino acid, negative nitrogen balance, and failure to gain or loss of weight. In addition there may be skin rashes, anaemia, and hypoglycaemia. At least two children have died as a result of phenylalanine deficiency. The skin rash has been cured by the administration of milk. There is also a misconception that a patient with PKU must maintain a plasma concentration of phenylalanine much higher than normal in order to grow.

Snyderman has reported experience with 125 patients in whom excellent growth was obtained when their plasma concentrations of phenylalanine were maintained in the normal range (Snyderman, 1974). Out of this experience she reported a requirement for phenylalanine of 90 mg/kg in 2 to 4 month old patients with PKU. This is the same value that she had found in normal infants. The requirement for phenylalanine decreases sharply after 4 months of age, and after 12 months of age it has fallen to 20–30 mg/kg. Our experience with patients with abnormalities in the metabolism of other amino acids is nowhere near as great as with phenylalanine, but it is clear that 0·9 to 1·1 g/kg of protein derived from cow's milk amply provides the requirements of a child $1\frac{1}{2}$ to 5 years of age. This may be much more than adequate for most patients with inborn errors of metabolism.

TABLE 1

MANIFESTATIONS OF TRYPTOPHAN DE-
FICIENCY IN PATIENTS WITH PHENYL
KETONURIA

| |
| --- |
| Hyperphenylalaninaemia |
| Hypoproteinaemia |
| Anaemia |
| Alopecia |
| Skin lesions |
|     Oral and nasal rhagades |
|     Papular erythema |
|     Desquamation of thumbs |
| Anorexia, lethargy |
| Neurological abnormalities |
| Oedema |
| Diarrhoea |

A syndrome of tryptophan deficiency has been described in patients with PKU receiving therapeutic diets in which the nitrogen content was composed entirely of individual amino acids (Table 1). In the patient with an inborn error of metabolism, the first sign of the deficiency of an amino acid is an increase in the concentration of the amino acid proximal to the enzymatic block as a consequence of the catabolic breakdown of tissue. All the patients with PKU who developed tryptophan deficiency experienced unexpected elevations in the serum concentrations of phenylalanine. The syndrome was otherwise characterized by lethargy, alopecia and a variety of skin lesions including desquamation of the thumbs. In addition there were anorexia and neurological disturbances. A few had diarrhoea. Serum concentrations of protein decreased and a few patients became oedematous. Signs and symptoms disappeared with the addition of tryptophan to the diet.

An artificial diet may lack some essential component, either because of a human failure to include a component such as tryptophan or because of instability during processing. For this reason we have tried to provide minimal quantities of essential amino acids in the form of a natural food such as milk protein.

## Disorders of Propionate Metabolism

We first began to consider the possibility that the patient with an inborn error of metabolism might teach us what the requirements of amino acids are for normal infants out of experience with the treatment of an infant with methylmalonic acidaemia (Nyhan *et al.*, 1973). Patients with methylmalonic acidaemia, and those with propionic acidaemia, cannot effectively catabolize isoleucine, valine, threonine and methionine, which are normally converted to propionyl CoA. This compound is converted to D-methylmalonyl CoA in a reaction catalysed by propionyl CoA carboxylase, the site of the defect in propionic acidaemia. D-methylmalonyl CoA is racemized to L-methylmalonyl CoA, and then converted to succinyl CoA in a reaction catalysed by methylmalonyl CoA mutase, the site of the defect in methylmalonic acidaemia. The quantitation of methylmalonate in the blood or urine provides a unique marker by which the course of a patient with methylmalonic acidaemia under treatment can be followed. Its elevation over some baseline level should reflect the provision of the relevant amino acids in amounts in excess of those required for growth.

Our original patient had a defect in the apoenzyme and was not responsive to vitamin $B_{12}$. When first seen at 13 months she had not grown or developed at all in the past 10 months. Nevertheless, normal infant catchup growth and development occurred on a diet devised to keep the concentrations of methylmalonate as low as possible and yet avoid nutritional oedema. The diet contained whole milk protein in amounts calculated to restrict the content of isoleucine, valine, threonine and methionine. Lipid, carbohydrate, and salts were added to provide adequate amounts of calories and other nutrients. It appeared that an optimal rate of growth and minimal quantities of methylmalonate were obtained when the patient was offered a mixture containing $0.75$ g/kg of protein. The infant took only 75 to 88 % of the formula. Anorexia is the rule in methylmalonic acidaemia. Many patients must be tube fed, but this patient took 75 to 90 kcal/kg and 0·6 to 0·7 g/kg of protein. From this we calculated that her intake was 34 mg/kg of isoleucine, 16 mg/kg of methionine, 29 mg/kg threonine, 40 mg/kg of valine, as well as 1 mEq/kg of Na. It seemed to us that these values were probably close to the minimal requirements of these nutrients. In Table 2 these values are compared with the Holt and Snyderman values.

TABLE 2

REQÚIREMENTS OF SOME ESSENTIAL AMINO ACIDS AS INDICATED BY AN INFANT WITH METHYLMALONIC ACIDURIA AND THE GENERALLY ACCEPTED VALUES

| Amino acid | Accepted values (Snyderman (1974)) | Values indicated by patient's response |
|---|---|---|
|  | (mg/kg/day) | (mg/kg/day) |
| Isoleucine | 100 | 34 |
| Methionine | 45 | 16 |
| Threonine | 87 | 29 |
| Valine | 105 | 40 |

The results of treatment in this patient were rewarding because in spite of the fact that diagnosis was not early, in terms of her development she ultimately achieved a normal level of physical growth, intellectual development and IQ. I do not believe that it was considered likely previously that an infant could exist, as it were, in suspended animation for 10 months of what is generally considered a critical period for the development of the brain and then exhibit catchup growth and intellectual development to achieve normality. I hope that it is not overly optimistic to consider that this example may indicate that a similar outcome is possible in other clinical states of nutrition.

We have more recently begun to study the issue systematically in the hope that the patient with an inborn error of metabolism will be able to teach us just what are the real requirements of essential amino acids for growth and development. This information can ultimately be useful in the management of normal infants and those with other types of illness. It is our intention to study each of the appropriate essential amino acids through the exploration of a wide variety of patients with inborn errors of amino acid metabolism. Initially we have focused on the disorders of propionate metabolism. Methylmalonic acidaemia is particularly useful because of its unique urinary metabolite which provides a sensitive index to show when excess quantities of the amino acid precursors of methylmalonic acid are being ingested. There is some likelihood that in propionic acidaemia the excretion of methylcitrate (Ando *et al.*, 1972*b*) or hydroxypropionate (Ando *et al.*, 1972*a*), or the blood concentrations of propionic acid will serve a similar purpose. In each infant currently under study we are also measuring nitrogen balance, as well as growth in weight and length and the

TABLE 3

REQUIREMENTS OF SOME ESSENTIAL AMINO ACIDS IN TWO INFANTS WITH DISORDERS OF PROPIONATE METABOLISM

| *Amino acid* | *Patient with methylmalonic acidaemia* | *Patient with propionic acidaemia* |
|---|---|---|
| | *(mg/kg/day)* | *(mg/kg/day)* |
| Isoleucine | 32–64 | 48 |
| Methionine | 13–26 | 20 |
| Threonine | 23–46 | 35 |
| Valine | 35–69 | 52 |

concentrations of proteins and amino acids in the serum. Data are available on an additional patient with methylmalonic acidaemia, also treated at 13 months, at which time his dimensions approximated those at birth, and a patient with propionic acidaemia treated from the first days of life.

The requirements for isoleucine, methionine, threonine and valine determined in this way are shown in Table 3. The data for the patient with methylmalonic acidaemia are shown as a range because of uncertainty in the second cycle tested, at a level intermediate between the first and third. Positive nitrogen balance was achieved at every level tested, including the first at 0·5 g/kg protein. The rate of gain in weight was virtually identical at this level and at 1·0 g/kg and the excretion of methylmalonic acid was virtually the same at both of these levels as well as at 1·25 g/kg. However, during the period of testing with 0·75 g/kg the rate of gain in weight fell to about half the previous value and methylmalonate excretion increased sharply. This is interpreted to represent a transient intercurrent illness, and therefore the lower figures probably more nearly represent the requirements. The relative constancy of the excretion of methylmalonate on the other hand probably reflects the validity of the ranges given. At levels of intake of 1·5 g/kg of protein and above, excretion of methylmalonate was considerably greater; at 2·0 g/kg the mean value was almost four times the baseline value. In the infant with propionic acidaemia, testing with 0·25 g/kg and 0·5 g/kg of protein yielded negative nitrogen balances.

Positive balances were achieved with 0·75 g/kg and above. Concomitantly, gain in weight was initiated. The excretion of methylcitrate was relatively constant at levels of intake from 0·5 to 1·5 g/kg of protein, but the values were about one-third of those observed on admission in acidosis. The excretion of hydroxypropionate appeared more closely to mirror the intake of precursors of propionate. The requirements as suggested by this infant ranged from 40 to 49 % of the Holt–Snyderman values (Snyderman, 1974).

## Discussion

*Harper* (*University of Wisconsin-Madison, USA*): Amino acid intakes of children with genetic defects should be below maximum, perhaps minimum to average. If it is assumed that 50 % above the average would give a value for the upper end of the requirement range, your values plus 50 % would be close to the values estimated by NAS/NRC (National Academy of Sciences, 1974) and FAO/WHO (1973) from minimum human milk intakes required to support satisfactory rates of growth.

*Kerr* (*Case Western Reserve University, Cleveland, USA*): We have been interested in the same relationship of inborn errors and amino acid requirements, and also find requirements lower than those previously described. Two situations should perhaps be distinguished. In phenyl-ketonuria or maple syrup urine disease, for example, the primary defect is in the first catabolic step, a lack of phenylalanine hydroxylase or α-ketoacid decarboxylase. What is enough for a child with one of these defects may be less than the normal requirement, because there is no obligatory catabolism. How do you compensate for the well known phenomenon of obligatory catabolism?

*Nyhan:* I think we do not, and it may be that we will have to find a way of extrapolating our data to normal children.

*Kerr:* The second situation is when the defect is several steps removed from the initial catabolic step. I wonder whether in this situation it is possible to achieve a state in which enough of the amino acid is provided for normal growth, and yet the level of the end product is reduced to normal?

*Nyhan:* I do not think it makes much difference where the level of a block is. The problem is more a function of the relative leakiness of the enzyme and the relative completeness of the block than where it is in the metabolic pathway.

*Pratt (Institute of Psychiatry, London, UK):* Is it the methylmalonic acid itself which causes the damage?

*Nyhan:* Again, I do not think we really know, but there is no question that it is the accumulation of organic acids. We have been trying to identify some of these things chemically. The accumulation of propionate is more toxic even than methylmalonate.

*Pratt:* The reason I asked is because it is important to distinguish between inborn errors in which damage is caused by such toxic metabolites and those like PKU in which the damage is more likely to be due to altered amino acid balance interfering with amino acid transport, especially across the blood–brain barrier. Treatment will differ according to the type of defect. As I have shown (p. 28), in PKU the levels of amino acids whose transport is affected can be raised to overcome this inhibition, so that the phenylalanine level can be allowed to rise higher, with a less severe diet. On the other hand, if there is a toxic by-product, the approach has to be different.

*Nyhan:* I do not think it is necessarily clear that the PKU model is any different from the others. I think that the problem in PKU results from the accumulation of something toxic rather than from the imbalance.

*Young (Massachusetts Institute of Technology, USA):* The criterion which is used to determine nitrogen balance can have a very important influence on the actual estimate of requirements as derived from nitrogen balance measurements. What specifically was your criterion of the minimum requirement for these various amino acids? Was it nitrogen equilibrium or was an allowance made for nitrogen retention?

*Nyhan:* I have not yet analysed our data on nitrogen balance. Our requirement was that the child should be in positive rather than in negative balance and should be growing at a rate consistent with that of normal individuals.

# References

ANDO, T., RASMUSSEN, K., NYHAN, W. L. and HULL, D. (1972*a*). 3-Hydroxy-propionate: Significance of *β*-oxidation of propionate in patients with propionic acidemia and methylmalonic acidemia, *Proc. Nat. Acad. Sci. USA*, **69**, 2807–11.

ANDO, T., RASMUSSEN, K., WRIGHT, J. M. and NYHAN, W. L. (1972*b*). Isolation and identification of methylcitrate, a major metabolic product of propionate in patients with propionic acidemia, *J. Biol. Chem.*, **247**, 2200–4.

FAO/WHO (1973). *Energy and protein requirements*, WHO Tech. Rep. Ser. No. 522, World Health Organization, Geneva.

GARROD, A. E. (1909). *Inborn Errors of Metabolism*. The Croonian lectures delivered before the Royal College of Physicians of London, June, 1908. Frowde, Hodder and Stoughton, London.

GIBLETT, E. R., ANDERSON, J. E., COHEN, F., POLLARA, B. and MEUWISSEN, H. J. (1972). Adenosine deaminase deficiency in two patients with severely impaired cellular immunity, *Lancet*, **2**, 1067–9.

MEDES, G. (1932). A new error of tyrosine metabolism: tyrosinosis. The intermediary metabolism of tyrosine and phenylalanine, *Biochem. J.*, **26**, 917–40.

NATIONAL ACADEMY OF SCIENCES (1974). *Improvement of protein nutriture*, Committee on amino acids, food and nutrition board, NAS/NRC, Washington, D.C.

NYHAN, W. L., FAWCETT, N., ANDO, T., RENNERT, O. M. and JULIUS, R. L. (1973). Response to dietary therapy in $B_{12}$ unresponsive methylmalonic acidemia, *Pediatrics*, **51**, 539–48.

SNYDERMAN, S. E. (1974). The amino acid requirements of the infant. in *Heritable Disorders of Amino Acid Metabolism* ((Ed.) W. L. Nyhan), John Wiley and Sons, Inc., New York, pp. 641–51.

# 7

# Contrasts in Nitrogen Metabolism between Animals and Plants

L. FOWDEN

*Rothamsted Experimental Station, Harpenden, Herts, UK*

## Introduction

Plants are autotrophs, animals heterotrophs. Green plants take up carbon dioxide from the air, and a range of simple inorganic materials through their roots, and produce all their complex organic materials from these in biosynthetic processes that depend upon solar radiation as the ultimate energy source. The assimilation of nitrogen in plants begins with the uptake of nitrate by root systems; its reduction to ammonia follows and subsequently this ammonia-N is introduced into the gamut of organic nitrogenous compounds encountered in members of the plant kingdom, including all twenty amino acid components of proteins. In contrast, animals depend upon their food sources for many pre-formed organic materials; they lack the complement of enzymes necessary for many biosynthetic processes, including those leading to at least eight of the twenty amino acids of proteins. On one view, plants can be regarded as being far more versatile synthesizers than animals, but an opposite view gives credit to animals for displaying conservation in their enzyme complements. In relation to amino acid synthesis, this conservation is quite evident; the 'essential' amino acids invariably represent compounds that are formed in plants only as a result of long and complex biosynthesis— pathways that animals have forsaken.

With many plant species, particularly annuals, growth and morphological development continue throughout a large part of their lifespan, and plants continue to take up nitrogen through their roots and increase their total nitrogen content. In contrast, a considerable fraction of the life of man and animals is represented by the adult phase, during which there is no net gain in body nitrogen content. Amino acid and protein metabolism then consists of a balanced turnover, ingested nitrogen equalling excreted nitrogen, and so nitrogen metabolism in the animal emphasizes repair and excretory processes. There is no evidence for active excretion of nitrogen by plants, but some reduction of total nitrogen in ageing plants may occur by shedding of leaves or other physical processes resulting in tissue loss.

Amino acid synthesis in plants is organized and regulated to provide plants at the various phases of their growth with the appropriate balance of individual compounds necessary for the elaboration of the structural, catalytic (enzymic) and reserve proteins characteristic of different species and their distinctive organs. The overall amino acid composition of the complex of proteins present in leaves does not differ very greatly between species, but large differences of composition are encountered between the seed protein of different classes of plants, between, for example, cereal and legume grains. We must recognize also that the *raison d'être* of plant proteins is to ensure the metabolic livelihood of the plant itself and we should not be surprised to find that many plant proteins, especially those of seed origin, have amino acid compositions that confer low nutritional quality. For example, the major storage protein of maize seeds (zein) has a low protein value because glutamic acid, glutamine and proline residues represent a very high proportion of the protein whilst several essential amino acids, particularly lysine, are present at very low levels, i.e. they are deficient in a nutritional sense. Traditional methods of plant breeding coupled with newer genetic engineering procedures may eventually lead to cereal or legume varieties whose seeds exhibit improved protein quality.

## Nitrogen Assimilation: A Comparative Assessment

### The Primary Route of Assimilation
Under situations of high nutrient-N flux into their root systems, plants respond by producing increased amounts of glutamine and/or asparagine. Amide production was once regarded as a metabolic response made by the plant to ensure that ammonia, the initial product of nitrate reduction, did not accumulate in amounts that could cause damage to its cells and tissues. Now glutamine synthesis is recognized to have much greater significance because reaction (1), catalysed by glutamine synthetase, is accepted as the predominant channel for the initial assimilation of ammonia-N into organic combination in higher plants (Miflin and Lea, 1977).

Nitrogen incorporated into the amide-N of glutamine is then transferred mainly to $\alpha$-oxoglutarate in a reductive amination (reaction (2)) catalysed

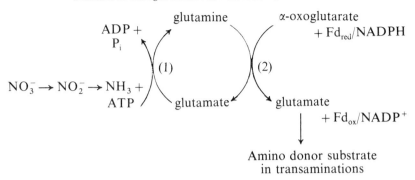

$$\text{Amino donor substrate in transaminations}$$

by glutamine-$\alpha$-oxoglutarate amido-transferase (GOGAT); two molecules of glutamic acid result, one of which is recycled through reaction (1), whilst the other is available as the principal amino group donor in a range of transaminations involved in the production of many of the remaining protein amino acids. All the reactions in the above scheme, with the exception of the initial reduction of $NO_3^-$ to $NO_2^-$, take place in leaves in the chloroplasts. ATP and reduced ferredoxin are continuously produced in these organelles as a direct result of photosynthetic reactions, and so nitrogen assimilation is coupled to light energy. In non-green tissues, e.g. roots, NADPH acts as an alternative source of reducing power for reaction (2).

This coupled system of glutamine synthetase and GOGAT enzymes has replaced the earlier concept in which glutamate dehydrogenase (GDH) was considered to provide the principal pathway for ammonia assimilation in green plants (reaction (3)).

$$2\text{-oxoglutarate} + NH_3 + NAD(P)H + H^+ \longrightarrow$$

$$\text{glutamate} + NAD(P)^+ + H_2O \quad (3)$$

We now recognize that the affinity of GDH for ammonia is too low (by comparison with that of glutamine synthetase) for the enzyme to operate efficiently at the physiological concentrations of ammonia normally present in plant cells. The two-stage cyclic process, however, requires the GOGAT enzyme and, at present, there is no evidence for its existence in animal tissues, although it is an important component of the ammonia assimilation system in bacteria (Tempest *et al.*, 1970). Therefore, the GDH-mediated reaction (3) apparently remains the accepted route for incorporation of ammonia into glutamate in animals, and the synthesis of glutamine follows, rather than precedes, that of glutamic acid. The GDH pathway remains satisfactory in animals because their tissues generally have higher concentrations of ammonia than those encountered in plants.

## Nitrogen Cycling

Plants and animals employ a series of similar reactions that facilitate the transfer of the amido-N from glutamine to various receptor molecules. Such amidotransferase-mediated reactions provide N atoms required for the synthesis of the purine skeleton, for the production of carbamoyl phosphate as a precursor of citrulline, arginine and the pyrimidines, for the heterocyclic amino acids histidine and tryptophan, and lastly for asparagine. Asparagine is formed by reaction (4), which is dependent upon $Mg^{2+}$ and ATP (Lea and Fowden, 1975).

$$\text{aspartate} + \text{ATP} + \text{glutamine} \xrightarrow{Mg^{2+}} \text{glutamate} + \text{asparagine} + \text{AMP} + \text{PP}_i$$

$$(4)$$

Under normal physiological conditions and concentrations of ammonia, little if any asparagine is synthesized by a counterpart of reaction (1), i.e. $NH_3$ cannot substitute effectively for glutamine as a source of the amido-N of asparagine.

Neither plants nor animals seem to possess an enzyme system akin to GOGAT to facilitate the direct transfer of the amido-N of asparagine to receptor molecules. Instead, ammonia is regenerated by the action of an amidohydrolase, asparaginase. It is then still reasonable to regard asparagine, in one role, as a temporary, non-toxic storage form of ammonia; it is, of course, an indispensable component of the proteins of both life forms. Meister (1965) records that asparagine can donate, by transamination, its $\alpha$-amino group to certain oxo-acid receptors in liver; the initial product, $\alpha$-oxosuccinamic acid, subsequently undergoes hydrolysis to yield oxaloacetic acid and ammonia. Although evidence has been sought for similar transamination of asparagine in plant systems, no positive findings are recorded.

Plants can be divided into two main groups based on the biochemical mechanism used for the initial fixation of $CO_2$ during photosynthesis. The majority of plants, and almost all temperate species, effect the light-mediated combination of ribulose bisphosphate (a $C_5$ sugar phosphate) with $CO_2$ to produce two molecules of 3-phosphoglyceric acid (a $C_3$ compound): such plants are termed $C_3$ species. Many tropical species photosynthesize more efficiently, especially at high light intensities and temperatures; they fix $CO_2$ initially with the formation of $C_4$ organic acids, compounds that are then translocated within the leaf to specialized cells, rich in chloroplasts, surrounding the vascular conducting tissues where decarboxylation and re-fixation of released $CO_2$ into ribulose bisphosphate (the normal mechanism) occur. Plants employing this mechanism are known as $C_4$ species; photosynthesis is increased because the mechanism ensures that $CO_2$ concentrations are enhanced within those cells abundantly endowed with chloroplasts. Furthermore, $C_4$ species do not exhibit the photorespiratory process typical of $C_3$ plants: photorespiration is counter-productive to photosynthesis, and consists of a series of reactions beginning with an oxidative cleavage of ribulose bisphosphate, continuing through several intermediates, and ending with the production of equimolar amounts of $CO_2$ and $NH_3$. These products are formed together with serine from two molecules of glycine (summary reaction (5)): initially one molecule of glycine is used to form $N^5,N^{10}$-tetrahydrofolate (and $NH_3$ plus $CO_2$) and this then donates a $C_1$ unit to the second glycine molecule to yield serine.

$$2CH_2(NH_2)COOH + [O] \rightarrow CH_2OHCH(NH_2)COOH + NH_3 + CO_2$$
(5)

Recent studies (Keys *et al.*, 1978) have indicated that whilst $NH_3$ liberated in this manner is rapidly re-assimilated by the glutamine synthetase–GOGAT pathway in chloroplasts, the flux of ammonia through this system may be an order of magnitude greater than that produced as a result of nitrate reduction in roots. The same reactions linking glycine with serine feature in animals, and can lead to high rates of ammonia formation especially in liver.

### Synthesis and Degradation of Other Amino Acids

As mentioned earlier, plants synthesize all the amino acids encountered in their proteins. Often the α-amino group is introduced at a late stage in the biosynthetic sequence and is frequently transferred from glutamate by transamination processes. When additional N atoms form part of the structure of an amino acid, these frequently originate from the amido-N of glutamine. The acceptor carbon skeletons can be traced metabolically, with varying degrees of directness, to the initial photosynthetic carbon fixation process. Although animals are deficient in the full enzyme complement necessary for the synthesis of the group of essential amino acids, sometimes they are able to effect certain stages in a synthesis; for instance, animal systems can form the branched chain amino acids (valine, leucine and isoleucine) if the corresponding oxo-acids are provided, i.e. they can effect the ultimate transamination step.

Evidence is increasing to show that the rates of nitrogen assimilatory reactions and the levels of nitrogenous constituents in plant cells are subject to regulation. One school of opinion suggests that the vigour and protein yield of a plant is governed by the level of nitrate reductase present in its tissues (Hageman *et al.*, 1976). This is obviously a key enzyme whose activity controls the initial utilization of nutrient-N. Furthermore, we know that the enzyme is induced by nitrate (Hewitt and Notton, 1967) and presumably its activity adjusts to changes in nitrate available to the plant. Experimental support for regulation of biosynthesis is available for many individual amino acids. Generally, end products of biosynthetic pathways play an important role in controlling the rates of their own syntheses, primarily by inhibiting the activity of key enzymes involved in early stages of the pathway. The ability of end products to influence enzyme formation (i.e. end-product repression of enzyme synthesis), which is a common feature of bacterial metabolism, appears unimportant in plant systems and probably is of little consequence in the regulation of amino acid metabolism in animal tissues. An end product may inhibit more than one enzyme implicated in its synthesis, but in situations where several protein amino acids are formed by a bifurcated metabolic pathway from a single precursor molecule, it is clearly important that the dominant inhibitory effect should occur at a point beyond the metabolic branch point. This situation is encountered in lysine biosynthesis, where threonine, methionine and lysine are all formed from aspartate. The first two reactions leading from aspartate are common to all three products, so it is at the site of the third enzyme (the first specific for lysine formation) that major end-product inhibition by lysine is encountered, although the first common-pathway enzyme is affected less strongly.

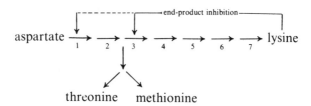

Structural analogues may mimic the role of end products as regulators of biosynthesis and have been used to select mutant plant cell lines capable of overproducing, to different extents, certain normal protein amino acids, e.g. methionine or tryptophan.

The concentrations of amino acids in the cells and tissues of plants are also influenced by physiological processes, especially by transport phenomena encountered at the whole plant, inter-cellular and sub-cellular levels. Furthermore, metabolites exist in distinct metabolic pools; for example, vacuoles make up a large fraction of the volume of fully expanded plant cells and contain significant proportions of the low molecular weight metabolites. Materials, however, may equilibrate between these separate pools somewhat slowly and so average tissue concentrations of a particular metabolite then provide only a crude measure of its potential, as an end-product inhibitor, at the active site of an enzyme.

Transport of amino acids between different tissues is also a dominant feature of metabolism in animals, but knowledge concerning their fluxes through different organs of the body is still inadequate. Following a meal of protein, the absorbed amino acids pass into the liver, where extensive catabolism occurs. At the cellular level, the liver responds to the increased influx by an accelerated protein synthesis leading to the accumulation of enzymes, with short half-lives, involved in the metabolism of the incoming nutrients and the formation of significant quantities of urea, i.e. catabolic enzymes are induced in response to metabolic needs (see Munro, 1976). Das and Waterlow (1974), working with rats, recorded the changes in activity of six hepatic enzymes involved in amino acid metabolism during adaptation from one level of dietary protein to another. They found that all six enzymes reacted similarly following a dietary change, a result suggesting that a common control mechanism was operative; half-lives of about 7 h were calculated for the enzymes. Perhaps it is then fair to conclude from the available experimental evidence that the activities of enzyme systems in animals are controlled more positively by inductive regulation via their substrates than by feedback inhibition processes associated with pathway end products—a situation that seems reversed in plants.

## Discussion

*Garlick* (*London School of Hygiene and Tropical Medicine, UK*): When plants are induced to make a lot more of an amino acid such as lysine than they would have been doing otherwise, what would they do with it, since they cannot make the excess into protein?

*Fowden:* I will comment not so much on lysine because at present levels of lysine have only been increased by 20 or 30 %. I am talking of free lysine, as you recognized. In the case of methionine, certain plants have been selected where the free methionine level is enhanced six-fold. In another system, using an analogue of tryptophan, the free tryptophan level has been increased 25-fold. I appreciate that it makes some difference in a nutritional sense whether the animal receives its amino acid bound or free, but it is better to have it free than not to have it at all.

*Harper* (*University of Wisconsin-Madison, USA*)*:* I should like to follow up your comment that plants produce proteins to meet their own needs and not to meet the needs of animals. On the other hand, plants are available and animals must adapt to what they find as a source of food. Cereal grains are the major source of calories and eventually the major source of protein for a great many animals, particularly man, and it is interesting that the ability has evolved, at least in the adult, to conserve lysine especially well, seeing that lysine is an amino acid which is most likely to be deficient.

*Fowden:* Yes, I fully accept your concept that animals have evolved and adapted; that is what I suppose any organism must and does do. And of course it is common for an animal, not simply to receive cereal protein as its sole form of intake, but as part of a mixture of proteins. The situations where one might encounter lysine stress are in intensive livestock production, when animals are raised on diets that are based principally on cereals. In pig breeding, for instance, there could be an advantage in having somewhat higher levels of lysine in some of the cereal grain proteins.

*Waterlow* (*London School of Hygiene and Tropical Medicine, UK*)*:* This is a question out of the depths of ignorance: do seed proteins in plants turn over actively?

*Fowden:* No, not actively at all in the mature dormant seed. During the growth and development of a seed there is clearly some turnover, but once the seed starts to germinate there is a major turnover because now new tissues are developing rapidly and have quite a different spectrum of proteins.

# References

Das T. K. and Waterlow, J. C. (1974). The rate of adaptation of urea cycle enzymes, aminotransferases and glutamic dehydrogenase to changes in dietary protein intake. *Br. J. Nutr.*, **32**, 353–73.
Hageman, R. H., Lambert, R. J., Loussaert, D., Dalling, M. and Klepper, L. A. (1976). Genetic improvement of seed proteins. *Proc. Int. Workshop*, Nat. Res. Council, Washington, D.C., 103–31.

HEWITT, E. J. and NOTTON, B. A. (1967). Inhibition by L-azetidine-2-carboxylic acid of induction of nitrate reductase in plants and its reversal by L-proline. *Phytochemistry*, **6**, 1329–35.

KEYS, A. J., BIRD, I. F., CORNELIUS, M. J., LEA, P. J., WALLSGROVE, R. M. and MIFLIN, B. J. (1978). Photorespiratory nitrogen cycle. *Nature, London*, **275**, 741–3.

LEA, P. J. and FOWDEN, L. (1975). The purification and properties of glutamine-dependent asparagine synthetase isolated from *Lupinus albus*. *Proc. R. Soc. Lond. B.*, **192**, 13–26.

MEISTER, A. (1965). *Biochemistry of the amino acids*, 2nd edition, vol. 1, p. 351. Academic Press, New York.

MIFLIN, B. J. and LEA, P. J. (1977). Amino acid metabolism. *Ann. Rev. Plant Physiol.*, **28**, 299–329.

MUNRO, H. N. (1976). Regulation of body protein metabolism in relation to diet. *Pro. Nutr. Soc.*, **35**, 297–308.

TEMPEST, D. W., MEERS, J. L. and BROWN, C. M. (1970). Synthesis of glutamate in *Aerobacter aerogenes* by a hitherto unknown route. *Biochem. J.*, **117**, 405–7.

# 8

# Metabolism of Branched Chain Amino Acids

A. E. HARPER and CAROL ZAPALOWSKI

*Departments of Nutritional Sciences and Biochemistry,*
*University of Wisconsin-Madison, Madison, Wisconsin, USA*

## Introduction

The first two steps in the degradation of the branched chain amino acids (BCAA)—leucine, isoleucine, and valine—are transamination of the BCAA to branched chain α-ketoacids (BCKA), which is reversible, followed by oxidative decarboxylation of the BCKA, which is irreversible. The coenzyme A (CoA) derivatives formed after decarboxylation of the BCKA are further oxidized in a series of reactions of which the end products are from leucine, acetyl CoA and acetoacetic acid; from isoleucine, acetyl CoA and succinyl CoA; and from valine, succinyl CoA. Thus, leucine is ketogenic, isoleucine is both ketogenic and glucogenic, and valine is glucogenic. The details of these pathways were elaborated largely by isotopic tracer studies (White *et al.*, 1978). In this paper we are concerned largely with the initial two steps of the degradative pathways in relation to regulation of catabolism of the BCAA. Recent work has shown that steps in the pathway subsequent to α-decarboxylation may also be regulated (Dancis and Levitz, 1972; Landaas, 1977).

## Enzymes of BCAA Catabolism

The BCAA aminotransferase, the catalyst for the first step, exists in three forms: two isozymes of a general aminotransferase which are found in the tissues of most species and which accept as substrates all three BCAA (Taylor and Jenkins, 1966; Ichihara *et al.*, 1975) and another enzyme, specific for leucine, which is found only in rat liver (Goto *et al.*, 1977). The subcellular distribution of the general BCAA aminotransferase varies from organ to organ in the rat, with 25 to 50 % being in the cytosol (Ichihara *et al.*, 1975). BCAA aminotransferase activities in liver are altered by some dietary and hormonal treatments of animals but there appear to be no unique regulators of BCAA transamination.

The BCKAs, $\alpha$-ketoisocaproic acid (KIC), $\alpha$-keto-$\beta$-methylvaleric acid (KMV) and $\alpha$-ketoisovaleric acid (KIV), products of the transaminations of leucine, isoleucine and valine respectively, undergo oxidative decarboxylation. This reaction is catalysed by BCKA dehydrogenase complex, a mitochondrial enzyme which requires $Mg^{++}$, coenzyme A, thiamin pyrophosphate, lipoic acid and $NAD^+$ as cofactors (Dancis and Levitz, 1972) and which accepts all three BCKAs as substrates (Connelly *et al.*, 1968). A second decarboxylase which appears to be a cytosolic oxygenase specific for leucine represents about 15 % of KIC oxidation by rat liver (Grant and Connelly, 1975).

The general BCAA aminotransferases (Ichihara *et al.*, 1975) and BCKA dehydrogenase (Connelly *et al.*, 1968; Wohlhueter and Harper, 1970) are distributed ubiquitously but non-uniformly throughout the body. BCAA aminotransferase activity is low in liver, intermediate in muscle and brain, and high in heart and kidney. Skeletal muscle, because it represents such a large proportion of body mass, contains most of the aminotransferase in the body (Shinnick and Harper, 1976). BCKA dehydrogenase activity is high in liver, intermediate in kidney and comparatively low in other organs, including muscle. In heart and muscle the enzyme appears to be regulated by adenine nucleotides, possibly by a phosphorylation–dephosphorylation mechanism (Parker and Randle, 1980; Waymack *et al.*, 1980). In heart, liver and muscle the enzyme is activated by BCAA and BCKA (Wohlhueter and Harper, 1970; Shinnick and Harper, 1977; Khatra *et al.*, 1977a,b; Hauschildt and Brand, 1980). Knowledge about the activation process and about the physiological importance of activators in many tissues and organs is incomplete. Because the enzyme exists in active and inactive forms (Parker and Randle, 1980; Waymack *et al.*, 1980), published values vary considerably depending upon conditions under which assays were done. Earlier literature values for BCKA dehydrogenase in liver (Sketcher *et al.*, 1974; Odessey and Goldberg, 1972) were also complicated by the fact that BCAA were used as substrates; low values were obtained because the aminotransferase is rate-limiting in liver (Shinnick and Harper, 1976) and thus the concentrations of BCKA, the substrates for the dehydrogenase, were limiting. Because dehydrogenase activity in muscle and heart was usually measured under conditions in which activity was lost, or inactivators were present, the percentage of dehydrogenase in these tissues remains to be established. Nonetheless, in the rat, the liver probably contains most of the BCKA dehydrogenase in the body.

The isolated perfused liver releases $^{14}CO_2$ from $^{14}C$-BCAA at rates below those observed for other amino acids (Miller, 1962). Although $^{14}CO_2$ was released from $^{14}C$-BCAA by liver preparations, $^{14}CO_2$ release from diaphragm (Noda and Ichihara, 1976) and kidney (Dawson *et al.*, 1967) was much greater. Observations on splanchnic amino acid balance indicate that BCAA are not taken up appreciably by visceral organs. After human subjects ingested a protein meal, BCAA accumulated in peripheral blood to a greater extent than other amino acids (Felig, 1975, 1976). Measurement of the enzyme activities for BCAA degradation revealed that, in rat (Ichihara *et al.*, 1975) and human (Goto *et al.*, 1977) liver BCAA aminotransferase activity is much lower than BCKA dehydrogenase activity (Connelly *et al.*, 1968; Shinnick and Harper, 1976). This information indicates that the aminotransferase is the rate-limiting enzyme for BCAA degradation in liver. Lindsay (1980) has calculated that if appreciable oxidation of leucine and isoleucine occur in sheep liver, the amounts of these amino acids taken up would be insufficient to support the amount of protein synthesis observed. The results of both the *in vivo* and *in vitro* studies on hepatic removal of BCAA indicate that the liver has limited capacity for oxidation of BCAA directly.

The high activity of BCKA dehydrogenase in rat liver (Shinnick and Harper, 1976) indicates that the capacity of the liver for BCKA oxidation is great. Liver mitochondria oxidize leucine to ketone bodies to a limited extent but KIC is converted to acetoacetate highly efficiently (Noda and Ichihara, 1976; Williamson *et al.*, 1979). Krebs and Lund (1977) have reported that KIC is converted to acetoacetate in liver at the same rate as butyrate and oleate. In view of the low rate of transamination of leucine to KIC in liver, ketogenesis from BCAA in this organ will be limited by the supply of substrate.

## Metabolism of BCAA in Different Tissues

As enzymes for degradation of the BCAA are distributed ubiquitously, it would be surprising if all tissues and organs did not degrade BCAA taken up in excess of the amounts required for protein synthesis. Many observations from several laboratories (Rowsell, 1956; Manchester, 1965; Dawson *et al.*, 1967; Goodman, 1963; Ichihara and Koyama, 1966; Dancis and Levitz, 1972) established, usually with tissue preparations *in vitro* and with leucine as the substrate, that BCAA can be degraded by heart, kidney, diaphragm, brain, leukocytes, skin fibroblasts, adipose tissue and liver. It has become evident from studies with the isolated perfused heart that this organ can be a highly active site of BCAA degradation (Barakat *et al.*, 1977; Buse *et al.*, 1972, 1973; Chua and Morgan, 1978) and that BCKA dehydrogenase in heart is well regulated (Buffington *et al.*, 1979; Snell, 1980). Adipose tissue has been recognized for some time as having the ability to oxidize BCAA (Goodman, 1963, 1964) and, at least in the rat, has been proposed as an important contributor to total BCAA catabolism of the body (Goodman, 1977; Tischler and Goldberg, 1980). Although kidney was among the organs first recognized as having a highly active enzyme system for BCAA degradation (Dawson *et al.*, 1967; Shinnick and Harper, 1976; Odessey and Goldberg, 1972), Noda and Ichihara (1976) concluded, after observing that BCAA oxidation by the rat was not greatly impaired after ligation of the blood vessels to the kidney, that oxidation of BCAA by the kidney was less than would be predicted from enzyme measurements *in vitro*. Although many organs and tissues can oxidize BCAA, it is not possible to assess their quantitative contributions to BCAA catabolism *in vivo*.

Skeletal muscle mass represents the greatest proportion of body mass (40–45 %) and the largest pool of free amino acids, including the BCAA, in the body. It is the major source of amino acids for gluconeogenesis during starvation and it has most of the BCAA aminotransferase of the body. Studies by Miller (1962) and Manchester (1965) had shown that peripheral tissues, particularly muscle, oxidized BCAA as extensively or more so than liver. Pozefsky *et al.* (1969) reported that BCAA represented an unexpectedly large proportion of amino acids taken up by muscle and that muscle released a much smaller proportion of BCAA than would be expected from the amino acid composition of muscle protein. They concluded that BCAA must be degraded in this tissue.

The high BCAA aminotransferase and low BCKA dehydrogenase activity of muscle and the low BCAA aminotransferase and high BCKA dehydrogenase activity of liver suggested to us that after transamination of BCAA, the carbon skeleton might circulate back as BCKA to other organs, particularly liver, for degradation. BCKA had been found in blood by Käser *et al.* (1960) and by McMenamy *et al.* (1965). Subsequently, Cree and Harper (1977) demonstrated that all three BCKA accumulated in the perfusate when the isolated rat hindquarter was perfused with a complete mixture of amino acids containing only 0·1 mM leucine, isoleucine and valine. In further studies (Hutson *et al.*, 1978, 1980), we have found that when this preparation is perfused with concentrations of $1-^{14}C$-leucine in the physiological range, the amount of KIC released can equal or exceed the amount oxidized to carbon dioxide. Noda and Ichihara (1976) observed that the capacity of the hepatectomized rat for $U-^{14}C$-leucine oxidation was reduced, suggesting to them that the liver contributed to oxidation of the BCAA carbon skeleton. Other BCAA metabolites are also released from perfused organs. Spydevold (1979) has observed that after perfusion of the isolated hindquarter with valine, both KIV and $\beta$-hydroxyisobutyrate were released into the perfusate and Pearce and Baptista (1980) have reported that the latter compound is released into the medium when heart is perfused with KIV. They would presumably be degraded in other organs.

BCKA are present in higher concentrations in blood than in skeletal muscle indicating that they must be readily transported out of muscle cells (Hutson *et al.*, 1978, 1980). Krebs and Lund (1977) calculated on the basis of kinetic considerations that BCKA concentrations should be only one fifteenth to one twenty-fifth those of BCAA in the cell. Values we have obtained are in this range and those for heart and liver are even lower (Livesey and Lund, 1980; Hutson and Harper, 1980). Thus, concentrations of BCKA are far too low for the BCKA dehydrogenase to be saturated with substrate (Parker and Randle, 1980) as is indicated by observations showing that when the leucine concentration of the perfusion medium is increased from $0\cdot1$ to $0\cdot5$ mM, the rate of oxidation of leucine by the rat hindquarter increased approximately fivefold but BCKA release did not increase proportionately to the increased leucine concentrations.

These observations indicate that BCKA arising from transamination of BCAA in skeletal muscle are released into the perfusion medium whether or not the BCKA dehydrogenase reaction is limiting the rate of BCKA oxidation. It would appear then that substrate supply limits BCAA oxidation by muscle, due first to the unfavourable equilibrium of the BCAA aminotransferase reaction, a result of the high glutamate concentration in muscle cells (Krebs and Lund, 1977), and second, to the release of BCKA into blood, even when their concentrations in muscle are low. It would seem, then, that a portion of the BCKA that arise from transamination of the BCAA will be released into the blood and be transported to other organs where they may be either degraded or reaminated (Nissen and Haymond, 1980). Observations on blood and tissue BCKA concentrations would indicate that BCKA in blood originate largely from skeletal muscle as the pools in other organs are very small (Livesey and Lund, 1980; Hutson and Harper, 1980). The high activity of BCKA dehydrogenase in liver should ensure that any KIC released into blood will be removed rapidly.

From measurements of concentrations of BCKA in the hepatic portal vein and hepatic vein, Livesey and Lund (1980) have shown that, in fact, the liver can oxidize all BCKA released by the periphery.

## Inter-relationships of Alanine and Glutamine in the Metabolism of BCAA

Measurements of arterio-venous differences of amino acids across muscle, of amino acid release during perfusion of rat hindquarter and release of amino acids during incubation of isolated muscle preparations revealed that not only are the amounts of BCAA released into the medium much less than would be anticipated, but those of alanine and glutamine are much greater (reviewed by Lindsay, 1980; Snell, 1980).

These observations, together with those of Mallette et al. (1969) that the reverse was true in liver provided evidence that transamination of BCAA was highly active in muscle in vivo and revealed the mechanism, the 'glucose–alanine cycle', by which the nitrogen removed from BCAA in muscle could be incorporated into a molecule that served as an efficient transport vehicle for transferring BCAA nitrogen to liver for conversion to urea. After Wahren et al. (1976) observed that a high protein intake, which elevates BCAA concentrations in peripheral blood, stimulated alanine efflux from muscle in normal and diabetic man and Odessey et al. (1974) showed that addition of BCAA to the incubation medium increased release of alanine from diaphragm while other amino acids failed to do this, there could be little doubt that muscle was a major site of BCAA transamination in vivo in both man and the rat.

It should also be noted that muscle releases glutamine in amounts greater than would be anticipated (Felig, 1975; Odessey et al., 1974) and that Garber et al. (1976) observed an increase in glutamine release into the medium when muscle was incubated with 0·5 mM leucine. As glutamate is the immediate acceptor of BCAA nitrogen, conversion of it to glutamine provides another readily transportable molecule for disposing of extra α-amino nitrogen and ammonia. Glutamine is taken up by both kidney and intestine which subsequently release alanine (Matsutaka et al., 1973; Felig, 1975; Snell, 1980).

Alanine is released into the circulation also by heart, small intestine, kidney, adipose tissue and brain, organs and tissue which also contain BCAA aminotransferase (Snell, 1980). Nevertheless, as skeletal muscle is such a large proportion of body mass and contains such a high proportion

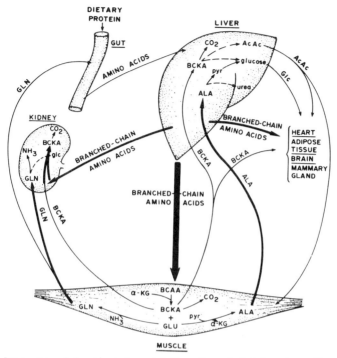

FIG. 1.    Inter-organ relationships in the metabolism of branched chain amino and ketoacids.

of total aminotransferase activity, it undoubtedly is the predominant site for removal of nitrogen from BCAA and for formation of BCKA.

The multi-organ relationships in the metabolism of BCAA in the animal body which have been discussed are portrayed in Fig. 1.

## Rate of BCAA Oxidation in vivo

Evidence that the BCKA concentrations in blood are responsive to precursor supply is given in Fig. 2, which shows changes in blood BCKA concentrations in rats fed graded levels of leucine in an amino acid diet containing slightly above adequate quantities of all of the indispensable amino acids (Benjamin, 1978). When leucine was limiting for growth and valine and isoleucine were in excess, KIC concentrations were low and KIV and KMV concentrations were elevated. As the leucine content of the diet was increased and valine and isoleucine were required in larger quantities

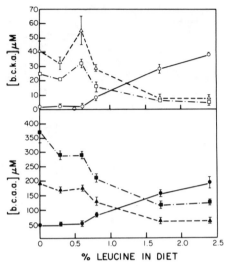

% LEUCINE IN DIET

FIG. 2.    Top: plasma BCKA concentrations of rats fed *ad libitum* for 12 h on diets graded in leucine content containing 1-[14]C-leucine. Values are the mean of two pooled samples each from three rats. Vertical lines indicate the range. O——O, α-ketoisocaproate (ketoleucine); □——□, α-ketoisovalerate (ketovaline); △－－－△, α-keto-β-methylvalerate (ketoisoleucine). Bottom: plasma BCAA concentrations of rats fed *ad libitum* for 12 h on diets graded in leucine content containing 1-[14]C-leucine. Values are the mean of two pooled samples each from three rats. Vertical lines indicate the range. ●——●, leucine; ■——■, valine; ▲－－－▲, isoleucine.

for growth, KIV and KMV concentrations fell and KIC concentration began to rise. Oxidation of leucine was low when leucine intake was low (Fig. 3) and rose rapidly when leucine intake exceeded the requirement.

Sketcher *et al.* (1974) estimated that skeletal muscle of a 100 g rat would be capable of oxidizing 0·52 mmol leucine/day and that with an adequate protein intake it would be required to oxidize only about 0·3 mmol/day. Kidney is capable of oxidizing about 0·2 mmol leucine/day. Goodman (1977) and Tischler and Goldberg (1980) have estimated that adipose tissue is capable of oxidizing 15 % of the daily leucine requirement, or about 0·12 mmol/day. Liver is capable of oxidizing in the order of 3 mmol KIC/day. Based on figures from Buffington *et al.* (1979), heart could oxidize about 0·6 mmol/day. Without taking into account the capabilities of other organs, these figures indicate that capacity of the rat for BCAA oxidation must far exceed its need. However, these figures for capacity are based on enzyme measurements made with saturating or, at least, high concentrations of a single BCAA, usually between 5 and 20 mM, well above those that occur in body fluids and tissues.

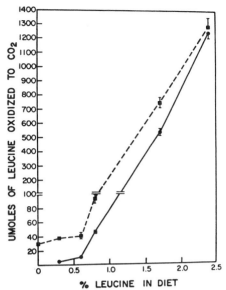

FIG. 3. Percent of total expired radioactivity recovered as $^{14}CO_2$ per hour from rats described in Fig. 2. Values are the means of six rats. Vertical lines indicate standard errors of the means. ●——●, based on specific radioactivity of dietary leucine; ■– – –■, based on specific radioactivity of plasma leucine.

Tissue concentrations of BCAA rarely exceed 0·5 mM (Livesey and Lund, 1980; Hutson and Harper, 1980), close to the lowest estimates of $K_m$ values of the aminotransferases for BCAA (Cree, 1980). Also, the BCAA compete with one another for the active site on the aminotransferase, hence competition among the BCAA will further reduce the rate of transamination of any one of them in the cell (Dawson *et al.*, 1967; Cree, 1980). As Krebs and Lund (1977) have pointed out from kinetic considerations, the high intracellular glutamate concentration of tissues should result in low steady-state concentrations of the BCKA participating in transamination reactions and favour their conversion to BCAA. Actual rates of transamination *in vivo* will, therefore, be well below the estimated enzymatic capacity and the probability that BCKA will be produced in quantities that will support the maximum rate of oxidative decarboxylation is equally low. Values for $K_m$s of the dehydrogenase for BCKA are in the range of 10–20 $\mu$M (Parker and Randle, 1978) or higher. BCKA concentrations in muscle are from 15–25 $\mu$M, in the range of the $K_m$, and would therefore support at best about half the maximum rate of oxidative decarboxylation of BCKA. With the dehydrogenase also, there will be competition among substrates (Buffington *et al.*, 1979). In other tissues that have been examined, BCKA concentrations are much lower than in muscle. It therefore seems highly probable that the rate of formation of BCKA through transamination becomes a major factor in control of the rate of BCAA oxidation. This is essentially what we observed in studies of BCAA oxidation by the perfused hindquarter in which the rate of leucine oxidation increased as BCAA concentrations in the perfusate were increased.

## *Nutritional and Hormonal Effects on BCAA Oxidation*

Both the rate of oxidation of BCAA by intact organisms and the activities of the BCAA aminotransferases and BCKA dehydrogenases are influenced by hormonal and dietary treatments, but most of the responses observed have been relatively small and often they have not been consistent. Observations on factors influencing the metabolism of BCAA have been reviewed by Adibi (1976, 1980) and by Felig (1975).

McFarlane and von Holt (1969) reported that severe and prolonged protein deficiency in the rat reduced oxidation of both leucine and KIC to 25–35 % of control values. In subsequent studies of effects of protein depletion, however, Reeds (1974) and Neale and Waterlow (1974*a,b*) did not observe depressed leucine oxidation *in vivo*, although Reeds (1974), Sketcher and James (1974) and Sketcher *et al.* (1974) found that protein depletion depressed BCKA dehydrogenase activity in muscle as Wohlhueter and Harper (1970) had found for liver. Also, although Mimura *et al.* (1968) had observed that BCAA aminotransferase was elevated in muscle in protein-depleted rats, Reeds (1974) and Sketcher *et al.* (1974) found the aminotransferase to be unchanged. It is evident from the results reported by Reeds (1974) and Sketcher *et al.* (1974) that the BCAA aminotransferase reaction was not limiting in muscle of protein-depleted rats in view of the accumulation of BCKA that occurred when they used valine as the substrate for measuring oxidative decarboxylation by muscle preparations. In a later study Sketcher and James (1976) reported that oxidation of leucine by the hind-limb from protein-depleted rats was depressed—providing further evidence that low BCKA dehydrogenase activity is limiting for BCAA degradation in muscle in protein-depleted rats.

Both starvation and diabetes increase rates of oxidation of BCAA by muscle but in diabetes increases were observed with other tissues as well (Paul and Adibi, 1978; May *et al.*, 1980). As plasma concentrations of BCAA are elevated in starvation and diabetes, it would be expected that oxidation rates would increase owing to increased substrate supply. But increased rates of oxidation have also been observed in isolated hindquarters from starved and diabetic rats (Hutson *et al.*, 1980 and unpublished) under conditions in which perfusate concentration was not elevated, suggesting that these conditions may have a specific effect on the dehydrogenase.

Several investigators have shown that substrates which can be used as alternate energy sources for various organs and tissues affect the metabolism of the BCAA. Since concentrations of glucose ketone bodies, pyruvate and fatty acids are altered in clinical situations, it is important to determine what effects these substrates have on the control of BCAA metabolism *in vivo*.

## Directions in Research on BCAA Metabolism

Other nutritional conditions, e.g. altered protein content of diet, and endocrine treatments, e.g. cortisol, insulin, have not produced either large or consistent effects on the activities of the enzymes of BCAA metabolism (Adibi, 1976, 1980), with the possible exception of the stimulation by cortisol and high protein feeding of the leucine specific aminotransferase in rat liver. Hormonal changes would be expected to exert effects on BCAA metabolism indirectly by altering substrate supply through mechanisms such as control of protein synthesis and degradation in muscle by insulin. In general, however, there is a distinct need for methodical examination and clarification of effects on BCAA metabolism of factors that are known to influence the metabolism of other amino acids.

Organ differences in the metabolism of BCAA have been established, but quantitative measurements to determine the extent of the differences are limited so far. This is an important direction for future research if the significance of inter-organ relationships in the metabolism of this group of amino acids is to be understood. Information about mechanisms by which the BCKA dehydrogenase is regulated is accumulating gradually, but extensive effort will be required in this area to establish not only the role of specific mechanisms such as regulation through covalent modification of the enzyme, but also the significance of dietary and hormonal factors. BCAA metabolism has clinical implications in malnutrition, renal disease, liver disease and trauma. Effective use of BCAA as therapeutic agents will depend on better understanding of factors that regulate BCAA metabolism (Adibi, 1980).

## Discussion

*Lund* (*Radcliffe Infirmary, Oxford, UK*): In our measurements of A–V differences across various organs in the rat, we were struck, as was Harper, by the fact that the concentrations of the branched chain keto acids in plasma were higher than the concentrations in all tissues except possibly skeletal muscle. The fact that liver removes branched chain keto acids, yet they were virtually undetectable in the freeze-clamped tissue, whereas their plasma concentration was about 40 nmol/ml, suggested to us that the plasma membrane of the liver may limit uptake. This provides another possible mechanism for regulation of branched chain keto acid metabolism, other than activity of the dehydrogenase.

*Harper:* At a recent meeting on branched chain amino acids evidence was presented for carrier-mediated transport of the branched chain keto acids across liver mitochondrial membranes (MacKay and Robinson, 1980). The authors did not discuss transport across the plasma membrane.

*Young* (*Massachusetts Institute of Technology, USA*): I would like to raise the question of the metabolic specificity of leucine and valine, because in much of the discussion it seems to be taken for granted that they behave in a uniform way. As Fig. 4 shows, with a reduction in leucine intake in young adults there is a reduction in plasma leucine, which is to be expected, but this is accompanied by a marked rise in valine concentration. This kind of relationship does not hold when the intake of valine is altered (Young *et al.*, 1972). A reduction in valine intake is associated with a fall in plasma valine but there is no change in plasma leucine or isoleucine concentrations. Thus, it appears as though leucine is regulating the plasma concentrations of the other two branched chain amino acids, but the latter do not regulate the level of plasma leucine.

*Felig* (*Yale University School of Medicine, USA*): In agreement with Young, Sherwin (1978) found in studies on human subjects that raising the plasma concentration of leucine by leucine infusion caused a marked reduction in valine and isoleucine; but if either of the other two branched chain amino acids is infused, there is no effect on leucine concentration. There seems to be some unique mechanism by which leucine controls the concentrations of the other branched chain amino acids, but which does not work in the opposite direction. This may well have clinical implications.

*Harper:* In support of your point, in animals we have found antagonisms among the branched chain amino acids (Harper *et al.*, 1970). For example,

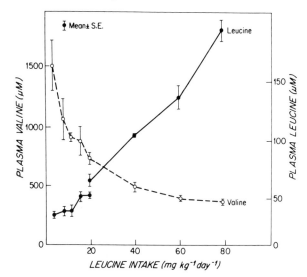

Fig. 4.   Plasma concentrations of valine and leucine in healthy young men receiving an
amino acid diet supplying graded intakes of leucine. (M. Meguid *et al.*, unpublished MIT
data).

if the leucine concentration is greatly elevated, either in the whole animal or
in the isolated perfused hindquarter, the concentrations of isoleucine and
valine fall. On the other hand, the leucine concentration does not fall if we
increase valine and isoleucine.

*Golden (University of the West Indies, Kingston, Jamaica):* We have some
more evidence of the different metabolic behaviour of leucine and valine
(Golden, Taruvinga and Jackson, unpublished). We were looking at the
relation between the nitrogen of the branched chain amino acids going into
muscle and that of the glutamine and alanine coming out. It has been shown
in studies *in vitro* that additions of different branched chain amino acids
and inhibitors lead to different relative amounts of glutamine and alanine
production (Ruderman and Berger, 1974; Garber *et al.*, 1976; Chang and
Goldberg, 1978*a,b*). We infused [15]N-valine or [15]N-leucine into rats for 6 h,
and looked at the labelling of alanine, glutamic acid, glutamine and free
ammonia in several tissues. The results are shown in Table 1. Both labelled
amino acids produced enrichment of ammonia and glutamine amide-N.
Valine produced a remarkably high enrichment in muscle alanine, and
moderate enrichment of the amino-N of glutamic acid and glutamine. With
leucine there was no enrichment in these amino acids. Harper pointed out

TABLE 1

ENRICHMENT OF AMMONIA, GLUTAMINE, GLUTAMIC ACID AND ALANINE IN VARIOUS TISSUES AFTER INFUSION OF [15]N-LABELLED VALINE AND LEUCINE

| [15]N-labelled amino acid | Tissue studied | Enrichment (atom % excess × $10^3$) | | | |
|---|---|---|---|---|---|
| | | Ammonia | Glutamine amide-N | Glutamine + glutamic acid amino-N | Alanine |
| L-valine | Liver | 66 + 3 | 50 + 7 | 40 + 11 | 35 + 7 |
| | Kidney | 88 + 8 | — | 7 + 2 | 8 + 2 |
| | Muscle | 74 + 17 | 39 + 7 | 25 + 4 | 179 + 45 |
| L-leucine | Liver | 49 + 4 | 43 + 5 | 4 + 1 | 3 + 1 |
| | Kidney | 96 + 7 | 38 + 2 | 1 + 0·3 | 2 + 1 |
| | Muscle | 38 + 5 | 34 + 5 | 4 + 1 | 6 + 1 |

Results are means ± SEM.

that most tissues contain a general aminotransferase which accepts all three branched chain amino acids. This difference in the behaviour of the N of valine and leucine suggests that we are missing something in our knowledge of the intermediary metabolism of these amino acids.

# References

ADIBI, S. A. (1976). Metabolism of branched-chain amino acids in altered nutrition, Metabolism, 25, 1287–302.

ADIBI, S. (1980). Roles of branched-chain amino acids in metabolic regulation, J. Lab. Clin. Med., 95, 475–84.

BARAKAT, H. A., BROWN, W. E. and DOHM, G. L. (1977). Oxidation of leucine by heart and muscle homogenates of the cardiomyopathic hamster, Biochem. Med., 18, 152–7.

BENJAMIN, E. (1978). In vivo studies of branched-chain amino acid and keto acid metabolism in the rat, M.S. Thesis, University of Wisconsin-Madison, Wisconsin.

BUFFINGTON, C. K., DEBUYSERE, M. S. and OLSON, M. S. (1979). Studies on the regulation of the branched chain α-keto acid dehydrogenase in the perfused rat heart, J. Biol. Chem., 254, 10453–8.

BUSE, M. G., BIGGERS, J. F., DRIER, C. and BUSE, J. F. (1973). The effect of epinephrine, glucagon, and the nutritional state on the oxidation of branched chain amino acids and pyruvate by isolated hearts and diaphragms of the rat, J. Biol. Chem., 248, 697–706.

BUSE, M. G., BIGGERS, J. F., FRIDERICI, K. H. and BUSE, J. F. (1972). Oxidation of branched chain amino acids by isolated hearts and diaphragms of the rat. The effect of fatty acids, glucose, and pyruvate respiration, *J. Biol. Chem.*, **247**, 8085–96.

CHANG, T. W. and GOLDBERG, A. L. (1978a). The origin of alanine produced in skeletal muscle, *J. Biol. Chem.*, **253**, 3677–84.

CHANG, T. W. and Goldberg, A. L. (1978b). The metabolic fates of amino acids and the formation of glutamine in skeletal muscle, *J. Biol. Chem.*, **253**, 3685–95.

CHUA, B. L. and MORGAN, H. E. (1978). Effect of branched-chain amino acids on protein turnover in perfused rat heart, *Fed. Proc.*, **37**, 540 (Abstr. 1723).

CONNELLY, J., DANNER, D. and BOWDEN, J. (1968). Branched chain α-keto acid metabolism. I. Isolation, purification, and partial characterization of bovine liver α-ketoisocaproic:α-keto-β-methylvaleric acid dehydrogenase, *J. Biol. Chem.*, **243**, 1198–203.

CREE, T. C. (1980). Studies on the degradation of the branched-chain amino acids in the perfused hindquarter and rat skeletal muscle extracts, Ph.D. Thesis, University of Wisconsin-Madison, Wisconsin.

CREE, T. C. and HARPER, A. E. (1977). Conversion of leucine to α-ketoisocaproic acid during perfusion of isolated rat hindquarter, *Fed. Proc.*, **36**, 528 (Abstr. 1355).

DANCIS, J. and LEVITZ, M. (1972). Abnormalities of branched-chain amino acid metabolism. in *The Metabolic Basis of Inherited Disease* (Eds. J. B. Stanbury, J. B. Wyngaarden and D. S. Fredrickson), 3rd Edition, McGraw-Hill Book Co., New York, pp. 426–39.

DAWSON, A. G., HIRD, F. J. R. and MORTON, D. J. (1967). Oxidation of leucine by rat liver and kidney, *Arch. Biochem. Biophys.*, **122**, 426–33.

FELIG, P. (1975). Amino acid metabolism in man, *Ann. Rev. Biochem.*, **44**, 933–55.

FELIG, P. (1976). Recent developments in body fuel metabolism. in *Year in Metabolism—1975–1976* (Ed. N. Freinkel), Plenum Medical Book Co., New York, pp. 113–36.

GARBER, A. J., KARL, I. E. and KIPNIS, D. M. (1976). Alanine and glutamine synthesis and release from skeletal muscle. II. The precursor role of amino acids in alanine and glutamine synthesis, *J. Biol. Chem.*, **251**, 836–43.

GOODMAN, H. M. (1963). The effects of growth hormone on leucine metabolism in adipose tissue *in vitro*, *Endocrinology*, **73**, 421–6.

GOODMAN, H. M. (1964). Stimulatory action of insulin on leucine uptake and metabolism in adipose tissue, *Am. J. Physiol.*, **206**, 129–32.

GOODMAN, H. M. (1977). Site of action of insulin in promoting leucine utilization in adipose tissue, *Am. J. Physiol.*, **233**, E97–E103.

GOTO, M., SHINNO, H. and ICHIHARA, A. (1977). Isozyme patterns of branched-chain amino acid transaminase in human tissues and tumors, *Gann*, **68**, 663–7.

GRANT, W. and CONNELLY, J. (1975). Mammalian cytosolic α-keto acid decarboxylase, *Fed. Proc.*, **34**, 640 (Abstr. 2390).

HARPER, A. E., BENEVENGA, N. J. and WOHLHUETER, R. M. (1970). Effects of ingestion of disproportionate amounts of amino acids, *Physiol. Rev.*, **50**, 480–97.

HAUSCHILDT, S. and BRAND, K. (1980). Effects of branched-chain α-keto acids on enzymes involved in branched-chain α-keto acid metabolism in rat tissues, *J. Nutr.* **110**, 1709–16.

HUTSON, S. M., CREE, T. C. and HARPER, A. E. (1978). Regulation of leucine and α-ketoisocaproate metabolism in skeletal muscle, *J. Biol. Chem.*, **253**, 8126–33.

HUTSON, S. M. and HARPER, A. E. (1980). Blood and tissue branched-chain amino and α-keto acid concentrations. Effect of diet, starvation and disease, *Am. J. Clin. Nutr.* **34**, 173–83.

HUTSON, S. M., ZAPALOWSKI, C., CREE, T. C. and HARPER, A. E. (1980). Regulation of leucine and α-ketoisocaproic acid metabolism in skeletal muscle. Effects of starvation and insulin. *J. Biol. Chem.*, **255**, 2418–26.

ICHIHARA, A. and KOYAMA, E. (1966). Transaminase of branched chain amino acids. I. Branched chain amino acids-α-ketoglutarate transaminase. *J. Biochem.*, **59**, 160–9.

ICHIHARA, A., YAMASAKI, Y., MASUGI, H. and SATO, J. (1975). Isozyme patterns of branched-chain amino acid transaminase during cellular differentiation and carcinogenesis, in *Isozymes III. Developmental Biology*, Academic Press, New York, pp. 875–89.

KÄSER, H., KÄSER, R. and LESTRADET, H. (1960). Separation and quantitative estimation of new alpha-keto-acids in human blood for paper chromatography, *Metabolism*, **9**, 926–31.

KHATRA, B. S., CHAWLA, R. K., SEWELL, C. W. and RUDMAN, D. (1977a). Distribution of branched-chain α-keto acid dehydrogenases in primate tissues, *J. Clin. Invest.*, **59**, 558–64.

KHATRA, B. S., CHAWLA, R. K., WADSWORTH, A. D. and RUDMAN, D. (1977b). Effect of dietary branched-chain α-keto acids on hepatic branched-chain α-keto acid dehydrogenase in the rat, *J. Nutr.*, **107**, 1528–36.

KREBS, H. A. and LUND, P. (1977). Aspects of the regulation of the metabolism of branched-chain amino acids, in *Advances in Enzyme Regulation* (Ed. G. Weber), Vol. 15. Pergamon Press, Oxford, pp. 375–94.

LANDAAS, S. (1977). Inhibition of branched-chain amino acid degradation by ketone bodies, *Scand. J. Clin. Lab. Invest.*, **37**, 411–18.

LINDSAY, D. B. (1980). Amino acids as energy sources, *Proc. Nutr. Soc.*, **39**, 53–9.

LIVESEY, G. and LUND, P. (1980). Enzymic determination of branched-chain amino acids and 2-oxoacids in rat tissues. Transfer of 2-oxoacids from skeletal muscle to liver *in vivo*, *Biochem. J.*, **188**, 705–13.

MACKAY, N. and ROBINSON, B. H. (1980). Transport of branched chain keto-acids across mitochondrial membranes, *Abst. Internat. Symp. on Metab. and Clin. Implications of Branched Chain Amino and Ketoacids*, Kiawah Island, S.C., USA, November 1980, p. 11.

MALLETTE, L. E., EXTON, J. H. and PARK, C. R. (1969). Effects of glucagon on amino acid transport and utilization in the perfused rat liver, *J. Biol. Chem.*, **244**, 5724–8.

MANCHESTER, K. L. (1965). Oxidation of amino acids by isolated rat diaphragm and the influence of insulin, *Biochim. Biophys. Acta*, **100**, 295–8.

MATSUTAKA, H., AIKAWA, T., YAMAMOTO, H. and ISHIKAWA, E. (1973). Gluconeogenesis and amino acid metabolism. III. Uptake of glutamine and output of alanine and ammonia by non-hepatic splanchnic organs of fasted rats and their metabolic significance, *J. Biochem.*, **74**, 1019–29.

MAY, M. E., MANCUSI, V. J., AFTRING, R. P. and BUSE, M. G. (1980). Effects of diabetes on oxidative decarboxylation of branched-chain keto acids, *Am. J. Physiol.*, **239**, E215–22.

MCFARLANE, I. G. and VON HOLT, C. (1969). Metabolism of amino acids in protein-calorie-deficient rats, *Biochem. J.*, **111**, 557–63.

MCMENAMY, R. H., VANG, J. and DRAPANAS, T. (1965). Amino acid and α-ketoacid concentrations in plasma and blood of the liverless dog, *Am. J. Physiol.*, **209**, 1046–52.

MILLER, L. (1962). The role of the liver and the nonhepatic tissues in the regulation of free amino acid levels in the blood, in *Amino Acid Pools*, Proceedings of the Symposium on Free Amino Acids, City of Hope Medical Center, 1961. (Ed. J. T. Holden), Elsevier, Amsterdam, pp. 708–21.

MIMURA, T., YAMADA, C. and SWENDSEID, M. E. (1968). Influence of dietary protein levels and hydro-cortisone administration on the branched chain amino acid transaminase activity in rat tissues. *J. Nutr.*, **95**, 493–7.

NEALE, R. J. and WATERLOW, J. C. (1974a). The metabolism of $^{14}$C-labelled essential amino acids given by intragastric or intravenous infusion to rats on normal and protein-free diets, *Brit. J. Nutr.*, **32**, 11–25.

NEALE, R. J. and WATERLOW, J. C. (1974b). Critical evaluation of a method for estimating amino acid requirements for maintenance in the rat by measurement of the rate of $^{14}$C-labelled amino acid oxidation *in vivo*, *Brit. J. Nutr.*, **32**, 257–72.

NISSEN, S. and HAYMOND, M. W. (1980). Effect of fasting on the interconversion of leucine and α-ketoisocaproate *in vivo*, *Proceedings Inter. Symp. Metab. Clinical Implications of Branched-Chain Amino Acids* (in press).

NODA, C. and ICHIHARA, A. (1974). Control of ketogenesis from amino acids. II. Ketone bodies formation from α-ketoisocaproate, the keto-analogue of leucine, by rat liver mitochondria, *J. Biochem.*, **76**, 1123–30.

NODA, C. and ICHIHARA, A. (1976). Control of ketogenesis from amino acids. IV. Tissue specificity in oxidation of leucine, tyrosine, and lysine, *J. Biochem.*, **80**, 1159–64.

ODESSEY, R. and GOLDBERG, A. (1972). Oxidation of leucine by rat skeletal muscle, *Am. J. Physiol.*, **223**, 1376–83.

ODESSEY, R. and GOLDBERG, A. L. (1979). Leucine degradation in cell free extracts of skeletal muscle, *Biochem. J.*, **178**, 475–89.

ODESSEY, R., KHAIRALLAH, E. A. and GOLDBERG, A. L. (1974). Origin and possible significance of alanine production by skeletal muscle, *J. Biol. Chem.*, **249**, 7623–9.

PARKER, P. J. and RANDLE, P. J. (1978). Branched-chain 2-oxoacid dehydrogenase complex of rat liver, *FEBS Letters*, **90**, 183–6.

PARKER, P. J. and RANDLE, P. J. (1980). Active and inactive forms of branched-chain 2-oxoacid dehydrogenase complex in rat heart and skeletal muscle, *FEBS Letters*, **112**, 186–190.

PAUL, H. S. and ADIBI, S. A. (1978). Leucine oxidation in diabetes and starvation: Effects of ketone bodies on branched-chain amino acid oxidation *in vitro*, *Metabolism*, **27**, 185–200.

PEARCE, F. J. and BAPTISTA, M. (1980). Branched chain ketoacid metabolism in the perfused rat heart, *Fed. Proc.*, **39**, 2140 (Abstr. 2813).

POZEFSKY, T., FELIG, P., TOBIN, J. D., SOELDNER, J. S. and CAHILL, G. F. JR. (1969). Amino acid balance across tissues of the forearm in postabsorptive man. Effects of insulin at two dose levels, *J. Clin. Invest.*, **48**, 2273–82.

REEDS, P. J. (1974). The catabolism of valine in the malnourished rat. Studies *in vivo* and *in vitro* with different labelled forms of valine, *Brit. J. Nutr.*, **31**, 259–70.

ROWSELL, E. V. (1956). Transaminations with L-glutamate and α-oxoglutarate in fresh extracts of animal tissues, *Biochem. J.*, **64**, 235–45.

RUDERMAN, N. B. and BERGER, M. (1974). The formation of glutamine and alanine in skeletal muscle, *J. Biol. Chem.*, **249**, 5500–6.

SANS, R. M., JOLLY, W. W. and HARRIS, R. A. (1980). Studies on the regulation of leucine catabolism. Mechanism responsible for oxidizable substrate inhibition and dichloroacetate stimulation of leucine oxidation by the heart, *Arch. Biochem. Biophys.*, **200**, 336–45.

SHERWIN, R. S. (1978). Effect of starvation on the turnover and metabolic response to leucine, *J. Clin. Invest.*, **61**, 1471–6.

SHINNICK, F. L. and HARPER, A. E. (1976). Branched-chain amino acid oxidation by isolated rat tissue preparations, *Biochim. Biophys. Acta*, **437**, 477–86.

SHINNICK, F. L. and HARPER, A. E. (1977). Effects of branched-chain amino acid antagonism in the rat on tissue amino acid and keto acid concentrations, *J. Nutr.*, **107**, 887–95.

SKETCHER, R., FERN, E. and JAMES, W. (1974). The adaptation in muscle oxidation of leucine to dietary protein and energy intake, *Brit. J. Nutr.*, **31**, 333–42.

SKETCHER, R. D. and JAMES, W. P. T. (1974). Branched-chain amino acid oxidation in relation to catabolic enzyme activities in rats given a protein-free diet at different stages of development, *Brit. J. Nutr.*, **32**, 615–23.

SKETCHER, R. D. and JAMES, W. P. T. (1976). Estimation of muscle leucine oxidation in the perfused hind-limb of the rat: Effect of feeding with a protein-free diet, *Proc. Nutr. Soc.*, **35**, 49A.

SNELL, K. (1980). Muscle alanine synthesis and hepatic gluconeogenesis, *Biochem. Soc. Trans.*, **8**, 205–13.

SPYDEVOLD, Ø. (1979). The effect of octanoate and palmitate on the metabolism of valine in perfused hindquarter of rat, *Eur. J. Biochem.*, **97**, 389–94.

TAYLOR, R. T. and JENKINS, W. T. (1966). Leucine aminotransferase. II. Purification and characterization, *J. Biol. Chem.*, **241**, 4396–405.

TISCHLER, M. E. and GOLDBERG, A. L. (1980). Leucine degradation and release of glutamine and alanine by adipose tissue, *J. Biol. Chem.*, **255**, 8074–81.

WAHREN, J., FELIG, P. and HAGENFELDT, L. (1976). Effect of protein ingestion on splanchnic and leg metabolism in normal man and in patients with diabetes mellitus, *J. Clin. Invest.*, **57**, 987–99.

WAYMACK, P. P., DEBUYSERE, M. S. and OLSON, M. S. (1980). Studies on the activation and inactivation of the branched chain α-keto acid dehydrogenase in the perfused rat heart, *J. Biol. Chem.*, **255**, 9773–81.

WHITE, A., HANDLER, P., SMITH, E. L., HILL, R. L. and LEHMAN, I. R. (1978). *Principles of Biochemistry*, 6th Edition, McGraw-Hill Book Co., New York, pp. 739–40.

WILLIAMSON, J. R., WALAJTYS-RODE, E. and COLL, K. E. (1979). Effects of branched chain α-ketoacids on the metabolism of isolated rat liver cells. I. Regulation of branched chain α-ketoacid metabolism, *J. Biol. Chem.*, **254**, 11511–20.

WOHLHUETER, R. M. and HARPER, A. E. (1970). Coinduction of rat liver branched chain α-keto acid dehydrogenase activities, *J. Biol. Chem.*, **245**, 2391–401.
YOUNG, V. R., TONTISIRIN, K., ÖZALP, I., LAKSHMANAN, F. and SCRIMSHAW, N. S. (1972). Plasma amino acid response curve and amino acid requirements in young men: valine and lysine, *J. Nutr.*, **102**,1159–70.

# 9

# A Model for Leucine Metabolism in the Human

D. E. MATTHEWS,† M. J. RENNIE‡ and D. M. BIER†

† Department of Internal Medicine,
Washington University School of Medicine,
St. Louis, Missouri, USA
‡ Department of Human Metabolism,
University College Hospital Medical School,
London, UK

## Introduction

Because the branched chain amino acids may play an important role as a muscle fuel source under physiological conditions such as fasting and may be an important nitrogen source for muscle alanine and glutamine production, much interest and research have been focused on the oxidation of the branched chain amino acids, particularly leucine (Adibi, 1976; Hutson et al., 1978; Odessey and Goldberg, 1979). Control of leucine oxidation occurs at either or both of the first two steps of catabolism: transamination of the leucine to α-ketoisocaproate (KIC) or decarboxylation of KIC. In muscle the latter step is thought to be controlling, although transamination activity may also decrease (Adibi, 1976; Hutson et al., 1978; Odessey and Goldberg, 1979). In previous work, $^{14}C$ has been used to measure the activity of both the transamination and decarboxylation steps. While this technique works well for measuring net transamination, it does not provide information about the reversibility of the transamination reaction or the equilibrium in vivo. In this work we describe a method for measuring the rates of both directions of the transamination reaction (deamination to KIC and reamination to leucine) and the rate of KIC oxidation.

## Model

Leucine transamination (formation of KIC) can be followed by the loss of the leucine amino-N (Matthews et al., 1980a). Leucine oxidation can be

117

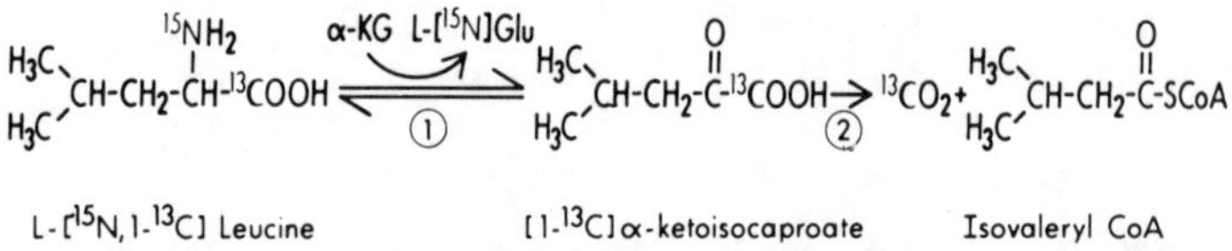

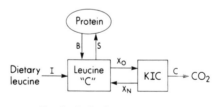

FIG. 1. Metabolism of L-[$^{15}$N,1-$^{13}$C]leucine. Step 1: transamination of leucine with α-ketoglutarate (α-KG) via leucine aminotransferase (branched chain amino acid transaminase); the $^{15}$N label is removed and incorporated into a [$^{15}$N]glutamate. Step 2: decarboxylation of KIC via branched chain keto acid dehydrogenase; the carboxyl-$^{13}$C is released as $^{13}CO_2$.

determined from the release of $CO_2$ from decarboxylation of the carboxyl-carbon (Matthews *et al.*, 1980*b*). Therefore, a $^{15}$N-labelled leucine is chosen for measurement of transamination and a carboxyl-$^{13}$C-labelled leucine for measurement of leucine oxidation (decarboxylation). In fact, matters are simplified if the two labels are incorporated into a single tracer molecule (Fig. 1).

During a primed, continuous infusion of L-[$^{15}$N,1-$^{13}$C]leucine, dilution of the dilabelled leucine in plasma reflects the appearance of unlabelled leucine derived primarily from leucine release from protein breakdown and from dietary leucine intake. [$^{15}$N,1-$^{13}$C]-leucine is removed by leucine oxidation and by incorporation of leucine into protein (protein synthesis). It is the former process that causes the differential release of the labels: removal of $^{15}$N by leucine transamination and production of $^{13}CO_2$ by 1-$^{13}$C-KIC decarboxylation (Fig. 1). When the 1-$^{13}$C-KIC is formed, it has two fates: (1) decarboxylation or (2) reamination to leucine. The latter route produces a 1-$^{13}$C-leucine.†

The process of dilabelled leucine catabolism in Fig. 1 can be viewed in

$$Q_C = I + B = S + C$$

FIG. 2. Model for leucine carboxyl-carbon metabolism. Symbols explained in text.

† Some $^{15}$N-leucine may be formed from the transamination of a KIC and $^{15}$N-glutamate, but the probability of a $^{15}$N from glutamate being incorporated back into a 1-$^{13}$C-KIC when it is reaminated to reform [$^{15}$N,-1-$^{13}$C]leucine is very low. Thus, transamination of [$^{15}$N,1-$^{13}$C]leucine irreversibly removes the dilabelled leucine.

terms of net leucine carbon oxidation and of leucine nitrogen transamination. Figure 2 presents a model for leucine C oxidation. Leucine enters the free leucine pool by dietary intake (I) and leucine release from protein breakdown (B). Net leucine removal occurs by incorporation of leucine into protein (S) and leucine oxidation (C). At steady state, appearance of leucine will equal disappearance of leucine and this is net leucine C flux ($Q_C$). Leucine transamination to and from KIC does not cause a loss of a leucine

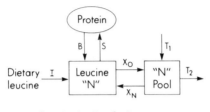

$$Q_N = I + B + X_N = S + X_O$$

FIG. 3.    Model for leucine N metabolism. Symbols explained in text.

carboxyl-$^{13}$C label; this only occurs as a result of the subsequent decarboxylation of the KIC.

In contrast, leucine N flux ($Q_N$) will occur at a faster rate (Fig. 3). First, leucine N formation includes reamination of KIC ($X_N$). Second, leucine N is lost immediately upon transamination ($X_O$). The leucine N enters the transaminating N-pool (primarily glutamate) which of course has other inflows and outflows ($T_1$ and $T_2$). The only difference in treatment of the $^{15}$N- and $^{13}$C-leucine labels occur in catabolism: the $^{15}$N-label will be removed faster by transamination. In other words, when the leucine N and C models are combined:

$$Q_N - Q_C = X_N = X_0 - C$$

The rate of KIC reamination ($X_N$) is the difference between leucine N and C fluxes. Decarboxylation (C) is measured from $^{13}$CO$_2$ release (Matthews *et al.*, 1980*b*); the sum of C and $X_N$ is leucine transamination to KIC ($X_0$). Whether leucine transamination, KIC decarboxylation, or both control leucine oxidation can be inferred from the magnitudes of C, $X_0$, and $X_N$.

## Methods

Three healthy, adult men were primed after an overnight fast with sodium $^{13}$C-bicarbonate and L-[$^{15}$N,1-$^{13}$C]leucine and then infused for seven hours with L-$[^{15}N,1-^{13}C]$leucine. The prime and infusion doses were similar to those used before (Matthews *et al.*, 1980*a*). Blood and breath samples were collected throughout the study. Plasma $[^{15}N,1-^{13}C]$leucine, $1-^{13}C$-leucine, and $^{15}N$-leucine enrichments were determined by chemical ionization gas chromatography–mass spectrometry (for more details of the method, see Bier, p. 289), and expired $^{13}CO_2$ was measured by conventional isotope ratio–mass spectrometry (Matthews *et al.*, 1980*b*). Rates of leucine metabolism were computed from the data obtained at isotopic steady state.

## Results

The parameters of leucine metabolism for the three subjects are presented in Table 1. The leucine C flux is slower than net leucine N flux. This is due to the faster removal of $^{15}N$ by leucine transamination compared to decarboxylation. However, the ratio of leucine C flux to leucine N flux ($Q_C/Q_N = 77\%$) is not nearly as small as the ratio of KIC decarboxylation to leucine transamination ($C/X_0 = 40\%$). Because the leucine oxidation

TABLE 1

LEUCINE METABOLISM MEASURED WITH A PRIMED, CONTINUOUS INFUSION OF L-$[^{15}N, 1-^{13}C]$LEUCINE

| Subjects | | PE | AK | JM | Mean ± SD |
|---|---|---|---|---|---|
| | | | *(μmol/kg/h)* | | |
| Leucine C flux | $Q_C$ | 119·7 | 95·4 | 145·0 | 120·0 ± 24·8 |
| Leucine N flux | $Q_N$ | 175·8 | 116·2 | 173·4 | 155·1 ± 33·7 |
| Leucine oxidation | C | 25·2 | 18·5 | 20·0 | 21·2 ± 3·5 |
| Leucine transamination to KIC | $X_0$ | 81·4 | 39·3 | 48·5 | 56·4 ± 22·1 |
| KIC reamination to leucine | $X_N$ | 56·2 | 20·8 | 28·4 | 35·1 ± 18·6 |
| % leucine flux oxidized | $C/Q_C$ | 21·1 | 19·4 | 13·9 | 18·1 ± 3·8 |
| % leucine N transaminated | $X_0/Q_N$ | 46·3 | 33·8 | 28·0 | 36·0 ± 9·4 |
| % KIC formed that is oxidized | $C/X_0$ | 31·0 | 47·0 | 41·4 | 39·8 ± 8·1 |
| % KIC formed that is reaminated | $X_N/X_0$ | 69·0 | 53·0 | 58·6 | 60·2 ± 8·1 |

component is a minor component of leucine flux ($C/Q_C = 18\%$) compared to leucine incorporation and release from protein synthesis and breakdown, a major reduction in oxidation is required for a moderate reduction in leucine flux.

The values of net leucine flux and oxidation are similar to values we have previously obtained for young adult men studied post-absorptively (Motil *et al.*, 1979). The rate of leucine oxidation compared to leucine C flux is attenuated in these three subjects ($C/Q_C = 18\%$). This attenuation occurs primarily by a reduction in KIC decarboxylation, not leucine transamination. Of the KIC formed from leucine, most is reaminated to leucine ($X_N/X_0 = 60\%$), rather than decarboxylated ($C/X_0 = 40\%$). The transamination reaction is operating about 2·5 times faster than decarboxylation in these subjects.

These results demonstrate that L-[$^{15}$N,1-$^{13}$C]leucine can be used to measure the whole body rates of leucine transamination to KIC, reamination of KIC to leucine, and KIC decarboxylation in man.

## Acknowledgements

The authors thank Drs D. Halliday, D. J. Millward, G. A. Clugston, and R. H. T. Edwards for assistance and helpful discussions.

## Discussion

*Millward* (*London School of Hygiene and Tropical Medicine, UK*)*:* In the studies which Matthews has described, in which $^{15}$N-leucine was infused, we looked at the distribution of label between plasma amino acids to see if we could determine the relative importance of alanine and glutamine as carriers to the liver of the nitrogen removed from leucine in muscle. In fact, at the level of detection which is possible with the gas chromatograph–mass spectrometer (0·1 atom %), no labelling of plasma alanine was found. Measurement of the labelling of glutamine by Halliday showed substantial enrichment of the amide-N. This supports the findings which Golden reported (see p.109).

*Bier:* Unfortunately, the data on this point are conflicting. We published a study in which a dog infused with $^{15}$N-leucine clearly transferred label to alanine-N (Matthews *et al.*, 1979). In addition, Matthews and I were part of a study in which alanine and leucine carbon and nitrogen balances were measured across the hind-limb of the dog (Ben Galim *et al.*, 1980). Both from stoichiometry and from isotopic measurements of nitrogen transfer it was clear that a substantial amount of leucine-N was in fact transferred to alanine. Haymond has some as yet unpublished results in man which are very similar to those obtained in the dog. Thus, the fact that leucine-N does appear in alanine-N seems certain. Of course, however, isotopic studies *per se* do not answer the question of whether there is *net* transfer of leucine-N to alanine-N. Here, stoichiometry as well is required.

*Matthews:* I have been involved in both [15]N-leucine infusion studies in which very little labelled plasma [15]N-alanine was found, mentioned by Millward, and in those described by Bier, in which plasma alanine was labelled with [15]N. One technical point may be important. We are discussing the incorporation of [15]N into large intracellular pools of free alanine and glutamine. In order to detect significant labelling with our gas chromatograph–mass spectrometer we have to give a larger dose of labelled leucine. With the isotope ratio mass spectrometer used by Halliday and the group in Jamaica much lower levels of [15]N abundance can be measured. Although we have one set of experiments with results different from another set, the doses of isotope and conditions of the experiments were quite different between the two studies, and this will largely affect the results and explain the differences. I suppose one could postulate that British subjects on a freely chosen diet dispose of their nitrogen in a different way from Americans.

*Felig (Yale University School of Medicine, USA):* There may be more than nitrogen metabolism at stake—perhaps Anglo-American relations.

*Jackson (University of the West Indies, Kingston, Jamaica):* In the light of what Waterlow said about the turnover rates of the free amino acid pools, I have difficulty in accepting the concept that the amino acid pools are not turning over rapidly enough to allow the label to be distributed through various metabolic pathways. Secondly, as has been pointed out, the amount of isotope given is different, according to whether the measurements are made by gas chromatograph– or isotope ratio–mass spectrometry. In view of the information which has been presented on the metabolic effects of leucine loads, is it possible that we are looking at two different situations, in one of which leucine, given in more than a tracer dose, is producing a disturbance in metabolic state?

*Matthews:* We share your concern about the amount of isotope infused, and have tried to keep the dose infusion rate below 5 % of the flux. This will not amount to a significant leucine load. Secondly, the glutamine and alanine pools are large. $^{15}$N-leucine is being infused at a fixed rate, and until enough of it has been transferred to these pools, there will not be significant enrichment.

*Jackson:* True, but the enrichment at any given time is determined both by the size of the pool and by its turnover rate.

# References

ADIBI, S. A. (1976). Metabolism of branched-chain amino acids in altered nutrition, *Metabolism*, **25**, 1287–1302.

BEN GALIM, E., HRUSKA, K., BIER, D. M., MATTHEWS, D. E. and HAYMOND, M. W. (1980). Branched-chain amino acid nitrogen transfer to alanine *in vivo*: Direct isotopic determination with [$^{15}$N] leucine, *J. Clin. Invest.*, **66**, 1295–307.

HUTSON, S. M., CREE, T. C. and HARPER, A. E. (1978). Regulation of leucine and α-ketoisocaproate metabolism in skeletal muscle, *J. Biol. Chem.*, **253**, 8126–33.

MATTHEWS, D. E., BEN GALIM, E. and BIER, D. M. (1979). Determination of stable isotope enrichment in plasma amino acids by chemical ionisation mass spectrometry. *Analytical Chemistry*, **51**, 80–4.

MATTHEWS, D. E., BEN-GALIM, E., HAYMOND, M. W. and BIER, D. M. (1980*a*). Alloisoleucine formation in maple syrup urine disease: Isotopic evidence for the mechanism, *Pediatr. Res.*, **14**, 854–7.

MATTHEWS, D. E., MOTIL, K. J., ROHRBAUGH, D. K., BURKE, J. F., YOUNG, V. R. and BIER, D. M. (1980*b*). Measurement of leucine metabolism in man from a primed, continuous infusion of L-[1-$^{13}$C]leucine, *Am. J. Physiol.*, **238**, E473–9.

MOTIL, K. J., MATTHEWS, D. E., ROHRBAUGH, D., BIER, D. M., BURKE, J. F. and YOUNG, V. R. (1979). Simultaneous estimates of whole body leucine and lysine flux in young men: Effect of reduced protein intake, *Fed. Proc.*, **38**, 708.

ODESSEY, R. and GOLDBERG, A. L. (1979). Leucine degradation in cell-free extracts of skeletal muscle, *Biochem. J.*, **178**, 475–89.

# 10

# Does Leucine Regulate Muscle Protein Synthesis *in vivo*?

P. J. GARLICK, M. A. MCNURLAN and E. B. FERN

*Clinical Nutrition and Metabolism Unit,*
*London School of Hygiene and Tropical Medicine,*
*London, UK*

A feature of the branched chain amino acids, particularly leucine, that has attracted much attention in recent years is their apparent involvement in the regulation of protein turnover in muscle. This has been shown most clearly in isolated muscles *in vitro*. Table 1 shows data from Fulks *et al.* (1975) who incubated small pieces of diaphragm with various mixtures of amino acids *in vitro* and estimated rates of protein synthesis and breakdown. Their conclusion was that protein synthesis was increased when all amino acids were included at five times the levels normally found in plasma. This effect could also be achieved by addition of only the branched chain amino acids, and particularly by leucine. There was in addition a decrease in the rate of protein breakdown. At the same time, and by very similar methods, Buse and Reid (1975) also reported a stimulatory effect of leucine, but not of valine and isoleucine, on protein synthesis in muscle. These authors also showed that the effect could be observed with lower concentrations of leucine, within the physiological range.

This property of leucine has been related to the fact that branched chain amino acids are extensively oxidized by muscle at rates comparable with their rates of incorporation into protein (Fulks *et al.*, 1975). Similar results have been observed in preparations of perfused skeletal muscle (Li and Jefferson, 1978) and in perfused cardiac muscle (Rannels *et al.*, 1974; Chua *et al.*, 1979).

Experiments in man have also led to the belief that the single amino acid leucine might have a specific regulatory role *in vivo*. Sherwin (1978) gave intravenous infusions of leucine to fasting patients and showed that an

## TABLE 1

EFFECT ON RATES OF PROTEIN SYNTHESIS OF INCUBATING RAT DIAPHRAGM
MUSCLE WITH VARIOUS AMINO ACIDS AT FIVE TIMES THE NORMAL PLASMA
CONCENTRATION

|  | *Protein synthesis (% of control)* | |
| --- | --- | --- |
|  | *Experiment 1* | *Experiment 2* |
| Control | 100 | 100 |
| + leucine | — | 125 |
| + leucine + isoleucine + valine | 124 | 147 |
| + isoleucine + valine | — | 116 |
| + all amino acids | 123 | — |
| + all minus BCAAs | 102 | — |

Tissues were taken from normal rats in experiment 1 and from
hypophysectomized rats in experiment 2. Data from Fulks *et al.*
(1975).

improvement in nitrogen balance resulted on the day of infusion. This
occurred after either 3 or 28 days of starvation. Similarly, Sapir and Walser
(1977) gave the α-keto analogues of the three branched chain amino acids to
starved patients and also observed an improvement in nitrogen balance. A
connection between this effect and a direct action on muscle *in vivo*,
however, has not yet been conclusively demonstrated. Buse *et al.* (1979)
showed that the injection of leucine into starving rats (which had also been
given an injection of glucose plus insulin) resulted in more aggregated
polysomes in muscle, but polysome profiles are not a reliable indicator of
the rate of protein synthesis, particularly in muscle from which the yield is
usually low (Pain, 1978). Measurements *in vivo* of the arterio-venous
difference across the human leg did not reveal any effect of leucine infusion
on the net outflow of phenylalanine or tyrosine (Hagenfeldt *et al.*, 1980). In
addition, changes in the levels of the branched chain amino acids in plasma
do not correlate with measured rates of protein synthesis in the muscle of
rats (Millward *et al.*, 1976).

In view of these somewhat conflicting observations we have attempted to
discover whether the concentration of leucine in the blood has any part to
play in the normal regulation of protein synthesis in muscle of the live
animal. Rats were injected with 100 μmol leucine/100 g body weight and the
rate of protein synthesis in tissues measured by intravenous injection of a
flooding dose of $^3$H-phenylalanine (Garlick *et al.*, 1980). The advantage of

this technique is that the large dose of phenylalanine (150 $\mu$mol/100 g body weight) floods the free phenylalanine pools of the tissues and plasma to almost the same specific activity. This largely avoids problems resulting from compartmentation of precursor pools and minimizes any effect that the increased concentration of leucine may have on this. Rates of protein synthesis were estimated from the incorporation of $^3$H-phenylalanine into protein during the 10 min following injection.

The effect of injected leucine on the rates of protein synthesis in gastrocnemius muscle, heart and jejunal serosa (smooth muscle) of fed rats (male, 100 g body weight) is shown in part A of Fig. 1. In no instance was there any evidence that leucine significantly stimulated protein synthesis. However, it is possible that in a fed rat protein synthesis is already maximal or that leucine concentrations are optimal. The experiment was therefore repeated in rats starved for 2 days and in rats given a protein-free diet for 9 days (parts B and C of Fig. 1). It can be seen that both these treatments resulted in a decline in protein synthesis in all three muscles compared with the fed state, yet a significant stimulation by leucine was still not detected.

In these experiments the leucine was given intravenously along with the labelled phenylalanine and protein synthesis was measured during the following 10 min. Because of the possibility that 10 min might not be long enough for leucine to exert an effect, protein synthesis was measured in starved rats 1 h after an intraperitoneal injection of 100 $\mu$mol leucine. This procedure is similar to that adopted by Buse *et al.* (1979), when reaggregation of polysomes was observed. However, as the results show (part D of Fig. 1) there was again no effect of leucine on the rate of protein synthesis in the three muscles.

A possible explanation for the lack of effect of leucine *in vivo* is that it was masked by a similar effect of the high dose of $^3$H-phenylalanine used to measure protein synthesis. However, *in vitro* studies have not demonstrated an influence of any amino acid except the branched chains on muscle protein synthesis (Fulks *et al.*, 1975; Buse and Reid, 1975), and high concentrations of labelled phenylalanine were also used for measuring the rates of protein synthesis in the studies on perfused hind limbs (Li and Jefferson, 1978) and perfused heart (Rannels *et al.*, 1974; Chua *et al.*, 1979). Secretion of insulin in response to the injection of leucine and phenylalanine might also mask an effect of leucine. Although the concentration of insulin does rise transiently at 2 min after injection of phenylalanine, the concentration is much higher after phenylalanine plus leucine, particularly in the starved group, yet there was no change in protein synthesis.

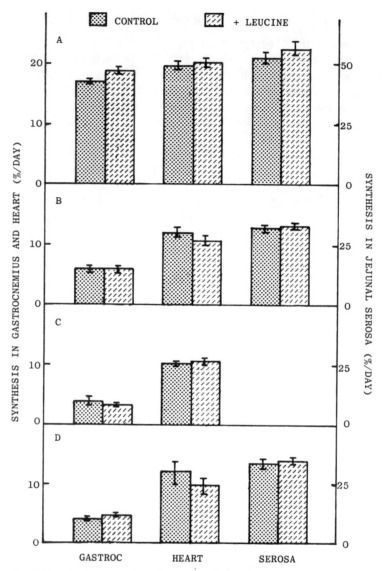

FIG. 1. Effect of an injection of leucine (100 $\mu$mol/100 g body weight) on rates of protein synthesis in gastrocnemius muscle, heart and jejunal serosa of 100 g male rats. A, leucine injected together with the [3]H-phenylalanine in fed rats; B, as in A but with rats starved for 2 days; C, as in A but with rats given a protein-free diet for 9 days; D, as in B but the leucine was given by intraperitoneal injection 1 h before injection of [3]H-phenylalanine. Rates of synthesis (% of protein pool renewed per day ± SEM) were not significantly altered by leucine injection under any of the conditions studied. (M. A. McNurlan, P. J. Garlick and E. B. Fern, unpublished).

It is difficult to reconcile these data with evidence obtained *in vitro* that leucine specifically exerts a stimulatory effect on muscle protein synthesis. It is possible that the conditions of incubation or perfusion *in vitro* are never achieved *in vivo*; or that they occur in man but not in the growing rat, whether fed, starved, or deprived of protein, as in our experiments. If so, there may be no direct connection between the change in nitrogen balance brought about by leucine (Sherwin, 1978) and the changes observed in muscle *in vitro*. Indeed, Sakamoto *et al.*, (1980) have observed that in fasting rats with pancreatitis alanine has the same effect as branched chain amino acids in sparing body nitrogen, although this amino acid does not appear to affect muscle protein synthesis *in vitro* (Fulks *et al.*, 1975). Our results, however, do suggest that whatever the effect of leucine on whole body nitrogen metabolism, it is probably not by means of an immediate and direct stimulation of the rate of protein synthesis in muscle.

# *Discussion*

*Walser* (*Johns Hopkins University, Baltimore, USA*): I would like to suggest another interpretation of the interesting data reported by Garlick. Effects on body protein stores are, of course, the net result of changes in protein synthesis and changes in degradation. Goldberg and Tischler (1981) have shown by using inhibitors of transamination that the stimulating effect on protein synthesis is a property of leucine and not of α-keto-isocaproate (KIC); on the other hand, an inhibitory effect on protein degradation is a property of KIC and not of leucine. Furthermore, Mitch, Sapir and I have examined the nitrogen-sparing effects of leucine as compared with KIC in early starvation in man (Mitch *et al.*, 1981). In contrast to the report of Sherwin (1978) we find no nitrogen-sparing effect of leucine infusions. However, KIC induced significant sparing of nitrogen in the same subjects during another fast. It is also possible that the nitrogen-sparing effect observed with leucine late in starvation by Sherwin (1978) could have been caused by an inhibitory effect on protein degradation exerted by the KIC derived from leucine. In any case, would you consider the possibility that an experiment designed to measure protein degradation rather than synthesis, along the same lines as the experiment which you reported, might yield different results?

*Garlick:* I agree that in Sherwin's experiment it may be breakdown which was affected. On the other hand, he reported that there was no change in methylhistidine production when he gave leucine, suggesting that there was no change in muscle protein breakdown.

It would indeed be interesting to measure protein breakdown, but we have no way of doing it in animals over short periods of time such as 10 min or 1 h, and over longer periods the results may be influenced by secondary effects rather than by direct effects of leucine.

*Felig (Yale University School of Medicine, USA):* I would like to raise another possibility. An interesting finding in Sherwin's study (1978) was that when leucine was infused after prolonged starvation, there was a rise in plasma glucose. This was totally unanticipated; here is a ketogenic amino acid which is causing a rise in blood sugar. When we looked at glucose turnover, we found that there was no stimulation of glucose production, but glucose utilization was falling. This suggests that leucine may have some glucose-sparing effect. Could the nitrogen-sparing effect of leucine that we find in the intact individual after prolonged fasting be related, if not to insulin production, to altered glucose kinetics?

*Garlick:* We have no idea what happens to glucose within the 10 min of this particular experiment. I think you are emphasizing the point that to try to relate results from experiments on bits of muscle *in vitro* to something occurring in the whole body is perhaps unrealistic.

*Felig:* I quite agree.

*Alleyne (University of the West Indies, Kingston, Jamaica):* Your studies were done with leucine given as a bolus. What would happen if you gave it as a constant infusion before the bolus of phenylalanine?

*Garlick:* That is a good suggestion. The single dose which we gave before the measurement of protein synthesis was given as an intraperitoneal injection, and therefore would be absorbed rather slowly over that period. I do not know what would happen if one gave it as an infusion.

# References

Buse, M. G. and Reid, S. S. (1975). Leucine: a possible regulator of protein turnover in muscle, *J. Clin. Invest.*, **56**, 1250–61.

Buse, M. G., Atwell, R. and Mancusi, V. (1979). In vitro effect of branched chain amino acids on the ribosomal cycle in muscles of fasted rats, *Horm. Metab. Res.*, **11**, 289–92.

CHUA, B., SIEHL, D. L. and MORGAN, H. E. (1979). Effect of leucine and metabolites of branched chain amino acids on protein turnover in heart, *J. Biol. Chem.*, **254**, 8358–62.

FULKS, R. H., LI, J. B. and GOLDBERG, A. L. (1975). Effects of insulin, glucose and amino acids on protein turnover in rat diaphragms, *J. Biol. Chem.*, **250**, 290–8.

GARLICK, P. J., MCNURLAN, M. A. and PREEDY, V. R. (1980). A rapid and convenient technique for measuring the rate of protein synthesis in tissues by injection of ($^3$H)phenylalanine, *Biochem. J.*, **192**, 719–23.

GOLDBERG, A. and TISCHLER, M. E. (1981). Regulatory effects of leucine on carbohydrate and protein metabolism, in *Metabolism and Clinical Implications of Branched Chain Amino and Ketoacids* (Eds. M. Walser and J. R. Williamson), Elsevier North Holland, Amsterdam.

HAGENFELDT, L., ERIKSSON, S. and WAHREN, J. (1980). Influence of leucine on arterial concentrations and regional exchange of amino acids in healthy subjects, *Clin. Sci.*, **59**, 173–81.

LI, J. B. and JEFFERSON, L. S. (1978). Influence of amino acid availability on protein turnover in perfused skeletal muscle, *Biochim. Biophys. Acta.*, **544**, 351–9.

MILLWARD, D. J., GARLICK, P. J., NNANYELUGO, D. O. and WATERLOW, J. C. (1976). The relative importance of muscle protein synthesis and breakdown in the regulation of muscle mass, *Biochem. J.*, **156**, 185–8.

MITCH, W. E., WALSER, M. and SAPIR, D. G. (1981). Nitrogen sparing induced by leucine compared with that induced by its keto analogue, α-ketoisocaproate, in fasting obese man, *J. Clin. Invest.*, **67**, 553–62.

PAIN, V. M. (1978). In *Protein Turnover in Mammalian Tissues and in the Whole Body* (Eds. J. C. Waterlow, P. J. Garlick and D. J. Millward), Chapter 2, Elsevier North Holland, Amsterdam.

RANNELS, D. E., HJALMARSON, A. C. and MORGAN, H. E. (1974). Effects of noncarbohydrate substances on protein synthesis in muscle, *Am. J. Physiol.*, **226**, 528–39.

SAKAMOTO, A., MOLDAWER, L. L., BOTHE, A., JR, BISTRIAN, B. R. and BLACKBURN, G. L. (1980). Nitrogen sparing mechanism of branched chain amino acid administration in acute experimental pancreatitis, *Proc. 2nd European Congress of Parenteral and Enteral Nutrition, Newcastle-upon-Tyne, UK.*

SAPIR, D. G. and WALSER, M. (1977). Nitrogen sparing induced in starvation by infusion of branched chain ketoacids, *Metab. Clin. Exp.*, **26**, 301–8.

SHERWIN, R. S. (1978). Effect of starvation on the turnover and metabolic response to leucine, *J. Clin. Invest.*, **61**, 1471–81.

# 11

# Recent Developments in Knowledge of Human Amino Acid Requirements

V. R. YOUNG, M. MEGUID and C. MEREDITH

*Department of Nutrition and Food Science,
Massachusetts Institute of Technology,
Cambridge, Massachusetts, USA*

and

D. E. MATTHEWS and D. M. BIER

*Department of Internal Medicine,
Washington University School of Medicine, St. Louis, Missouri, USA*

## *Introduction*

In this symposium we recognize the extensive contributions made by investigators at the Tropical Metabolism Research Unit, University of the West Indies, during the past 25 years. Thus, in a chapter that is devoted to recent knowledge of human amino acid requirements it is appropriate to begin with the state of knowledge in 1955 about these requirements and some major advances that occurred in the following 20 years. Because there are a number of extensive reviews which can be consulted on the amino acid requirements of human subjects (Rose, 1957; Leverton, 1959; Holt and Snyderman, 1965; Irwin and Hegsted, 1971; FAO/WHO, 1973; NRC, 1974; Young and Scrimshaw, 1978) this paper will focus only on a few topics of research largely carried out in our laboratories.

## *Summary of Advances Between 1955 and 1975*

In the early studies of Rose (1957) in young men and of Leverton and co-workers in young women (1959) nitrogen balance was used as the main criterion of requirements. Rose (1957) developed a classification in which eight amino acids were defined as essential: leucine, valine, isoleucine, threonine, methionine, phenylalanine, lysine and tryptophan. Holt and Snyderman (1965) subsequently showed that histidine is essential for the young infant. More recent studies in uraemic patients (e.g. Fürst, 1972) and 'long-term' nitrogen balance studies in normal healthy subjects indicate that this amino acid should now be classified as an indispensable amino acid in adult nutrition. Thus, Kopple and Swendseid (1975) observed that healthy adults fed a histidine-deficient diet developed a negative nitrogen balance after 20–30 days and this could be corrected by dietary supplementation with histidine. However, the absolute requirement for histidine is apparently quite low. This is supported by the findings of Sheng *et al.* (1977) which showed that, from an oral dose of $^{15}NH_4Cl$, $^{15}N$ was incorporated into the imidazole moiety of histidine isolated from globin. In addition Sheng *et al.* (1977) were able to detect $^{15}N$ in urinary $N^\tau$-methylhistidine, an amino acid originating through methylation of a histidine residue in muscle actin and myosin which is subsequently released and excreted in urine following breakdown of these myofibrillar proteins (Young and Munro, 1978). These findings confirm that synthesis of histidine occurs within the body. The slow rate at which a negative nitrogen balance develops in the absence of histidine suggests that endogenous synthesis of histidine can cover a significant proportion of the requirement for this amino acid.

Reports on the minimum needs in age groups other than infants and young adults are sparse, contradictory or non-existent. In the late 1950s and early 1960s, Tuttle, Swendseid and collaborators reported a series of studies on the amino acid requirements of elderly subjects and concluded that elderly men have markedly higher requirements for lysine and methionine than young adults (Tuttle *et al.*, 1965). However, the studies of Watts *et al.* (1964) and our own nitrogen balance studies (Tontisirin *et al.*, 1973, 1974) have not confirmed these conclusions. Nevertheless, in view of the changes in the distribution of body protein metabolism that occur with advancing age (e.g. Young *et al.*, 1981) and the increasing numbers of elderly in the populations of technically advanced regions, reliable and more extensive data on the effects of increasing age in adults on amino acid requirements are desirable.

Recent improvements in techniques for the nutritional support of pre-term infants also bring into question whether the established qualitative pattern of amino acids required by young infants and pre-school children (e.g. FAO/WHO, 1973; NRC, 1974) applies equally to prematures, in which many enzyme systems associated with the metabolism and

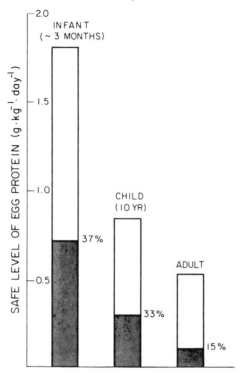

FIG. 1. Requirements for total protein and essential amino acids by human subjects of various ages. Drawn from FAO/WHO (1973).

interconversion of amino acids are biochemically immature. Thus the limited *in vitro* enzymatic capacity to metabolize tyrosine (Kretchmer *et al.*, 1957) and methionine (Gaull *et al.*, 1972) may have important consequences for evaluation of amino acid requirements and protein quality in the nutrition of pre-term infants (Gaull *et al.*, 1977). Furthermore, the absence or low level of cystathionase in the liver has raised the suggestion that cystine might be an essential amino acid for the fetus and new-born infant (Sturman *et al.*, 1970) and the lower hepatic activity of cysteine sulphinic acid decarboxylase (Rassin and Gaull, 1977) further suggests that conversion of cystine to taurine may be insufficient to maintain normal urine and plasma taurine concentrations. Because taurine may function as a neurotransmitter (Oja and Kontro, 1978) and a dietary taurine deficiency in the cat is associated with a retinal degeneration which may progress to blindness (Hayes *et al.*, 1975), it can be questioned whether there is a taurine requirement for rapid growth in the pre-term infant. The effects of a cystine-free diet and of the sparing effect of cystine on the methionine requirement of the pre-term and newborn infant have yet to be explored. This approach is necessary because the body's capacity for conversion of methionine to cystine and, possibly, the dietary need for the latter cannot be judged solely from determination of enzyme activity of samples from isolated organs *in vitro*. The same applies to the effect of tyrosine intake on the phenylalanine requirement.

A striking feature of the current estimates of the amino acid requirements of humans is the marked decrease in these needs which occurs with increasing maturity and development and, thus, the relatively low requirement for indispensable amino acids by the time adulthood is reached (Fig. 1).

The level and pattern of amino acids required for optimal nutritional support of sick patients receiving parenteral nutrition has been reviewed by Shenkin and Wretlind (1978). It may be concluded that in some pathological states amino acids that are normally regarded as dispensable for healthy subjects should be supplied in the nutritional regimens of sick patients to achieve adequate protein status. However, a major purpose here is to assess the more general concepts and state of knowledge about these needs.

## Shortcomings of Methods for Establishing Amino Acid Requirements

Estimates of the needs for individual essential amino acids are based on methods similar to those used for establishing total protein needs. For individual essential amino acids, most experimental diets in the classical studies already referred to have been based on mixtures of synthetic L- or DL-amino acids. However, in some studies intact proteins or mixtures of proteins have been used (Fomon *et al.*, 1973). Several observations suggest the possibility that amino acid mixtures may be utilized less efficiently than equivalent amino acid intakes when consumed in the form of intact proteins. Studies in rats (Itoh *et al.*, 1973) indicate that the nutritional equivalence of a mixture of amino acids patterned as in casein depends upon the pattern of food ingestion, but comparable critical studies in man are lacking. Thus, it would be highly worthwhile to explore further the factors that might determine the efficiency of utilization of amino acids when obtained from a dietary synthetic L-amino acid mixture as compared with equivalent intakes from intact proteins. This should be undertaken in adults because differences in the efficiency of utilization of nitrogen from amino acid mixtures and intact proteins do not appear too large in the case of young infants (e.g. Harper, 1977).

Secondly, many of the estimates of human amino acid requirements are based on results obtained from metabolic nitrogen balance studies. The limitations of balance studies are well recognized and have been discussed in a number of reviews (e.g. Hegsted, 1963, 1976; Young and Scrimshaw, 1978; Young *et al.*, 1980) and include the following: (a) neglect of integumental and other routes of loss may lead to substantial errors; (b) on high nitrogen intakes many authors have found unrealistic positive nitrogen balances which are difficult to explain by cumulative technical errors; (c) the excessive energy intakes used in many studies promote a positive balance which would not occur with normal everyday intakes; (d) high nitrogen intakes *per se* may reduce apparent minimum requirements for the indispensable amino acids (Kies *et al.*, 1965), although the mechanisms responsible for this effect have not been identified. In a recent study we gave an intravenous infusion of $1$-$^{13}$C-leucine to young men (Meredith, Bier, Matthews and Young, unpublished data) to explore the effect of addition of non-specific nitrogen, in the form of a mixture of dispensable amino acids, to an L-amino diet supplying a limiting level of leucine. This dietary treatment resulted in a reduced rate of leucine oxidation and a lowered rate of entry of leucine into the plasma compartment via endogenous protein breakdown (Table 1).

TABLE 1

EFFECT OF ADDITION OF NON-SPECIFIC NITROGEN (NSN) ON
WHOLE BODY LEUCINE DYNAMICS IN FOUR YOUNG MEN

|  | *Diet group* | | |
|---|---|---|---|
|  | *1* | *2* | *3* |
| Leucine intake (mg/kg/day) | 36 | 36 | 36 |
| Added NSN ($\equiv$ g protein/kg/day) | 0 | 0·6 | 1·1 |
| Leucine dynamics[a] ($\mu$mol/kg/h) |  |  |  |
| Oxidation | 15 | 14 | 11 |
| Flux | 138 | 138 | 83 |
| Endogenous inflow | 127 | 127 | 72 |

[a] Determined with $1$-$^{13}$C-leucine by the method of Matthews *et al.* (1980).
Unpublished MIT data of Meredith *et al.*

These preliminary observations, therefore, support the concept that high intakes of non-specific nitrogen 'affect| the dynamic status of whole body amino acid metabolism and, possibly, the apparent minimum requirement for indispensable amino acids.

For these and other reasons, the nutritional value of current estimates for amino acid requirements in adults remains highly uncertain. Possibly they have been grossly underestimated. This problem does not appear to apply to infants and young children, however, because there is generally good agreement between the requirements for indispensable amino acids, as judged from experiments involving specific manipulation of the level of test amino acid in L-amino acid diets (Holt and Snyderman, 1965), and those derived from nitrogen balance and growth data obtained with varying intakes of high quality protein (Fomon *et al.*, 1973). This is probably due to the fact that changes in body weight or nitrogen balance are sensitive criteria of dietary amino acid adequacy in the growing organism, whereas short-term nitrogen balance is an insensitive index of dietary amino acid adequacy for protein nutritional maintenance in the healthy adult. Recent studies at INCAP by Pineda *et al.* (1981) and Torun *et al.* (1981), in which plasma amino acid responses and biochemical data in addition to nitrogen balance determinations were used to arrive at requirements of 2-year-old children, support the conclusion that the published amino acid needs for the young are not in serious error.

There is a need, therefore, to develop new approaches and more sensitive indices for assessment of dietary amino acid adequacy in healthy adults.

## *Dynamic Aspects of Amino Acid and Nitrogen Metabolism in Relation to Amino Acid Requirements*

To meet this need our laboratory has extended recently the lead made with animal models and has conducted experiments in humans to explore the relationships between the intake of a specific indispensable amino acid, the quantitative status of whole body amino acid dynamics and the amino acid requirement. Furthermore, there is growing evidence that the biochemical mechanisms responsible for maintaining body protein and amino acid homoeostasis are closely integrated with the individual's requirements for these nutrients (Young and Bier, 1980). Also, we previously observed that hepatic tryptophan oxygenase in rats showed a cyclic change in activity in response to ingestion of adequate protein-containing meals (Young and Munro, 1973), but this pattern was abolished when tryptophan intake was below requirements. This suggested that the liver monitors the adequacy of the tryptophan intake in relation to the body's need. In addition, rates of oxidation of lysine (Brookes *et al.*, 1972), threonine (Kang-Lee and Harper, 1978) and histidine (Kang-Lee and Harper, 1977) are low and constant when intakes of these amino acids are below requirements, but the oxidation rate rises linearly when amino acid intakes are raised above the minimum level required for maximal growth in young rats.

TABLE 2

LEUCINE, VALINE AND LYSINE OXIDATION RATES IN YOUNG MEN DURING FED STATE RECEIVING ADEQUATE AND VERY LOW INTAKES OF THE SPECIFIC AMINO ACID AND AN OTHERWISE ADEQUATE [L-AMINO ACID DIET

| | *Adequate intake* | | | *Low intake* | | |
|---|---|---|---|---|---|---|
| | *Leucine* | *Valine* | *Lysine* | *Leucine* | *Valine* | *Lysine* |
| Intake (mg/kg/day) | 79 | 68 | 100 | 4 | 4 | 2 |
| Oxidation rate | | | | | | |
| μmol/kg/h | 32 | 24 | — | 7·4 | 4·3 | — |
| % of flux | 27 | 17 | 20 | 13 | 11 | 8 |

Unpublished MIT data of Meguid, Meredith, Matthews, Bier and Young.

The quantitative whole body function of the indispensable amino acids is to support an adequate rate of protein synthesis, and their major fate is a subsequent liberation from proteins during protein breakdown and either reutilization or irreversible oxidation-via catabolic pathways. Hence, it is reasonable to assume that quantitative estimates of these processes and a knowledge of how they change under varying conditions of amino acid intake and adequacy might establish a useful basis for computing the dietary amino acid requirement level. Briefly, therefore, we have shown recently, as summarized in Table 2, that when the intake of a specific indispensable amino acid is reduced to very low levels in an otherwise adequate diet, the oxidation of that amino acid is markedly reduced. Thus, about 10% of the amino acid entering the metabolic pool, largely from protein breakdown, is oxidized compared with much higher values when the intake exceeds known requirements (Table 2). These oxidation studies were carried out in young men while they were receiving small meals and the oxidation rate of the limiting amino acid would be expected to be even lower during the post-absorptive phase (e.g. Motil *et al.*, 1981; Garlick *et al.*, 1980).

With adequate intakes of total nitrogen and indispensable amino acids, whole body protein synthesis rates approximate 3–4 g/kg/day in healthy adult men (see review by Waterlow et al., 1978). The rate of protein synthesis is higher during the day-time when meals are usually consumed and falls during the period of an overnight fast (e.g. Motil et al., 1981; Garlick et al., 1980). In contrast the rate of whole body protein breakdown is higher in the post-absorptive period. Thus, our data based on [13]C-leucine suggest that in young men receiving an adequate protein diet whole body protein breakdown proceeds at a rate of about 1·3 g/kg for the 12-h period of the day when meals are consumed and this increases to about 2·5 g/kg during a 12-h overnight fasting period (Motil et al., 1981). Rhythms in protein and amino acid metabolism are well-documented (e.g. Wurtman, 1970; Buckley and Marquardt, 1980) and appear to be generated largely by the inflow of exogenous energy-yielding substrate and the amino acids provided by protein-containing meals.

Now it can reasonably be assumed that an overall protein turnover of about 3–4 g/kg/day in adult men is consistent with maintenance of a healthy state and that there is low but obligatory loss, via oxidation, of indispensable amino acids when they are consumed in amounts just sufficient to support this rate of protein synthesis and breakdown. Therefore, it should be possible to arrive at a theoretical estimate of the minimum dietary requirement level for each indispensable amino acid. With the above information, this can be approached if the amino acid composition of whole body protein and the differences between the rates of oxidation of the specific amino acid during the fed and post-absorptive states are known for intakes that just meet physiological needs. Although precise knowledge is not available, whole body protein contains approximately 8 % leucine and 8 % lysine (Waterlow et al., 1978) and amino acid oxidation rates in the post-absorptive phase may be reduced to about one half of those occurring during the period when meals supply amino acids in amounts approximating physiological needs (e.g. Motil et al., 1981; Garlick et al., 1980).

Table 3 combines this information and presents the steps used to compute a minimum requirement for leucine, or lysine, in healthy young men, that is based on considerations of whole body protein and amino acid dynamics. Approached in this way, the estimated mean requirement for both amino acids is approximately 20 mg/kg/day. This compares with average values of 11 mg/kg/day for lysine as derived by Munro (1972) from the data of Rose. Whether the computed requirement represents a more adequate estimate than those obtained in Rose's nitrogen balance studies is

## TABLE 3

A THEORETICAL APPROXIMATION OF THE LEUCINE AND LYSINE REQUIREMENT IN ADULT
MEN, BASED ON CONSIDERATIONS OF AMINO ACID DYNAMICS

1. *Assume whole body protein breakdown:* 1·3 per kg during day-time (fed state);
   2·5 per kg during night-time (fasted state)
2. Assume *body protein contains* 8 % leucine and 8 % lysine
3. Assume that % *of endogenous inflow oxidized* = 10 % during day-time and 5 %
   during night-time
4. Therefore, endogenous leucine (or lysine) entering plasma pool = 104 mg/kg in
   daytime and 200 mg/kg during night-time
5. Thus, leucine (or lysine) oxidation = 10 mg/kg in day-time and 10 mg/kg during
   night-time
6. Hence, daily leucine (or lysine) requirement (per kg/day) = 10 + 10 = 20

not known. However, the evidence reviewed earlier indicates that the latter
are probably too low.

If adult human subjects show a similar pattern of response in dynamics of
whole body amino acid and nitrogen metabolism to graded intakes of
specific amino acids as has been observed for growing rats (Kang-Lee and
Harper, 1977, 1978), then marked changes in amino acid oxidation rates
and/or in whole body protein turnover would be expected to occur when
dietary intakes change in the region round about the minimal physiological
need for maintenance of normal protein turnover.

A series of experiments currently underway in our laboratories provides
an initial opportunity to evaluate this possibility. Our results are shown in
Fig. 2 for lysine oxidation in young men receiving graded intakes of lysine
and in Fig. 3 for the rate of incorporation of leucine into whole body
protein, in young men receiving graded leucine intakes. From these data it
is apparent that marked changes in the response curves occur in the range of
30 mg/kg/day for lysine and about 23 mg/kg/day for leucine. These intake
levels, especially for leucine, more closely correspond to the computed
mean requirement (Table 3) than to the much lower estimates that have
been based on nitrogen balance data. A comparison of requirements for
lysine and leucine as estimated in these ways is summarized in Table 4. It
should be pointed out that these observations apply to the mean
requirement and do not take into account the problem of individual
variation in these needs. Our data are not sufficiently extensive to assess the
magnitude of this variation.

This evidence further suggests that Rose's values and the comparable
estimates made by others, based on use of nitrogen balance techniques (see

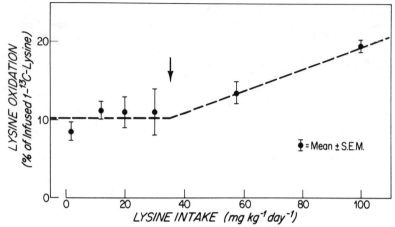

FIG. 2.    Lysine oxidation, determined with a primed constant intravenous infusion of 1-$^{13}$C-lysine (see Matthews *et al.*, 1980) in young men receiving graded intakes of lysine from an L-amino acid diet. Preliminary data of Meredith *et al.* Lines drawn by inspection.

Irwin and Hegsted, 1971; Munro, 1972), are probably deficient and that a more likely mean adequate level for protein nutritional maintenance is that predicted from a knowledge of whole body amino acid dynamics. However, we recognize that this conclusion is based on still very limited and preliminary results, but the few, difficult studies that we have conducted so far offer support for our concept and underscore the reservations that we have expressed about the uncertainty of published requirements for indispensable amino acids in adult human nutrition. Furthermore, in view of the relationships between concentrations of amino acids in the

TABLE 4

COMPARISON OF APPROXIMATIONS OF MEAN REQUIREMENTS FOR LEUCINE AND LYSINE, AS EVALUATED IN DIFFERENT WAYS, FOR YOUNG MEN

| Approach | Requirement (mg/kg/day) | |
|---|---|---|
| | Leucine | Lysine |
| Nitrogen balance (Rose)[a] | 11 | 9 |
| Theoretical | 20 | 20 |
| $^{13}$C oxidation studies | $\simeq 20$ | $\simeq 30$ |

[a] Midpoint of range of values taken for average need, as suggested by Munro (1972).

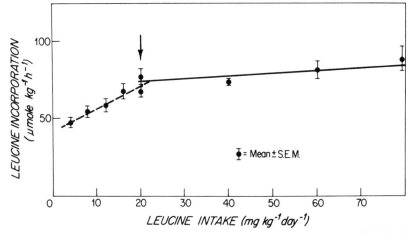

FIG. 3.   Rate of incorporation of leucine into whole body protein, determined with aid of a primed, constant intravenous infusion of 1-$^{13}$C-leucine (e.g. Matthews *et al.,* 1980) in young men receiving graded intakes of leucine from an L-amino acid diet. Preliminary data of Meguid *et al.* Lines drawn by inspection.

circulating blood plasma and rates of formation of catecholamines and serotonin from tyrosine and tryptophan, respectively, in the central nervous system (Growdon, 1979), it would be of great interest to know whether the requirement, as computed from whole body amino acid dynamics, would result in the generation of plasma amino acid levels and patterns that are more favourable for meeting this particular function of amino acids than the currently accepted and lower requirement values.

We do not yet know whether the appropriate criterion to evaluate this concept critically should be based on a measure of oxidation rate, whole body amino acid flux and/or rate of whole body protein synthesis and breakdown. It is conceivable that the criterion might depend upon the specific amino acid under investigation. For example, the oxidation rate of methionine in rats is approximately linearly related throughout a broad range of methionine intake both above and below the requirement level (Aguilar *et al.,* 1974) compared with the 'break-point' type of response observed for the oxidation of lysine, histidine and threonine (Brookes *et al.,* 1972, Kang-Lee and Harper, 1977, 1978) to graded intakes of these specific amino acids. Therefore, the relationship between methionine intake, whole body amino acid dynamics and the methionine requirement might be explored more usefully by measurement of the effect of changes in methionine intake on whole body protein synthesis or breakdown rates. On the other hand, for the case of lysine, a determination of the rate of lysine oxidation might be quite appropriate. This problem can be resolved for human subjects only with more extensive knowledge of the comparative and quantitative aspects of whole body metabolism for the various indispensable amino acids, when explored under various conditions of protein and amino acid intake.

A final consideration of our concept can be made, however, by comparison of amino acid requirements for infants with those computed by the approach suggested above from whole body protein and amino acid dynamics. In a previous study, we estimated the rate of whole body protein synthesis in premature infants and also measured nitrogen balance and weight gain in these babies to determine a protein intake that would support adequate growth in this age group (Pencharz et al., 1977). Assuming, as in the case of term-infants (e.g. see Harper, 1977), that total essential amino acid requirements approximate the intake furnished by an adequate, but not excessive, level of high quality protein, a computed requirement for these premature infants may then be compared with that based on growth and nitrogen balance data. These calculations are shown in Table 5, and there is

TABLE 5

REQUIREMENTS FOR AMINO ACIDS IN PREMATURE INFANTS DETERMINED IN TWO DIFFERENT WAYS

| Amino acid | Requirement (mg/kg/day) | | B/A |
|---|---|---|---|
| | Based on milk protein intake[a] of 3·8 g/kg/day (A) | Computed[b] (B) | |
| Isoleucine | 159 | 142 | 0·9 |
| Leucine | 258 | 271 | 1·1 |
| Lysine | 140 | 230 | 1·6 |
| S-amino acid | 95 | 115 | 1·2 |
| Aromatic | 266 | 272 | 1·0 |
| Threonine | 129 | 126 | 1·0 |
| Valine | 175 | 176 | 1·0 |
| Total | 1 222 | 1 332 | 1·0 |

[a] Calculated from Harper (1977) and based on N balance data of Pencharz et al. (1977).

[b] Assuming body amino acid composition similar to that of pig liver (Munro and Fleck, 1969), that protein synthesis approximates 20 g/kg/day and that % of indispensable amino acid flux oxidized is 14 % (Pencharz et al., 1977).

generally close agreement between the two methods of estimation. This strengthens the argument that a knowledge of whole body amino acid dynamics may offer a novel approach to the determination of dietary amino acid requirements in humans.

## Acknowledgement

We thank our colleagues and volunteer subjects who have helped us explore many exciting aspects of whole body amino acid metabolism. The unpublished results discussed herein were obtained with the support of NIH grants AM15856, RR88 and USDA Grant 5901-0410-8-0066-0.

## Discussion

*Waterlow* (*London School of Hygiene and Tropical Medicine, UK*)*:* Neale and I adopted a somewhat different approach in rats. Three weeks after giving a labelled amino acid all body proteins are more or less uniformly labelled. At this point we measured the rate of loss of radioactivity from the body, either by collecting $^{14}CO_2$ or by carcass analysis. If the rat is on a protein-free diet, the value so obtained represents the rate of obligatory loss of the amino acid. In the case of lysine this came to about 3 % of body lysine per day (Neale and Waterlow, 1974). It is much more difficult to get this kind of information from measurements of $^{14}CO_2$ output immediately after administration of a labelled amino acid. In that situation, in order to determine the rate of amino acid oxidation, it is necessary to know the specific activity at the site of oxidation.

*Munro* (*Tufts University, Boston, USA*)*:* Motil *et al.* (1981) carried out a study in young men in which oxidation of $1\text{-}^{13}C$-leucine was followed at different levels of protein intake from very low to 0·6 g/kg daily, which is about the accepted safe level of protein intake (FAO/WHO, 1973), and up to a much higher level (1·5 g/kg). At the highest level oxidation was greatly stimulated, but in addition the evidence from whole body leucine flux suggested that protein synthesis was also stimulated at the same time. Therefore, there may be advantages in exceeding the point of intake at which oxidation goes up. There is no reason to believe that simply because oxidation begins to climb incrementally you are necessarily or immediately moving into a toxic phase of excess amino acid intake.

*Young:* I think that the nutritional interpretation of the amino acid intake–oxidation response curve is uncertain and the point you make is a good one. We have interpreted the oxidation studies on the assumption that our young adults were responding in a similar fashion to the responses observed in growing rats (Kang-Lee and Harper, 1977, 1978) with reference to intake, oxidation and requirement relationships. Obviously, further studies are required before firm conclusions regarding amino acid requirements can be drawn from our data on whole body amino acid dynamics.

*Walser* (*Johns Hopkins University, Baltimore, USA*): I should like to ask whether you can extrapolate from these data to a protein requirement and also to a non-essential nitrogen requirement. These two values were inferred from the earlier work of Rose (Rose *et al.*, 1955) and perhaps are subject to criticisms similar to those you have raised with regard to the observations on requirements of essential amino acids.

*Young:* No, I don't think I would be prepared to go any further than I have at the moment, particularly in relation to a total protein or nitrogen requirement.

*Kerr* (*Case Western Reserve University, Cleveland, USA*): In relation to Nyhan's observations reported earlier about essential amino acid requirements in children with inborn errors of metabolism and the observations which you have presented in non-growing adults, I should like to report our own observations of essential amino acid requirements in the absence of catabolism (Ruch and Kerr, submitted for publication). The intakes of phenylalanine and leucine required to produce slightly raised blood levels and normal growth were determined in infants with classical phenylketonuria (PKU, phenylalanine hydroxylase deficiency) and maple syrup urine disease (MSUD, branched chain α-keto acid decarboxylase deficiency) respectively. These values were compared with estimates of amounts of these two amino acids provided by the minimal requirement for human milk protein in normal infants (Fomon, 1973). Intakes were expressed in relation to energy.

The amount of phenylalanine tolerated by infants with classical PKU was found to be considerably less than the estimated normal requirement (Fig. 4). This difference persisted over two years of life. These findings are in agreement with the lower range of estimated phenylalanine tolerance found in a more heterogeneous population in the PKU collaborative study (Acosta *et al.*, 1977). Likewise, the amount of dietary leucine tolerated by infants with classical MSUD was considerably less than the normal

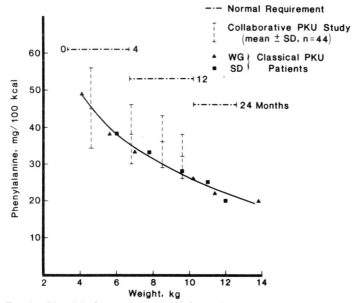

FIG. 4.  Phenylalanine requirement of infants with phenylketonuria (PKU).

requirement estimated from human milk protein intake over the first two years of life (Fig. 5). When these data are normalized and related to the percentage of total intake from 0 to 4 months a good correspondence is found (Fig. 6). The sum of previous fractional estimates of protein requirements for growth (accretion), catabolism and other losses (skin, hair, etc.) is very similar to the total protein requirement based on intake studies. The decreasing total requirement can be related to decreasing protein accretion. The percentage of the estimated normal total requirement for phenylalanine and leucine which is tolerated by infants with PKU and MSUD is similar to the percentage of the normal requirement which is due to accretion plus other losses. The difference from the total can be accounted for by lack of obligatory catabolism. Therefore, this type of data demonstrates the significant contribution of obligatory catabolism to the normal requirement for essential amino acids and emphasizes the relatively greater reliability of estimating amino acid requirements from long term growth studies with complete proteins.

*Reeds* (*Rowett Research Institute, Aberdeen, UK*): I have always been in favour of this metabolic approach to amino acid requirements. There are three aspects which concern me a bit with respect to amino acid

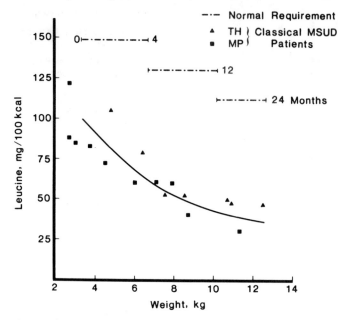

FIG. 5.   Leucine requirements of infants with maple syrup urine disease (MSUD).

requirements for maintenance. First, I think maintenance requirements cannot be interpreted solely in terms of maintaining protein synthesis and amino acid oxidation, because some amino acids have specific metabolic requirements, for example the utilization of histidine for 3-methylhistidine and of glycine for creatine synthesis. Secondly, is egg protein, with its imbalance with respect to essential/non-essential amino acids, a good 'reference' protein? The third point which I hope to make tomorrow concerns the use of 8 % leucine as a value for whole body protein when it is based on muscle or liver, when the inclusion of collagenous tissues has quite a marked effect on the amino acid composition of the whole body protein; one tends to forget about the requirements for collagen synthesis.

*Young:* The assumed value of 8 % for the leucine content of body protein is of course a debatable issue and the point you make is well taken. What I was trying to do was to offer a new conceptual and metabolic approach to the determination of amino acid requirements. I think that there are serious issues with respect to the specific values that I have chosen to state the case and a small change in some of the values could make quite a difference to the computed estimation of the requirement. I appreciate the limitations of our

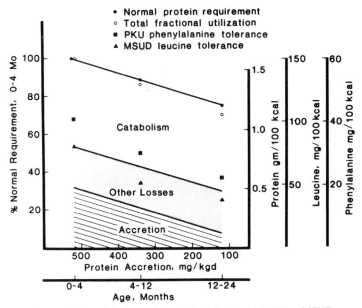

FIG. 6.   Tolerance of phenylalanine and leucine by infants with PKU or MSUD compared with normal protein requirements.

approach, but nevertheless, on the basis of the arguments presented and available data, I think it provides a reasonable basis for additional investigation.

*Harper* (*University of Wisconsin-Madison, USA*): How do your estimates fit with the values that are obtained from the low protein or amino acid diets or infusions that have been used to maintain renal patients? It seems to me that the amounts used to maintain patients with renal disease would come out considerably lower than your estimates.

*Young:* I agree that the estimates based on studies in such patient populations would indicate a lower requirement for specific amino acids. I am not convinced that the available estimates made on unhealthy or hospitalized patients are any better established than in the case of healthy adults.

# References

ACOSTA, P. B., WENZ, E. and WILLIAMSON, M. (1977). Nutrient intake of treated infants with phenylketonuria, *Am. J. Clin. Nutr.*, **30**, 198–208.

150    V. R. Young, M. Meguid, C. Meredith, D. E. Matthews and D. M. Bier

AGUILAR, T. S., BENEVENGA, N. J. and HARPER, A. E. (1974). Effect of dietary methionine level on its metabolism in rats, *J. Nutr.*, **104**, 761–71.

BROOKES, I. M., OWENS, F. N. and GARRIGUS, U. S. (1972). Influence of amino acid level in the diet upon amino acid oxidation by the rat, *J. Nutr.*, **102**, 27–36.

BUCKLEY, W. T. and MARQUARDT, R. R. (1980). Diurnal variation in oxidation of L-[$^{14}$C]leucine and L-[1-$^{14}$C]lysine in rats, *J. Nutr.*, **108**, 974–81.

CONWAY, J. M., BIER, D. M., MOTIL, J. J., BURKE, J. F. and YOUNG, V. R. (1980). Whole body lysine flux in young adult men: effects of reduced total protein and of lysine intake, *Am. J. Physiol.*, **239**, E192–E200.

FAO/WHO (1973). *Energy and Protein Requirements*, WHO Technical Report Ser. No. 522, World Health Organization, Geneva.

FOMON, S. J. (1978). *Infant Nutrition*, Saunders, Philadelphia.

FOMON, S. J., THOMAS, L. N., FILER, L. J., JR., ANDERSON, T. A. and BERGMANN, K. E. (1973). Requirements for protein and essential amino acids in early infancy, *Acta Paediat. Scand.*, **62**, 33–45.

FÜRST, P. (1972). $^{15}$N studies in severe renal failure. II. Evidence for essentiality of histidine, *Scand. J. Clin. Lab. Invest.*, **30**, 307–12.

GARLICK, P. J., GLUGSTON, G. A., SWICK, R. W. and WATERLOW, J. C. (1980). Diurnal pattern of protein and energy metabolism in man, *Am. J. Clin. Nutr.*, **33**, 1983–6.

GARZA, C., SCRIMSHAW, N. S. and YOUNG, V. R. (1976). Human protein requirements: the effect of variations in energy intake within the maintenance range, *Am. J. Clin. Nutr.*, **29**, 280–7.

GAULL, G. E., STURMAN, J. A. and RÄIHA, N. C. R. (1972). Development of mammalian sulfur metabolism: absence of cystathionase in human fetal tissues, *Pediatr. Res.*, **6**, 538–47.

GAULL, G. E., RASSIN, D. K., RÄIHA, N. C. R. and HEINONEN, K. (1977). Milk protein quality and quantity in low-birth weight infants. III. Effects on sulfur amino acids in plasma and urine, *J. Pediatr.*, **90**, 348–55.

GROWDON, J. H. (1979). Neurotransmitter precursors in the diet: their use in the treatment of brain diseases, in *Nutrition and the Brain* (Ed. R. J. Wurtman), Vol. 3, pp. 117–81, Raven, New York.

HARPER, A. E. (1977). Amino acid requirements—general, in *Clinical Nutrition Update: Amino Acids* (Eds. H. L. Greene, M. A. Holliday and H. N. Munro), pp. 58–65, American Medical Association, Chicago.

HAYES, K. C., CAREY, R. E. and SCHMIDT, S. Y. (1975). Retinal degeneration associated with taurine deficiency in the cat, *Science*, **188**, 949–51.

HEGSTED, D. M. (1963). Variation in requirements of nutrients—amino acids, *Fed. Proc.*, **22**, 1424–30.

HEGSTED, D. M. (1976). Balance studies, *J. Nutr.*, **106**, 307–11.

HOLT, L. E., JR. and SNYDERMAN, S. E. (1965). Protein and amino acid requirements of infants and children, *Nutr. Abstr. Rev.*, **35**, 1–13.

IRWIN, M. I. and HEGSTED, D. M. (1971). A conspectus of research on amino acid requirements of man, *J. Nutr.*, **101**, 539–66.

ITOH, H., KISHI, T. and CHIBATA, I. (1973). Comparative effects of casein and amino acid mixture simulating casein on growth and food intake in rats, *J. Nutr.* **103**, 1709–15.

KANG-LEE, T. A. and HARPER, A. E. (1977). Effect of histidine intake and hepatic histidase activity on the metabolism of histidine *in vivo*, *J. Nutr.*, **107**, 1427–43.

KANG-LEE, T. A. and HARPER, A. E. (1978). Threonine metabolism *in vivo*: effect of threonine intake and prior induction of threonine dehydratase in rats, *J. Nutr.*, **108**, 163–75.

KIES, C. F., SHORTRIDGE, L. and REYNOLDS, N. S. (1965). Effect on nitrogen retention of men of varying the total dietary nitrogen with essential amino acid intake kept constant, *J. Nutr.*, **85**, 260–4.

KOPPLE, J. D. and SWENDSEID, M. E. (1975). Evidence that histidine is an essential amino acid in normal and chronically uremic men, *J. Clin. Invest.*, **55**, 881–91.

KRETCHMER, N., LEVINE, S. Z. and MCNAMARA, H. (1957). The *in vitro* metabolism of tyrosine and its intermediates in the liver of the premature infant, *A.M.A. J. Dis. Child.*, **93**, 19–20.

LEVERTON, R. M. (1959). Amino acid requirements of young adults, in *Protein and Amino Acid Nutrition* (Ed. A. A. Albanese), pp. 477–506, Academic Press, New York.

MATTHEWS, D. E., MOTIL, K. J., ROHRBAUGH, D. K., BURKE, J. F., YOUNG, V. R. and BIER, D. M. (1980). Measurement of leucine metabolism in man from a primed, continuous infusion of L-[1-$^{13}$C]leucine, *Am. J. Physiol.*, **238**, E473–E479.

MOTIL, K. J., MATTHEWS, D. E., BIER, D. M., BURKE, J. F., MUNRO, H. N. and YOUNG, V. R. (1981). Whole body leucine and lysine metabolism: response to dietary protein intake in young men, *Am. J. Physiol.*, in press.

MUNRO, H. N. and FLECK, A. (1969). Analysis of tissues and body fluids for nitrogenous constituents, in *Mammalian Protein Metabolism* (Ed. H. N. Munro), Vol. III, Chapter 30, pp. 423–525, Academic Press, New York.

MUNRO, H. N. (1972). Amino acid requirements and metabolism and their relevance to parenteral nutrition, in *Parenteral Nutrition* (Ed. A. W. Wilkinson), pp. 34–67, Churchill Livingstone, London.

NEALE, R. J. and WATERLOW, J. C. (1974). Critical evaluation of a method for estimating amino acid requirements for maintenance in the rat by measurement of the rate of $^{14}$C-labelled amino acid oxidation *in vivo*, *Br. J. Nutr.*, **32**, 257–72.

NRC (1974). *Improvement of Protein Nutriture*, National Academy of Sciences, Washington, DC.

OJA, S. S. and KONTRO, P. (1978). Neurotransmitter actions of taurine in the central nervous system, in *Taurine and Neurological Disorders* (Ed. A. Barbeau and R. J. Huxtable), pp. 181–200, Raven, New York.

PENCHARZ, P. B., STEFFEE, W. P., COCHRAN, W., SCRIMSHAW, N. S., RAND, W. M. and YOUNG, V. R. (1977). Protein metabolism in human neonates: nitrogen balance studies, estimated obligatory loss of nitrogen and whole body turnover of nitrogen, *Clin. Sci. Mol. Med.*, **52**, 485–98.

PINEDA, O., TORUN, B., VITERI, F. and ARROYAVE, G. (1981). Protein quality in relation to estimates of essential amino acid requirements, in *Protein Quality in Humans: Assessment and* in vitro *Estimation* (Eds. C. E. Bodwell, J. D. Adkins and D. Hopkins), Avi, Westport, Connecticut.

RASSIN, D. K. and GAULL, G. E. (1977). Protein requirements and the development of amino acid metabolism in the preterm infant, in *Clinical Nutrition Update: Amino Acids* (Eds. H. L. Green, M. A. Holliday and H. N. Munro), pp. 84–95, American Medical Association, Chicago.

ROSE, W. C. (1957). The amino acid requirements of adult man, *Nutr. Abstr. Rev.*, **27**, 631–47.

ROSE, W. C., WIXOM, R. L., LOCKHART, H. B. and LAMBERT, G. F. (1955). The amino acid requirements of man. XV. The valine requirement; summary and final observations, *J. Biol. Chem.*, **217**, 987.

SHENG, Y-B., BADGER, T. M., ASPLUND, J. M. and WIXAM, R. L. (1977). Incorporation of $^{15}NH_4Cl$ into histidine in adult man, *J. Nutr.*, **107**, 621–30.

SHENKIN, A. and WRETLIND, A. (1978). Parenteral nutrition, *World Rev. Nutr. Diet.*, **28**, 1–111.

STURMAN, J. A., GAULL, G. E. and RÄIHA, N. C. R. (1970). Absence of cystathionase in human fetal liver: is cystine essential? *Science*, **169**, 74–6.

TONTISIRIN, K., YOUNG, V. R., MILLER, M. and SCRIMSHAW, N. S. (1973). Plasma tryptophan response curve and tryptophan requirements of elderly people, *J. Nutr.*, **103**, 1220–8.

TONTISIRIN, K., YOUNG, V. R., RAND, W. M. and SCRIMSHAW, N. S. (1974). Plasma threonine response curve and threonine requirements of young men and elderly women, *J. Nutr.*, **104**, 495–505.

TORUN, B., PINEDA, O., VITERI, F. E. and ARROYAVE, G. (1981). Use of amino acid composition data to predict protein nutritive value for children with specific reference to new estimates of their essential amino acid requirements, in *Protein Quality in Humans: Assessment and* in vitro *Estimation* (Eds. C. E. Bodwell, J. D. Adkins and D. Hopkins), Avi, Westport, Connecticut.

TUTTLE, S. G., BASSETT, S. H., GRIFFITH, W. H., MULCARE, D. B. and SWENDSEID, M. E. (1965). Further observations on the amino acid requirements of older men. II. Methionine and lysine, *Am. J. Clin. Nutr.*, **16**, 229–31.

WATERLOW, J. C., GARLICK, P. J. and MILLWARD, D. J. (1978). *Protein Turnover in Mammalian Tissues and in the Whole Body*, p. 804. Elsevier North Holland, Amsterdam.

WATTS, J. H., MANN, A. N., BRADLEY, L. and THOMPSON, D. J. (1964). Nitrogen balances of men over 65 fed the FAO and milk patterns of essential amino acids, *J. Gerontol.*, **19**, 360–74.

WURTMAN, R. M. (1970). Diurnal rhythms in mammalian protein metabolism, in *Mammalian Protein Metabolism* (Ed. H. N. Munro), Vol. 4, pp. 445–79, Academic Press, New York.

YOUNG, V. R. and BIER, D. M. (1980). Stable isotopes ($^{13}C$ and $^{15}N$) in the study of human protein and amino acid metabolism and requirements, in *Nutritional Factors: Modulating Effects on Metabolic Processes* (Eds. R. F. Beers and E. G. Bassett), pp. 267–308, Raven, New York.

YOUNG, V. R., GERSOVITZ, M. and MUNRO, H. N. (1981). Human aging: protein and amino acid metabolism and implications for protein and amino acid requirements, in *Nutritional Approaches to Aging Research* (Ed. G. B. Moment), CRC, Boca Raton, Florida.

YOUNG, V. R. and MUNRO, H. N. (1973). Plasma and tissue tryptophan levels in relation to tryptophan requirements of weanling and adult rats, *J. Nutr.*, **103**, 1756–63.

YOUNG, V. R. and MUNRO, H. N. (1978). N$^\tau$-methylhistidine (3-methylhistidine) and muscle protein turnover: an overview, *Fed. Proc.*, **37**: 2291–300.

YOUNG, V. R., PERERA, W. D., WINTERER, J. C. and SCRIMSHAW, N. S. (1976). Protein and amino acid requirements of the elderly, in: *Nutrition and Aging* (Ed. M. Winick), pp. 77–118, John Wiley, New York.

YOUNG, V. R. and SCRIMSHAW, N. S. (1978). Nutritional evaluation of proteins and protein requirements, in *Protein Resources and Technology* (Eds. M. Milner, N. S. Scrimshaw and D. I. C. Wang), pp. 136–73, Avi, Westport, Connecticut.

YOUNG, V. R., SCRIMSHAW, N. S. and BIER, D. M. (1980). Whole body protein and amino acid metabolism: relation to protein quality evaluation in human nutrition, *J. Agric. Food Chem.*, **29**, 440–7.

PART II

# End Products of Nitrogen Metabolism

# 12

# Metabolism of Glutamine, Glutamate and Aspartate

Patricia Lund

*Metabolic Research Laboratory,*
*Nuffield Department of Clinical Medicine, Radcliffe Infirmary, Oxford, UK*

## Introduction

Glutamine, glutamate and aspartate are a unique group of amino acids in that they undergo extensive interconversion in the small intestinal mucosa during absorption, with alanine and ammonia as the main nitrogenous products released into the portal vein (Windmueller and Spaeth, 1980). The concentrations of these three amino acids are therefore maintained in the body from endogenous sources; moreover, glutamine and aspartate can only be formed via glutamate. The supply of glutamate presents no real problem under normal conditions since there is a virtually indefinite supply from transamination and catabolism of other amino acids. Consequently aspartate concentrations are maintained through the near-equilibrium existing in the aspartate aminotransferase reaction in most tissues. By contrast, regulation of the synthesis and metabolism of glutamine is poorly understood and biochemical observations made in man are most frequently followed up at the cellular and subcellular level in the rat. One aim of this paper is to draw attention to differences between man and rat in distribution and properties of enzymes, particularly of those involving glutamine, which may make the rat an unsuitable experimental model for man.

## Glutamate Metabolism

Interest in glutamate metabolism has been revived because of the introduction of parenteral nutrition. On the one hand the cheapness of glutamate has meant that it is included as a major nitrogenous component of these amino acid mixtures, while on the other there has been the

155

*Patricia Lund*

awareness of the possible danger of bypassing the normal 'filter-system' of the small intestine and liver on administering a load of glutamate intravenously. The 'toxic' effects of glutamate (Chinese Restaurant Syndrome) are well known. Furthermore, several tissues, notably brain, liver and erythrocytes, appear to be relatively impermeable to dicarboxylic amino acids. Therefore it is surprising that a load of glutamate administered intravenously is removed extremely rapidly in man (Kingsland *et al.*, 1980), more rapidly than a similar load of leucine (Elia *et al.*, 1980*a*). The different rates are presented for comparison (Table 1). In

TABLE 1

RATES OF REMOVAL OF INFUSED AMINO ACID LOADS IN NORMAL MAN AFTER AN OVERNIGHT FAST

| Amino acid | Amount infused (mmol) | Period of infusion (min) | $t_{0.5}$ (min) |
|---|---|---|---|
| L-glutamate[a] | 24·5 | 10 | $10·7 \pm 1·67$ (7) |
| L-leucine[c] | 29 | 10 | $27·8 \pm 1·6$ (4) |
| L-alanine[c] | 133 | 5 | $31·5 \pm 1·2$ (6) |

[a] Kingsland *et al.* (1980).
[b] Elia *et al.* (1980*a*).
[c] Elia *et al.* (1980*b*).
Figures in parentheses are the number of observations.

spite of the very high rate of glutamate removal and metabolism, plasma concentrations of glutamine, aspartate and alanine are not increased (Kingsland *et al.*, 1980).

## Aspartate Metabolism

Aspartate is used mainly in argininosuccinate synthesis in liver and kidney, in the 'purine nucleotide cycle' for the resynthesis of AMP from IMP, and in pyrimidine synthesis. The quantitative requirements are not known. As already mentioned, the supply of aspartate presents no problem, and depends only on availability of glutamate.

An outstanding question, which applies to glutamate, aspartate and glutamine (and alanine and glycine) in particular, is how the concentrations, characteristic for individual tissues, are 'set' and how the high tissue/plasma gradients for these amino acids are maintained.

## Glutamine Metabolism

### Links between Branched Chain Amino Acids and Glutamine

The observations that large amounts of alanine (London *et al.*, 1965; Pozefsky *et al.*, 1969) and glutamine (Marliss *et al.*, 1971) are released from skeletal muscle during the post-absorptive state in man, largely as a result of the metabolism of branched chain amino acids, means that the metabolism of these amino acids is inextricably linked. During the last 10 years a good deal of effort has gone into research, mainly in the rat, to determine whether the carbon skeleton of alanine is also derived from branched chain or other amino acids (which would lead to a net synthesis of glucose in starvation) or from glucose (which would not). This is still a controversial question. Davis and Bremer (1973), Goldstein and Newsholme (1976), Garber *et al.* (1976), Spydevold (1976) and Snell and Duff (1977) present evidence that the carbon skeletons of branched chain amino acids do appear in alanine. Goldberg and co-workers (Goldberg and Chang, 1978) claim that only the carbon skeleton of glutamine is derived from other amino acids. Branched chain amino acids have been the focus of attention as possible precursors because they constitute about 20 % of the total amino acids of protein and, unlike other amino acids, they are not metabolized by the liver in rat. But there are many differences between rat and man (see Table 2) that make the rat a poor experimental model for man.

To comment in more detail on the points made in Table 2:

### Branched Chain Amino Acid Metabolism

Rate-limitation of the branched chain 2-oxoacid dehydrogenase in skeletal muscle in the rat leads to the release of large amounts of branched chain 2-oxoacids into the circulation which are removed and further metabolized by the liver. Thus, this amount (representing some 25 % of the dietary intake of leucine, isoleucine and valine) is lost as a potential carbon source for alanine or glutamine, and the contribution of these 2-oxoacids to alanine or glutamine formation must be reassessed in the rat.

In man, there is little release of branched chain 2-oxoacids from skeletal muscle, and conditions that might be expected to increase release (leucine infusion, a meal of steak) do not do so. As expected, both leucine infusion and a meal of steak increase the proportion of glutamine in the venous blood draining skeletal muscle. Surprisingly, though, neither increases alanine release. The preoccupation with branched chain amino acid metabolism in skeletal muscle has meant that an interesting point from the original arterio-venous measurements of the Cahill group tends to be

## TABLE 2

SOME DIFFERENCES BETWEEN RAT AND MAN IN DISTRIBUTION AND PROPERTIES OF ENZYMES RELATED TO BRANCHED CHAIN AMINO ACID AND GLUTAMINE METABOLISM

| Rat | | Man | |
|---|---|---|---|
| Observation | Consequence | Observation | Consequence |
| Uneven distribution of branched chain aminotransferase and 2-oxoacid dehydrogenase between skeletal muscle and liver (Ichihara et al., 1973; Shinnick and Harper, 1976) | Release of branched chain 2-oxoacids from skeletal muscle and removal for further metabolism by liver (Livesey and Lund, 1980) | Branched chain amino acid aminotransferase and 2-oxoacid dehydrogenase more evenly distributed between skeletal muscle and liver than in rat (Khatra et al., 1977) | Little release of branched chain 2-oxoacids from skeletal muscle, even after leucine infusion or a meal of steak which increase glutamine, but not alanine, release (Elia and Livesey, personal communication) |
| Low activity of branched chain amino acid aminotransferase in liver. Predominantly high $K_m$ isoenzyme II specific for leucine (Ichihara et al., 1973) | Branched chain amino acids not metabolized by liver (see Mendes-Mourão et al., 1975) | Low $K_m$ isoenzyme I of branched chain amino acid aminotransferase present in liver; no isoenzyme II (Goto et al., 1977) | Possible that branched chain amino acids metabolized by liver |
| Both 'phosphate-dependent' glutaminase and glutamine synthetase present in kidney (Krebs, 1935) | No net uptake of glutamine by kidney of normal non-acidotic rat (Lund and Watford, 1976; Squires et al., 1976; Reiman and Yablon, 1978) | Glutamine synthetase absent from kidney (Lemieux et al., 1976) | Large amounts of glutamine (about 67 mmol/day) removed by kidney of normal man (Owen and Robinson, 1963) |
| Liver 'phosphate-dependent' glutaminase unique (Horowitz and Knox, 1968). Activated by Pi, $NH_4^+$, $HCO_3^-$; possibly also by cyclic AMP and branched chain amino acids (for review see Lund, 1980) | Physiologically-occurring activators give flexibility in extent to which glutamine is metabolized by liver | 'Phosphate-dependent' glutaminase of liver different from rat liver enzyme (Linder-Horowitz et al., 1970) | Unknown |

overlooked. Although the greater part of the branched chain amino acids arising from proteolysis do undergo transamination there is still a net release from the tissue, with removal by the splanchnic bed (Marliss *et al.*, 1971) and brain (Felig *et al.*, 1973*a*). The possibility exists, therefore, that branched chain amino acids are metabolized by the liver in man, especially as the aminotransferase of human liver, in contrast to that of rat liver, is the isoenzyme I that predominates in the metabolism of branched chain amino acids in other tissues and has approximately the same activity per unit weight as the enzyme in skeletal muscle (Goto *et al.*, 1977). There is some support for this suggestion: on infusion of leucine in man, a considerable proportion of the leucine is metabolized by liver (Hagenfeldt *et al.*, 1980).

*Quantitative Aspects of Glutamine Metabolism in Man and Rat*

The large output of glutamine from skeletal muscle in man fasted overnight appears to be virtually balanced by uptake by the splanchnic bed (mainly intestine), brain and kidney (Marliss *et al.*, 1971; Owen and Robinson, 1963; Felig *et al.*, 1973*b*; Tizianello *et al.*, 1978), although the possible contribution of amino acid exchange via erythrocytes (Felig *et al.*, 1973*c*) has not always been taken into account. The kidneys of normal man use an enormous amount of glutamine: 47 $\mu$mol/min, which amounts to 67·7 mmol/day. During mild $NH_4Cl$ acidosis (6–8 g/day for 6–9 days) the demand increases to about 147 mmol/day (Owen and Robinson, 1963). The question is, where does the extra glutamine come from? Interestingly, in a footnote, Owen and Robinson (1963) report that 1-methyl- and 3-methylhistidine amount to about 20 % of the total nitrogen excretion in acidosis. This is an indication that increased muscle protein degradation and/or turnover occur in acidosis in the face of an adequate dietary intake of protein; it also suggests that skeletal muscle is involved in the increased synthesis of glutamine. The work of Tizianello *et al.* (1978) and Finley *et al.* (1979) gives some evidence of this. Weisswange *et al.* (1973), however, found no increased glutamine release from muscle in acidosis. In any event, skeletal muscle almost certainly does not provide the extra 80 mmol/day required by the kidney.

By analogy with man it is taken for granted that muscle is the main source of glutamine under normal conditions in the rat, but the evidence is not overwhelming. The kidney makes no demand on the glutamine supply in the normal rat (Lund and Watford, 1976; Squires *et al.*, 1976; Relman and Yablon, 1978), which leaves the mucosa of the small intestine as the main glutamine-utilizing tissue with a requirement of about 1·8 mmol/day for a 280 g rat starved overnight (Windmueller and Spaeth, 1974). Some measurements of arteriovenous differences across the hindlimbs suggest that glutamine release from skeletal muscle (including skin and adipose tissue) is more than adequate to supply the requirement of the small intestine (Ruderman and Berger, 1974; Squires, 1977; Rémésy *et al.*, 1978; Ishikawa and co-workers (see Ishikawa, 1976)), while others indicate no significant release (Lund and Watford, 1976; Lemieux *et al.*, 1980; Schröck *et al.*, 1980). These experiments are not strictly comparable in that physiological states were different, flow rates were rarely measured, and some data refer to whole blood, others to plasma.

In metabolic acidosis in the rat the glutamine requirement, calculated from data of Squires *et al.* (1976) and flow rate given by Schröck *et al.* (1980), becomes about 0·5 mmol/day/100 g body weight, an amount that should easily be detectable as an output from another tissue. This may be an underestimate because of the discrepancy between this value and the total renal release of ammonia (2 mmol/day/100 g body weight) into urine (see Parry and Brosnan, 1978) plus a similar amount into the renal vein). No significant increase in release from skeletal muscle has been found in chronic $NH_4Cl$ acidosis (Lund and Watford, 1976; Squires, 1977; Lemieux *et al.*, 1980), although a very large release in acute HCl acidosis was found by Schröck *et al.* (1980). The apparent failure of skeletal muscle to increase glutamine production and release in chronic $NH_4Cl$ acidosis is in keeping with the unexpected finding that glutamine turnover is not increased (Squires, 1977). The implication is that synthesis continues, as in the normal rat, whereas utilization by extra-renal tissues is decreased. The arterial concentration of glutamine is also decreased in chronic $NH_4Cl$ acidosis (Squires *et al.*, 1976; Lund and Watford, 1976; Lemieux *et al.*, 1980). As glutamine utilization by the intestinal mucosa is concentration-dependent over the range 0·6–0·2 mM (Windmueller and Spaeth, 1974) a decreased uptake by the small intestine seems very likely. Data we obtained some years ago (Welt and Lund, unpublished) provided evidence of this. The aorta-portal vein difference for glutamine decreased in 200 g rats from a mean control value of −0·134 to −0·048 μmol/ml whole blood in chronic $NH_4Cl$ acidosis. Using the portal flow rate given by Pardridge (1977), the rate of glutamine utilization decreases from about 1·4 mmol/day to about 0·5 mmol/day in a 200 g rat—an amount of the same order as that required by the kidney. At some time between 3 and 6 days of $NH_4Cl$ acidosis in man a similar adaptation of the splanchnic uptake of glutamine appears to occur (see Tizianello *et al.*, 1978).

*Liver Phosphate-dependent Glutaminase Isoenzymes in Man and Rat*
    The function of the liver in glutamine metabolism is by no means clear in

either man or rat. Mammalian liver generally contains both glutamine synthetase and 'phosphate-dependent' glutaminase, theoretically allowing the liver flexibility to synthesize or degrade glutamine, depending on the dietary or hormonal 'signals'. Pardridge (1977) reports a high clearance rate of $^{14}C$-glutamine in rat liver *in vivo* which, as it is not paralleled by an equivalent net uptake, is indicative of rapid synthesis and degradation in a metabolic cycle, but a cycle which does not involve the same molecules of glutamate. A similar interpretation may apply in the experiments of Haussinger and Sies (1979). This raises the question of how the relative activities of the two enzymes, located in different compartments of the cell, are regulated. It is very likely that regulation is different in man and rat in that Linder-Horowitz *et al.* (1970) suggested that the 'phosphate-dependent' glutaminase of human liver has the same properties as the widely distributed 'kidney-type' isoenzyme of the rat and not those of the 'liver-type' isoenzyme described by Horowitz and Knox (1968). At that time the complexity of rat liver glutaminase had not been recognized. In fact, until recently, it appeared that rat liver *in vivo* was unlikely to be involved in glutamine metabolism because the intact cell *in vitro* neither degraded glutamine when incubated in the presence of near-physiological plasma concentrations of the amino acid, nor synthesized it from suitable amino acid precursors (see Lund and Watford, 1976). The well-established property of glutamine as precursor of both glucose and urea appeared to be true only at 5–10 mM. Now a number of physiologically-occurring activators of rat liver glutaminase have been recognized which alter this interpretation. The recent work has been reviewed (see Lund, 1980).

In summary, in addition to the activation by phosphate, the enzyme has been reported to be activated by $NH_4^+$ (*in vivo*, in isolated perfused liver, isolated hepatocytes, mitochondria and mitochondria disrupted by freeze-thawing), by $HCO_3^-$ (isolated hepatocytes, mitochondria, disrupted mitochondria), by glucagon or dibutyryl cyclic AMP (isolated hepatocytes, mitochondria isolated from glucagon-treated rats), by branched chain amino acids (isolated hepatocytes) and by ATP (mitochondria, disrupted mitochondria). Thus, rat liver glutaminase is an allosteric protein possessing multiple regulatory sites.

In an attempt to compare the rat and human liver enzymes in more detail, it soon became clear that results obtained from isolated mitochondria must be interpreted with caution, and that no definite conclusions can be drawn regarding the kinetic properties of glutaminase. Enzyme activity and the quantitative effects of the effectors depend on (a) composition of the isolation medium, (b) composition of the incubation medium, (c) temperature, (d) whether mitochondria are incubated with or without

*Patricia Lund*

shaking and (e) age of mitochondrial suspension. In the author's opinion, changes in mitochondrial volume are a major factor in the variability of the results because many of the components, of the incubation medium in particular, cause swelling (i.e. phosphate, succinate) or prevent swelling (i.e. EDTA, $Mg^{++}$, rotenone); ATP appears to counteract the effects of $Mg^{++}$. The rat enzyme exhibits the phenomenon of latency in that the effects of phosphate, $HCO_3^-$ and $NH_4^+$ are more pronounced under hypotonic conditions. Sigmoid kinetics are shown under all conditions; the effects of $HCO_3^-$ and $NH_4^+$ are to decrease the sigmoidicity. The requirement for phosphate appears to decrease on storage of the suspensions at 4°C. Similarly, the activating effects of $HCO_3^-$ and $NH_4^+$ are lost after storage for 48 h. Thus, the properties of the mitochondrial glutaminase in rat liver are highly dependent on the conformation of the mitochondrion, and no meaningful information on its kinetic properties in the environment of the cell can be obtained from experiments in isolated mitochondria or in solubilized preparations. This probably applies to the glutaminase isoenzyme of kidney which also exhibits latency (Kovačević *et al.*, 1979). In view of the sensitivity of the inner mitochondrial membrane to external factors, reports that metabolic acidosis increases transport of glutamine into mitochondria should also be treated with caution. But this very sensitivity to ionic changes could conceivably lead to regulation of glutamine metabolism *in vivo* through changes in mitochondrial volume. Many of the well-established events in the kidney in metabolic acidosis would automatically follow activation of glutamine transport and/or glutaminase activity by this mechanism.

In human liver biopsy specimens, preliminary data obtained from freshly prepared mitochondria confirm that the pH optimum of the enzyme is different from that of the rat liver enzyme. Activity increases with increasing pH between pH 7·4 and pH 8·6 (see also Linder-Horowitz *et al.*, 1970). At pH 8·6 there is no activation by $HCO_3^-$ or by $NH_4^+$.

In view of the fact that Simpson (1980) reports that $HCO_3^-$ is an inhibitor of glutaminase in dog kidney mitochondria, different isoenzymes may exist from those which have so far been identified in rat tissues (see Horowitz and Knox, 1968; Huang and Knox, 1976). As nothing is known of the regulation of glutamine synthetase in the intact cell, no conclusions can be drawn in either rat or human liver on how synthesis and degradation of glutamine are controlled. Data from the literature suggest that rat liver can function as either a glutamine-producing or a glutamine-utilizing tissue, depending on physiological state (see Lund, 1980), so that it may play a fundamental role in glutamine metabolism in this species. So far, there appears to be no evidence that human liver functions as a glutamine-producing tissue.

The activation of rat liver glutaminase by $NH_4^+$ raises an important point in relation to metabolic acidosis. Clearly, $NH_4Cl$ should not be used as the inducing agent when studying the mechanism by which the extra-renal tissues respond to meet the increased demand of the kidney for glutamine in acidosis. Activation of liver glutaminase by a load of $NH_4Cl$ will override the normal response of the liver and, indirectly, possibly that of other tissues. For instance, the presence of excess $NH_4^+$ may increase the production of glutamine by muscle for reasons unrelated to acidosis.

### Source of the Amide-N of Glutamine Released from Muscle
A point which has not been considered to any extent is the stoichiometry of glutamine release from muscle—i.e. the fact that the synthesis of 1 molecule of glutamine requires 1 molecule of ammonia in addition to an amino-group. There are two possible sources of free ammonia (except when $NH_4Cl$ is used to induce metabolic acidosis), namely formation via glutamate dehydrogenase and the purine nucleotide cycle. From glutamate, via glutamate dehydrogenase, the sum of reactions (ignoring cofactors) would be:

$$2 \text{ monoaminoacid} + 2 \text{ } \alpha\text{-oxoglutarate} \rightarrow 2 \text{ glutamate} + 2 \text{ } \alpha\text{-oxoacid}$$
$$\text{glutamate} \rightarrow \alpha\text{-oxoglutarate} + NH_3$$
$$\text{glutamate} + NH_3 \rightarrow \text{glutamine}$$

*Sum:* 2 monoaminoacid + $\alpha$-oxoglutarate $\rightarrow$ 2 $\alpha$-oxoacid + glutamine

If, as suggested by Goldberg and co-workers (Goldberg and Chang, 1978), the carbon skeletons of glutamine are derived from glutamate, valine, isoleucine, aspartate and asparagine, then 2 molecules of any of the latter four amino acids would be required for the formation of 1 molecule of glutamine, as 2 molecules of $\alpha$-oxoacid would be required to regenerate 1 molecule of $\alpha$-oxoglutarate except when glutamate is the precursor.

From the purine nucleotide cycle the reactions would be:

$$2 \text{ monoaminoacid} + 2 \text{ } \alpha\text{-oxoglutarate} \rightarrow 2 \text{ glutamate} + 2 \text{ } \alpha\text{-oxoacid}$$
$$\text{glutamate} + \text{oxaloacetate} \rightarrow \text{aspartate} + \alpha\text{-oxoglutarate}$$
$$AMP \rightarrow IMP + NH_3$$
$$IMP + \text{aspartate} \rightarrow AMP + \text{fumarate}$$
$$\text{fumarate} \rightarrow \text{oxaloacetate}$$
$$\text{glutamate} + NH_3 \rightarrow \text{glutamine}$$

*Sum:* 2 monoaminoacid + $\alpha$-oxoglutarate $\rightarrow$ 2 $\alpha$-oxoacid + glutamine

Here, although the sum of reactions is the same, the ammonia arises from AMP deaminase. It is difficult to accept that this could be the only pathway for the formation of free ammonia in muscle as it would demand an obligatory link between the rate of amino acid degradation and the rate of the purine nucleotide cycle. On the other hand, the extremely low activity of glutamate dehydrogenase in skeletal muscle (see Williamson *et al.*, 1967) has provided the main argument against this route. It is clear from these reactions that any net loss of $\alpha$-oxoacid from muscle—for instance, of branched chain 2-oxoacids as occurs in the rat—depletes the available precursors for the regeneration of $\alpha$-oxoglutarate. Admittedly these reactions represent a highly artificial situation in which the entire carbon and nitrogen of glutamine is provided by these particular amino acids. Leucine, for example, is probably an important precursor of glutamine-N. Furthermore, acetyl-CoA formed from leucine, isoleucine, glucose and fatty acids could also contribute to the carbon skeleton through incorporation into $\alpha$-oxoglutarate.

## Concluding Remarks

In pointing to differences in glutamine metabolism between man and rat the emphasis has been on the inter-organ relationships in the maintenance of the glutamine supply and on glutamine as a transport form of ammonia. The importance of the mucosa of the small intestine as a glutamine-consuming tissue has been stressed but, in addition to being the most important respiratory fuel for this tissue, glutamine also serves as the precursor of citrulline which is synthesized and released into the portal vein (Windmueller and Spaeth, 1974). This is interesting from the nutritional point of view in that it provides an explanation as to why arginine is not an essential amino acid in the adult. It does not mean, however, that arginine should not be included in amino acid mixtures for parenteral nutrition—loss of mucosa, as occurs in patients needing parenteral nutrition, will lead to arginine becoming an essential amino acid. Glutamine also plays a part in the formation of other essential amino acids in its role as donor of nitrogen for the amination of several of the $\alpha$-oxoacid analogues of essential amino acids which are administered clinically as protein-sparing agents. The importance of the liver in this, as in other aspects of the regulation of glutamine metabolism, remains to be explored.

# References

DAVIS, E. J. and BREMER, J. (1973). Studies with isolated surviving rat hearts. Interdependence of free amino acids and citric acid cycle intermediates, *Eur. J. Biochem.*, **38**, 86–97.

ELIA, M., FARRELL, R., ILIC, V., SMITH, R. and WILLIAMSON, D. H. (1980*a*). The removal of infused leucine after injury, starvation and other conditions in man, *Clin. Sci.*, **59**, 275–83.

ELIA, M., ILIC, V., BACON, S., WILLIAMSON, D. H. and SMITH, R. (1980*b*). Relationship between basal blood alanine concentration and the removal of an alanine load in various clinical situations in man, *Clin. Sci.*, **58**, 301–9.

FELIG, P., WAHREN, J. and AHLBORG, G. (1973*a*). Uptake of individual amino acids by the human brain, *Proc. Soc. Exp. Biol. Med.*, **142**, 230–1.

FELIG, P., WAHREN, J., KARL, I., CERASI, E., LUFT, R. and KIPNIS, D. M. (1973*b*). Glutamine and glutamate metabolism in normal and diabetic subjects, *Diabetes*, **22**, 573–6.

FELIG, P., WAHREN, J. and RÄF, L. (1973*c*). Evidence of inter-organ amino acid transport by blood cells in humans, *Proc. Nat. Acad. Sci. USA*, **70**, 1775–9.

FINLEY, R. J., AOKI, T. T., BECKMAN, C., CALLAWAY, C. W. and CAHILL, G. F. (1979). Glutamine metabolism during ammonium chloride loading in man, *Clin. Res.*, **27**, 484A.

GARBER, A. J., KARL, I. E. and KIPNIS, D. M. (1976). Alanine and glutamine synthesis and release from skeletal muscle. II. The precursor role of amino acids in alanine and glutamine synthesis, *J. Biol. Chem.*, **251**, 836–43.

GOLDBERG, A. L. and CHANG, T. W. (1978). Regulation and significance of amino acid metabolism in skeletal muscle, *Fed. Proc.*, **37**, 2301–7.

GOLDSTEIN, L. and NEWSHOLME, E. A. (1976). The formation of alanine from amino acids in diaphragm muscle of the rat, *Biochem. J.*, **154**, 555–8.

GOTO, M., SHINNO, H. and ICHIHARA, A. (1977). Isozyme patterns of branched-chain amino acid transaminase in human tissues and tumors, *Gann*, **68**, 663–7.

HAGENFELDT, L., ERIKSSON, S. and WAHREN, J. (1980). Influence of leucine in arterial concentrations and regional exchange of amino acids in healthy subjects, *Clin. Sci.*, **59**, 173–81.

HAUSSINGER, D. and SIES, H. (1979). Hepatic glutamine metabolism under the influence of the portal ammonia concentration in the perfused rat liver, *Eur. J. Biochem.*, **101**, 179–84.

HOROWITZ, M. L. and KNOX, W. E. (1968). A phosphate activated glutaminase in rat liver different from that in kidney and other tissues, *Enzym. biol. clin.*, **9**, 241–55.

HUANG, Y.-Z. and KNOX, W. E. (1976). A comparative study of glutaminase isozymes in rat tissues, *Enzyme*, **21**, 408–26.

ICHIHARA, A., NODA, C. and OGAWA, K. (1973). Control of leucine metabolism with special reference to branched-chain amino acid transaminase isoenzymes, *Adv. Enz. Reg.*, **11**, 155–66.

ISHIKAWA, E. (1976). The regulation and output of amino acids by rat tissues, *Adv. Enz. Reg.*, **14**, 117–36.

KHATRA, B. S., CHAWLA, R. K., SEWELL, C. W. and RUDMAN, D. (1977).

Distribution of branched-chain α-keto acid dehydrogenase in primate tissues, *J. Clin. Invest.*, **59**, 558–64.

KINGSLAND, P. A., KINGSNORTH, A., ROYLE, G. T., KETTLEWELL, M. G. W. and ROSS, B. D. (1980). Glutamate metabolism in malnutrition and sepsis in man, *Br.|J. Surg.*, **68**, 234–7.

KOVAČEVIĆ, Z., BREBERINA, M., PAVLOVIĆ, M. and BAJIN, K. (1979). Molecular form and kinetic properties of phosphate-dependent glutaminase in the mitochondria isolated from the kidneys of normal and acidotic rats, *Biochim. Biophys. Acta*, **567**, 216–24.

KREBS, H. A. (1935). Metabolism of amino acids. IV. The synthesis of glutamine from glutamic acid and ammonia and the enzymic hydrolysis of glutamine in animal tissues, *Biochem. J.*, **29**, 1951–69.

LEMIEUX, G., BAVEREL, G., VINAY, P. and WADOUX, P. (1976). Glutamine synthetase and γ-glutamyltransferase in the kidney of man, dog and rat, *Am. J. Physiol*, **231**, 1068–73.

LEMIEUX, G., WATFORD, M., VINAY, P. and GOUGOUX, A. (1980). Metabolic changes in skeletal muscle during chronic metabolic acidosis, *Int. J. Biochem.*, **12**, 75–83.

LINDER-HOROWITZ, N., O'TOOLE, W. F. and KNOX, W. E. (1970). Glutaminase in normal human tissues and in lung carcinomata, *Enzym. Biol. Clin.*, **11**, 154–61.

LIVESEY, G. and LUND, P. (1980). Enzymic determination of branched-chain amino acids and 2-oxoacids in rat tissues. Transfer of 2-oxoacids from skeletal muscle to liver *in vivo*, *Biochem. J.*, **188**, 705–13.

LONDON, D. R., FOLEY, T. H. and WEBB, C. G. (1965). Evidence for the release of individual amino acids from resting human forearm, *Nature*, **208**, 588–9.

LUND, P. and WATFORD, M. (1976). Glutamine as a precursor of urea, in *The Urea Cycle* (Eds. S. Grisolia, R. Baguena and F. Mayor), pp. 479–88, John Wiley, New York.

LUND, P. (1980). Glutamine metabolism in the rat, *FEBS Lett. 117 Suppl.*, K86–K92.

MARLISS, E. B., AOKI, T. T., POZEFSKY, T., MOST, A. S. and CAHILL, G. F. (1971). Muscle and splanchnic glutamine and glutamate metabolism in postabsorptive and starved man, *J. Clin. Invest.*, **50**, 814–7.

MENDES-MOURÃO, J., McGIVAN, J. D. and CHAPPELL, J. B. (1975). The effects of L-leucine on the synthesis of urea, glutamate and glutamine by isolated rat liver cells, *Biochem. J.*, **146**, 457–64.

OWEN, E. E. and ROBINSON, R. R. (1963). Amino acid extraction and ammonia metabolism by the human kidney during the prolonged administration of ammonium chloride, *J. Clin. Invest.*, **42**, 263–76.

PARDRIDGE, W. M. (1977). Unidirectional influx of glutamine and other neutral amino acids into liver of fed and fasted rats *in vivo*, *Am. J. Physiol*, **232**, E492–E496.

PARRY, D. M. and BROSNAN, J. T. (1978). Glutamine metabolism in the kidney during induction of, and recovery from, metabolic acidosis in the rat, *Biochem. J.*, **174**, 387–96.

POZEFSKY, T., FELIG, P., TOBIN, J. D., SOELDNER, J. S. and CAHILL, G. F. (1969). Amino acid balance across tissues of the forearm in post-absorptive man. Effects of insulin at two dose levels, *J. Clin. Invest.*, **48**, 2273–82.

RELMAN, A. S. and YABLON, S. (1978). The regulation of ammonia production in the rat. *Curr. Prob. Clin. Biochem.*, **8**, 198–200.

RÉMÉSY, C., DEMIGNÉ, C. and AUFRÈRE, J. (1978). Inter-organ relationships between glucose, lactate and amino acids in rats fed on high-carbohydrate or high-protein diets, *Biochem. J.*, **170**, 321–9.

RUDERMAN, N. B. and BERGER, M. (1974). The formation of glutamine and alanine in skeletal muscle, *J. Biol. Chem.*, **249**, 5500–6.

SCHRÖCK, H., CHA, C.-J. M. and GOLDSTEIN, L. (1980). Glutamine release from hindlimb and uptake by kidney in the acutely acidotic rat, *Biochem. J.*, **188**, 557–60.

SHINNICK, F. L. and HARPER, A. E. (1976). Branched-chain amino acid oxidation by isolated rat tissue preparations, *Biochim. Biophys. Acta*, **437**, 477–86.

SIMPSON, D. P. (1980). Modulation of glutamine transport and metabolism in mitochondria from dog renal cortex. Influence of pH and bicarbonate, *J. Biol. Chem.*, **255**, 7123–8.

SNELL, K. and DUFF, D. A. (1977). The release of alanine by rat diaphragm muscle *in vitro*, *Biochem. J.*, **162**, 399–403.

SPYDEVOLD, O. (1976). Sources of carbon skeleton of alanine released from skeletal muscle, *Acta Physiol. Scand.*, **97**, 273–80.

SQUIRES, E. J., HALL, D. E. and BROSNAN, J. T. (1976). Arterio-venous differences for amino acids and lactate across the kidneys of normal and acidotic rats, *Biochem. J.*, **160**, 125–8.

SQUIRES, E. J. (1977). Glutamine metabolism in normal and acidotic rats *in vivo*, MSc Thesis, Memorial University of Newfoundland, Canada.

TIZIANELLO, A., DE FERRARI, G., GORIBOTTO, G. and GURRERI, G. (1978). Effects of chronic renal insufficiency and metabolic acidosis on glutamine metabolism in man, *Clin. Sci. Mol. Med.*, **55**, 391–7.

WEISSWANGE, A., CUENDET, G. S., BOPP, P., STAUFFACHER, W. and MARLISS, E. B. (1973). Effect of acid-base alterations upon arterial levels and forearm arteriovenous differences of glutamine, glutamate and alanine, *Clin. Res.*, **21**, 642.

WILLIAMSON, D. H., LUND, P. and KREBS, H. A. (1967). The redox state of free nicotinamide adenine dinucleotide in the cytoplasm and mitochondria of rat liver, *Biochem. J.*, **103**, 514–27.

WINDMUELLER, H. G. and SPAETH, A. E. (1974). Uptake and metabolism of plasma glutamine by the small intestine, *J. Biol. Chem.*, **249**, 5070–9.

WINDMUELLER, H. G. and SPAETH, A. E. (1980). Respiratory fuels and nitrogen metabolism *in vivo* in small intestine of fed rats, *J. Biol. Chem.*, **255**, 107–12.

For Discussion of this paper see p. 185.

# 13

## Metabolism of Ammonia

G. A. O. Alleyne, N. McFarlane-Anderson,
*Department of Medicine,*
*University of the West Indies, Kingston, Jamaica*

and

J. A. Lupianez, P. Hortelano and F. Sanchez-Medina
*Department of Biochemistry,*
*University of Granada, Granada, Spain*

### Introduction

Only a small fraction of the body's amino nitrogen is present as ammonia but its importance in terms of clinical disease and metabolic regulation is being increasingly recognized by clinical investigators. Because of the ready diffusibility of ammonia it is found in virtually all tissues. The concentration in arterial blood is relatively low, approximately 0·1 mM in rat and 0·04 mM in fasted man. It is produced in significant amounts by the intestine and the kidney—the latter organ contributing most to the arterial ammonia levels in health. Hyperammonaemia however is relatively uncommon, occurring most often in patients with liver disease or with rare errors of urea biosynthesis (Hsia, 1974). Elevated levels of ammonia play a significant role in the alteration of cerebral metabolism of hepatic encephalopathy (Phillips *et al.*, 1952; McDermott and Adams, 1954) and may also play a minor regulatory role in control of enzymatic processes, for example, ammonia is an activator of phosphofructokinase and pyruvate kinase (Passonneau and Lowry, 1964; Pogson, 1968). There is also a possibility that in those tissues in which, because the deamination of glutamate by glutamate dehydrogenase is at near equilibrium, the concentration of ammonia may effect the cell redox state and thus alter metabolism (Krebs and Veech, 1969). This review will not focus on the role

169

of ammonia as an intracellular regulator of metabolism, but will deal mainly with the problem of ammonia production in the kidney and intestine and to a lesser extent with the dynamics of its metabolism.

## Renal Ammoniagenesis

**Urinary Ammonia**

Renal ammonia is found in the renal vein and the urine and it is believed that the pH gradient across the renal cell determines the distribution between blood and urine. Ammonia (as net base) has to be generated from amino nitrogen:

$$\text{Protein} \rightarrow R.CH(NH_2)COOH \rightarrow R.CO.COOH + NH_3$$

This base will neutralize acid derived from diet or excess acid added to body fluids. Thus it is commonly held that the major function of the production and excretion of ammonia is the preservation of acid-base homoeostasis. The classical mechanism through which urinary ammonia neutralizes acidosis is as follows. There is increased production of ammonia in the kidney cell and as the urinary pH falls there is intraluminal trapping of ammonia, thus permitting excretion of protons. Acidification of the urine is thus central to proton excretion. In acute acidosis in man, rat and dog, when urine pH falls, urinary ammonia increases. The increased urinary ammonia of starvation is related to the acidosis produced by ketosis. This critical role of urinary acidification is also borne out in the isolated perfused kidney (Tannen and Ross, 1979). When HCl was added to the perfusate, the urine pH fell and the increase in total ammonia production was linearly correlated with the fall in urine pH.

Recently, however, it has been proposed that in chronic metabolic acidosis the situation may be different and urinary ammonia may be the determinant of urine pH (Schloeder and Stinebaugh, 1977). In chronically acidotic man a minimum urine pH is achieved as the plasma bicarbonate falls and then as the urinary ammonia increases, the urine pH steadily rises. A similar finding has also been observed when adult males were given glutamine orally—as urine ammonia increased, urine pH rose (Welbourne et al., 1972). It is proposed that as proximal tubular production of ammonia increased with acidosis, ammonia diffused readily through the kidney, was trapped in the collecting duct and thereby raised urinary pH.

The concept that urinary ammonia per se can contribute directly to acid-base homoeostasis has been challenged by Oliver and Bourke (1975). They claim that at physiological pH, glutamine, the main precursor of urinary ammonia is a zwitterion and its metabolism yields $NH_4^+$ and not $NH_3$, thus excretion of the ammonium ion does not change body proton status. It is the metabolism of the carbon skeleton remaining which produces two $HCO_3^-$ ions and thus eliminates excess protons. However, the urinary ammonia in their schema is still a quantitative estimate of the protons which are ultimately removed.

They showed also that rats given HCl had a decrease in urinary urea when urinary ammonia increased, resulting in no net loss or gain of nitrogen (Table 1). Thus acidosis decreased urea synthesis, leading to a 'saving' of

TABLE 1

RATES OF URINARY EXCRETION OF AMMONIUM, UREA NITROGEN AND AMMONIUM PLUS UREA NITROGEN IN RATS MADE ACIDOTIC

| Treatment | Rates (mmol/24 h) | | |
|---|---|---|---|
| | Urea N | Ammonia N | Total |
| Control | $12\cdot47 \pm 0\cdot23$ | $0\cdot54 \pm 0\cdot01$ | $13\cdot07 \pm 0\cdot25$ |
| 4 mmol HCl | $9\cdot19 \pm 0\cdot31$ | $3\cdot17 \pm 0\cdot30$ | $12\cdot33 \pm 0\cdot51$ |
| 6 mmol HCl | $8\cdot47 \pm 0\cdot39$ | $4\cdot66 \pm 0\cdot23$ | $13\cdot09 \pm 0\cdot45$ |
| 4 mmol $NH_4Cl$ | $13\cdot44 \pm 0\cdot07$ | $3\cdot78 \pm 0\cdot15$ | $17\cdot22 \pm 0\cdot10$ |

Data from Oliver and Bourke (1975).
Means $\pm$ SEM.

bicarbonate ions while metabolism of the carbon skeleton of glutamine led to a production of $HCO_3^-$. In this sense, ureagenesis is an acid-producing mechanism and is balanced by ammoniagenesis—a bicarbonate-producing mechanism. Some confirmatory evidence for this comes from our experiments in the dog which show that acute acidosis induced by HCl reduced urinary urea as urinary ammonia increased (Fine *et al.*, 1978). In the perfused liver, acidosis also decreases urea production (Lueck and Miller, 1970).

The concentration of bicarbonate rather than acidosis may be the critical fact since in the isolated hepatocytes, reduction of $HCO_3^-$ in the medium at constant pH led to a fall in glutamine utilization and urea production (Bavarel and Lund, 1979).

It has been proposed that another major role of urinary ammonia is in sparing cations (Lotspeich, 1959); in man taking $NH_4Cl$, there is first an increase in urinary sodium which decreases as urinary ammonia increases.

There have been few studies to quantitate this relationship further, but in our laboratory we have never been able to demonstrate that animals on a sodium-free diet and thus avidly conserving sodium have any enhancement of renal ammoniagenic capacity.

## Precursors of Urinary Ammonia

Van Slyke *et al.* (1943) first proposed that glutamine nitrogen was the major precursor of urinary ammonia and subsequent studies showed that increasing plasma glutamine led to an increase in urinary ammonia. This was shown in the dog (Lotspeich, 1959) and then in man by Welbourne *et al.* (1972), who found that increasing plasma glutamine by oral administration in both controls and persons with chronic acidosis led to an increase in urinary ammonia (Table 2). It is of interest that increasing plasma glutamine in subjects with renal insufficiency did not increase urinary ammonia.

TABLE 2

EFFECT OF ORAL GLUTAMINE ON PLASMA GLUTAMINE, URINE pH AND URINE $NH_4^+$ IN NORMAL ADULTS AND IN ADULTS MADE ACIDOTIC WITH ORAL $NH_4Cl$

| *Subjects* | *Plasma glutamine* ($\mu mol/ml$) | *Urine* | |
|---|---|---|---|
| | | *pH* | $NH_4$ ($\mu mol/min$) |
| Normal | $0.544 \pm 0.056$ | $5.87 \pm 0.24$ | $27.2 \pm 3.3$ |
| + glutamine | $1.360 \pm 0.286$ | $6.22 \pm 0.22$ | $49.9 \pm 7.8$ |
| Acidotic | $0.367 \pm 0.066$ | $5.05 \pm 0.10$ | $104.8 \pm 14$ |
| + glutamine | $1.022 \pm 0.059$ | $5.31 \pm 0.10$ | $152.3 \pm 23$ |
| Renal disease | $0.439 \pm 0.031$ | $6.22 \pm 0.22$ | $16.7 \pm 8.9$ |
| + glutamine | $1.612 \pm 0.293$ | $6.10 \pm 0.15$ | $24.3 \pm 14.0$ |

Data from Welbourne *et al.* (1972).

Several investigators have measured glutamine extraction by the kidney in relation to total ammonia production and some of these data are shown in Table 3. Except for one study (Vinay *et al.*, 1980) the ratio of ammonia production to glutamine extracted is above 1·0 and less than 2·0 in rat, dog and man in normal acid-base state. With induction of acidosis, the ratio approaches 2·0 in the rat and the dog but in two studies in man it remains unchanged. The data of Vinay *et al.* (1980) are apparently anomalous and would have to be interpreted as showing that non-glutamine sources are major contributors to urinary ammonia in normal rats. The other data in

## TABLE 3

GLUTAMINE EXTRACTION AND AMMONIA PRODUCTION BY THE KIDNEY IN DIFFERENT
SPECIES AND DIFFERENT STATES

| Subject | Condition | Glutamine extraction ($\mu mol/min$) | $NH_4$ production ($\mu mol/min$) | Ratio |
|---|---|---|---|---|
| Rat[a] | Control | 0·122 | 0·722 | 5·92 |
|  | Acute acidosis | 0·643 | 1·119 | 1·74 |
| Rat[b] | Control | 0·510 | 0·555 | 1·09 |
|  | Chronic acidosis | 0·815 | 1·820 | 2·23 |
| Dog[c] | Control | 15·7[a], 18·4[c] | 25·2[a], 20·7[c] | 1·61[a], 1·13[c] |
|  | Acute acidosis | 14·9, 17·1 | 32·2, 20·3 | 2·17, 1·19 |
| Dog[d] | Chronic alkalosis | 8·0 | 8·8 | 1·1 |
|  | Chronic acidosis | 27·2 | 44·4 | 1·60 |
| Man[e] | Control | 47·0 | 77·4 | 1·65 |
|  | Chronic acidosis | 102·5 | 139·1 | 1·36 |
| Man[f] | Control | 34·8 | 50·3 | 1·4 |
|  | Chronic acidosis | 109·6 | 149·0 | 1·4 |
|  | Renal insufficiency | 2·7 | 22·4 | 8·3 |

[a] Vinay et al. (1980).
[b] Brennan and Alleyne (unpublished).
[c] Fine et al. (1978).
[d] Denis et al. (1964), Pitts et al. (1972).
[e] Owen and Robinson (1963).
[f] Tizianello et al. (1978).

the rat and the dog are consonant with the thesis that in chronic acidosis there is not only increased extraction of glutamine, but that there is utilization of both the amide-N and the amino-N. In the isolated kidney perfused with glutamine at a concentration of 1·0 mM the ratio of ammonia formed to glutamine extracted is indeed 2·0 indicating complete metabolism and recovery of both nitrogens (Ross and Bullock, 1978).

There are two major differences in man. There is increased extraction of glutamine in acidosis, but there is no evidence of more complete utilization of the amino-N. In patients with renal insufficiency the ratio of 8·3 (Table 3) indicates that there must be other sources of renal ammonia apart from glutamine. This is in keeping with the studies shown in Table 2 in which elevation of plasma glutamine did not lead to increase in urinary ammonia. The other sources of ammonia in renal failure are not yet clear, but there are now two definitive studies showing that they must exist. Tizianello et al. (1978, 1980) have postulated that intrarenal breakdown of glutathione, protein and peptides must be the source of at least half of the urinary ammonia in patients with renal insufficiency. In this context, the demonstration of increased activity of $\gamma$-glutamyl transpeptidase in acidosis (McFarlane-Anderson and Alleyne, 1977) may be of relevance as it has been proposed that the major function of this enzyme is intra-luminal degradation of glutathione (McIntyre and Curthoys, 1979).

None of the above studies provides direct data on the immediate precursor pool of the nitrogen of urinary ammonia. Pitts et al. (1965) used $^{15}$N-labelled amino acids and showed in the chronically acidotic dog that approximately 40 % of the urinary ammonia came from the amide-N of glutamine and 19 % from the amino-N.

There are as yet no similar data in man and this is perhaps one of the major deficiencies in this area of research especially when one considers that the basic methods are available and are relatively non-invasive.

The purine-nucleotide cycle exists in kidney and appears to show an adaptive increase in metabolic acidosis (Bogusky et al., 1976). It has been suggested (Tannen, 1978) that this pathway is of doubtful significance since inhibition of formation of aspartate through inhibition of transaminase activity by amino oxyacetate does not affect ammonia production (Relman and Narins, 1975). However, renal cortical slices incubated with glutamine alone or glutamine plus amino oxyacetate do not show any differences in levels of aspartate (Bennett and Alleyne, unpublished data), and Alleyne (1970) and Vinay et al. (1980) reported that in acute acidosis there was an increase in the levels of aspartate in the kidney. Further work is needed on the possible relevance of the purine nucleotide cycle to renal ammonia production.

**Origin of Glutamine**
Although other organs may synthesize glutamine most of it is derived from muscle (Marliss et al., 1971; Hills et al., 1972; Felig et al., 1973). Circulating ammonia probably contributes very little to muscle glutamine formation except when there is hyperammonaemia and the source of both amide-N and amino-N is probably through intracellular amino acid degradation (Ruderman and Berger, 1974). It has been suggested that branched chain amino acids may be the nitrogen source for glutamine synthesis but their levels and metabolism in muscle make this unlikely (Ruderman and Berger, 1974). Although the liver contains glutamine synthetase, it is agreed that at least in man, rat and dog there is no net release of glutamine by the liver (Tizianello et al., 1978; Lund, 1971; Fine, 1980, personal communication). Under normal conditions the brain takes up rather than releases glutamine (Tizianello et al., 1978).

**Enzymic Control of Ammonia Production from Glutamine**
Because metabolic acidosis is the major physiological stimulus for ammonia production, there continues to be considerable interest in the control of glutamine metabolism in acidosis. This has been focussed mainly on the enzymes responsible for glutamine degradation and the possible signals which might initiate the whole process. (For review see Tannen, 1978).

The major enzymes responsible for the formation of ammonia from glutamine are: (1) phosphate-dependent glutaminase, (2) glutamate dehydrogenase and (3) $\gamma$-glutamyl transpeptidase. Phosphate-dependent glutaminase has been studied extensively since its adaptation in chronic acidosis in the rat was first noted by Davies and Yudkin (1952). However, its critical role was disputed when it was shown that the enzyme in the canine kidney did not adapt during acidosis (Pollak *et al.*, 1965) and rats treated with actinomycin D showed increased ammoniagenesis when they were made acidotic but did not show an adaptive increase in glutaminase (Goldstein, 1965). The distribution of the enzyme in the kidney has also cast doubt on its primary role in increasing ammoniagenesis in acidosis. The highest levels are found in the distal tubule, while it is only the enzyme located in the proximal tubule which shows an adaptive increase in response to acidosis (Curthoys and Lowry, 1973). The enzyme also adapts long after there is an increase in urinary ammonia or an increase in the ammoniagenic capacity of the renal cortical cells (Alleyne, 1970).

In considering the role of glutaminase in ammoniagenesis, there has been a tendency to ignore some of the basic thermodynamic facts about the enzyme (Krebs and Vinay, 1975). The equilibrium in the glutaminase system favours virtually complete hydrolysis of glutamine: the standard free energy is $-3 \cdot 5$ kcal and the $\Delta G$ value about twice this (Benzinger and Hems, 1956). Krebs points out that it is thermodynamically impossible for the enzyme and its substrate to exist in the same cellular compartment (as they do) with glutamine at a stable value of approximately $1 \cdot 5$ mM unless the enzyme is strongly inhibited. The sole candidate for this inhibitor role so far is glutamate. Thus the three possible mechanisms for enhanced enzymatic hydrolysis of glutamine must be reduction of an inhibitor, a change in the characteristic of the enzyme and an increase in enzyme protein. It is definite, at least in the rat, that the last is the mechanism of enhanced glutamine utilization in chronic acidosis.

Recent studies in our laboratory have concentrated mainly on the acute response to acidosis, partly because an understanding of these changes may give a clue to the sequence of events in chronic acidosis. In addition, in the clinical situation, acute metabolic acidosis is more frequent and a more lethal condition than chronic metabolic acidosis.

In the rat, 2 h after administration of $NH_4Cl$, renal cortical slices show increased ammoniagenesis from glutamine and this occurs before there are changes in renal glutaminase (Alleyne, 1970). If the acidosis is severe enough, these changes may be seen after $\frac{1}{2}$ h.

This acute response is seen not only in rats made acidotic with $NH_4Cl$, but in rats which are exercised by swimming and develop an acute lactic acidosis. The severity of the acidosis and the other aspects of the renal response such as enhanced gluconeogenesis and increased activity of phosphoenol pyruvate carboxykinase have been extensively studied (Sanchez-Urrutia *et al.*, 1975).

We studied the ammoniagenic capacity of these rats, especially in relation to glutaminase activity (Table 4). After the rats had been swimming

TABLE 4

AMMONIA PRODUCTION BY RENAL CORTICAL SLICES AND RENAL GLUTA-MINASE ACTIVITY IN RATS WITH ACUTE METABOLIC ACIDOSIS AS A RESULT OF SWIMMING FOR 15 MIN

| | *Ammonia production* ($\mu mol/h/g\ dry\ wt$) *by slices from* $2\cdot0$ mM *glutamine* | *Glutaminase activity* ($nmol/min/mg\ protein$) *with substrate glutamine* | |
| --- | --- | --- | --- |
| | | $1\cdot0$ mM | $20\cdot0$ mM |
| Control | $321 \pm 6$ (14) | $9\cdot4 \pm 2\cdot8$ (9) | $70\cdot5 \pm 2\cdot8$ (11) |
| Exercised | $362 \pm 12$ (14) | $38\cdot9 \pm 5\cdot0$ (9) | $114\cdot0 \pm 8\cdot4$ (11) |

Values are means $\pm$ SEM.
Numbers in parentheses are the number of estimations.

for 15 min or 1 h, renal cortical slices showed an increase in ammoniagenesis from $2\cdot0$ mM glutamine. At this stage there is a 60 % increase in glutaminase activity when measured at saturating concentrations of substrate. However, at more physiological concentrations of glutamine ($1\cdot0$ mM) there is a 400 % increase in activity of the enzyme. This phenomenon may indicate a conformational change in the enzyme in acute acidosis which precedes the increase in enzyme protein that occurs in chronic acidosis. If this phenomenon proves to be widespread, it would be the ideal method of rapid control of glutamine metabolism. Conformational change in enzymes with resultant alteration in kinetic properties is a well described mechanism for regulation, but hitherto it has not been considered applicable to renal glutaminase.

There is also adaptation of glutamate dehydrogenase (GDH) in the kidneys of rats with chronic metabolic acidosis (Pitts, 1971, Seyama *et al.*, 1973). Studies with isolated mitochondria from acidotic rats have shown a four-fold increase in flux through GDH with only a doubling of flux through glutaminase. The increased flux through GDH is partly caused by increased deamination of glutamine and synthesis of new enzyme, but it is possible that there are other factors intrinsic or extrinsic to the enzyme which activate it in acidosis (Schoolwerth *et al.*, 1978).

## Glutamine Transport

We have also considered the possibility that glutamine transport might be limiting its metabolism. There is an intracellular to extracellular gradient for glutamine in renal cortex, clearly indicating active transport. We have described an increase in glutamine transport into brush border membrane vesicles in chronically acidotic rats and have suggested that this might be a mechanism of control of metabolism (McFarlane-Anderson and Alleyne, 1979). However, recent studies with free flow micropuncture and microperfusion of tubules have shown that end-proximal concentrations of glutamine are virtually zero in the normal rat, indicating almost complete reabsorption of glutamine in the proximal tubule (Silbernagl, 1980). This would suggest that increased transport across the luminal surface cannot occur in acidosis. However, since there is increased glutamine extraction by the kidney in acidosis, if it does not occur luminally, it should occur at the antiluminal surface. Studies in the dog have already established that antiluminal transport of glutamine can account for between 35 and 60 % of renal ammonia production (Lemieux *et al.*, 1974). It is possible that the brush border preparation used by us was contaminated by basolateral membranes and the latter would show the true adaptive increase in transport in acidosis.

## Modifiers of Glutamine Metabolism

Studies *in vitro* and *in vivo* have shown that almost every intermediate of the tricarboxylic acid cycle and several circulating metabolites may influence ammoniagenesis from glutamine. However, there has been no consistent change in the pattern of most of these in situations in which there is increased ammoniagenesis. As an example, 2-oxoglutarate has been shown to inhibit glutamine entry into mitochondria (Goldstein, 1976) and the fall in 2-oxoglutarate levels found in kidneys of acutely acidotic rats was interpreted as a regulatory mechanism for glutamine metabolism. However, we have shown that rat kidney slices metabolizing glutamine in the presence of maleate show increased glutamine uptake and ammonia-genesis even though the levels of 2-oxoglutarate rise in the slice (McFarlane-Anderson and Alleyne, 1981).

We have concentrated on the possibility that in acute acidosis a humoral factor may be produced which will increase glutamine uptake and ammonia production in the rat (Alleyne and Roobol, 1974). Within $\frac{1}{2}$ h of induction of

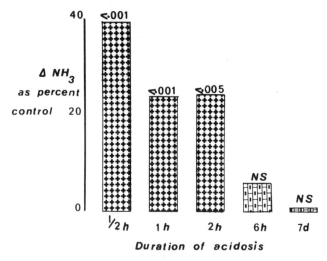

FIG. 1.   The presence of an ammoniagenic stimulating activity in plasma drawn from rats at different times after induction of acidosis. Rats sacrificed at $\frac{1}{2}$ h were given a single dose of 16 mmol $NH_4Cl/kg$; rats sacrificed at 1, 2 and 6 h were given a single dose of 10 mmol/kg, and the chronically acidotic rats sacrificed at 7 days were allowed to drink 280 mM $NH_4Cl$. All controls were fed equivalent amounts of NaCl.

acidosis there is a circulating plasma factor which will stimulate renal uptake of glutamine and ammoniagenesis. This factor is not a protein, is dialysable, is not of adrenal origin and appears to exert its effect by an α-adrenergic like mechanism. The factor is not detectable at 6 h after an acid load or in the plasma of chronically acidotic rats (Fig. 1). We are investigating whether this factor stimulates glutamine metabolism by enhancing transport or modifying the glutaminase enzyme.

# Ammonia Production in the Intestine

The gut produces ammonia, but very little appears in arterial blood if hepatic function is normal. This ammonia is derived from both intraluminal and intracellular metabolism.

## Intraluminal Metabolism

Dietary protein is not completely absorbed in the small intestine and on reaching the caecum is broken down by bacteria to produce ammonia (Dawson, 1978). Urea is the other source of ammonia and it has been known for years that blood ammonia is elevated in cirrhotic patients who are uraemic. Urea probably diffuses into the ileum, passes into the caecum and is there degraded by bacterial or mucosal ureases: approximately one third of the body urea is broken down in the gut every day (Gibson *et al.*, 1976). The entry of ammonia into the portal vein from the caecum is via the process of non-ionic diffusion (Bown *et al.*, 1975).

## Intracellular Metabolism

Most of the ammonia from intracellular metabolism is derived from glutamine. The studies of Windmueller and Spaeth (1974, 1975, 1980) have shown that in the rat 25–33 % of plasma glutamine is extracted in a single pass through the intestine. There is a single small intracellular pool of glutamine which is supplied by both intraluminal and intravascular sources; after metabolism, the glutamine-N appears in the portal vein as citrulline, alanine, proline and ammonia with ammonia accounting for 23 % of the total.

The pathways of glutamine metabolism in gut and kidney are similar but are perhaps subject to different control mechanisms. Increasing $HCO_3^-$ at constant pH increases glutamine removal by enterocytes but has no effect on kidney tubules (Bavarel and Lund, 1979).

We have been unable to discover any condition in which there is any adaptive increase in the enzymes of glutamine metabolism. When chronic metabolic acidosis was induced in rats and the kidneys showed increased ammoniagenesis and increased glutaminase activity, enterocytes and pieces of intestine showed no change in ammonia production or in glutaminase activity (Table 5). However, when glutamine is measured in arterial and portal venous blood, there is an increase in the A-V differences with acute or chronic acidosis, indicating that there was increased glutamine extraction by the gut (Table 5). When enterocytes from normal rats were exposed to plasma from acutely acidotic rats, there was an increase in ammoniagenesis

## TABLE 5

CHANGES IN GLUTAMINE METABOLISM BY THE INTESTINE *IN VITRO* AND *IN VIVO* AFTER INDUCTION OF CHRONIC METABOLIC ACIDOSIS

|  | Control | Acidosis |
|---|---|---|
| *Ammonia production* ($\mu$mol/h/g dry wt) |  |  |
| Slices of jejunum | $197 \pm 9$ | $187 \pm 10$ |
| Enterocytes | $279 \pm 25$ | $289 \pm 24$ |
| *Glutaminase activity* (nmol/min/mg protein) | $51 \pm 4$ | $52 \pm 4$ |
| *Artery—portal vein* Glutamine ($\mu$mol) | $+0.06 \pm 0.009$ | $+0.114 \pm 0.005$ |

from glutamine indicating that the ammoniagenic plasma factor has similar effects on gut and kidney.

### Utilization of Blood Ammonia

The hepatic metabolism of portal vein ammonia is well understood, but there is less information about the metabolism of ammonia in the systemic circulation. Studies in the intact animal have shown uptake of ammonia into brain and muscle (Berl *et al.*, 1962; Hills *et al.*, 1972) and there are several studies on the neurotoxicity of ammonia. More recently the availability of a $^{13}$N-radionuclide has made it possible to obtain quantitative data on the tissue uptake of ammonia in man (Lockwood *et al.*, 1979). The rate of ammonia clearance is a linear function of the arterial concentration and this was most strikingly seen in the irreversible uptake of ammonia by the brain probably with rapid fixation in the amide group of glutamine (Berl *et al.*, 1962). In normal subjects approximately 50% of arterial ammonia was utilized by muscle and it is possible that this route of metabolism assumes even greater significance in the presence of liver disease (Ganda and Ruderman, 1976). The detailed enzymology of the incorporation of ammonia into amino acids and the mechanism by which this is modified is beyond the scope of this review.

## Conclusion

This review has deliberately focussed mainly on the origin rather than the utilization of ammonia since the biochemistry and physiology of the latter are relatively clear. The major areas for future investigation would seem to be in the production of blood ammonia, particularly the role of the kidney and intestine, and the methods of physiological control. This will have clinical relevance in the control of hyperammonaemic states.

Urinary ammonia is used as a metabolic marker in studies on protein metabolism, yet there are few data on the precursors of ammonia and whether the precursors change, depending on the state of the organism. Glutamine-N is assumed to be the immediate precursor of urinary ammonia but some other nitrogen source may serve this function and glutamine-N entering the kidney may only indirectly contribute to ammonia in normal states.

Our own interest in a possible humoral control of glutamine metabolism and ammoniagenesis must be extended to other species beside the rat and the nature of the ammoniagenic factor must be elucidated. The possibility of a conformational change rather than an increase in enzyme protein as a means of control needs to be further evaluated with respect to the enzymes of ammoniagenesis. It is hoped that some of the answers to these problems will come from this symposium.

## *References*

ALLEYNE, G. A. O. (1970). Renal metabolic response to acid base changes. II. The early effects of metabolic acidosis on renal metabolism in the rat, *J. Clin. Invest*, **49**, 945–51.

ALLEYNE, G. A. O. and ROOBOL, A. (1974). Regulation of renal cortex ammoniagenesis. I. Stimulation of renal cortex ammoniagenesis *in vitro* by plasma isolated from acutely acidotic rats, *J. Clin. Invest.* **53**, 117–21.

BAVAREL, G. and LUND, P. (1979). A role for bicarbonate in the regulation of mammalian glutamine metabolism, *Biochem. J.* **184**, 599–606.

BENZINGER, T. H. and HEMS, R. (1956). Reversibility and equilibrium of the glutaminase reaction observed calorimetrically to find the free energy of adenosine triphosphate hydrolysis. *Proc. Natl. Acad. Sci.*, **42**, 896–900.

BERL, S., TAKAGAKI, G., CLARKE, D. D. and WAELSCH, H. (1962). Metabolic compartments *in vivo*: ammonia and glutamic acid metabolism in brain and liver, *J. Biol. Chem.*, **237**, 2562–9.

BOGUSKY, R. T., LOWENSTEIN, L. M. and LOWENSTEIN, J. M. (1976). The purine nucleotide cycle. A pathway for ammonia production in the rat kidney, *J. Clin. Invest.*, **58**, 326–35.

BOWN, R. L., GIBSON, J. A., FENTON, J. C. B., SNEDDEN, W., CLARKE, M. L. and SLADEN, G. E. (1975). Ammonia and urea transport by the excluded human colon, *Clin. Sci. Mol. Med.*, **48**, 279–87.

CURTHOYS, N. P. and LOWRY, O. H. (1973). Distribution of rat kidney glutaminase iso enzymes in the various structures of the nephron and their response to metabolic acidosis and alkalosis, *J. Biol. Chem.*, **248**, 162–8.

DAVIES, B. M. A. and YUDKIN, J. (1952). Studies in biochemical adaptation. The origin of urinary ammonia as indicated by the effects of chronic acidosis and alkalosis on some renal enzymes in the rat, *Biochem. J.*, **52**, 407–12.

DAWSON, A. M. (1978). Regulation of blood ammonia, *Gut.*, **19**, 504–9.

DENIS, G., PREUSS, H. and PITTS, R. (1964). The $pNH_3$ of renal tubular cells, *J. Clin. Invest.*, **43**, 571–82.

FELIG, P., WAHREN, J., KARL, I., CERASI, E., LUFT, R. and KIPNIS, D. M. (1973). Glutamine and glutamate metabolism in normal and diabetic subjects, *Diabetes*, **22**, 573–6.

FINE, A., BENNETT, F. I. and ALLEYNE, G. A. O. (1978). Effects of acute acid base alterations on glutamine metabolism and renal ammoniagenesis in the dog, *Clin. Sci. Mol. Med.*, **54**, 503–8.

GANDA, O. P. and RUDERMAN, N. B. (1976). Muscle nitrogen metabolism in chronic hepatic insufficiency, *Metabolism*, **25**, 427–35.

GIBSON, J. A., PARK, N. J., SLADEN, G. E. and DAWSON, A. M. (1976). The role of the colon in urea metabolism in man, *Clin. Sci. Mol. Med.*, **48**, 279–87.

GOLDSTEIN, L. (1965). Actinomycin D inhibition of the adaptation of renal glutamine deamidating enzymes in the rat, *Nature (London)* **205**, 1330–1.

GOLDSTEIN, L. (1976). $\alpha$-Ketoglutarate regulation of glutamine transport and deamidation by renal mitochondria, *Biochem. Biophys. Res. Comm.* **70**, 1136–41.

HILLS, A. G., REID, E. L. and KERR, W. D. (1972). Circulatory transport of L-glutamine in fasted mammals: cellular sources of urine ammonia, *Amer. J. Physiol*, **223**, 1470–6.

HSIA, Y. E. (1974). Inherited hyperammonemic syndromes, *Gastroenterology*, **67**, 347–74.

KREBS, H. A. and VEECH, R. L. (1969). Equilibrium relations between pyridine nucleotides and adenine nucleotides and their role in the regulation of metabolic processes, *Advances in Enzyme Regulation*, **7**, 397–413.

KREBS, H. A. and VINAY, P. (1975). Regulation of renal ammonia production, *Med. Clin. Amer.*, **59**, 595–603.

LEMIEUX, G., VINAY, P. and CARTIER, P. (1974). Renal haemodynamics and ammoniagenesis. Characteristics of the antiluminal site for glutamine extraction, *J. Clin. Invest*, **53**, 884–94.

LOCKWOOD, A. H., McDONALD, J. M., REIMAN, R. E., GELBARD, A. S., LAUGHLIN, J. S., DUFFY, T. E. and PLUM, F. (1979). The dynamics of ammonia metabolism in man. Effects of liver disease and hyperammonemia, *J. Clin. Invest.*, **63**, 449–60.

LOTSPEICH, W. I. (1959). The synthesis and secretion of ammonia, in *Metabolic Aspects of Renal Function*, pp. 89–122, Charles C. Thomas, Illinois.

LUECK, J. D. and MILLER, L. L. (1970). The effect of perfusate pH on glutamine metabolism in the isolated perfused rat liver, *J. Biol. Chem.*, **245**, 5491–7.

LUND, P. (1971). Control of glutamine synthesis in rat liver, *Biochem. J.* **124**, 653–60.

McDERMOTT, W. V. JR. and ADAMS, R. D. (1954). Episodic stupor associated with an ECK fistula in the human with special reference to the metabolism of ammonia, *J. Clin. Invest.*, **33**, 1–9.

McFARLANE-ANDERSON, N. and ALLEYNE, G. A. O. (1977). The effect of metabolic acidosis on γ-glutamyl transpeptidase activity in the rat kidney, *FEBS Letters*, **79**, 51–3.

MCFARLANE-ANDERSON, N. and ALLEYNE, G. A. O. (1979). Transport of glutamine by rat kidney brush border membrane vesicles, *Biochem. J.*, **182**, 295–300.

MCFARLANE-ANDERSON, N. and ALLEYNE, G. A. O. (1981). The effect of maleate on glutamine metabolism by rat kidney, *Biochemical Medicine*, **25**, 149–59.

MCFARLANE-ANDERSON, N., BENNETT, F. I. and ALLEYNE, G. A. O. (1976). Ammonia production by the small intestine of the rat, *Biochim. Biophys. Acta.*, **437**, 238–43.

MCINTYRE, T. M. and CURTHOYS, N. P. (1979). Comparison of the hydrolytic and transfer activities of rat renal γ-glutamyl transpeptidase, *J. Biol. Chem.*, **254**, 6499–504.

MARLISS, E. B., AOKI, T. T., POZEFSKY, T., MOST, A. S. and CAHILL, G. F. JR. (1971). Muscle and splanchnic glutamine and glutamate metabolism in post absorptive and starved man, *J. Clin. Invest.*, **50**, 814–17.

OLIVER, J. and BOURKE, E. (1975). Adaptations in urea ammonium excretion in metabolic acidosis in the rat: a reinterpretation, *Clin. Sci. Mol. Med.*, **48**, 515–20.

OWEN, E. C. and ROBINSON, R. R. (1963). Amino acid extraction and ammonia metabolism by the human kidney during the prolonged administration of ammonium chloride, *J. Clin. Invest.*, **42**, 263–76.

PASSONNEAU, J. V. and LOWRY, O. H. (1964). The role of phosphofructokinase in metabolic regulation, *Adv. Enzyme Reg.*, **2**, 265–74.

PHILLIPS, G. B., SCHWARTZ, R., GABUZDA, G. J. JR. and DAVIDSON, C. S. (1952). The syndrome of impending hepatic coma in patients with cirrhosis of the liver given certain nitrogenous substances, *New Eng. J. Med.*, **247**, 239–46.

PITTS, R. F. (1971). Metabolism of amino acids by the perfused rat kidney, *Amer. J. Physiol.*, **220**, 862–7.

PITTS, R. F., PILKINGTON, L. A. and DEHAAS, J. (1965). N[15] tracer studies on the origin of urinary ammonia in the acidotic dog with notes on the enzymatic synthesis of labelled glutamic acid and glutamine, *J. Clin. Invest.*, **44**, 731.

PITTS, R. F., PILKINGTON, R. A., MACLEOD, M. D. and LEAL-PINTO, E. (1972). Metabolism of glutamine by the intact functioning kidney of the dog, *J. Clin. Invest.*, **51**, 557–65.

POGSON, C. I. (1968). Adipose tissue pyruvate kinase. Properties and interconversion of two active forms, *Biochem. J.*, **110**, 67–77.

POLLAK, V. E., MATTENHEIMER, H., DEBRUIN, H. and WEINMANN, K. J. (1965). Experimental metabolic acidosis. The enzymatic basis of ammonia production by the dog kidney, *J. Clin. Invest.*, **44**, 169–81.

RELMAN, A. S. and NARINS, R. G. (1975). The control of ammonia production in the rat, *Med. Clin. N. America*, **59**, 583–93.

ROSS, B. D. and BULLOCK, S. (1978). The metabolic fate of glutamine nitrogen in the perfused rat kidney, *Biochem. J.*, **170**, 177–9.

RUDERMAN, N. B. and BERGER, M. (1974). The formation of glutamine and alanine in skeletal muscle, *J. Biol. Chem.*, **249**, 5500–6.

SANCHEZ-URRUTIA, L., GARCIA-RUIZ, J. P., SANCHEZ-MEDINA, F. and MAYOR, F. (1975). Lactic acidosis and renal phosphoenol pyruvate carboxykinase activity during exercise, *Biochem. Med.*, **14**, 355–67.

SCHLOEDER, F. X. and STINEBAUGH, B. J. (1977). Urinary ammonia content as a determinant of urinary pH during chronic metabolic acidosis, *Metabolism*, **26**, 1321–31.
SCHOOLWERTH, A. C., NAZAR, B. L. and LANOUE, K. F. (1978). Glutamate dehydrogenase activation and ammonia formation by rat kidney mitochondria, *J. Biol. Chem.*, **253**, 6177–83.
SEYAMA, S., SAEKI, T. and KATANUMA, N. (1973). Comparison of properties and inducibility of glutamate dehydrogenases in rat kidney and liver, *J. Biochem.*, **73**, 39–45.
SILBERNAGL, S. (1980). Tubular reabsorption of L-Glutamine studies by free flow micropuncture and microperfusion of rat kidney, *Internat. J. Biochem.*, **12**, 9–16.
TANNEN, R. L. (1978). Ammonia metabolism, *Amer. J. Physiol.*, **235**, F265–F277.
TANNEN, R. L. and ROSS, B. D. (1979). Ammoniagenesis by the isolated perfused rat kidney: the critical role of urinary acidification, *Clin. Sci.*, **56**, 353–64.
TIZIANELLO, A., DEFERRARI, G., GARIBOTTO, G. and GURRERI, G. (1978). Effects of chronic renal insufficiency and metabolic acidosis on glutamine metabolism in man, *Clin. Sci. Mol. Med.*, **55**, 391–7.
TIZIANELLO, A., DEFARRARI, G., GARIBOTTO, G., GURRERI, G. and ROBAUDO, C. (1980). Renal metabolism of amino acids and ammonia in subjects with normal renal function and in patients with chronic renal insufficiency, *J. Clin. Invest.*, **65**, 1162–73.
VAN SLYKE, D. D., PHILLIPS, R. A., HAMILTON, P. B., ARCHIBALD, R. M., FUTCHER, P. H. and MILLER, A. (1943). Glutamine as source material of urinary ammonia, *J. Biol. Chem.*, **150**, 481–3.
VINAY, P., ALLIGNET, E., PICHETTE, C., WATFORD, M., LEMIEUX, G. and GOURGOUX, A. (1980). Changes in renal metabolite profiles and ammoniagenesis during acute and chronic metabolic acidosis in dog and rat, *Kidney International*, **17**, 312–25.
WELBOURNE, T., WEBER, M. and BANK, N. (1972). The effect of glutamine administration on urinary ammonium excretion in normal subjects and patients with renal disease, *J. Clin. Invest.*, **51**, 1852–60.
WINDMUELLER, H. G. and SPAETH, A. E. (1974). Uptake and metabolism of plasma glutamine by the small intestine, *J. Biol. Chem.*, **249**, 5070–9.
WINDMUELLER, H. G. and SPAETH, A. E. (1975). Intestinal metabolism of glutamine and glutamate from the lumen as compared to glutamine from blood. *Arch. Biochem. Biophys.*, **171**, 662–72.
WINDMUELLER, H. G. and SPAETH, A. E. (1980). Respiratory fuels and nitrogen metabolism in vivo in small intestine of fed rats. Quantitative importance of glutamine, glutamate and aspartate, *J. Biol. Chem.*, **255**, 107–12.

# DISCUSSION ON PAPERS 12 AND 13

*Walser* (*Johns Hopkins University, Baltimore, USA*): I would like to comment on the roles of ammonia excretion and urea formation in acid–base balance as proposed by Alleyne. I think we can all agree that the removal of a proton from an aqueous system is equivalent to the addition of a hydroxyl ion. Furthermore, the addition of a hydroxyl ion in an open system containing $CO_2$ is equivalent to the addition of a bicarbonate ion. Thus the excretion of ammonia both removes a proton and adds a bicarbonate ion to the body. Oliver and Bourke (1975) have made some interesting points, but their distinction between proton excretion and bicarbonate reclamation is artificial; the two processes are simply different ways of looking at the same event. Furthermore, excretion of $NH_4^+$ certainly is equivalent to excretion of $H^+$.

Secondly, while it is true that bicarbonate ions are consumed in urea synthesis, the stoichiometry of the reaction is such that the number of

$$2NH_4^+ - H^+ + HCO_3^- \rightarrow H_2NCONH_2 + 2H_2O$$

protons removed is exactly equal to the number of bicarbonate ions removed. Consequently, ureagenesis has no effect whatsoever on acid–base balance.

*Alleyne:* I am not sure that I agree with your last point. What is the site of removal of the protons? We agree that bicarbonate is removed in urea synthesis, the problem is to identify the manner in which a stoichiometrically identical number of protons is removed in the process of ureagenesis. I do not know if you accept the thesis of Oliver and Bourke (1975) that, since glutamine is a zwitterion, excretion of $NH_4$ does not change the body's proton state.

*Felig* (*Yale University School of Medicine, USA*): I am intrigued by the plasma factor. Are you suggesting that it appears in any condition where there is an increased acid load?

*Alleyne:* The plasma factor is found in the rat given ammonium chloride or HCl, and it is also found in the swimming rat.

*Felig:* I want to interject a caution about the forced-swimming rat. We have used it as a model for exercise and training. However, these rats are swimming for their lives, and it is a model for extremely severe stress. Catecholamines are enormously elevated in these animals, well beyond what is seen in any kind of exercise.

*Waterlow (London School of Hygiene and Tropical Medicine, UK):* One of the main points raised was the source of the extra glutamine needed by the kidney in acidosis. Lund suggested that in the rat there is evidence for a compensatory decrease in uptake by the splanchnic bed. Does that apply in all kinds of acidosis?

*Alleyne:* I think most of the results on glutamine metabolism are the same whether acidosis is produced by HCl or ammonium chloride. We have shown clearly that in acute acidosis in the dog and the rat, levels of plasma glutamine rise (Fine *et al.*, 1978). In some new work, Fine (personal communication) has confirmed this and has found, surprisingly, that when plasma glutamine levels rise the uptake of glutamine by the gut decreases.

*Waterlow:* Since Lund has emphasized the species differences, is the source of extra glutamine the same in man as in the rat?

*Lund:* I think there is too little information available to be able to say. It would not surprise me if there are differences. For instance, the most important source may be liver in the rat and skeletal muscle in man.

*Millward (London School of Hygiene and Tropical Medicine, UK):* One difference between rat and man, which may be important, is that the glutamine concentration in muscle is very much higher in man, e.g. 20 mM, than in the rat, e.g. 5 mM. There seems to be no obvious reason for this, and I wonder if we should treat muscle as a store, so that glutamine can be released when needed, without removing other amino acids from the pool.

*Pratt (Institute of Psychiatry, London, UK):* We also have some data on the release of glutamine from rat muscle. When we rapidly varied the levels of both glutamine and alanine in the blood, we found that the extent to which both of these amino acids, and indeed gluconeogenic amino acids generally, are released correlates well with the blood alanine level but not at all with the blood glutamine level (Daniel *et al.*, 1979). Our findings suggest that there must be compartmentation of alanine and probably also of glutamine within muscle.

*Felig:* There is another possible mechanism for the increased provision of glutamine in metabolic acidosis. Lund mentioned a change in the utilization of glutamine in the gut. To what extent is glutaminase in the gut pH-sensitive? There is a lot of clinical evidence that by giving lactulose you can lower the blood ammonia levels in cirrhotic patients. The presumed explanation is that acidification of the gut changes the relationships of $NH_3$ to $NH_4^+$, and that through non-ionic diffusion $NH_3$ is got rid of. Is it conceivable that there is a change in gut glutaminase activity which is pH-dependent, and allows more glutamine to be available to the kidney? Are there any data on the pH-sensitivity of intestinal glutaminase?

*Lund:* The pH optimum of the intestinal enzyme is around 8·6, but activity rapidly decreases below pH 7·6. In chronic acidosis, of course, pH is usually back to near normal.

*James (Dunn Clinical Nutrition Centre, Cambridge, UK):* I am not convinced by Felig's suggestion that a fall in blood ammonia can be explained by an effect of lactulose on the enzymes of glutamine metabolism in the gut. Stephen and Cummings (1980) have shown that there may be two mechanisms by which lactulose could affect ammonia metabolism. First, microbial metabolism of lactulose generates volatile fatty acids in the colon, with a fall in pH. This will reduce the reabsorption of ammonia through the colonic mucosa. The second mechanism depends on the promotion of microbial growth which will then entrap ammonia for synthesis of bacterial protein in the colon, thereby reducing the availability of ammonia.

*Golden (University of the West Indies, Kingston, Jamaica):* In order to get more information about glutamine production in man, and to estimate the proportion of glutamine flux consumed in renal ammonia excretion, we infused [15]N-amide-labelled glutamine (Prochem, London, 97 atom % excess) into a normal adult subject at the rate of 0·56 $\mu$mol/kg/h for 6 h. Enrichment was measured in plasma and RBC glutamine amide and in plasma ammonia at 1 h intervals, and in urinary ammonia every $\frac{1}{2}$ h (Golden *et al.*, unpublished).

A constant enrichment (plateau) was reached in 1 h, except for RBC glutamine, which took 3 h (Fig. D1). At plateau the enrichments were:

| | |
|---|---|
| plasma glutamine amide | 0·080 $\pm$ 0·009 |
| plasma ammonia | 0·043 $\pm$ 0·006 |
| RBC glutamine amide | 0·046 $\pm$ 0·004 |
| urinary ammonia | 0·056 $\pm$ 0·003 |

From the formula, flux = dose/plateau enrichment, the plasma glutamine flux is estimated to be 41·6 mmol/h(0·67 mmol/kg/h). The rate of urinary

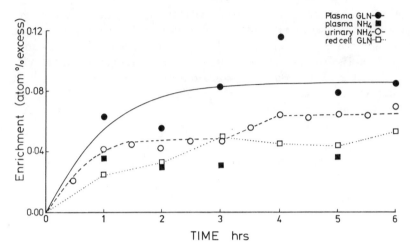

FIG. D1.    Enrichment of nitrogen in plasma and red cell glutamine and in plasma and urinary ammonia during constant infusion of glutamine amide-$^{15}$N in an adult subject.

ammonia excretion was $0.88 \pm 0.06$ mmol/h. From the relative plateau enrichments of plasma glutamine and urinary ammonia one can calculate that 69 % of urinary ammonia was derived from plasma glutamine. Thus the contribution of glutamine to urinary $NH_3$ excretion was only 0.62 mmol/h, which represents only 1.5 % of the glutamine flux. Even if ammonia production by the kidney is 10 times greater than the amount of ammonia excreted in the urine, renal consumption of glutamine would still only account for 15 % of the glutamine flux. Consumption of glutamine by the gut far outweighs consumption by the kidney. The corollary of this is that when the animal is acidotic and the kidney requires more glutamine for renal ammonia formation, the existing production rate of glutamine is quite sufficient to accommodate the extra requirement.

*Jackson (University of the West Indies, Kingston, Jamaica):* The question of the production rate of glutamine can be approached in another way. If you take the figures given by Marliss *et al.* (1971) for the A-V difference in glutamine concentration across the forearm, make an extrapolation to the rest of the body and assume a blood flow of 4 litre/min, the result is a figure of 40 mmol/h for glutamine production by the carcass—a figure almost identical with our estimate of the total flux. From the figures given by Lund (68 mmol/day or 2.8 mmol/h in normal, 147 mmol/day or 6.1 mmol/h in acidotic man), the utilization of glutamine by the kidney works out at about

7 % of the flux in normal man and 15 % in acidosis. Therefore, it is not necessary to postulate any other source than muscle or carcass as a source of glutamine.

*Lund:* You have to remember that there is a lot of ammonia released into the renal vein as well. Is that included in your calculation?

*Jackson:* Yes, that is included in the 7 % of the glutamine flux which is consumed by the kidney. Since, as Golden pointed out, less than 2 % of the flux goes into urinary ammonia, the rest of the ammonia produced from glutamine in the kidney must pass into the renal vein.

*Waterlow:* Alleyne in his paper raised the question of the sources of urinary ammonia and whether these sources alter in acidosis. The results presented by Golden and Jackson, if they are representative, go a long way towards answering this question. Not only have they shown that under normal conditions about 70 % of urinary $NH_3$ is derived from glutamine, but according to their argument, the supply of glutamine is so large that there is no need for any change in the precursors of urinary ammonia in acidosis. This is a very important point in relation to our work, in which we calculate protein turnover rates from the labelling of urinary ammonia (Waterlow *et al.*, 1978). It would produce great problems if the pattern of ammonia precursors alters in different metabolic states. So far we have obtained no evidence that the level of labelling is related in any way to the amount of ammonia excreted. This fits in with the arguments we have just heard.

*Alleyne:* If no one has any more information about rates of glutamine production and consumption by different tissues, I suggest that we should give some consideration to the enzymatic mechanisms of glutamine synthesis and degradation, taking up the points raised by Lund.

*Cohen* (*University of Wisconsin-Madison, USA*): The persistent view that glutaminase in liver is essentially non-functioning raises the question whether glutamine aminotransferase might be the mechanism for breaking down glutamine. Has that been looked into? Some of the properties of the glutaminase which were worrying you might be consistent with a different enzyme system, namely the glutamine aminotransferase, which requires other factors than the substrates provided for the ordinary glutaminase.

*Lund:* I agree, but one cannot rule out glutaminase in view of recent work showing that glucagon and ammonium ions increase metabolism of low concentrations of glutamine by activating glutaminase (Joseph and McGivan, 1978). At one time we looked to see if the aminotransferase pathway could be important. The only ketoacids that increased metabolism of plasma concentrations of glutamine (0·5 mM) were the ketoacid analogues of methionine and phenylalanine. More physiologically important ketoacids, such as pyruvate, had no effect (Lund and Watford, 1976). Meister suggests a salvage function for the aminotransferase in re-aminating small amounts of ketoacids formed during normal metabolism of essential amino acids (Cooper and Meister, 1974).

*Munro* (*Tufts University, Boston, USA*): Some years ago Katunuma *et al.* (1967) showed with isolated glutaminase that glutamate was an allosteric feedback inhibitor at physiological levels of glutaminase action. The idea was that when the glutamate level fell in acidosis, then ammonia production was increased. This allosteric control mechanism is further supported by the fact that when glutaminase is prevented from having an adaptive increase by giving inhibitors of protein synthesis, you still get increased amounts of ammonia in acidosis. My question is whether the original observations of Katunuma are superseded by subsequent information, and whether glutamate levels were measured in the experiments you described.

*Alleyne:* There is no doubt that Katunuma's data on the inhibition of glutaminase by glutamate have been confirmed over and over again. However, glutamate levels may be unchanged at the time when there is increased ammoniagenesis. In our studies on freeze-clamped kidneys at various stages of acidosis, the whole tissue levels of glutamate have not changed in our system at a time when there is an increase in ammonia production, but we cannot be sure that intramitochondrial glutamate does not fall.

*Lund:* Steady state concentrations can give you no indication of flux; for example, one can have an increased flux of glutamate through glutamate dehydrogenase without any change in glutamate concentration.

*Alleyne:* I would not dispute that, but Munro's point was whether the steady state levels of glutamate had fallen at the time when there was enhancement of glutaminase activity. Regardless of the flux, if the steady state level is not changed, then presumably there will not be an allosteric effect on the enzyme.

*Coore* (*University of the West Indies, Kingston, Jamaica*): Munro suggested an allosteric effect on glutaminase. Have you ruled out a covalent change of the enzyme protein? If you contrasted the $V_{max}$ of the extracted enzyme with its activity *in situ*, and it turned out that the $V_{max}$ was greater than *in situ*, then maybe you have a covalent change and your strategy would be a little bit different. You would have to think of freeze-clamping, rapid assay and perhaps incubating semi-purified enzyme with pyruvate kinase and cyclic AMP.

*Alleyne:* I agree. Our data showed a conformational change in the enzyme.

*Munro:* Does not the location of the glutaminase in the proximal tubule, which Curthoys and Weiss (1974) demonstrated was mitochondrial, make it extremely difficult to know what the local glutamate concentration is?

*Alleyne:* I agree. It depends on the site of action of glutaminase. Curthoys and Lowry (1973) showed that at the site where there was a maximum amount of glutaminase, there was apparently no adaptation at all. Your point is whether glutamate levels fell at the site in the proximal tubule where there was increased adaptation of glutaminase. We have not studied dissected tubules, but Curthoys and Lowry showed an increase in enzyme protein in the proximal tubule, not just activation of the enzyme.

# *References*

COOPER, A. J. L. and MEISTER, A. (1974). Isolation and properties of a new glutamine transaminase from rat kidney, *J. Biol. Chem.*, **249**, 2554–61.

CURTHOYS, N. P. and LOWRY, O. H. (1973). Distribution of rat kidney glutaminase iso enzymes in the various structures of the nephron and their response to metabolic acidosis and alkalosis, *J. Biol. Chem.*, **248**, 162–8.

CURTHOYS, N. P. and WEISS, R. F. (1974). Regulation of renal ammoniagenesis, *J. Biol. Chem.*, **249**, 3261–6.

DANIEL, P. M., PRATT, O. E. and SPARGO, E. (1979). Blood alanine as a regulator of amino acid release from muscle in rats, *J. Physiol.*, **295**, 12–13P.

FINE, A., BENNETT, F. I. and ALLEYNE, G. A. O. (1978). Effects of acute acid base alterations on glutamine metabolism and renal ammoniagenesis in the dog, *Clin. Sci. Mol. Med.*, **54**, 503–8.

JOSEPH, S. K. and MCGIVAN, J. D. (1978). The effect of ammonium chloride and glucagon on the metabolism of glutamine in isolated liver cells from starved rats, *Biochim. Biophys. Acta*, **543**, 16–28.

KATUNUMA, N., HÜZINO, A. and TOMINO, I. (1967). Organ specific control of glutamine metabolism, *Adv. Enzyme Reg.*, **5**, 55–69.

LUND, P. and WATFORD, M. (1976). Glutamine as a precursor of urea, in *The Urea Cycle* (Eds S. Grisolia, R. Baguena and F. Mayor), pp. 479–88, John Wiley, New York.

MARLISS, E. B., AOKI, T. T., POZEFSKY, T., MOST, A. S. and CAHILL, G. F., JR. (1971). Muscle and splanchnic glutamine and glutamate metabolism in post absorptive and starved man, *J. Clin. Invest.*, **50**,|814–17.

OLIVER, J. and BOURKE, E. (1975). Adaptations in urea ammonium excretion in metabolic acidosis in the rat: a reinterpretation, *Clin. Sci. Mol. Med.*, **48**, 515–20.

STEPHEN, A. M. and CUMMINGS, J. H. (1980). Mechanism of action of dietary fibre in the human colon, *Nature*, **284**, 283–4.

WATERLOW, J. C., GOLDEN, M. H. N. and GARLICK, P. J. (1978). Protein turnover in man measured with $^{15}N$: comparison of end-products and dose regimes, *Am. J. Physiol.*, **253**(2), E165–E174.

# 14

# Ammonia Metabolism *in vivo* in the Rat

F. JAHOOR

*Tropical Metabolism Research Unit,*
*University of the West Indies, Kingston, Jamaica*

We have been discussing the metabolism of ammonia, its production from glutamine and its excretion by the kidney. One aspect which has not been considered is the uptake of ammonia into amino acids. Schoenheimer (1942) and Sprinson and Rittenberg (1949) clearly demonstrated that when [15]N-labelled ammonium salts were given, a substantial fraction of [15]N was retained in the body, bound in most amino acids isolated from mixed body proteins. We thought it would be interesting to get some quantitative information about the distribution *in vivo* of [15]N from ammonia into the free amino acids of the body pools. We therefore infused adult rats with [15]$NH_4Cl$, and assumed that after six hours the precursors and products of nitrogen transfer had reached isotopic equilibrium. The results in Table 1 show that $NH_3$ is well taken up by most tissues. The tissues in which $NH_3$ is normally generated had low [15]N enrichment of $NH_3$ relative to tissues which fix $NH_3$, the enrichment of gut $NH_3$ being only half that of liver.

The second column of the table shows the enrichment of glutamine amide-N. Muscle is a major site of glutamine synthesis. The low enrichment of glutamine amide-N in muscle relative to that of $NH_3$ shows that muscle glutamine derives about 60 % of its amide-N from a pool other than free $NH_3$. The enrichment of glutamine amide-N in gut and kidneys, the major users of glutamine, reflects the low enrichment in muscle, from which they derive most of their own glutamine. Hepatic glutamine amide-N is relatively highly enriched, indicating substantial glutamine synthesis from $NH_3$ in the liver; again, however, about 40 % of the amide-N comes from sources other than free $NH_3$.

The major site for amination of amino acids from $NH_3$ is the liver. Almost 20 % of alanine-N is derived from free $NH_3$, indicating substantial synthesis of this amino acid in the liver. Unfortunately, we are not able to

TABLE 1

$^{15}$N ENRICHMENTS (ATOMS % EXCESS × $10^3$) OF METABOLITES IN THE FREE POOLS (MEANS ± SEM)

| Tissue | $NH_4$ | Glutamine amide-N | Urea | Glutamine + glutamic acid α-amino-N | Aspartate | Alanine |
|---|---|---|---|---|---|---|
| Liver | 211 ± 22 | 123 ± 18 | 45 ± 4 | 15 ± 3 | 11 ± 1 | 35 ± 4 |
| Kidneys | 153 ± 17 | 32 ± 5 | 51 ± 4 | NE[a] | 4 ± 0·5 | NE |
| Muscle | 145 ± 17 | 48 ± 11 | 32 ± 4 | NE | 7 ± 1 | NE |
| Gut | 109 ± 12 | 70 ± 15 | 25 ± 3 | 5 ± 2 | — | NE |

[a] NE not enriched.

Six rats weighing 427–571 g were infused through a tail vein with $^{15}$NH$_4$Cl (97 atoms % excess) at a dose of about 2 mg $^{15}$N/kg/day for 6 h. After decapitation, the tissues were homogenized in ice-cold 10 % PCA. Free NH$_3$, glutamine amide-N, alanine-N, glutamine + glutamic acid α-amino-N, aspartate-N and urea-N were sequentially extracted by macro modifications of specific enzyme reactions (Bergmeyer, 1976) and measured for isotopic enrichment as described by Jackson *et al.* (1980). During isolation and extraction about 16 % of glutamine amide-N is degraded to NH$_3$, so that the actual enrichment of free NH$_3$ is probably higher than that shown (unpublished results of Jahoor, Jackson and Golden).

determine separately the enrichment of the α-amino-N of glutamine and glutamic acid, so the immediate precursor of alanine cannot be specified. The enrichment of aspartate-N in kidney and muscle without any enrichment in glutamate or alanine may indicate synthesis of aspartate in these tissues by a pathway which does not involve glutamate or alanine as an intermediate.

## Discussion

*Waterlow (London School of Hygiene and Tropical Medicine, UK):* I think this is a very interesting contribution, and I would like to make two points about it. First, this kind of quantitative information about the proportion of a product which is derived from a particular precursor, etc., could only be obtained with a constant infusion, and I hope we shall see more of such studies. Secondly, I found when I was working in Jamaica 15 years ago that when $^{15}$NH$_4$Cl is given by mouth, about 60 % of the $^{15}$N was excreted in urea, whereas if $^{15}$N-glycine was given under the same conditions, only some 30 % of the $^{15}$N came out in urea. This only confirmed what Sprinson and Rittenberg showed in 1949; nevertheless, it seems rather paradoxical that $NH_3$ should preferentially form urea, and that very little of the $^{15}N$ should be found in urinary $NH_3$. We can, of course, always fall back on the explanation of compartmentation in the liver cell, but this really tells us nothing new.

## References

BERGMEYER, H. U. (1976). *Methods of Enzymatic Analysis*, 2nd Edition, Vol. IV, Academic Press, New York.

JACKSON, A. A., GOLDEN, M. H. N., JAHOOR, P. F. and LANDMAN, J. P. (1980). The isolation of urea nitrogen and ammonia nitrogen from biological samples for mass spectrometry, *Analyt. Biochem.*, **105**, 14–17.

SCHOENHEIMER, R. (1942). *The Dynamic State of Body Constituents*, Harvard University Press, Cambridge, Mass.

SPRINSON, D. B. and RITTENBERG, D. (1949). The rate of utilization of ammonia for protein synthesis, *J. Biol. Chem.*, **180**, 707–14.

# 15

# Precursors of Urea

PATRICIA LUND

*Metabolic Research Laboratory,
Nuffield Department of Clinical Medicine,
Radcliffe Infirmary, Oxford, UK*

The principal precursor delivered to the liver is alanine. The quantitative importance of alanine derives from the fact that it is the main nitrogenous end product of the extensive interconversion that occurs during absorption of dietary glutamate, aspartate and glutamine. Alanine is also formed and released in large amounts as an end product of the metabolism of arterial glutamine by the mucosa of the small intestine (Windmueller and Spaeth, 1974) and as an end product of amino acid metabolism in heart, diaphragm and skeletal muscle (Felig, 1975). So, although glutamine and alanine are considered to be the transport forms of nitrogen and carbon from the periphery to liver, glutamine-N is delivered to the liver largely as alanine, ammonia, proline and citrulline. The citrulline, however, does not appear to contribute directly to urea production, as there is no net uptake by the liver. Instead it is diverted to the kidney, where it forms arginine (Rogers *et al.*, 1972) which is used in the synthesis of total body protein, and in the formation of creatine (Reeds, p. 263).

The overall balance of nitrogen removal by the liver is given in Table 1. There are several possible explanations for the discrepancy between amino-N uptake and urea excretion. First, no account is taken of influx of amino acids or ammonia arising from loss of protein from the intestinal mucosa, or of influx of ammonia resulting from degradation of urea in the gastrointestinal tract. Secondly, no account is taken of degradation of amino acids arising endogenously from liver proteolysis. Thirdly, some amino acids, notably histidine, lysine, ornithine and arginine, were omitted in the measurements made by Felig *et al.* (1973). Fourthly, the calculations depend on the accuracy of the total blood flow measurement. In spite of these sources of error, the importance of glutamine-N and alanine-N as precursors of urea is clear cut.

TABLE 1

GLUTAMINE-N AND ALANINE-N AS MAJOR PRECURSORS OF UREA IN MAN

|  | *Total removed mmol/day* | *Urea equivalent mmol/day* |
|---|---|---|
| Glutamine[a] | 106 | 106 |
| Alanine | 149 | 75 |
| Citrulline | −30 | −45 |
| Other amino acids | 162 | 81 |
| Total | 387 | 217 = 13 g |
| Urea excretion during first day of fasting (Cathcart, 1907) |  | 19 g |

[a] Less than 50 % removed directly by liver (Felig *et al.*, 1973).
Rates are calculated from arterio-hepatic venous differences for whole blood in overnight fasted man, taking the total liver blood flow as 1124 ml/min (Felig *et al.*, 1973). Values of arterio-portal venous differences from the same authors show that glutamine is largely metabolized by the small intestine. By analogy with the rat (Windmueller and Spaeth, 1974) the nitrogenous end products are alanine, ammonia, proline and citrulline. The assumption is made that loss of amino-N into other tissues drained by the splanchnic bed is negligible.

In the fed state excess dietary amino acids are rapidly degraded by the liver to form urea. But approximately 30 % of dietary protein is in the form of glutamine, glutamate, aspartate and alanine, and metabolism of arterial glutamine continues even during absorption of dietary protein (Windmueller and Spaeth, 1980) so that alanine remains an important precursor.

The synthesis of urea requires stoichiometric amounts of ammonia and aspartate. As alanine is the major precursor, transfer of alanine-N to aspartate can occur only via the alanine- and aspartate-aminotransferases. Ammonia needed in addition to that arriving at the liver in the portal vein, or formed directly from histidine, asparagine, threonine, glutamine, serine and glycine, is formed via glutamate dehydrogenase and not via the purine nucleotide cycle (Krebs *et al.*, 1978).

An interesting problem is how glutamine-N is diverted from urea synthesis in the liver to ammonia production by the kidney in starvation and other metabolic acidoses. There is some evidence that glutamine metabolism by the small intestinal mucosa may be decreased (Tizianello *et al.*, 1978) thus decreasing the supply of urea precursors to the liver and allowing glutamine to be diverted to the kidney for ammoniagenesis. As urea synthesis is obligatorily linked to gluconeogenesis (Krebs *et al.*, 1976) through fumarate generated at the argininosuccinase reaction, the switch to ammoniagenesis involves also the diversion of glucogenic precursor to the kidney.

## Discussion

*Munro (Tufts University, Boston, USA):* What is the contribution within the liver cell from essential amino acids which are primarily or exclusively degraded by enzymes in the liver? To what extent do they account for the missing nitrogen?

*Lund:* The calculations in the table are very crude. They assume no change in amino acid concentration inside the liver cell and no change in A–V difference over 24 h. But if you are thinking of amino acids coming from proteolysis during the first day of fasting, then these may go a long way to explaining the missing nitrogen, especially if losses of protein from the intestinal mucosa are also included.

*Harper (University of Wisconsin-Madison, USA):* If one adds up the amino acids which are deaminated, such as histidine, threonine, tryptophan and serine, and those that are transaminated, there is roughly a 40:60 balance between ammonia production from the deamination reactions and aspartate formation from transamination reactions. This represents a rather higher proportion of ammonia from deamination than you suggested.

*Lund:* The correction for citrulline synthesis may have been confusing. In fact, the values for individual amino acids (Felig *et al.*, 1973) confirm that ratio when citrulline is subtracted from the glutamine value—assuming that two glutamine molecules are needed to form citrulline plus ammonia in the small intestinal mucosa. The citrulline is lost as a potential source of urea because it is not removed by the liver.

*Jackson (University of the West Indies, Kingston, Jamaica):* I have done a similar calculation using more or less the same data and I think that the difference between the amount of ammonia formed by deamination and the amount of nitrogen entering urea via aspartate can largely be accounted for by ammonia coming from urea recycled through the bowel. If this is added to the ammonia side of the equation the result is an almost exact balance.

*Harper:* It seems to me that the contribution of glutamate to urea synthesis should not be overlooked. Most transamination reactions in liver give rise to glutamate, and the amino group of glutamate can be transferred to form aspartate. Glutamate can also be oxidized by glutamic dehydrogenase to yield ammonia. It therefore has the potential for providing nitrogen in whichever form is needed to balance the substrate supply.

*Lund:* I agree with that.

*Walser (Johns Hopkins University, Baltimore, USA):* Dr Lund, why do you say that the citrulline is lost to ureagenesis? I though that it was converted to arginine in the kidney, which obviously can then be converted to urea.

*Lund:* That is debatable. I would say that it goes to protein synthesis. Freedland has given evidence for this (Featherston *et al.*, 1973).

*Walser:* In fact, however, there is no need for the body to synthesize arginine on a normal diet, since protein intake is in excess of need, and except when most unusual diets are consumed the proportion of arginine in the food is about the same as in body protein. Consequently I have always had difficulty with the idea that the arginine made from citrulline in the kidney is used for protein synthesis. The diet already contains plenty of arginine for this purpose.

*Reeds (Rowett Research Institute, Aberdeen, UK):* Later this morning I shall be talking about creatine synthesis. One must not forget that the guanido group of creatine is derived from arginine in the kidney, and that in the adult there is a net synthesis of 1–2 g creatine/day, requiring 7–15 mmol arginine.

*Harper:* Is not the problem that, because of the high activity of arginase, urea formation from arginine in the liver is so rapid that the liver free arginine pool is very low. Thus only a small amount circulates to other organs and tissues, which may therefore depend on production in the kidney to provide arginine for protein synthesis (Rogers *et al.*, 1972).

*Lund:* I agree with Harper.

*Cohen (University of Wisconsin-Madison, USA):* I cannot accept that. After all, if one includes the plasma proteins, the liver makes more protein per unit weight than almost any other organ. The point is that the specific activating enzyme for the formation of arginine-tRNA has a $K_m$ which is much more favourable than the $K_m$ for arginase, and so the liver is able to trap all the arginine it needs to make liver and plasma proteins. There is evidence, already quoted by Lund, that kidney can provide some arginine (by synthesis from citrulline formed in the liver) to some tissues, especially muscle (Featherston *et al.*, 1973). However, the argument that liver cannot provide arginine for protein synthesis because of the high arginase content, and thus must depend on the kidney, is fallacious. There is no evidence that arginine formed by the kidney and transported to the liver is any less susceptible to degradation by arginase than that formed by the liver.

*Lund:* The suggestion is that arginine is used for extra-hepatic protein synthesis.

*Harper:* I agree. The free arginine pool in the liver is very small, and organs other than the liver may depend on arginine formed in the kidney.

*Millward (London School of Hygiene and Tropical Medicine, UK):* Walser's paper (p. 237) contains some information which is relevant to this point. He quotes the case of a patient with a deficiency of one of the urea cycle enzymes, who therefore could not make urea, but growth continued. If citrulline in the kidney were an important source of arginine needed for protein synthesis, then growth could not occur because the patient could not make citrulline. Is that not so?

*Walser:* You are right. These infants cannot synthesize arginine, and therefore need arginine supplements; in fact, they need a disproportionate excess, in part for the reason just mentioned.

# References

CATHCART, E. P. (1907). Uber die Zusammensetzung des Hungerharns, *Biochem. Z.*, **6**, 109–48.

FEATHERSTON, W. R., ROGERS, Q. R. and FREEDLAND, R. A. (1973). Relative importance of kidney and liver in synthesis of arginine by the rat, *Am. J. Physiol.*, **224**, 127–9.

FELIG, P., WAHREN, J. and RÄF, L. (1973). Evidence of interorgan transport by blood cells in humans, *Proc. Soc. Nat. Acad. Sci. USA*, **70**, 1775–9.

FELIG, P. (1975). Amino acid metabolism in man, *Ann. Rev. Biochem.*, **44**, 933–55.

KREBS, H. A., LUND, P. and STUBBS, M. (1976). Interrelations between gluconeogenesis and urea synthesis, in *Gluconeogenesis* (Eds. Hanson, R. W. and Mehlman, M. A.), John Wiley and Sons, Inc., New York, pp. 269–91.

KREBS, H. A., HEMS, R., LUND, P., HALLIDAY, D. and READ, W. W. C. (1978). Sources of ammonia for mammalian urea synthesis, *Biochem. J.*, **176**, 733–7.

ROGERS, Q. R., FREEDLAND, R. A. and SYMMONS, R. A. (1972). *In vivo* synthesis and utilization of arginine in the rat, *Am. J. Physiol.*, **223**, 236–40.

TIZIANELLO, A., DE FERRARI, G., GARIBOTTO, G. and GURRERI, G. (1978). Effects of chronic renal insufficiency and metabolic acidosis on glutamine metabolism in man, *Clin. Sci. Mol. Med.*, **55**, 391–7.

WINDMUELLER, H. G. and SPAETH, A. E. (1974). Uptake and metabolism of plasma glutamine by the small intestine, *J. Biol. Chem.*, **249**, 5070–9.

WINDMUELLER, H. G. and SPAETH, A. E. (1980). Respiratory fuels and nitrogen metabolism *in vivo* in small intestine of fed rats, *J. Biol. Chem.*, **255**, 107–12.

# 16

## Deamination versus Transamination

A. A. JACKSON and M. H. N. GOLDEN

*Tropical Metabolism Research Unit,*
*University of the West Indies, Kingston, Jamaica*

The pioneering work carried out by the Columbia group in the 1940s with [15]N demonstrated the specificity with which nitrogen moves between different nitrogen-containing compounds. At the same time they made the conceptual leap of appreciating the dynamic relationships of nitrogen metabolism (Schoenheimer, 1942) and it is this finding that has come to dominate our thinking over the succeeding decades. During this period there have been occasional reports of the use of [15]N to study the movement of nitrogen between different compounds in the body, but for a variety of reasons this area of work has been relatively neglected. In part this is a result of the inherent technical difficulties. Probably of more importance is the fact that, in general, diets are relatively well balanced in terms of their amino acid composition. Although it has been clearly demonstrated that amino acid imbalances can have profoundly deleterious effects upon metabolism (Harper *et al.*, 1970), this knowledge has had little practical application in medical practice until the more recent advent and wide use of elemental diets, given either orally or intravenously.

Our initial interest in intermediary nitrogen metabolism arose out of a desire to identify the specific route whereby the label from a tracer dose of [15]N-glycine reached the end products of nitrogen metabolism, urea and ammonia (Jackson and Golden, 1980). During a continuous infusion of [15]N-glycine we followed the rise to plateau in urinary urea and ammonia and at the same time measured the enrichment in blood of those amino acids which are recognized as playing a major role in transporting nitrogen from muscle in the periphery to the visceral organs for excretion, namely alanine, glutamic acid and glutamine. We were surprised to be unable to identify any label in the $\alpha$-amino-N of these amino acids, and only a low level of enrichment in the amide group of glutamine relative to that in urinary ammonia. We were unable to alter this relationship by changing the metabolic state from fed to fasting, or by producing an acute metabolic acidosis with hydrochloric acid. We reasoned that on a normal diet there may not be any requirement for the nitrogen from glycine to pass to these amino acids, recognizing that glycine is both synthesized and degraded in the liver and kidneys. Therefore, we tried to create a situation where glycine would distribute its nitrogen around other amino acids by feeding a diet containing sufficient energy to place the subject in an anabolic state. Essential amino acids were given to the recommended allowance and all non-essential nitrogen was given in the form of glycine. Hence to maintain balanced synthesis of new tissue non-essential amino acids would perforce have to be synthesized to requirements utilizing the predominant source of nitrogen available, namely glycine-N. However, we were still unable to measure any significant enrichment in the $\alpha$-amino-N of the three amino acids.

Studies in rats confirmed the failure of $^{15}$N-glycine to send label to these amino acids in blood (Taruvinga, 1978) but did show some enrichment in glutamic acid and aspartate in liver. This finding of a modest enrichment in the free amino acid pool agrees with the relative levels of enrichment found in mixed liver protein of rats after large single doses of $^{15}$N-glycine (Ratner et al., 1940; Shemin and Rittenberg, 1944; Aqvist, 1951; Vitti and Gaebler, 1963). The findings of Aqvist (1951) are representative of the results obtained. He looked at the distribution of nitrogen from a wide range of amino acids into the individual amino acids of mixed liver protein and demonstrated a common pattern of distribution with reference to glutamic acid. However, threonine, histidine and lysine neither received or disposed of label. Glycine and serine disposed of label to a limited extent. As these two amino acids interchange readily with the release and fixation of ammonia, they will give label to and receive label from the free ammonia pool without any net synthesis of glycine plus serine.

Cammarata and Cohen (1950) demonstrated a wide range of transaminating activity for individual amino acids in heart, liver and kidney. The amino acids reacted with different degrees of intensity, which the authors ascribed to technical factors. However, it is worth noting that serine, histidine, glycine, threonine, lysine and cystine showed the lowest levels of activity. This contrasts strikingly with the activity of the deaminating enzymes, which are high in liver for all this group of amino acids except lysine (Cedrangolo, 1975). We would conclude that this group of amino acids are normally catabolized by deamination, and do not freely exchange nitrogen with the other transaminating amino acids. Quantitatively they form as much as one-third to one-half of the $\alpha$-amino-N pool in a variety of tissues in different metabolic states in both man and rat (Jackson and Golden, 1980).

Rudman *et al.* (1973) measured the effect of an oral load of one of eighteen amino acids on the concentration of blood ammonia in normal individuals and patients with impaired hepatic function. The normal subjects were able to handle all the amino acids without developing hyperammonaemia. The patients with hepatic disorders handled the different amino acids with variable effectiveness. Rudman *et al.* divided the amino acids into groups, in relation to the degree of hyperammonaemia they produced. Group A contains those amino acids which tend to cause a rise in blood ammonia and includes glycine, serine, threonine, histidine and lysine. The authors felt that deamidation of these amino acids exceeded the capacity of the urea cycle to handle the ammonia generated. Glycine and serine are the only two non-essential amino acids which consistently cause hyperammonaemia when given in excess (Harper *et al.*, 1970).

Therefore, the amino acids have been divided into two categories with respect to their degradative pathways; those that lose nitrogen through transamination and provide nitrogen for urea synthesis through aspartate and those that lose nitrogen by deamination to generate 'free ammonia' which has been considered to enter urea through carbamyl phosphate. The urea cycle has to be balanced stoichiometrically and this is thought to take place through glutamate dehydrogenase (GDH) (Krebs *et al.*, 1973). Krebs *et al.* have calculated the theoretical proportions of glutamate-N and ammonia that would be expected to be liberated from the catabolism of bovine $\alpha$-casein, on the basis of known degradative pathways. The relative yield of glutamate-N to ammonia is 182:69, leading to the conclusion that ammonia must be generated from glutamate, through the action of GDH to maintain the stoichiometry. We have recalculated these results on the basis of a number of considerations; the evidence would suggest that lysine cannot be considered a transaminating amino acid (*vide supra*); dietary aspartate is deaminated in the gut, and so the nitrogen would reach the liver as ammonia (Windmueller and Spaeth, 1980); dietary glutamine-N arrives in the portal vein as ammonia (38 %), citrulline (28 %) and alanine (24 %) (Windmueller and Spaeth, 1980); urea is degraded in the bowel to ammonia, some of which is recycled to urea synthesis. In an adult on a normal diet this recycling amounts to about 20 % of urea production (Paper 19 of this volume). When these corrections are made the ratio of glutamate-N to ammonia-N is 145:147, giving a balanced stoichiometry to urea cycle precursors. Therefore in the fed state on a normal mixed intake the urea cycle should be in balance.

We considered the same problem from a different perspective. In postabsorptive man the net release across the forearm is glutamine (170 µmol/litre) and alanine (110 µmol/litre) (Marliss et al., 1971). Assuming a carcass blood flow of about 4 litre/min, then 40 mmol/h of glutamine and 27 mmol/h of alanine (total 107 mmol nitrogen/h) would be added to the circulation by the periphery for urea production. The alanine represents transaminating nitrogen. The glutamine reaches the liver predominantly via the portal tract, where the nitrogen is released as alanine, ammonia, and citrulline in the amounts shown in Table 1 (Windmueller and Spaeth, 1980). Recycled ammonia from urea synthesis will be about 18 mmol (20 % of urea production). Therefore the balance of transaminator to deaminator nitrogen is 47 to 48.

TABLE 1
SOURCES OF NITROGEN FOR UREA FORMATION

|  |  | Source of nitrogen (mmol N/h) |  |
| --- | --- | --- | --- |
| From periphery | alanine 27 | glutamine 80 |  |
|  | alanine 20 | NH₃ 30 | citrulline 22 |
| From gut |  | NH₃ 18 |  |
| Total | transaminators 47 | deaminators 48 |  |

From Windmueller and Spaeth, 1980.

These calculations are obviously only approximate but they serve to emphasize one important point, that the urea recycled through the gut has a critical role to play in balancing the stoichiometry of the urea cycle. In effect this system acts to deaminate transaminating amino acids. Hence in the normal individual in the postprandial state or on a normally balanced diet, the precursors for urea synthesis are naturally balanced and there is no need to invoke a major role for GDH in maintaining the stoichiometry of the reaction. Rather it would seem that GDH acts as a last resort in situations where imbalance is created. Thus a general diminution in hepatic function limits the capacity of the urea cycle (Rypins et al., 1980) to a greater extent than the capacity of the liver to deaminate amino acids (Rudman et al., 1973). A sufficiently large load of a deaminating amino acid can lead to hyperammonaemia in normal animals (Kamin and Handler, 1951). The hepatic deaminase enzymes are particularly sensitive to dietary protein intakes (Cedrangolo, 1975), and the ability with which the ammonia formed can be handled by the urea cycle would seem to be a rate-limiting step. This would indicate a definite limit to the extent to which GDH can fix free ammonia.

It is standard for amino acids to be thought of as forming two groups, essential and non-essential, based upon the ease with which the carbon skeleton can be synthesized in the body. We would suggest that the identification in terms of deaminators or transaminators is also of importance in relation to the way in which the nitrogen is handled. On a normal mixed diet in normal man this distinction is not likely to manifest itself in any way, but in certain clinical conditions, especially when special diets are being formulated, it may be a matter for serious consideration.

Glycine and serine are the only two of the deaminators that are considered to be non-essential; however, this may not be true for all situations (Giordano *et al.*, 1972; Jackson *et al.*, 1981). Lysine, threonine and probably histidine are essential, and because these amino acids are not transaminated, and because the deamination they undergo is irreversible, it is difficult if not impossible to substitute an α-keto analogue for any of them. The concept of non-essential nitrogen, and the utilization of endogenously degraded urea nitrogen as a source for non-essential nitrogen, has played a major role in the development of low protein diets. These diets, after an initial success have recently lost favour somewhat. Consideration of them in terms of the balance of transaminators and deaminators may allow for better formulations. Elemental diets of widely varying composition are given both orally and intravenously in diverse clinical conditions. Although of great therapeutic value, they are not without problems, which may possibly be associated with the balance of amino acids.

## Acknowledgement

The support of the Wellcome Trust is gratefully acknowledged.

## Discussion

*Bier (Washington University School of Medicine, St. Louis, USA):* We have carried out a series of 60 h infusions of labelled glycine for the purpose of measuring protein synthesis, as many people in this room have done. We measured the enrichment of $^{15}N$ in plasma amino acids from 45 to 60 h in about 15 studies. In our procedure, the amide-N of glutamine is lost in preparation and, thus, glutamine is measured as glutamate. Since the plasma content of glutamine is much greater than that of glutamate, the combined effects of amide-N loss and co-measurement have the result that $^{15}N$ enrichment in glutamate is underestimated. Yet, in every case we found $^{15}N$ enrichment in glutamate and alanine similar in relative magnitude to that found in the liver studies which you quoted. Our findings are therefore very distinctly different from the data you have just presented.

*Jackson:* Distinctly different—yes and no. There is a quantitative difference and a qualitative difference. We are not trying to suggest that there is absolutely no exchange of nitrogen between the deaminators and the transaminators; what we are suggesting is that under normal physiological conditions and probably under a wide range of abnormal conditions the extent to which this happens is distinctly limited, and the extent to which nitrogen actually flows from the free pools into glutamate, and then through glutamate dehydrogenase, is limited, although this flow does take place. The question is not whether it happens or not, but to what extent it happens.

*Bier:* I accept that glycine is primarily handled by a pathway other than transamination. However, I think our results are quite surprisingly different qualitatively as well as quantitatively, in terms of the relative amount of labelled nitrogen in alanine and glutamate.

*Stein (University of Pennsylvania School of Medicine, Philadelphia, USA):* There are two distinct mechanisms by which $^{15}N$ can be transferred from glycine to other amino acids: (1) by conventional nitrogen interchange reactions, and (2) via bacterial reactions in the gut, the first step of which is urea hydrolysis. In short term experiments, the latter is likely to be negligible, but this may not be the case after a 45 or 60 h infusion.

*Jackson:* That is a very important point, and I am not aware that anyone has considered it specifically. One question is the extent to which urea nitrogen is returned as ammonia and introduced into amino acids in the body pool. A second question is the extent to which the colon actually functions as a rumen, so that nitrogen is fixed by bacteria in amino acids, perhaps essential as well as non-essential, and made available for metabolism. That is one potential way in which differences between results in different populations might be resolved, in that there is a different colonic flora functioning in a different way, depending on diet and other considerations.

*Felig* (*Yale University School of Medicine, USA*): This is perhaps a very simple-minded question. If we accept that the nitrogen coming from the periphery to the liver in the overnight fasted individual is primarily alanine and glutamine, how is it getting into the urea cycle?

*Lund* (*Radcliffe Infirmary, Oxford, UK*): Theoretically glutamine is the ideal urea precursor because its amide-N is released as free ammonia and its amino-N appears in aspartate (after transamination of glutamate) giving the 1:1 stoichiometry required for urea synthesis. These reactions may occur to some extent in liver. Most of the glutamine is metabolized in the small intestine with release of ammonia and alanine into the portal vein. In this case the alanine-N is transferred to form aspartate via glutamate. For urea synthesis from alanine alone, 2 mol alanine are needed to form 1 mol aspartate (via glutamate) and 1 mol free ammonia.

*Felig:* So the key enzyme is glutamic dehydrogenase?

*Lund:* Yes, for maintaining the balance of free ammonia and aspartate through glutamate.

*Millward* (*London School of Hygiene and Tropical Medicine, UK*): Jackson has discussed the concept of separating the amino acids into those deaminated and those transaminated. One of the differences between the two groups is that some of the deaminating reactions are not linked to NAD reduction. Therefore there is likely to be a considerable amount of heat produced in these steps. In some cases this will mean that the oxidation of amino acids in terms of the removal of nitrogen is an inefficient process. I know that serine/threonine dehydratase is not linked to NAD reduction, but is the deamination of histidine and lysine also an inefficient process?

*Neuberger* (*Charing Cross Hospital Medical School, London, UK*): As far as glycine and serine are concerned, they probably go through the glycine cleavage system which will yield ammonia and generate ATP. That is a reaction which yields quite a lot of useful energy.

*Matthews* (*Washington University School of Medicine, St. Louis, USA*): Professor Neuberger, what do you think is the importance of serine aminotransferase in serine metabolism?

*Neuberger:* I do not know any data which would allow me to give an answer. I can only say that it seems from isotope experiments that the transformation of serine to glycine occurs very readily. The glycine cleavage complex is a most efficient system for glycine metabolism. Therefore one does not know what proportion of serine is deaminated by the dehydratase. This is an inducible enzyme which increases enormously in amount on a high carbohydrate diet, whereas in other conditions it is present in very small amounts. It has a very high $K_m$, and therefore I am inclined to believe that serine degradation mostly goes through conversion to glycine and the glycine cleavage complex, but we do not have any quantitative data to be very sure about it.

*Jackson:* In so far as the serine–pyruvate aminotransferase is concerned, it is my understanding that the kinetics of this reaction are very much in the direction of serine synthesis, so that in normal physiological situations this is unlikely to be a pathway for serine degradation. The same thing applies to the glycine–pyruvate aminotransferase.

*Matthews:* I have a second question to Jackson: lysine is an amino acid whose degradative pathway has been up in the air for some time. Why do you not expect nitrogen from lysine to go into glutamate?

*Jackson:* I can only say that both we and others have found that when labelled lysine is infused, not very much of the label goes to α-amino-N of other amino acids. We have tried to study the degradative pathway of lysine and we are unable to identify where the nitrogen goes to. It appears in urea without going through any of the transaminating amino acids. This would suggest, on the basis of our thesis, that somewhere, somehow it must be deaminated.

*Lund:* On the question of deamination versus transamination: in the case of histidine, which has been mentioned, the first step is a simple hydrolytic splitting off of the $NH_3$, so there is no NAD involved. Histidine is converted directly to urocanic acid. In the case of lysine, I agree that the first step is not a transamination. It is generally accepted that lysine reacts with ketoglutarate to give saccharopine, but the net result is the equivalent of two transamination reactions in the degradation of lysine. Thirdly, serine is generally considered to be a glucogenic amino acid, so that it cannot all be metabolized via glycine. Serine transaminates with pyruvate to give hydroxypyruvate or forms pyruvate directly via serine dehydratase in the glucogenic pathway.

*Jackson:* These reactions which have been elucidated *in vitro* and the identification of the enzymes which catalyse them tell us what pathways are possible, but they do not necessarily describe what happens in the whole animal *in vivo*. Jahoor in our laboratory has tried to test experimentally whether in the whole animal we can demonstrate the expected balance between deaminators and transaminators as precursors of the two nitrogen atoms of urea.

*Jahoor (University of the West Indies, Kingston, Jamaica):* We have been trying to test experimentally the theoretical requirement that the nitrogen atoms of urea are derived in equal amounts from free $NH_3$ and aspartate. According to the theory, one would expect that if the nitrogen in the precursor pool for urea formation is labelled, then the abundance in urea should be the mean of the abundances in $NH_3$ and aspartate.

TABLE 2

$^{15}N$ ENRICHMENT (ATOMS % EXCESS $\times 10^3$) OF NITROGEN METABOLITES IN THE FREE POOLS OF LIVER RELATIVE TO THAT OF UREA (MEANS $\pm$ SEM)

| Isotope source | | Relative enrichment (urea = 100%) | | |
| | $NH_3$ | Aspartate | $(NH_3 + Aspartate)/2$ | Amide-N |
| --- | --- | --- | --- | --- |
| Glycine | $94 \pm 6$ | $27 \pm 4$ | 61 | $126 \pm 26$ |
| Aspartic acid | $97 \pm 6$ | $33 \pm 3$ | 65 | $117 \pm 3$ |
| Lysine | $51 \pm 4$ | $17 \pm 1$ | 34 | $52 \pm 1$ |
| Glutamine | $361 \pm 25$ | $60 \pm 12$ | 211 | $663 \pm 29$ |
| $NH_4Cl$ | $468 \pm 18$ | $35 \pm 2$ | 248 | $272 \pm 27$ |

Five groups of six adult male rats weighing 407–571 g were infused intravenously after a meal. Each group received a different tracer: $^{15}N$-glycine, $^{15}N$-α-amino-lysine, $^{15}N$-aspartic acid, $^{15}N$-amide glutamine, or $^{15}NH_4Cl$. The dose of each was about 2 mg $^{15}N$/kg/day. After 6 h the animals were decapitated and the livers rapidly homogenized in ice-cold 10% PCA, centrifuged and neutralized in the cold. The nitrogen from $NH_3$, glutamine-amide, aspartic acid and urea was sequentially removed by macro-modification of specific enzyme reactions (Bergmeyer, 1976) and the resulting $NH_3$ extracted and measured for isotopic enrichment as described by Jackson et al. (1980).

The results in Table 2 show that this relationship does not hold when total hepatic free $NH_3$ and aspartate are considered as the precursors. Thus with infusions of $^{15}N$-labelled glycine, lysine and aspartic acid the mean abundance in ($NH_3$ + aspartate) is much less than the abundance in urea. With our methodology about 16% of glutamine amide-N is degraded to free $NH_3$ during isolation. This admixture could not be the cause of the low labelling of $NH_3$, since in each experiment the labelling of glutamine amide-N was higher than that of $NH_3$. Without the admixture the mean abundance in ($NH_3$ + aspartate) would be even lower. With infusions of glutamine and $NH_4Cl$ the reverse situation occurred; ($NH_3$ + aspartate) has more than twice the mean abundance of urea, and free $NH_3$ several times more. Thus this pool cannot be the direct precursor pool for urea. Two explanations seem possible: (1) the assumption often made that $NH_3$ is freely diffusible is wrong and the abundance in intramitochondrial $NH_3$ differs greatly from that in total liver $NH_3$; or (2) carbamyl phosphate synthesis is linked to ammoniagenesis from specific amino acids and total free $NH_3$ is not normally the major *in vivo* precursor for carbamyl phosphate synthesis. These results support the contention of Krebs *et al.* (1978) that aspartate is not a major source of hepatic $NH_3$ through operation of the adenine nucleotide cycle, since aspartate invariably had a far lower abundance than free $NH_3$.

*Millward:* Is it possible that aspartate can derive nitrogen from $NH_3$ by operation of the purine nucleotide cycle in reverse?

*Jahoor:* I cannot answer that question.

*Kerr (Case Western Reserve University, Cleveland, USA):* When you gave $^{15}N$-labelled aspartate you found extremely low labelling of the aspartate pool. How can you be sure that isotopic equilibrium had been reached? If it had not been reached, how can you then conclude, from the infusions of $^{15}NH_3$, that 40 % of the amide-N of glutamine is coming from some other source?

*Jahoor:* We think there is good evidence that in the rat isotopic equilibrium is reached in the free pool of the liver with infusions lasting 6 h. The low enrichment of liver aspartate, when labelled aspartic acid was infused, is probably due to the fact that liver cells do not take up dicarboxylic amino acids (Kamin and Handler, 1951).

# References

AQVIST, S. E. G. (1951). Metabolic interrelationships among amino acids studied with isotopic nitrogen, *Acta. Chem. Scand.*, **5**, 1046–64.

BERGMEYER, H. U. (1976). *Methods of Enzymatic Analysis.* 2nd Edition, Vols I–IV. Academic Press, New York.

CAMMARATA, P. S. and COHEN, P. P. (1950). The scope of the transamination reaction in animal tissues, *J. Biol. Chem.*, **187**, 439–52.

CEDRANGOLO, F. (1975). in *The Urea Cycle* (Eds. S. Grisolia, R. Bagnena and F. Mayor), John Wiley and Sons, New York, 551–60.

GIORDANO, C., DE SANTO, N. G., RINALDI, S., DE PASCALE, C. and PLUVIO, M. (1972). Histidine and glycine essential amino acids in uraemia. in *Uraemia: pathogenesis, diagnosis and treatment* (Ed. R. Kluthe, G. Berlyne and B. Burton), Verlag, K.G., Stuttgart, 138–43.

HARPER, A. E., BENEVENGA, N. J. and WOHLHUETER, R. M. (1970). Effects of ingestion of disproportionate amounts of amino acids, *Physiol. Rev.*, 50, 428–558.

JACKSON, A. A. and GOLDEN, M. H. N. (1980). $^{15}N$ glycine metabolism in normal man: The metabolic $\alpha$-amino-nitrogen pool, *Clin. Sci.*, 58, 517–22.

JACKSON, A. A., GOLDEN, M. H. N., JAHOOR, P. F. and LANDMAN, J. P. (1980). The isolation of urea nitrogen and ammonia nitrogen from biological samples for mass spectrometry, *Analyt. Biochem.*, 105, 14–17.

JACKSON, A. A., SHAW, J. C. L., BARBER, A. and GOLDEN, M. H. N. (1981). Nitrogen metabolism in pre-term infants fed human donor breast milk: the possible essentiality of glycine, *Pediat. Res.*, in press.

KAMIN, H. and HANDLER, P. (1951). The metabolism of parentally administered amino acids. II. Urea synthesis, *J. Biol. Chem.*, 188, 193–205.

KAMIN, H. and HANDLER, P. (1951). The metabolism of parenterally administered amino acids. III. Ammonia formation, *J. Biol. Chem.*, 193, 873–80.

KREBS, H. A., HEMS, R. and LUND, P. (1973). Some regulatory mechanisms in the synthesis of urea in the mammalian liver, *Adv. Enz. Regulation*, 11, 361–77.

KREBS, H. A., HEMS, R., Lund, P., HALLIDAY, D. and READ, W. W. C. (1978). Sources of ammonia for mammalian urea synthesis, *Biochem. J.*, 176, 733–7.

MARLISS, E. B., AOKI, T. T., POZEFSKY, T., MOST, A. S. and CAHILL JR, G. F. (1971). Muscle and splanchnic glutamine and glutamate metabolism in postabsorptive and starved man, *J. Clin. Invest.*, 50, 814–17.

RATNER, S., RITTENBERG, D., KESTON, A. S. and SCHOENHEIMER, R. (1940). Studies in protein metabolism. XIV. The chemical interaction of dietary glycine and body proteins in rats, *J. Biol. Chem.*, 134, 665–76.

RUDMAN, D., GALAMBOS, J. T., SMITH, III, R. B., SALAM, A. A. and WARREN, W. D. (1973). Comparison of the effect of various amino acids upon the blood ammonia concentration of patients with liver disease, *Amer. J. Clin. Nutr.*, 26, 916–25.

RYPINS, E. B., HENDERSON, J. M., FULENIWIDER, J. T., MOFFITT, S., GALAMBOS, J. T., WARREN, W. D and RUDMAN, D. (1980). A tracer method for measuring rate of urea synthesis in normal and cirrhotic subjects, *Gastroenterology*, 78, 1419–24.

SCHOENHEIMER, R. (1942). *The Dynamic State of Body Constituents*, Harvard University Press, Massachusetts.

SHEMIN, D. and RITTENBERG, D. (1944). Some interrelationships in general nitrogen metabolism, *J. Biol. Chem.*, 153, 401–21.

TARUVINGA, M. (1978). M.Sc. project, University of London.

VITTI, T. G. and GAEBLER, O. H. (1963). Effects of growth hormone on metabolism of nitrogen from several amino acids and ammonia, *Arch. Biochem. Biophys.*, 101, 292–8.

WINDMUELLER, H. G. and SPAETH, A. E. (1980). Respiratory fuels and nitrogen metabolism *in vivo* in small intestine of fed rats, *J. Biol. Chem.*, 255, 107–12.

# 17

# Regulation of the Ornithine–Urea Cycle Enzymes

P. P. COHEN

*Department of Physiological Chemistry,*
*University of Wisconsin-Madison, Madison, Wisconsin, USA*
*and Centro Universitario de Profesores Visitantes,*
*Universidad Nacional Autonoma de Mexico,*
*Mexico City, Mexico*

The enzymes involved in urea biosynthesis in ureotelic animals are listed in Table 1. All of the enzymes have been purified to a degree which has permitted detailed studies of their physical and kinetic properties.

The enzymes of urea biosynthesis provide a metabolic pathway for conversion of potentially toxic ammonia to non-toxic metabolites and ultimately to arginine and the metabolically inert excretory product, urea.

As can be seen from Fig. 1, the enzymes are compartmentalized in the liver. The enzymes involved in the formation of citrulline from ammonia, bicarbonate and ornithine, carbamylphosphate synthetase I (CPS I) and ornithine transcarbamylase (OTC), are confined to the mitochondria (see Cohen (1981) for a recent review). The remaining three enzymes, argininosuccinate synthetase, argininosuccinase and arginase, are extra-mitochondrial (see Ratner (1976), and Soberon and Palacios (1976) for recent reviews). Low, but metabolically significant levels of CPS I and OTC are present in mitochondria of the intestinal mucosa of ureotelic animals (Hall *et al.*, 1960; Jones *et al.*, 1961; Windmueller and Spaeth, 1980). Metabolically significant levels of argininosuccinate synthetase, argininosuccinase and arginase are present in kidney (Ratner, 1976).

As can be seen from Table 1 and Fig. 1, CPS I is the only enzyme involved in urea biosynthesis which has a regulatory cofactor, N-acetyl-L-glutamate (AG). Liver mitochondria of ureotelic animals contain an enzyme, acetylglutamate synthetase, which synthesizes AG from acetyl CoA plus glutamate (Tatibana *et al.*, 1976; Shigesada and Tatibana, 1978).

The sequestration of CPS I, OTC and AG synthetase in liver mitochondria of ureotelic animals provides the opportunity for a number

## TABLE 1

ENZYMES AND METABOLITES INVOLVED IN UREA BIOSYNTHESIS IN LIVER OF UREOTELIC
ANIMALS

(1) *Carbamylphosphate synthetase I* [EC 6.3.4.16] (CPS I):

$$NH_4^+ + HCO_3^- + 2ATP^{4-} \xrightarrow[Mg^{2+}, K^+]{AG} NH_2CO_2PO_3^{2-} + 2ADP^{3-} + HPO_4^{2-} + 2H^+$$

(2) *Ornithine transcarbamylase* [EC 2.1.3.3] (OTC):

   Carbamyl phosphate + ornithine $\longrightarrow$ citrulline + $P_i$

(3) *Argininosuccinate synthetase* [EC 6.3.4.5]

   Citrulline + aspartate + ATP $\underset{}{\overset{Mg^{2+}}{\rightleftharpoons}}$ argininosuccinate + AMP + $PP_i$

(4) *Argininosuccinase* [EC 4.3.2.1]:

   Argininosuccinate $\rightleftharpoons$ arginine + fumarate

(5) *Arginase* [EC 3.5.3.10]:

   Arginine $\xrightarrow{Mn^{++}}$ ornithine + urea

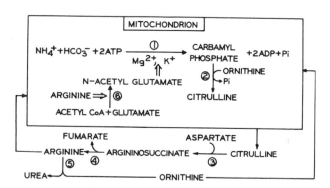

FIG. 1.   Numbers shown correspond to reactions listed in Table 1. Reaction (6) represents *N*-acetyl-L-glutamate synthetase. Double arrows indicate activation of carbamylphosphate synthetase I by *N*-acetyl-L-glutamate (1) and acetylglutamate synthetase by arginine (6).

of metabolic systems to influence the capacity of the liver to regulate the conversion of potentially toxic levels of ammonia to non-toxic metabolites and ultimately to urea.

## Regulatory Factors in Urea Biosynthesis Involving the Mitochondrial Enzymes CPS I, OTC and AG Synthetase

### Role of N-Acetyl-L-glutamate

N-Acetylglutamate (AG) is an essential cofactor which functions as an allosteric regulator of CPS I (Marshall *et al.*, 1961; Fahien *et al.*, 1964; Marshall, 1976). Recent studies by Hensgens *et al.* (1980) have demonstrated the important role of intramitochondrial AG in the short term regulation of urea biosynthesis under different nutritional and hormonal conditions.

### Mitochondrial Metabolites and Fluxes

The activity of CPS I is dependent on mitochondrial concentrations of ATP, ammonia, bicarbonate, $Mg^{2+}$, $K^+$, and AG (see reaction (1), Table 1, and Fig. 1). Because the synthesis of AG will be affected by the availability of acetyl CoA and glutamate, the intramitochondrial concentration of AG will be dependent on acetyl CoA generation from fatty acid and pyruvate oxidation, and on glutamate dehydrogenase and transaminases for maintaining glutamate levels. The effect of acetyl CoA levels on the synthesis of AG and in turn on the synthesis of citrulline and urea has been reported (Aoyagi *et al.*, 1979). A further regulatory factor for the synthesis of AG is arginine which is a positive effector of AG synthetase (Shigesada and Tatibana, 1978). The effects of a number of metabolites on citrulline synthesis in liver mitochondria and urea synthesis in isolated liver cells have been investigated (Meijer *et al.*, 1975; Stubbs, 1976; Meijer and Van Woerkom, 1977; Wojtczak *et al.*, 1978; Raijman and Bartulis, 1979; and Bryla and Niedzwiecka, 1979). Meijer (1979) has recently reviewed the effect of mitochondrial metabolic alterations on CPS I activity. The importance of $K^+$ for the CPS I system is that it decreases the $K_m$ for $NH_4^+$, ATP and AG (Marshall *et al.*, 1961).

In addition to the above regulatory influences by mitochondrial metabolites on CPS I activity, glutamate and α-ketoglutarate have been shown to act as competitive inhibitors of AG binding by CPS I ($K_m$, $1 \times 10^{-3}$M) with $K_i$ values of $1 \times 10^{-3}$M and $3 \times 10^{-3}$M respectively (Fahien *et al.*, 1964).

It is thus clear that CPS I represents a uniquely poised system responsive to relatively rapid regulation by metabolic events in liver mitochondria.

## Hormonal Effects on Levels or Activity of Enzymes

Effects on the levels of activities of the enzymes involved in urea biosynthesis have been reported for the following hormones:

(1)  Corticosteroids (Schimke, 1963; McLean and Gurney, 1963; Räihä, 1976; Edkins and Räihä, 1976; Gautier and Vaillant, 1978; Gebhardt and Mecke, 1979; Lamers and Mooren, 1980).

(2)  Glucagon (Gebhardt and Mecke, 1979; Yamazaki and Graetz, 1977; Triebwasser and Freedland, 1977; Snodgrass et al., 1978; Halestrap et al., 1980; Titheradge and Haynes, 1980; Hensgens et al., 1980).

(3)  Growth hormone (McLean and Gurney, 1963).

(4)  Cyclic AMP (Räihä, 1976; Edkins and Räihä, 1976; Gebhardt and Mecke, 1979; Halestrap et al., 1980).

(5)  Thyroid hormone (Cohen, 1970; Cohen et al., 1978; Lamers and Mooren, 1980).

The systems used have included fetal rat liver, human fetal liver, isolated liver cells and mitochondria and intact animals (rats and tadpoles). With the exception of the effect of thyroid hormones on amphibian systems, the other hormonal effects appear to be secondary to effects on gluconeogenesis, ATP production, or other metabolic processes.

In the case of amphibian metamorphosis one of the earliest detectable effects of thyroid hormone is the induction of CPS I (see Cohen, 1970), with a coordinate increase in the levels and activities of the other enzymes involved in urea biosynthesis (Wixom et al., 1972).

## Effects of Dietary Protein on the Levels of the Ornithine–Urea Cycle Enzymes

Schimke (1962a, 1962b, 1963) observed that the total content of the enzymes involved in urea biosynthesis increased coordinately as dietary protein intake was increased in the rat. While rats fed a protein-free diet, but adequate in calories, showed a decrease in the levels of all of the enzymes, starvation resulted in an increase (Schimke, 1962b). Schimke (1963) concluded from his studies that 'all conditions studied which lead to protein breakdown and resultant increased urea excretion, whether it be feeding a protein-rich diet, starvation or corticosteroid administration, were associated with increases in all five urea cycle enzymes proportionate to the increase in urea synthesis'. A similar adaptation to dietary protein intake has been reported for primates, including man (Nuzum and Snodgrass, 1971). In a study by Das and Waterlow (1974) it was shown that increases and decreases in dietary protein in rats resulted in corresponding increases and decreases within 30 h of the extramitochondrial enzymes argininosuccinate synthetase, argininosuccinase and arginase. They further observed that the quality of the protein fed was of importance. Thus, substituting gelatin for casein resulted in a doubling of the nitrogen output with no change in the levels of enzyme activity.

It should be noted that feeding a high protein diet increases the level of AG, the activator of CPS I (Tatibana *et al.*, 1976; Tatibana and Shigesada, 1976; Shigesada *et al.*, 1978).

## Synthesis and Degradation of Enzymes

In view of the effect of dietary protein on the levels of the enzymes involved in urea biosynthesis, data on the relative rates of synthesis and degradation of the enzymes are of interest. Schimke (1964) studied the effect of starvation and of different levels of protein intake on arginase turnover in the rat. He observed that the increased level of arginase in rats fed a high protein diet was the result of an increased rate of synthesis, while in rats whose diet was changed from a high to a low protein intake, the resulting decreased level of arginase was the result of an increased rate of degradation. On the other hand, Schimke found that the increased level of arginase in rats which were starved was the result of a decreased rate of degradation.

In a study of the effect of dietary protein intake on the turnover of rat liver argininosuccinate synthetase by Tsuda *et al.* (1979), it was found that while both the rates of synthesis and degradation were involved in the regulation of the enzyme level during dietary transition, under steady state conditions, the rate of degradation was found to be unchanged while the rate of synthesis of the enzyme was found to be the determining factor.

In the case of the rat liver mitochondrial enzyme CPS I, Nicoletti *et al.* (1977) estimated a half life of 7·7 days on a normal diet, 3·3 days on a high protein diet and 4·6 days on a protein-free diet. The finding of a shorter half life for CPS I on a high protein diet associated with an elevated level of the enzyme would indicate that there is both an increased rate of synthesis and of degradation, with the former predominating.

Studies have recently been carried out in rats on the effects of high and low protein diets, and of starvation, on the *m*RNA levels and synthesis of CPS I and OTC (Mori *et al.*, 1981). The translatable levels of hepatic *m*RNAs for CPS I and OTC were found to be 4·2- and 2·2-fold higher, respectively, in the case of rats fed a 60 % casein diet as compared with rats fed a 5 % casein diet. The differences in *m*RNA levels were found to be slightly higher than the differences in the levels of the enzyme activities (3·3- and 1·9-fold, respectively) and of the enzyme proteins (3·3- and 2·1-fold, respectively). These results indicate that the dietary protein-dependent changes in the levels of CPS I and OTC are due mainly to changes in levels of translatable *m*RNAs for the enzymes.

On the other hand, it was found that while levels of CPS I and OTC were increased 1·7- and 1·3-fold, respectively, during a 7-day fasting period over those of control animals kept on a 24 % casein diet, the levels of translatable *m*RNAs for CPS I and OTC were decreased to 54 and 67 % respectively, of those of control animals. Thus, the increase of these two enzyme levels during fasting appears to be the result of a decreased rate of degradation of the enzymes.

## Concluding Remarks

The mitochondrial sequestration of CPS I, OTC and AG synthetase provides the liver cell with the metabolic opportunity for ready control and efficient fixation of ammonia and bicarbonate to form carbamyl phosphate and the utilization of its carbamyl group to form citrulline. While rat liver CPS I is present in relatively high concentration in mitochondria (20 % of the matrix proteins) with a relatively low specific activity (35 units/mg protein), OTC is present in relatively low concentration (0·5 %) but with a relatively high specific activity (900 units/mg protein). Lusty (1978) has estimated the relative maximal activities of CPS I and OTC in rat liver mitochondria to be 1:10. Thus, under normal metabolic conditions, CPS I activity is rate-limiting and dependent on AG concentration; on the other hand, the relative excess of OTC activity insures rapid utilization of carbamylphosphate for citrulline biosynthesis. Under normal conditions in ureotelic animals, reversal of the pathway to citrulline does not occur because (i) 2 mol ATP are required in the CPS I system and thus synthesis of carbamylphosphate is favoured and (ii) the mammalian OTC system has an equilibrium that lies overwhelmingly in the direction of citrulline synthesis (Marshall and Cohen, 1972).

In the case of the extramitochondrial enzymes, arginase catalyses an essentially irreversible step, but the argininosuccinate synthetase and argininosuccinase steps are reversible (Ratner, 1976). In addition, Ratner

(1976) has pointed out that the argininosuccinate synthetase step is subject to appreciable product inhibition by AMP and PPi. Because of these factors assessment of the maximal steady state rate of argininosuccinate synthesis *in vivo* is uncertain. While the evidence from a variety of experimental systems indicates that citrulline production in mitochondria is more rapid than its utilization for arginine and urea biosynthesis, it is possible that the rate-limiting factor in the overall system could be the transport rate of citrulline (and aspartate) from the mitochondria (see Gamble and Lehninger, 1973). The presence of significant levels of argininosuccinate synthetase, argininosuccinase and arginase in kidney (see Ratner, 1976) provides an auxiliary system for conversion of excess citrulline to arginine and urea (see Featherston *et al.*, 1973).

While the ultimate purpose of the ornithine–urea cycle is that of conversion of ammonia to urea, it is important to recognize that the conversion of potentially toxic levels of ammonia to the non-toxic metabolite citrulline may be of greater immediate importance to the animal. The mitochondrial enzymes CPS I and OTC thus must be responsive and efficient in dealing with sudden and potentially toxic levels of ammonia. While feeding or injection of AG is ineffective in influencing CPS I activity, because of its rapid hydrolysis by amino acid acylases, the analogue *N*-carbamylglutamate (which is not readily hydrolysed) has been shown to be effective. Thus, rats given lethal doses of ammonium acetate can be protected 100 % if given a mixture of *N*-carbamylglutamate plus arginine (Kim *et al.*, 1972).

The coordinated response of the two mitochondrial enzymes and the three extramitochondrial enzymes to changes in dietary protein intake, and to thyroxine in the metamorphosing tadpole, indicates that a regulatory mechanism exists for the coordinated regulation of the levels of all five enzymes. The nature of the factors involved in this regulatory process remains unknown. A possibility which merits exploration is that of the role of the 'signal' peptides (see Blobel, 1980), shown to be present in the precursor forms of CPS I (Shore *et al.*, 1979; Mori *et al.*, 1979) and of OTC (Conboy *et al.*, 1979; Mori *et al.*, 1980a) synthesized in rat liver. Since it has recently been found that the levels of CPS I and OTC in rats on high and low protein diets are correlated with levels of *m*RNAs for these enzymes (Mori *et al.*, 1981), it is possible that the peptides cleaved by a selective protease (see Mori *et al.*, 1980b) from the precursor forms of the enzymes play a role in the transcription of the genes involved, or affect an as yet to be demonstrated regulatory gene.

While the synthesis of all of the enzymes involved in urea biosynthesis is under nucleo-cytoplasmic control, little is known about the proteolytic systems (and their regulation) involved in their degradation, and in particular the basis for the selective degradation of liver protein during starvation.

## Acknowledgement

This study was supported in part by Grant CA-03571, awarded by the National Cancer Institute, National Institute of Health.

## Discussion

*Neuberger* (*Charing Cross Hospital Medical School, London, UK*): Do you think that the amount of messenger RNA gives an unequivocal indication of the rate of protein synthesis, or could transcription be less efficient under some conditions than others?

*Cohen:* We have tried to test that possibility in some experiments in which we mixed $m$RNA preparations from the livers of rats on high and low protein diets, and from fasted and normal rats, to see whether there could be any factor lacking as a result of a low protein intake or fasting, or conversely whether some activator was formed. We found that the mixture gave the mean value of the two $m$RNA levels. This does not answer your question directly, but at least it suggests that there is no gross change, under these dietary conditions, in the translatability of $m$RNA in the cell-free system.

*Jackson* (*University of the West Indies, Kingston, Jamaica*): Dr Cohen, can you comment on the build-up of citrulline which you mentioned?

*Cohen:* Under high loads of ammonia or protein intake, the mitochondrial enzymes are capable of producing more citrulline inside the mitochondria than the extra-mitochondrial enzymes are able to deal with. Because of the characteristics of argininosuccinate synthetase–potential reversibility and strong product inhibition–and because there is a barrier to the transport of ornithine into the mitochondrion and of citrulline out, in addition to the need for aspartate, formed mainly in the mitochondria, I contend that the bottleneck may be not at the enzyme level but at the level of intracellular transport. Under a moderate ammonia load, there is no significant accumulation of citrulline within the mitochondria. If citrulline formation by isolated mitochondria or hepatocytes is compared with urea formation there is a greater capacity for rapidly forming citrulline than for the overall reaction. I think that the major function of the mitochondrial system is the rapid regulation of potentially toxic ammonia concentrations. Some of the citrulline can even be transported out of the liver and utilized in the kidney, as has already been mentioned. Thus the cycle of urea biosynthesis is a totally effective cycle under most conditions and a partial cycle under extreme conditions.

*Felig* (*University of Wisconsin-Madison, USA*): Dr Cohen, would you comment on the regulation of the urea cycle enzymes by glucagon? There have been some studies in which patients with congenital deficiencies of urea cycle enzymes have been treated with large amounts of glucagon in order to increase the activity of ornithine transcarbamylase or other enzymes. It is important to distinguish between the pharmacological and physiological effects of glucagon. Is there evidence that physiological amounts of glucagon can regulate urea cycle enzymes?

*Cohen:* There has been a recent paper by Hensgens *et al.* (1980) in Holland on the glucagon effect in hepatocytes and mitochondria. They show clearly that a primary effect of glucagon is to raise the level of ATP. While an increase in ATP level might account in part for the stimulatory effect of glucagon on citrulline biosynthesis, Hensgens *et al.* also reported that glucagon caused an 8-fold increase in mitochondrial *N*-acetyl-glutamate concentration. The latter effect in particular could account for the short-term enhancement of citrulline biosynthesis by glucagon.

*Felig:* The question still remains, whether these effects can be demonstrated with the concentrations of glucagon that are known to be present under physiological conditions. For example, in starvation the glucagon concentration may go up from the basal level of 75 pg/ml to about 150 pg/ml. This is a different order of magnitude from the 1 mg doses which have been used in various studies.

*Cohen:* I have no comment on that.

*Young* (*Massachusetts Institute of Technology, USA*): Dr Cohen, do you consider the metabolic association between *N*-acetyl-glutamate and fatty acid and pyruvic acid utilization a possible way in which energy supply might influence ureagenesis? It seems to me an unlikely mechanism of action to explain the effect of energy intake on urea production, since we are dealing with an activator, and the specific generation of $N$-acetyl-glutamate would represent only a very small proportion of the total carbon flow associated with oxidation of fatty acids and pyruvate.

*Cohen:* I referred to a review by Meijer (1979) which deals with the metabolic factors which can influence not only the synthesis of $N$-acetyl-glutamate but also the activity of CPS I. With respect to the question of whether the level of acetyl CoA generated from the oxidation of pyruvate and fatty acids would so greatly exceed the amount needed for synthesis of $N$-acetyl-glutamate that it is not likely to be a factor involved in regulating the level of $N$-acetyl-glutamate, two points should be made. First, Aoyagi *et al.* (1979) demonstrated that the inhibition of urea biosynthesis in hepatocytes by pent-4-enoate was the result of a decrease in the level of acetyl CoA which in turn resulted in a decrease in the level of $N$-acetyl-glutamate. Secondly, the relatively large amount of acetyl CoA generated is utilized rapidly in many metabolic systems and thus is not likely to accumulate. Under most conditions the pool size and concentration of available acetyl CoA may be such as to be rate-limiting for one or more of the competing reactions which also may have different $K_m$ values for acetyl CoA.

I would like to make a few additional remarks. I commented on the possible role of the so-called signal peptides in regulating the synthesis of the enzymes involved in urea biosynthesis. I probably did not emphasize sufficiently the need for understanding the basis for the selective degradation of liver proteins. The problem is particularly challenging in the case of starvation as to why the enzymes of urea biosynthesis resist degradation while the total liver protein is decreased to a very low level.

Finally, reference has been made to the fact that while the relative rate of protein synthesis by the muscle may be low, the total amount of protein synthesis is very large. Since the source of amino acids released from muscle under certain conditions remains uncertain, I would propose the following possibility. With the synthesis of the large amount of protein by muscle there must be a relatively large pool of amino acids involved in the signal peptides associated with the synthesis of the many proteins making up the muscular system. These peptides are removed post-translationally and may represent a sequestered pool of peptides and/or amino acids. Many of the amino acids in the signal peptides have non-polar side chains which aid in transport through or incorporation into membranes. Many of the branched chain amino acids released by muscle under certain conditions may arise from this source.

Finally, the present difficulties in understanding the true precursor relationship of an isotopically labelled amino acid, administered for the purpose of studying protein synthesis and degradation in a given organ, may reflect our lack of knowledge about the characteristics of such an intracellular amino acid pool (see Waterlow, p. 8).

## References

AOYAGI, K., MORI, M. and TATIBANA, M. (1979). Inhibition of urea synthesis by pent-4-enoate associated with decrease in N-acetyl-L-glutamate concentration in isolated rat hepatocytes, *Biochim. Biophys. Acta*, **587**, 515–21.

BLOBEL, G. (1980). Intracellular protein topogenesis, *Proc. Natl. Acad. Sci. USA*, **77**, 1496–500.

BRYLA, J. and NIEDZWIECKA, A. (1979). Relationship between pyruvate carboxylation and citrulline synthesis in rat liver mitochondria: the effect of ammonia and energy, *Int. J. Biochem.*, **10**, 235–9.

COHEN, P. P. (1970). Biochemical differentiation during amphibian metamorphosis, *Science*, **168**, 533–43.

COHEN, P. P. (1981). The ornithine–urea cycle: biosynthesis and regulation of carbamyl phosphate synthetase I and ornithine transcarbamylase, in *Current Topics in Cellular Regulation: Biological Cycles* Vol. 18 (Eds B. L. Horecker, P. Stadtman-Srere and R. Estabrook), Academic Press, New York.

COHEN, P. P., BRUCKER, R. F. and MORRIS, S. M. (1978). Cellular and molecular aspects of thyroid hormone action during amphibian metamorphosis, in *Hormonal Proteins and Peptides*, Vol. VI: *Thyroid Hormones*, (Ed C. H. Li), pp. 273–381, Academic Press, New York.

CONBOY, J. G., KALOUSEK, F. and ROSENBERG, L. E. (1979). *In vitro* synthesis of a putative precursor of mitochondrial ornithine transcarbamylase, *Proc. Natl. Acad. Sci. USA*, **76**, 5724–7.

DAS, T. K. and WATERLOW, J. C. (1974). The rate of adaptation of urea cycle enzymes, aminotransferases and glutamic dehydrogenase to changes in dietary protein intake, *Brit. J. Nutr.*, **32**, 353–73.

EDKINS, E. and RÄIHÄ, N. C. R. (1976). Changes in the activities of urea synthesis caused by dexamethasone and dibutyryladenosine 3′:5′-cyclic monophosphate in foetal rat liver maintained in organ culture, *Biochem. J.*, **160**, 159–62.

FAHIEN, L. A., SCHOOLER, J. M., GEHRED, G. A. and COHEN, P. P. (1964). Studies on the mechanism of action of acetylglutamate as an activator of carbamyl phosphate synthetase, *J. Biol. Chem.*, **239**, 1935–41.

FEATHERSTON, W. R., ROGERS, Q. R. and FREEDLAND, R. A. (1973). Relative importance of kidney and liver in synthesis of arginine by the rat, *Amer. J. Physiol.*, **224**, 127–9.

GAMBLE, J. G. and LEHNINGER, A. L. (1973). Transport of ornithine and citrulline across the mitochondrial membrane, *J. Biol. Chem.*, **248**, 610–18.

GAUTIER, C. and VAILLANT, R. (1978). Effects of administration of hydrocortisone, actinomycin D and puromycin on carbamoylphosphate synthetase-I and ornithine carbamoyltransferase activities in fetal rat liver, *Biol. Neonate*, **33**, 289–96.

GEBHARDT, R. and MECKE, D. (1979). Permissive effect of dexamethasone on glucagon induction of urea-cycle enzymes in perfused primary monolayer cultures of rat hepatocytes, *Eur. J. Biochem.*, **97**, 29–35.

HALESTRAP, A. P., SCOTT, R. D. and THOMAS, A. P. (1980). Mitochondrial pyruvate transport and its hormonal regulation, *Int. J. Biochem.*, **11**, 97–105.

HALL, L. M., JOHNSON, R. C. and COHEN, P. P. (1960). Presence of carbamylphosphate synthetase in intestinal mucosa, *Biochim. Biophys. Acta*, **37**, 144–5.

HENSGENS, H. E. S. J., VERHOEVEN, A. J. and MEIJER, A. J. (1980). The relationship between intramitochondrial *N*-acetylglutamate and activity of carbamoyl-phosphate synthetase (ammonia). The effect of glucagon, *Eur. J. Biochem.*, **107**, 197–205.

JONES, M. E., ANDERSON, A. D., ANDERSON, C. and HODES, S. (1961). Citrulline synthesis in rat tissues, *Arch. Biochem. Biophys.*, **95**, 499–507.

KIM, S., PAIK, W. K. and COHEN, P. P. (1972). Ammonia intoxication in rats: protection by *N*-carbamoyl-L-glutamate plus L-arginine, *Proc. Natl. Acad. Sci. USA*, **69**, 3530–3.

LAMERS, W. H. and MOOREN, P. G. (1980). Role of glucocorticosteroid hormones on the levels of rat liver carbamoylphosphate synthase (ammonia) and arginase activity during ontogenesis, *Biol. Neonate*, **37**, 113–37.

LUSTY, C. (1978). Carbamoylphosphate synthetase I of rat liver mitochondria, purification, properties and polypeptide molecular weight, *Eur. J. Biochem.*, **85**, 373–83.

MARSHALL, M. (1976). Carbamylphosphate synthetase I from frog liver, in *The Urea Cycle* (Eds S. Grisolia, R. Baguena and F. Mayor), pp. 133–42, Wiley, New York.

MARSHALL, M. and COHEN, P. P. (1972). Ornithine transcarbamylase from *Streptococcus faecalis* and bovine liver. II. Multiple binding sites for carbamyl-P and L-norvaline, correlation with steady state kinetics, *J. Biol. Chem.*, **247**, 1654–68.

MARSHALL, M., METZENBERG, R. L. and COHEN, P. P. (1961). Physical and kinetic properties of carbamyl phosphate synthetase from frog liver, *J. Biol. Chem.*, **236**, 2229–37.

MCLEAN, P. and GURNEY, M. W. (1963). Effect of adrenalectomy and of growth hormone on enzymes concerned with urea synthesis in rat liver, *Biochem. J.*, **87**, 96–104.

MEIJER, A. J. (1979). Regulation of carbamoylphosphate-synthase (ammonia) in liver in relation to urea cycle activity, *Trends Biochem. Sci.*, **4**, 83–6.

MEIJER, A. J. and VAN WOERKOM (1977). Relationship between intramitochondrial citrate and the activity of carbamoylphosphate synthase (ammonia), *Biochim. Biophys. Acta*, **500**, 13–26.

MEIJER, A. J., GIMPEL, J. A., DELEEUW, G. A., TAGER, J. M. and WILLIAMSON, J. R. (1975). Role of anion translocation across the mitochondrial membrane in the regulation of urea synthesis from ammonia by isolated rat hepatocytes, *J. Biol. Chem.*, **250**, 7728–38.

MORI, M., MIURA, S., TATIBANA, M. and COHEN, P. P. (1979). Cell-free synthesis and processing of a putative precursor for mitochondrial carbamylphosphate synthetase I of rat liver, *Proc. Natl. Acad. Sci. USA*, **76**, 5071–5.

MORI, M., MIURA, S., TATIBANA, M. and COHEN, P. P. (1980*a*). Processing of a putative precursor of rat liver ornithine transcarbamylase, a mitochondrial matrix enzyme, *J. Biochem.* (*Tokyo*), **88**, 1829–36.

MORI, M., MIURA, S., TATIBANA, M. and COHEN, P. P. (1980*b*). Characterization of a protease apparently involved in processing of pre-ornithine transcarbamylase of rat liver, *Proc. Natl. Acad. Sci. USA,*|**77**, 7044–8.

MORI, M., MIURA, S., TATIBANA, M. and COHEN, P. P. (1981). Cell-free translation of carbamylphosphate synthetase I and ornithine transcarbamylase messenger RNAs in rat liver. Effect of dietary protein and fasting on translatable *m*RNA levels, *J. Biol. Chem.*, **256**, 4127–32.

NICOLETTI, M., GUERRI, C. and GRISOLIA, S. (1977). Turnover of carbamylphosphate synthase, of other mitochondrial enzymes and of rat tissues. Effect of diet and of thyroidectomy, *Eur. J. Biochem.*, **75**, 583–92.

NUZUM, C. T. and SNODGRASS, P. J. (1971). Urea cycle adaptation to dietary protein in primates, *Science*, **172**, 1042–3.

RÄIHÄ, N. C. R. (1976). Developmental changes of urea-cycle enzymes in mammalian liver, in *The Urea Cycle* (Eds S. Grisolia, R. Baguena and F. Mayor), pp. 261–72, Wiley, New York.

RAIJMAN, L. and BARTULIS, T. (1979). Effect of ATP translocation on citrulline and oxaloacetate synthesis by isolated rat liver mitochondria, *Arch. Biochem. Biophys.*, **195**, 188–97.

RATNER, S. (1976). Enzymes of arginine and urea synthesis, in *The Urea Cycle* (Eds S. Grisolia, R. Baguena and F. Mayor), pp. 181–219, Wiley, New York.

SHIGESADA, K. and TATIBANA, M. (1978). *N*-Acetylglutamate synthetase from rat-liver mitochondria. Partial purification and catalytic properties, *Eur. J. Biochem.*, **84**, 285–91.

SHIGESADA, K., AOYAGI, K. and TATIBANA, M. (1978). Role of acetylglutamate in ureotelism. Variations in acetylglutamate level and its possible significance in control of urea synthesis in mammalian liver, *Eur. J. Biochem.*, **85**, 385–91.

SCHIMKE, R. T. (1962*a*). Differential effects of fasting and protein-free diets on levels of urea cycle enzymes in rat liver, *J. Biol. Chem.*, **237**, 1921–4.

SCHIMKE, R. T. (1962*b*). Adaptive characteristics of urea cycle enzymes in the rat, *J. Biol. Chem.*, **237**, 459–68.

SCHIMKE, R. T. (1963). Studies on factors affecting the levels of urea cycle enzymes in rat liver, *J. Biol. Chem.*, **238**, 1012–8.

SCHIMKE, R. T. (1964). The importance of both synthesis and degradation in the control of arginase levels in rat liver, *J. Biol. Chem.*, **239**, 3808–17.

SHORE, G. C., CARIGNAN, P. and RAYMOND, Y. (1979). *In vitro* synthesis of a putative precursor to the mitochondrial enzyme, carbamylphosphate synthetase, *J. Biol. Chem.*, **254**, 3141–4.

SNODGRASS, P. J., LIN, R. C., MULLER, W. A. and AOKI, T. (1978). Induction of urea cycle enzymes of rat liver by glucagon, *J. Biol. Chem.*, **253**, 2748–53.

SOBERON, G. and PALACIOS, R. (1976). Arginase, in *The Urea Cycle* (Eds S. Grisolia, R. Baguena and F. Mayor), pp. 221–35, Wiley, New York.

STUBBS, M. (1976). The effect of ethanol on ammonia metabolism, in *The Urea Cycle* (Eds S. Grisolia, R. Baguena and F. Mayor), pp. 468–78, Wiley, New York.

Tatibana, M. and Shigesada, K. (1976). Regulation of urea biosynthesis by the acetylglutamate-arginine system, in *The Urea Cycle* (Eds S. Grisolia, R. Baguena and F. Mayor), pp. 301–13, Wiley, New York.

Tatibana, M., Shigesada, K. and Mori, M. (1976). Acetylglutamate synthetase, in *The Urea Cycle* (Eds S. Grisolia, R. Baguena and F. Mayor), pp. 95–105, Wiley, New York.

Titheradge, M. A. and Haynes, Jr., R. C. (1980). The hormonal stimulation of ureogenesis in isolated hepatocytes through increases in mitochondrial ATP production, *Archiv. Biochem. Biophys.*, **201**, 44–55.

Triebwasser, K. C. and Freedland, R. A. (1977). The effect of glucagon on ureogenesis from ammonia by isolated rat hepatocytes, *Biochem. Biophys. Res. Comm.*, **76**, 1159–65.

Tsuda, M., Shikata, Y. and Katsunuma, T. (1979). Effect of dietary proteins on the turnover of rat liver argininosuccinate synthetase, *J. Biochem. (Tokyo)*, **85**, 699–704.

Windmueller, H. G. and Spaeth, A. E. (1980). Respiratory fuels and nitrogen metabolism *in vivo* in small intestine of fed rats. Quantitative importance of glutamine, glutamate and aspartate, *J. Biol. Chem.*, **255**, 107–12.

Wixom, R. L., Reddy, M. K. and Cohen, P. P. (1972). A concerted response of the enzymes of urea biosynthesis during thyroxine-induced metamorphosis of *Rana catesbeiana*, *J. Biol. Chem.*, **247**, 3684–92.

Wojtczak, A. B., Walajtys-Rode, E. I. and Geelen, M. J. H. (1978). Interrelations between ureogenesis and gluconeogenesis in isolated hepatocytes. The role of anion transport and the competition for energy, *Biochem. J.*, **170**, 379–85.

Yamazaki, R. K. and Graetz, G. S. (1977). Glucagon stimulation of citrulline formation in isolated hepatic mitochondria, *Arch. Biochem. Biophys.*, **178**, 19–25.

# 18

# Urea Metabolism

M. Walser

*Johns Hopkins University School of Medicine,
Baltimore, Maryland, USA*

Urea is the major end product of nitrogen metabolism in man, except under exceptional circumstances. It is also the nitrogenous end product whose rate of formation is subject to the greatest changes in response to both physiological and pathological stimuli. Accordingly the rate of ureagenesis plays a major role in determining nitrogen balance.

It has often been assumed that the rate of ureagenesis under any given set of circumstances is the rate appropriate to the current needs of the organism and is determined by the substrate load (amino acids and ammonia) and by induction or repression of the levels of urea cycle enzymes. As indicated herein, this is an oversimplification: recent evidence indicates that the rate of ureagenesis is rapidly autoregulated, independently of induction or repression mechanisms. Furthermore, there are clearly situations in which ureagenesis is either excessive or inadequate in relation to the needs of the organism.

## The Rate of Urea Degradation in Normal Man

Some decades ago, it was also generally believed that urea was an inert end product of metabolism in monogastric animals. It has been known since the start of this century, however, that urea could serve as a nitrogen source for the synthesis of amino acids in ruminants, replacing a substantial portion of dietary protein nitrogen (Stangel, 1967). It became evident that the conversion of urea-N to ammonia-N by bacterial urease in ruminants was a prerequisite to the utilization of urea-N, and it was also shown that ammonium salts could be effectively utilized by ruminants as a source of dietary nitrogen.

*M. Walser*

The breakdown of urea to ammonia and carbon dioxide in monogastric animals was demonstrated in a series of investigations beginning in 1948 (Leifer *et al.*, 1948; Chao and Tarver, 1953; Dintzis and Hastings, 1953; Kornberg and Davies, 1955; Liu *et al.*, 1955; Levenson *et al.*, 1959). These studies also established that urea breakdown was attributable to the action of bacterial urease in the gastrointestinal tract. However, little utilization of urea-N was demonstrable in the pig (Hoefer, 1967) or the chicken (Featherston, 1967).

Degradation of urea in man was demonstrated by Walser and Bodenlos (1959), on the basis of studies of the disappearance of intravenously injected $^{15}$N-urea or $^{14}$C-urea. The rate of urea metabolism in normal subjects averaged $20\%$ of urea production. After intestinal bacteriostasis, metabolism was virtually eliminated. These results have in general been confirmed in several studies (Table 1), whether expressed as the fraction of produced urea that is degraded, as extrarenal clearance of urea, or as the amount of nitrogen released each day by intestinal ureolysis.

TABLE 1
UREA METABOLISM IN NORMAL HUMAN SUBJECTS

| Reference | No. of subjects | Degradation g N/day | Extrarenal clearance litres/day |
|---|---|---|---|
| Walser and Bodenlos, 1959 | 3 | 3·5 | 14 |
| Jones *et al.*, 1969 | 2 | 3·4 | 19 |
| Murdaugh, 1970 | 2 | 2·6 | 36 |
| Walser and Dlabl, 1974 | 6 | 4·9 | 27 |
| Varcoe *et al.*, 1975 | 6 | 1·5 | 18 |
| Gibson *et al.*, 1976 | | | |
| 40 g protein | 6 | 2·9 | 31 |
| 100 g protein | 6 | 5·1 | 31 |
| Long *et al.*, 1978 | 3 | 3·2 | 22 |

A technical problem that may require some minor modification of these results in a quantitative sense is the reincorporation of labelled nitrogen and possibly labelled carbon into urea. There is every reason to expect this to occur with labelled nitrogen, since the ammonia released by ureolysis is likely to be a nitrogen source for the synthesis of urea. Evidence for such recycling of urea-N in man was presented by Walser and Bodenlos (1959). They showed that, when urea is reacted directly with hypobromite (instead of being converted first to an ammonium salt), the resulting $N_2$ gas contains the three isotopic forms, $^{30}N_2$, $^{29}N_2$, and $^{28}N_2$ in the same proportions as the three isotopic forms of urea, $H_2^{15}NCO^{15}NH_2$, $H_2^{14}NCO^{15}NH_2$, and $H_2^{14}NCO^{14}NH_2$ in the sample of urea reacted (Walser *et al.*, 1955). Ordinarily, the proportions of these three forms of nitrogen are randomly distributed according to the simple expression: (1) $[^{29}N_2]^2/[^{28}N_2][^{30}N_2] = k = 4$, and the same proportionality describes the relative concentrations of the three forms of urea. However, when a highly enriched sample of urea (or nitrogen gas) is mixed with a sample containing normal or low enrichment, $k \neq 4$, as can be readily shown algebraically. When highly enriched $^{15}N$-urea is administered, urea isolated from the blood or urine and reacted directly with hypobromite will produce a $N_2$ sample in which $k \neq 4$. This can be demonstrated by measuring the peak heights at mass 30, 29, and 28 in the mass spectrometer. As ureolysis proceeds, releasing some $^{15}NH_4$ and some $^{14}NH_4$, the newly synthesized urea will become progressively enriched and $k$ will approach 4. However, as long as some of the injected $^{15}N$-urea remains intact, this alteration in proportions of the three forms of urea in the blood can be detected.

Walser and Bodenlos (1959) demonstrated this phenomenon in man and estimated that 7 h after injection of $^{15}N$-urea 15 % of the circulating $^{15}N$ in urea was derived from recirculated labelled material and 85 % from the originally injected material. Hence the use of $^{15}N$-urea to measure the rate of urea metabolism can lead to substantial underestimation of this rate unless reincorporation of labelled material is taken into account, as it was, for example, in a later report by Walser (1970).

Reincorporation of the $^{14}C$ label into urea should be slight, since the $^{14}C$ should be diluted by the large bicarbonate pool. Evidence against $^{14}C$ reincorporation was presented by Koj *et al.* (1964) and Regoeczi *et al.* (1965). A more direct test of this question is a comparison of the rate of disappearance from the blood of injected $^{15}N$-urea, corrected for reincorporation of labelled material by measuring peak heights at masses 30, 29 and 28, with the rate of disappearance of simultaneously injected $^{14}C$-urea. Long *et al.* (1978) recently reported experiments of this type in normal subjects and patients with sepsis. They found that the slope of $^{14}C$-urea disappearance was similar to the slope of total $^{15}N$-urea disappearance (contrary to the report of Regoeczi *et al.*, 1965), but slower than the slope of unmetabolized $^{15}N$-urea (measured as the disappearance of $^{30}N_2$ evolved from urea). They estimated from these data that 63 % of urea carbon was recycled to urea. To explain this finding, they postulated that carbamate, which may be the first product of the urease reaction (Gorin,

1959), may be converted to carbamyl phosphate and thence to urea, either in the liver or the gut wall.

This proposal seems untenable on chemical grounds, for several reasons. Carbamate is highly unstable at pHs near neutrality, and when at equilibrium with both $NH_3$ and $CO_2$ (Faurholt, 1925) will be present at only minute concentrations in physiological fluids. Carbonic anhydrase catalyses the hydrolysis of carbamate (Gorin, 1959), rather than promoting its formation, as maintained by Long et al. (1978). Carbamate is a poor substrate for carbamyl phosphate synthetase; indeed, it either does not bind to the enzyme or reacts so slowly that it cannot compete with $CO_2$ (Jones, 1976). Transport of carbon as carbamino haemoglobin in portal venous blood, also mentioned by Long et al. (1978), is irrelevant, since such carbon would be in isotopic equilibrium with bicarbonate and $CO_2$, and furthermore would have no access to carbamyl phosphate synthetase without first becoming $CO_2$. Finally, the extent of urea synthesis by the gut (if it occurs at all) would be far too little to account for 63 % of the rate of ureagenesis.

Thus some other explanation must be found for the results reported by Long et al. (1978). One possibility is a technical error in the measurement of the peak at mass 30,—a critical element of this technique. Contamination of $N_2$ with NO is a major problem in the analysis of the gas evolved by the hypobromite reaction, and recent workers have recommended the use of a liquid $N_2$ trap to freeze this gas out before admitting the sample into the mass spectrometer inlet (Ross and Martin, 1970). This was evidently not done by Long et al. (1978), and conceivably their results may have been affected by this omission. Similarly, the earlier study of Walser (1970), in which the $^{30}N_2$ peak was measured without such a trap, yielded results in a single uraemic patient far out of line with subsequent data (Walser, 1980). Thus the evidence for significant recycling of urea carbon in man must be considered tenuous at best.

Some of the factors affecting the amount of urea hydrolysed in man have been studied. Gibson et al. (1976) showed that degradation increased from 2·9 g N/day to 5·1 g N/day on changing from a 40 g protein diet to a 100 g protein diet. They also showed that colectomy reduced but did not abolish urea degradation. An extremely high protein diet (400 g/day), consumed for years by an eccentric patient studied by Richards and Brown (1975), was associated with a rate of urea degradation of only 1 g N/day. The effect of a protein-free diet or starvation on urea degradation does not appear to have been studied, nor has the effect of urea loading.

In chronic renal failure the amount of urea-N converted to ammonia-N per day is usually similar to the amount converted in normal subjects (Fig. 1). This is difficult to understand, because urea concentration in body fluids is increased many times, and furthermore, there appears to be an adaptive increase in the urease activity of bowel contents in patients with chronic uraemia (Brown *et al.*, 1971). On the other hand, there is some evidence that high urea concentrations inhibit bacterial urease (Smith and Bryant, 1979). Measurements of urea degradation in acute renal failure have apparently not been reported.

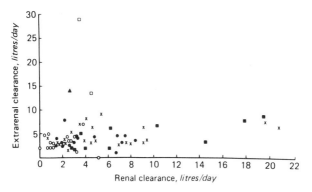

FIG. 1.  Extrarenal clearance of urea as a function of renal clearance of urea in patients with chronic renal failure, according to published reports. Normal values for these two quantities are approximately 25 litres/day (Table 1) and 80 litres/day. × Robson, 1964; □ Jones *et al.*, 1969; ▲ Walser, 1970; ○ Walser *et al.*, 1973; ■ Varcoe *et al.*, 1975; ● Mitch *et al.*, 1977. Reprinted with permission of author and publisher from Walser (1980).

Although the absolute rate of degradation of urea is unaltered in chronic renal failure, the fraction degraded may be greatly increased, to as much as 80 %. This is because renal clearance is reduced even more than is extrarenal clearance, so that the latter quantity becomes a larger fraction of total clearance.

In six patients with cirrhosis of the liver, urea degradation averaged 3·8 g N/day and extrarenal clearance averaged 17 litres/day—both values insignificantly different from normal (Weber, 1979).

From these results it is clear that most of the factors controlling the rate of urea degradation in man have yet to be elucidated. Even the common assumption that degradation should increase with rising blood urea concentration (and conversely) has not been substantiated.

# Biological Value of Urea Degradation in Man

The physiological significance of urea degradation has long been a subject of debate. Even before urea degradation in monogastric animals was established, it was known that ammonium salts could promote growth in animals fed low protein diets (Underhill and Goldschmidt, 1913; Foster *et al.*, 1939; Rittenberg *et al.*, 1939; Lardy and Feldott, 1949; Rose *et al.*, 1949; Sprinson and Rittenberg, 1949). Rose *et al.* (1949) also showed that urea had the same effect in rats. In man, urea supplements can provide a utilizable source of nitrogen for protein synthesis under certain circumstances, notably when diets adequate in essential amino acids but inadequate in non-essential nitrogen are fed (Watts *et al.*, 1964; Huang *et al.*, 1966; Scrimshaw *et al.*, 1966; Gallina and Dominguez, 1971; Jekat and Pabst, 1971). Positive nitrogen balance can be induced by urea supplements in children on low protein diets (Snyderman *et al.*, 1962) or in malnourished adults undergoing repletion (Tripathy *et al.*, 1970).

These and other observations led to the suggestion that urea-N could be reutilized (following degradation to ammonia-N) in uraemic patients, provided the diet contained adequate essential amino acids as such or as their N-free analogues, but inadequate total nitrogen (Schloerb, 1966; Richards *et al.*, 1967). The earlier suggestion of Giordano (1963) that urea-N could be used for protein synthesis in uraemic patients without prior hydrolysis is clearly inconsistent with our present knowledge of the reversibility of the urea cycle *in vivo*.

Studies of protein-N labelling after $^{15}$N-urea administration to patients or normal subjects (Richards *et al.*, 1967; Giordano *et al.*, 1968; Varcoe *et al.*, 1975) were initially interpreted as supporting this hypothesis until it became generally realized that such labelling merely demonstrates exchange, not net utilization (Richards, 1975).

When ureolysis is suppressed in uraemic patients on low protein diets appearance of urea-N ( = production minus degradation) should rise if the ammonia-N arising from urea breakdown was previously being used for protein synthesis to a significant degree. In fact, this does not occur (Mitch *et al.*, 1977) and nitrogen balance improves rather than worsens (Mitch and Walser, 1977). Furthermore, it should be possible to induce a negative urea appearance rate in uraemic patients by giving a diet inadequate in total nitrogen but adequate in carbon skeletons of essential amino acids. As reported recently by Abras and Walser (1981), this also fails to occur; urea-N appearance under these conditions becomes very low (about 0·6 g/day) but does not become negative.

Thus there is no evidence for reutilization of urea-N in uraemia and substantial evidence that it does not occur. Yet there are clearly situations in non-uraemic man in which urea supplements can improve nitrogen retention, as noted above.

There is also a substantial body of evidence indicating that antibiotics, which should suppress ureolysis, improve nitrogen retention not only in animals (Brause *et al.*, 1953; Hensley *et al.*, 1953; Rao and Sadhir, 1963), but also in malnourished children (Scrimshaw *et al.*, 1954), premature infants (Robinson, 1952; Coodin, 1953), normal children (Wetzel and Hopwood, 1954) and in young adults undergoing physical training (Haight and Pierce, 1954). The explanation of these observations is not yet at hand, but it is reasonably certain that antibiotics must exert their effects by some mechanism unrelated to bacterial urease.

From the above summary of studies of urea degradation, it is apparent that under most circumstances the enterohepatic cycling of urea-N, urea → ammonia → urea will neither result in loss of nitrogen from the body nor play a significant role in nitrogen balance.

## Urea Appearance

In contrast, when nitrogen derived from endogenous and exogenous protein is irreversibly converted to urea, it appears as such in urine and body fluids. Thus, the difference between urea production and recycled or metabolized urea has been termed urea appearance (Walser *et al.*, 1973) or 'net production'. This quantity, rather than total urea production, is one of the determinants of nitrogen balance. When blood urea concentration and volume of distribution is constant, urea appearance is equal to urea excretion. Steady-state blood urea concentration obviously varies with steady-state urea output. In renal failure urea can accumulate progressively or, when renal function improves, the body urea pool may diminish significantly. In anuria, accumulation obviously becomes equal to appearance. These considerations underscore why the definition of urea appearance must include both urea excretion and accumulation, whether the latter is a positive or negative quantity.

The time course of the response in the size of the urea pool (and thus in blood urea concentration) to a step change in the rate of urea appearance (as for example, when protein intake is changed) depends upon the turnover time of the pool. In normal subjects, the turnover time is measured in hours, so that blood urea concentration responds quite rapidly to a change in urea appearance. In renal failure, however, the turnover time is greatly prolonged and is measured in days. Consequently, a step change in urea appearance produces a progressive change in blood urea that approaches a new limiting value with a half-time that may be as long as a week. Hence accumulation, whether positive or negative, is quite commonly a significant fraction of total appearance in such patients, and thus may significantly affect nitrogen balance measurements.

Even in subjects with normal renal function, there are circumstances in which excretion may differ from appearance owing to changes in the urea pool. For example, several studies have documented the decline in urea excretion that follows the onset of fasting or ingestion of a protein-free diet (Adibi, 1971; Scrimshaw et al., 1972). The declining curve of urea excretion under such circumstances is usually taken as an accurate reflection of the rate of fall of net urea production. Actually, appearance falls much faster than these curves would suggest because the urea pool is decreasing, and hence urea accumulation is a negative quantity. Another source of error is the accumulation that occurs on intravenous amino acid loading. Under these circumstances, ignoring urea accumulation (as well as accumulation of amino acids themselves) makes nitrogen balance seem more positive than it is. Finally, accurate measurement of urea accumulation becomes the limiting factor in assessing hour-to-hour changes in urea appearance in normal subjects fed high-nitrogen loads. In one such study, Rafoth and Onstad (1975) showed that normal human subjects could produce at least 50 mg of urea/kg lean body mass/hour, with no evidence of a plateau being reached. Earlier work (Rudman et al., 1973), in which a maximal rate of urea synthesis in normal subjects was purportedly demonstrated, suffered from defects in experimental design such that synthesis appeared to reach a plateau before it was in fact maximal, as pointed out in later work from the same laboratory (Rypins et al., 1980).

## Determinants of Urea Appearance

The major determinant of the rate of urea appearance is the concentration of circulating amino acids. In one of the first studies of this relationship in man, Rafoth and Onstad (1975) administered protein loads to normal subjects, and measured the rate of urea formation as a function of the total plasma amino-N concentration. As Fig. 2 shows, their observations show an approximately linear relationship. A particularly interesting feature of this linear function is that when extrapolated, it has an intercept on the

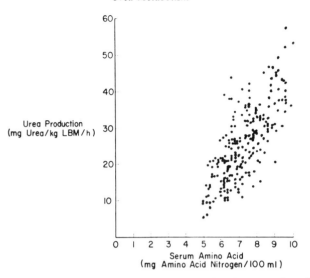

FIG. 2.   Urea appearance in normal subjects as a function of total serum amino acid nitrogen concentration. Hour-to-hour changes in urea excretion plus accumulation were measured following protein loads. As urea degradation was close to zero in these subjects, the results give production as well as appearance. Reprinted with permission of the authors and publisher from Rafoth and Onstad, (1975).

horizontal axis. This suggests that ureagenesis tends to autoregulate serum amino acid concentration, for the following reasons. An increase in amino acid concentration evidently produces a proportionately larger increase in the rate of ureagenesis, thereby restoring amino acid concentration towards normal. On the other hand, when protein intake is low, or carbohydrate intake is high, ureagenesis falls proportionately more than does the concentration of circulating amino acids, and appears to approach zero at an amino acid nitrogen concentration of about 4·5 mg/100 ml. In this way, progressive depletion of the circulating pool of amino acids during periods of protein deprivation or carbohydrate feeding may be minimized.

Indirect evidence that this intercept value for total amino acid concentration is nevertheless high enough to sustain protein synthesis has been obtained from observations of infants with complete congenital absence of one of the first two enzymes of the urea cycle (carbamyl phosphate synthetase or ornithine transcarbamylase). In these children, normal growth can be achieved by providing a low protein diet supplemented with a mixture of essential amino acids and their nitrogen-free analogues (ketoacids and hydroxyacids) (Brusilow et al., 1979). Total plasma amino acid concentration in these children is characteristically low, and ureagenesis is of course zero (except for urea derived from dietary arginine through the action of arginase). These patients evidently maintain high enough levels of circulating amino acids on this regimen to achieve growth without the necessity of converting any amino-N to urea (other than the guanidino-N of arginine). Obligatory ureagenesis is zero in these patients, in the sense that normal growth can be achieved and hyperammonaemia avoided as long as total circulating amino acid concentration remains within a certain (as yet ill-defined) range, with zero flux through the Krebs–Henseleit cycle.

Even in the presence of a normal urea cycle, 'obligatory' ureagenesis is nearly zero, as indicated by the studies in patients with chronic renal failure cited above (Abras and Walser, 1981). Total amino acid concentration in these patients is also low, though not as low as in children with urea cycle defects.

Data from these disparate groups of patients thus support, in a qualitative sense, the sort of relationship shown in Fig. 2 in that (1) the stimulus to ureagenesis virtually ceases at a low value of circulating amino acid concentration, and (2) an increment in circulating amino acid concentration induces a proportionately much larger increase in the rate of ureagenesis. At some point which is as yet undefined in normal man, ureagenesis becomes maximal. Presumably it is at this point that further increments in amino acid load induce clinically significant hyper-ammonaemia, as occurs in rats (Stewart and Walser, 1980a).

## Autoregulatory Mechanism of Ureagenesis

The biochemical mechanism by which this sort of autoregulation of amino acid levels is achieved has been clarified recently. N-acetyl-glutamate (NAG), an activator of mitochondrial carbamyl phosphate synthetase (CPS I), has long been suspected of playing a role in the regulation of ureagenesis (Tatibana and Shigesada, 1976; Hensgens et al., 1980). In the absence of NAG, CPS I is completely inactive. It is continuously synthesized intramitochondrially and evidently leaks through the mito-chondrial membrane to be destroyed by cytosolic enzymes. Saheki et al. (1978) and Hensgens et al. (1980) showed that its concentration can rise several-fold within minutes in isolated liver or hepatocytes following addition of $NH_4Cl$. Stewart and Walser (1980a) showed that complete amino acid loads injected intraperitoneally into rats induced a fivefold increase in liver NAG concentrations within 5 min. A resulting threefold increase in CPS I activity of intact mitochondria ensued. Presumably ureagenesis increased to the same degree (though this was not shown). When progressively larger amino acid loads were injected, a point was reached at which no further increase in CPS I activity occurred (even though NAG continued to increase). At this point, the CPS I activity of disrupted mitochondria, exposed to optimal concentrations of NAG, was the same as the CPS I activity of intact mitochondria. Further increments in amino acid dosage induced hyperammonaemia.

The possible role of arginine, a known activator of NAG synthetase (Tatibana and Shigesada, 1976), in these effects was examined by comparing the results in rats injected with a complete amino acid mixture with the results in rats given a mixture devoid of arginine. No differences were detected in the response of NAG levels or CPS I activity.

In searching for the stimulus to increased NAG synthesis, it was found that liver glutamate also rose within minutes in a dose-dependent fashion. Since glutamate is one of the substrates for NAG synthesis, the hypothesis was proposed that the urea cycle is activated following protein loads by the resulting increase in liver glutamate, which facilitates its own conversion to urea-N by this mechanism involving NAG and CPS I.

Additional evidence for the importance of this mechanism was adduced by showing that its blockade leads to hyperammonaemia following moderate amino acid loads (Stewart and Walser, 1980b). Propionyl CoA is known to be a competitive inhibitor of acetyl CoA in the synthesis of NAG (Coude *et al.*, 1979). Large doses (10–20 mmol/kg) of propionate were shown to induce hyperammonaemia in rats, especially when given along with moderate loads of amino acids. This effect was associated with a reduction in liver NAG and CPS I activity. Thus, blockade of amino acid-induced activation of ureagenesis can induce intolerance to amino acid loads which do not induce hyperammonaemia in normal animals. Presumably, the hyperammonaemia often observed in infants with propionic acidaemia (Rosenberg, 1978) is attributable to the same mechanism, as first proposed by Bachmann (1974).

In contrast to this rapidly activated mechanism, induction of the enzymes of the urea cycle is a slow process requiring days or at least hours (Das and Waterlow, 1974). It appears that this mechanism provides long-term modulation of ureagenesis in response to the demands of the organism, while the NAG mechanism facilitates the ureagenic response to single protein-containing meals. The fundamental function of both mechanisms may be the same—namely, to maintain circulating amino acid concentration within the optimal physiological range despite variations in protein intake or amino acid catabolism, whether the latter results from increased amino acid release from the periphery or from increased hepatic utilization of amino acids for gluconeogenesis.

## Acknowledgement

This work was supported by a grant (AM 18020) from the National Institute of Health.

## Discussion

*Lund* (*Radcliffe Infirmary, Oxford, UK*): For the control system which you have described there has to be a very rapid increase in mitochondrial glutamate. It occurs to me that the rapid switching on of glutaminase by ammonium ions might be the signal for increased synthesis of acetyl glutamate.

*Walser:* I like that idea.

*Cohen* (*University of Wisconsin-Madison, USA*): I am curious why you picked propionate as an inhibitor. When Grisolia and I were first studying the derivatives of glutamate as activators of CPS I, it turned out that propionyl glutamate was a reasonably good activator, with a $K_m$ roughly one tenth that of acetyl glutamate (Grisolia and Cohen, 1953). Pent-4-enoate, which forms a CoA derivative that is not utilizable, produces a very marked inhibition of ammonia utilization and urea biosynthesis (Aoyagi *et al.*, 1979).

*Walser:* I am very well aware of those experiments. Although I did not show the data, we did in fact measure the levels of $N$-propionyl glutamate in our experiments and found them to be very low. Therefore under these conditions propionyl glutamate was not in fact playing any role as an activator of ureagenesis. As you suggest, pent-4-enoate might have been a better inhibitor.

*Nyhan* (*University of California, San Diego, USA*): The effects of propionate are biologically important because it accumulates in children with propionicacidaemia. We have shown that propionyl CoA is a substrate for the synthesis of propionyl glutamate (Coude *et al.*, 1979) which, as you have indicated, is a relatively trivial activator of CPS, whereas propionyl CoA itself is a phenomenally active inhibitor. It fits in, therefore, that these children get lethal hyperammonaemia.

*Garlick* (*London School of Hygiene and Tropical Medicine, UK*): Looking at the question from the point of view of the whole animal, what do you think is the overall rate-limiting step in urea production? It seems to me that there might be two possible mechanisms. The first would be regulation at the stage of oxidation of individual amino acids, so that amino-N is merely 'driven down' to urea. In the whole animal this would require that the oxidation of all amino acids would have to be integrated in some way. The second possible rate-limiting step would be at the other end, by regulation of urea synthesis, so that amino-N is 'drawn out' of the amino acids.

*Walser:* If glycine is taken as an example of what Jackson and Golden call a deaminated amino acid (see p. 204), you are asking whether regulation of glycine catabolism is accomplished by the enzymes specific to this pathway, and the resulting ammonia is inevitably and quantitatively converted to urea; or whether ureagenesis from ammonia is regulated and glycine catabolism is secondarily affected. I cannot answer that particular question. However, I would expect the metabolism of the transaminating amino acids to be affected secondarily, since the transaminases are in general ubiquitous and probably are near equilibrium.

It is necessary to avoid confusion between 'rate-limiting' and 'rate-controlling'. The former term is applied only to the maximum rate. There is obviously a maximal rate for CPS; it may well be that argininosuccinate synthetase activity reaches a maximum at about the same point, but we do not know for certain.

*Cohen:* In response to Garlick, the enzyme involved in utilization of ammonia for urea biosynthesis (CPS I) is indifferent to whether the ammonia comes from amino acids or from ammonium chloride. A separate problem relates to the extent to which a given amino acid contributes its nitrogen to form aspartate via the ammonia pool, or by transamination.

*Reeds* (*Rowett Research Institute, Aberdeen, UK*): Dr Walser, have you any views on how, within your hypothesis, energy intake might influence the control of ureagenesis?

*Walser:* It is clear that the rate of ureagenesis is not simply a function of amino acid concentration. We know from the work of Felig and others that carbohydrate and insulin reduce the circulating concentrations of amino acids, so we can allow for an effect of energy intake without abandoning our thesis that amino acid concentration is the major determinant. Certainly there are also direct hormonal effects on ureagenesis, e.g. by glucagon. The level of acetyl CoA is probably also very important, by virtue of its being a substrate for the synthesis of *N*-acetyl glutamate. This may be another major link with energy metabolism.

*Reeds:* I was glad that you said that, because we have been doing some experiments in which we have been acutely altering the fat intake of pigs and looking at urea synthesis. Despite the fact that increasing fat intake causes an immediate fall in the amino acid concentrations in arterial blood, urea synthesis is, if anything, stimulated for the first 12 h after the change in diet. I have been wondering whether this was due to a change in the concentration of acetyl CoA.

*Walser:* It could well be.

# References

ABRAS, E. and WALSER, M. (1981). Optimization of N utilization in chronic renal failure by continuous enteral alimentation, *Clin. Res.* (in press).
ADIBI, S. A. (1971). Alteration in the urinary excretion rate of amino acids and nitrogen by dietary means in obese and normal human subjects, *J. Lab. Clin. Med.*, **77**, 278–89.
AOYAGI, K., MORI, M. and TATIBANA, M. (1979). Inhibition of urea synthesis by pent-4-enoate associated with decrease in *N*-acetyl-L-glutamate concentration in isolated rat hepatocytes, *Biochim. Biophys. Acta*, **587**, 515–21.
BACHMANN, C. (1974). Urea cycle, in *Heritable Disorders of Amino Acid Metabolism* (Ed. W. L. Nyhan), John Wiley and Sons, Inc., New York, pp. 361–86.
BRAUSE, R., WALLACE, H. C. and CUNHA, T. J. (1953). The value of antibiotics in the nutrition of swine: A review, *Antibiot. Chemother.*, **3**, 127–31.
BROWN, C. L., HILL, M. J. and RICHARDS, P. (1971). Bacterial ureases in uraemic men, *Lancet*, **2**, 406–7.
BRUSILOW, S. B., BATSHAW, M. L. and WALSER, M. (1979). Use of ketoacids in inborn errors of urea synthesis, in *Nutritional Management of Genetic Disorders*, (Ed. M. Winick), John Wiley and Sons, Inc., New York.
CHAO, F. C. and TARVER, H. (1953). Breakdown of urea in the rat, *Proc. Soc. Exp. Biol. (NY)*, **84**, 406.
COODIN, F. J. (1953). Studies of terramycin in premature infants, *Pediatrics*, **13**, 462–75.
COUDE, F. X., SWEETMAN, L. and NYHAN, W. L. (1979). Inhibition by propionyl-coenzyme A of *N*-acetylglutamate synthetase in rat liver mitochondria, *J. Clin. Invest.*, **64**, 1544–51.
DAS, T. K. and WATERLOW, J. C. (1974). The rate of adaptation of urea cycle

enzymes, aminotransferases and glutamic dehydrogenase to changes in dietary protein intake, *Br. J. Nutr.*, **32**, 353–73.

DINTZIS, R. Z. and HASTINGS, A. B. (1953). The effect of antibiotics on urea breakdown in mice, *Proc. Nat. Acad. Sci.*, **39**, 571.

FAURHOLT, C. (1925). *J. Chem. Phys.*, **22**, I.

FEATHERSTON, W. R. (1967). Utilization of urea and other sources of non-protein nitrogen by the chicken, in *Urea as a Protein Supplement* (Ed. Michael H. Briggs), Pergamon Press Ltd, London.

FOSTER, G. L., SCHOENHEIMER, R. and RITTENBERG, D. (1939). Studies in protein metabolism: V. The utilization of ammonia for amino acid and creatine formation in animals, *J. Biol. Chem.*, **127**, 319–27.

GALLINA, D. L. and DOMINGUEZ, J. M. (1971). Human utilization of urea nitrogen in low calorie diets, *J. Nutr.*, **101**, 1029–35.

GIBSON, J. A., PARK, N. J., SLADEN, G. E. and DAWSON, A. M. (1976). The role of the colon in urea metabolism in man, *Clin. Sci. Mol. Med.*, **50**, 51–9.

GIORDANO, C. (1963). Use of exogenous and endogenous urea for protein synthesis in normal and uremic subjects, *J. Lab. Clin. Med.*, **62**, 231–46.

GIORDANO, C., DEPASCUALE, C., BALLESTEIRI, C., CITTADINI. D. and CRESCENZI, A. (1968). Incorporation of urea [15]N in amino acids of patients with chronic renal failure on low nitrogen diets, *Am. J. Clin. Nutr.*, **21**, 394–404.

GORIN, G. (1959). On the mechanism of urease action, *Biochem. Biophys. Acta*, **34**, 268–70.

GRISOLIA, S. and COHEN, P. P. (1953). Catalytic rôle of glutamate derivatives in citrulline biosynthesis, *J. Biol. Chem.*, **204**, 753–7.

HAIGHT, T. H. and PIERCE, W. E. (1954). Influence of small doses of antibiotics on weight behaviour of young adults, *J. Lab. Clin. Med.*, **44**, 807–8.

HENSGENS, H. E. S. J., VERHOVEN, A. J. and MEIJER, A. J. (1980). The relationship between intramitochondrial *N*-acetylglutamate and activity of carbamoyl-phosphate synthetase (ammonia), *Eur. J. Biochem.*, **107**, 197–205.

HENSLEY, G. W., CARROL, R. W., WILCOX, E. L. and GRAHAM, W. R. (1953). The effect of aureomycin and methionine supplements fed to rats receiving soybean meals, *Arch. Biochem. Biophys.*, **45**, 270–4.

HOEFER, J. A. (1967). The effect of dietary urea in the pig, in *Urea as a Prótein Supplement*. (Ed. Michael J. Briggs), Pergamon Press Ltd., London.

HUANG, P. C., YOUNG, V. R., CHOLAKOS, B. and SCRIMSHAW, N. S. (1966). Determination of the minimum dietary essential amino acid-to-total ratio for beef protein fed to young men, *J. Nutr.*, **90**, 416–22.

JEKAT, F. and PABST, N. I. Unpublished data cited by Kofranyi, E. (1971). in *Proteins and Food Supplies in the Republic of South Africa* (Ed. J. W. Claasens and H. J. Potgeiter), A. A. Balkema, Capetown.

JONES, E. A., SMALLWOOD, R. A., CRAIGIE, A. and ROSENOER, V. M. (1969). The enterohepatic circulation of urea nitrogen, *Clin. Sci.*, **37**, 825–36.

JONES, M. E. (1976). Partial reactions of carbamyl-P synthetase: A review and an inquiry into the role of carbamate, in *The Urea Cycle*, (Ed. S. Grisolia, R. Baguena and F. Mayor), John Wiley and Sons, Inc., New York.

KOJ, A. E., REGOECZI, E. and MCFARLANE, A. S. (1964). Endogenous catabolism of urea in relation to measurement of plasma-protein synthesis, *Biochem. J.*, **91**, 26.

KORNBERG, H. L. and DAVIES, R. E. (1955). Gastric urease, *Physiol. Rev.*, **35**, 169.

LARDY, H. A. and FELDOTT, G. (1949). The net utilization of ammonium nitrogen by the growing rat., *J. Biol Chem.*, **179**, 509–10.

LEIFER, E., ROTH, L. J. and HEMPLEMAN, L. H. (1948). Metabolism of $C^{14}$ labelled urea, *Science*, **108**, 748.

LEVENSON, S. M., CROWLEY, L. V., HOROWITZ, R. E. and MALM, O. J. (1959). The metabolism of carbon labelled urea in the germfree rat, *J. Biol. Chem.*, **234**, 2061–2.

LIU, C. H., HAYS, V. W., SVEC, H. J., CATRON, D. V., ASHTON, G. C. and SPEER, V. C. (1955). The fate of urea in growing pigs. *J. Nutr.*, **57**, 241.

LONG, C. L., JEEVANANDAM, M. and KINNEY, J. M. (1978). Metabolism and recycling of urea in man, *Am. J. Clin. Nutr.*, **31**, 1367–82.

MEIJER, A. J. (1979). Regulation of carbamoyl-phosphate synthase (ammonia) in liver in relation to urea cycle activity, TIBS, 83–6.

MITCH, W. E., LEITMAN, P. S. and WALSER, M. (1977). Effects of oral neomycin and kanamycin in chronic uremic patients: I Urea metabolism, *Kidney Int.*, **11**, 116–22.

MITCH, W. E. and WALSER, M. (1977). Effects of oral neomycin and kanamycin in chronic uremic patients: II Nitrogen balance, *Kidney Int.*, **11**, 123–7.

MURDAUGH, H. V. (1970). Urea metabolism and low protein intake: Studies in man and dog, in *Urea and the Kidney* (Ed. B. Schmidt-Nielsen), Excerpta Medica Press, Amsterdam.

RAFOTH, R. J. and ONSTAD, G. R. (1975). Urea synthesis after oral protein loading in man, *J. Clin. Invest.*, **56**, 1170–4.

RAO, B. S. and SADHIR, P. P. (1963). Studies on growth and metabolism of albino rats with penicillin supplements, *Ann. Biochem. Exp. Med.*, **23**, 391–4.

REGOECZI, E., IRONS, L., KOJ, A. and MCFARLANE, A. S. (1965). Isotopic studies of urea metabolism in rabbits, *Biochem. J.*, **95**, 521–32.

RICHARDS, P. (1975). Nitrogen recycling in uremia: a reappraisal, *Clin. Nephrol.*, **3**, 166–7.

RICHARDS, P. and BROWN, C. L. (1975). Urea metabolism in an azotaemic woman with normal renal function, *Lancet*, **2**, 207–9.

RICHARDS, P., METCALF-GIBSON, A., WARD, E. E., WRONG, O. and HOUGHTON, B. J. (1967). Utilisation of ammonia nitrogen for protein synthesis in man, and the effect of protein restriction and uraemia, *Lancet*, **2**, 845–9.

RITTENBERG, D., SCHOENHEIMER, R. and KESTON, A. S. (1939). Studies in protein metabolism: IX. The utilization of ammonia by normal rats on a stock diet, *J. Biol Chem.*, **128**, 603–7.

ROBINSON, P. (1952). Controlled trial of aureomycin in premature twins and triplets, *Lancet*, **1**, 52.

ROBSON, A. M. (1964). Urea metabolism in chronic renal failure. *M.D. Thesis*, University of Newcastle upon Tyne.

ROSE, W. C., SMITH, L. C., WOMACK, M. and SHANE, M. (1949). The utilization of the nitrogen in ammonium salts, urea and certain other compounds *in vivo*, *J. Biol. Chem.*, **181**, 307–16.

ROSENBERG, L. E. (1978). Disorders of propionate, methylmalonate and cobalamin, in *The Metabolic Basis of Inherited Disease* (4th edn) (Ed. J. B. Stanbury, J. B. Wyngaarden and D. S. Fredrickson, McGraw-Hill, Inc., New York, 411–29.

Ross, P. J. and Martin, A. E. (1970). A rapid procedure for preparing gas samples for nitrogen-15 determination, *Analyst*, **95**, 817–22.

Rudman, D., Difulco, T., Galambos, J. T., Smith, R. B., Salam, A. and Warren, W. D. (1973). Maximal rates of urea synthesis in normal and cirrhotic subjects, *J. Clin. Invest.*, **52**, 2241–9.

Rypins, E. B., Henderson, J. M., Fulenwider, J. T., Moffitt, S., Galambos, J. T., Warren, W. D. and Rudman, D. (1980). A tracer method for measuring rate of urea synthesis in normal and cirrhotic subjects, *Gastroenterol.*, **78**, 1419–24.

Saheki, T., Ohkibo, T. and Katunuma, T. (1978). Regulation of urea synthesis in rat liver. Increase in the concentrations of ornithine and acetylglutamate in rat liver in response to urea synthesis stimulated by the injection of ammonium salt, *J. Biochem. (Tokyo)*, **84**, 1423–40.

Schloerb, P. R. (1966). Essential L-amino acid administration in uremia, *Am. J. Med. Sci.*, **252**, 650–9.

Scrimshaw, N. S., Guzman, M. A. and Tanden, O. B. (1954). Effect of aureomycin and penicillin on growth of Guatemalan school children, *Fed. Proc.*, **13**, 477.

Scrimshaw, N. S., Young, V. R., Schwartz, R., Piche, M. L. and Das, J. B. (1966). Minimum dietary essential amino acid-to-total nitrogen ratio for whole egg protein fed to young men, *J. Nutr.*, **89**, 9–18.

Scrimshaw, N. A., Hussein, M. A., Murray, E., Rand, W. M. and Young, V. R. (1972). Protein requirements of man: variations in obligatory urinary and fecal N losses in young men, *J. Nutr.*, **102**, 1595–1604.

Smith, C. J. and Bryant, M. P. (1979). Introduction to metabolic activities of intestinal bacteria, *Am. J. Clin. Nutr.*, **32**, 149–57.

Snyderman, S. E., Holt, L. E., Dancis, J., Roitman, E., Boyer, A. and Ballis, M. E. (1962). 'Unessential' nitrogen: a limiting factor for human growth, *J. Nutr.*, **78**, 57–71.

Sprinson, D. B. and Rittenberg, D. (1949). The rate of utilization of ammonia for protein synthesis, *J. Biol. Chem.*, **180**, 707–14.

Stangel, H. J. (1967). History of the use of urea in ruminant feeds, in *Urea as a Protein Supplement* (Ed. Michael H. Briggs), Pergamon Press Ltd., London.

Stewart, P. and Walser, M. (1980a). Short term regulation of ureagenesis, *J. Biol. Chem.*, **11**, 5270–80.

Stewart, P. and Walser, M. (1980b). Failure of the normal ureagenic response to amino acids in organic acid-loaded rats, *J. Clin. Invest.*, **65**, 484–92.

Tatibana, M. and Shigesada, K. (1976). Regulation of urea biosynthesis by the acetylglutamate-arginine system, in *The Urea Cycle* (Ed. S. Grisolia, R. Baguena, F. Mayor), John Wiley and Sons, Inc., New York.

Tripathy, K., Klahr, S. and Lotero, H. (1970). Utilization of exogenous urea nitrogen in malnourished adults, *Metabolism*, **19**, 253–62.

Underhill, F. P. and Goldschmidt, S. (1913). Studies on the metabolism of ammonium salts: III. The utilization of ammonium salts with a non-nitrogenic diet, *J. Biol. Chem.*, **15**, 341–55.

Varcoe, R., Halliday, D., Carson, E. R., Richards, P. and Tavill, A. S. (1975). Efficiency of utilisation of urea nitrogen for albumin synthesis by chronically uraemic and normal man, *Clin. Sci. Mol. Med.*, **48**, 379–90.

WALSER, M. (1970). Use of isotopic urea to study the distribution and degradation of urea in man, in *Urea and the Kidney* (Ed. B. Schmidt-Nielsen), Excerpta Medica Press, Amsterdam.

WALSER, M. (1980). Determinants of ureagenesis with particular reference to renal failure, *Kidney Int.*, **17**, 709–21.

WALSER, M. and BODENLOS, L. J. (1959). Urea metabolism in man, *J. Clin. Invest.*, **38**, 1617–26.

WALSER, M. and DLABL, P. (1974). Urea metabolism, in *Proc. Conference on Adequacy of Dialysis*, sponsored by the Artificial Kidney-Chronic Uremia Program, National Institute of Arthritis, Metabolism and Digestive Diseases, Monterey, California, March 20–22.

WALSER, M., GEORGE, J. and BODENLOS, L. J. (1955). Altered proportions of isotopes of molecular nitrogen as evidence for a monomolecular reaction, *J. Chem. Phys.*, **22**, 1146.

WALSER, M., COULTER, A. W., DIGHE, S. and CRANTZ, F. R. (1973). The effect of keto-analogues of essential amino acids in severe chronic uremia, *J. Clin. Invest.*, **52**, 678–90.

WATTS, J. H., TOLBERT, B. and RUFF, W. L. (1964). Nitrogen balances for young adult males fed two sources of total nitrogen intake, *Metabolism*, **13**, 172–80.

WEBER, F. L. JR. (1979). The effect of lactulose on urea and nitrogen metabolism in cirrhotic patients, *Gastroenterol.*, **77**, 518–23.

WETZEL, N. D. and HOPWOOD, H. H. (1954). Antibiotic supplements overcome growth failure in school children, *J. Agric. Food Chem.*, **2**, 1148–57.

# 19

# Urea Production Rates in Man in Relation to the Dietary Intake of Nitrogen and the Metabolic State of the Individual

A. A. Jackson, J. Landman

*Tropical Metabolism Research Unit,*
*University of the West Indies, Kingston, Jamaica*

and

D. Picou

*Mount Hope Medical Complex Task Force,*
*Port of Spain, Trinidad*

Urea nitrogen is the major end product of nitrogen metabolism in man. However, the urea is not excreted quantitatively, since a proportion of the urea produced is degraded in the bowel to ammonia, which is then available for further metabolic interaction (Walser and Bodenlos, 1959). This endogenous urea breakdown may act as a source of non-essential nitrogen. The extent to which the degraded urea nitrogen is recycled back to urea synthesis, or passes to other metabolic pathways can be measured using the method of Picou and Phillips (1972). They showed that the extent to which degraded urea nitrogen is recycled to urea production is directly related to the dietary intake of nitrogen.

This method has been used to measure urea nitrogen kinetics in three normal adult males on an intake of 240 mg N/kg/day and two uraemic patients on an intake of 80 mg N/kg/day and 40 mg N/kg/day. All diets provided a maintenance intake of energy. Table 1 shows that in the normal adult, in nitrogen balance on a normal protein intake, urea nitrogen production exceeds the dietary intake of nitrogen. The difference is accounted for by the urea nitrogen recycled to urea after degradation in the bowel. As nitrogen intake falls there is a decrease in the total urea production, but a greater proportion of the total production is degraded in

A. A. Jackson, J. Landman and D. Picou

## TABLE 1
### UREA PRODUCTION IN NORMAL AND URAEMIC ADULTS AND MALNOURISHED AND RECOVERED INFANTS

| Subjects | Diet | Intake (mg N/kg/day) | Produced | Degraded | Recycled to urea | Metabolized[a] | Excreted | Produced/ intake | Degraded/ produced | Metabolized/ degraded |
|---|---|---|---|---|---|---|---|---|---|---|
| | | | | | (mg urea N/kg/day) | | | | (%) | |
| *Infants* | | | | | | | | | | |
| Malnourished | HP | 592 | 293 | 117 | 40 | 77 | 176 | 49 | 40 | 66 |
| Malnourished | LP | 180 | 84 | 50 | 6 | 44 | 34 | 47 | 60 | 88 |
| Recovered | HP | 592 | 338 | 116 | 40 | 76 | 222 | 57 | 34 | 66 |
| Recovered | LP | 180 | 65 | 45 | 4 | 41 | 20 | 36 | 69 | 91 |
| *Adults* | | | | | | | | | | |
| Normal | | 240 | 285 | 77 | 47 | 30 | 208 | 119 | 27 | 39 |
| Uraemic | LP | 80 | 93 | 55 | 6 | 49 | 38 | 116 | 59 | 89 |
| Uraemic | VLP | 40 | 61 | 27 | 3 | 24 | 34 | 152 | 44 | 89 |

[a] Entering pathways other than urea synthesis.
LP, low protein; HP, high protein.

the bowel. This nitrogen, however, is not recycled to urea synthesis, but disappears to other metabolic pathways.

This picture can be compared with that seen in infants (see Table 1), all of whom were receiving supramaintenance intakes of energy and were therefore in a net anabolic state and positive nitrogen balance. Urea nitrogen production was only half the dietary nitrogen intake, and the proportion of the degraded urea nitrogen being recycled to urea synthesis, even on the higher protein intake, was less than 10% of the intake of nitrogen.

We would conclude therefore, that urea production responds to both the absolute dietary intake of nitrogen and to the overall metabolic state with respect to nitrogen balance. The proportion of urea nitrogen production which is degraded in the bowel is inversely related to the dietary intake of nitrogen provided the subject is not in negative nitrogen balance. This nitrogen is available for other metabolic pathways and on an adequate nitrogen intake tends to be resynthesized to urea.

## Discussion

*Walser (Johns Hopkins University, Baltimore, USA):* By what method did you measure urea recycled versus urea utilized?

*Jackson:* By the method of Picou and Phillips (1972). A constant infusion of doubly labelled urea was given and the production rate measured from the abundance of $^{15}N^{15}N$ urea in the urine. The amount recycled is determined from the abundance of $^{15}N^{14}N$ urea.

*Walser:* I am afraid I must disagree with your interpretation. Labelled ammonia released in the gut by urea breakdown becomes diluted in a large portal ammonia pool. It inevitably labels most transaminating amino acids, simply by exchange, thereby further lowering the ammonia specific activity. This will occur quite independently of any net utilization of ammonia-N for urea synthesis. Therefore the reappearance of the labelled material in newly synthesized urea underestimates the extent to which the ammonia derived from urea breakdown is simply reconverted to urea.

*Jackson:* Even though the labelled nitrogen from the gut is being diluted in a large ammonia pool, this does not prevent us from measuring the amount of urea synthesized from recycled nitrogen. These are the figures given in the sixth column of Table 1 of our paper.

*Walser:* Again I must point out the importance of exchange reactions. In a test tube containing glutamate dehydrogenase and its substrates and co-factors in concentrations such that the net flux is from glutamate to ammonia, the addition of labelled ammonia will result in labelled glutamate. This is not utilization of ammonia for glutamate synthesis, but simply a reflection of the fact that the reaction is reversible and therefore has both a forward and a backward flux.

*Jackson:* Yes, but the fact remains that some of the urea produced is unaccounted for. The most likely way for this nitrogen to disappear is that after liberation from urea in the gut it is taken up into amino acids and protein synthesis.

*Reeds (Rowett Research Institute, Aberdeen, UK):* Have you any idea what proportion of the urea recycled into the gut appears in faecal nitrogen due to fixation by bacteria in the hind-gut?

*Jackson:* Very little. In the constant infusion studies the amount of labelled nitrogen appearing in the faeces is less than 5 %, depending on the clinical state of the patient. In normal subjects it is virtually nil. In uraemic patients it may rise slightly, but it is a very small proportion of the flux.

*Waterlow (London School of Hygiene and Tropical Medicine, UK):* This discussion has been going on for years, and listening to it I have come to the conclusion that both are right. Obviously Walser is correct when he says that exchange of labelled material is not evidence of *net* transfer. On the other hand, I believe that Jackson is also justified in claiming that some of the ammonia produced by hydrolysis of urea in the gut is taken up into protein. The key question, as I see it, is whether or not this has any importance for the nitrogen economy of the animal. It seems to me that in the steady state it can have no importance. If ammonia derived from urea is taken up into non-essential amino acids (NEAA), an equivalent amount of endogenous NEAA must be catabolized to urea, otherwise NEAA would accumulate, and this does not happen. However, the uptake into amino acids of ammonia derived from urea might be important in a situation such as that of the malnourished child trying to grow on a marginal protein intake, if one postulates that there is a fixed obligatory rate of urea synthesis, which cannot fall to zero. In this situation the uptake of urea-N into amino acids and thence into protein would represent a genuine saving of nitrogen if, and only if, the supply of carbon skeletons was not limiting. This brings us back to the question of whether the rate of urea formation is ultimately determined by the activity of the enzymes which irreversibly oxidize the carbon|skeletons.

## References

PICOU, D. and PHILLIPS, M. (1972). Urea metabolism in malnourished and recovered children receiving a high or low protein diet, *Amer. J. Clin. Nutr.*, **25**, 1261–6.

WALSER, M. and BODENLOS, L. J. (1959). Urea metabolism in man, *J. Clin. Invest.*, **38**, 1617–26.

# 20

# Pathways of Glycine Metabolism

A. Neuberger

*Department of Biochemistry,*
*Charing Cross Hospital Medical School, London, UK*

Glycine is the most simple amino acid, it has no centre of asymmetry, and the $\alpha$-carbon atom carries no bulky substituents. This is of particular importance in protein structure since the glycine residue imposes a minimum of steric restraint on the construction of a complex protein molecule. Thus, in collagen the fact that every third residue is a glycine makes it possible to construct the coiled triple helix which is such a characteristic of this group of proteins. Various other examples could be given demonstrating the special properties of glycine which are derived from its lack of a bulky substituent.

## The Biosynthesis of Serine

The metabolism of glycine is intimately connected with that of serine, and at least in animals one can say that almost all the glycine is derived from serine. There are two important reactions by which serine can be formed— one from D-glycerate and one from 3-phosphoglycerate, both arising from glycolysis. 3-Phosphoglycerate is formed from 1,3-diphospho-D-glycerate; it also arises from photosynthesis in the operation of the Calvin cycle. D-Glycerate is of course readily formed from the phosphate derivative by dephosphorylation.

The relevant enzymes, the dehydrogenase, the transaminase and the enzyme which catalyses the dephosphorylation are widely distributed and have been studied in detail. In the glycerate pathway the dephosphorylation occurs first, but otherwise the enzymic processes are similar to the one described. Similar mechanisms operate in plants and in micro-organisms. Glycine in almost all animals is a non-essential amino acid, and from experiments which I carried out with Arnstein many years ago (Arnstein and Neuberger, 1953*a,b*), it could be calculated that the amount of glycine formed in animals and man is probably 10–50 times greater than the amount present in the average diet. In birds glycine improves growth under certain conditions, and its presence in the diet in significant amounts is thus desirable.

## The Formation of Glycine and Serine Associated with Photosynthesis

As already mentioned, in the Calvin cycle 3-phosphoglycerate is produced which is an effective precursor of serine, which then gives rise to glycine. Glycine is also produced in photosynthetic tissue especially in connection with photorespiration. The latter is defined as 'respiration', i.e. the evolution of carbon dioxide which differs from dark respiration and is specifically associated with the mechanisms characteristic of photosynthesis. Thus, photorespiration is closely linked with the so-called C3 plants. During photosynthesis glycollic acid is produced, the formation of

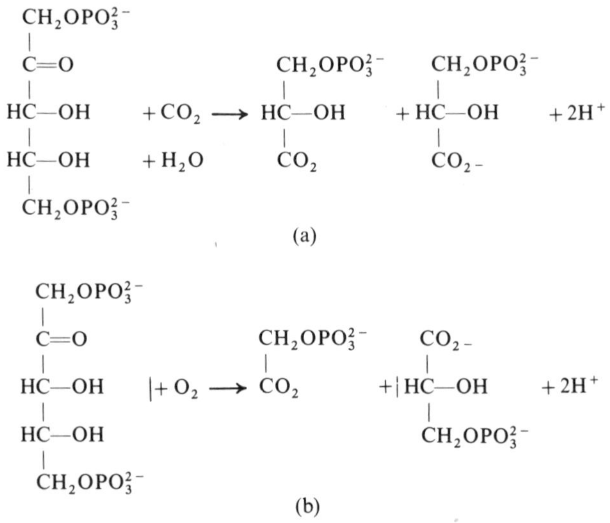

$$
\begin{array}{c}
CH_2OPO_3^{2-} \\
| \\
C{=}O \\
| \\
HC{-}OH \\
| \\
HC{-}OH \\
| \\
CH_2OPO_3^{2-}
\end{array}
\;+\,CO_2 \;\longrightarrow\;
\begin{array}{c}
CH_2OPO_3^{2-} \\
| \\
HC{-}OH \\
| \\
CO_2
\end{array}
\;+\;
\begin{array}{c}
CH_2OPO_3^{2-} \\
| \\
HC{-}OH \\
| \\
CO_2{-}
\end{array}
\;+\,2H^+
$$

(a)

$$
\begin{array}{c}
CH_2OPO_3^{2-} \\
| \\
C{=}O \\
| \\
HC{-}OH \\
| \\
HC{-}OH \\
| \\
CH_2OPO_3^{2-}
\end{array}
\Big|\,+\,O_2 \;\longrightarrow\;
\begin{array}{c}
CH_2OPO_3^{2-} \\
| \\
CO_2
\end{array}
\;+\Big|
\begin{array}{c}
CO_2{-} \\
| \\
HC{-}OH \\
| \\
CH_2OPO_3^{2-}
\end{array}
\;+\,2H^+
$$

(b)

FIG. 1.   The reactions of ribulose 1,5-diphosphate in photosynthesis. (a) The conversion of the ribulose derivative to two molecules of phosphoglycerate with the uptake of one molecule of $CO_2$. (b) The oxidative conversion of ribulose diphosphate to one molecule of 3-phosphoglycerate and one molecule of phosphoglycollate.

which is associated with the action of ribulose 1,5-diphosphate carboxylase. This important enzyme has two active sites: one which binds $CO_2$ and cleaves the ribulose diphosphate molecule with the incorporation of $CO_2$ giving rise to two molecules of 3-phosphoglycerate (Fig. 1a); the second site has an affinity for oxygen, and this oxidative reaction produces one molecule of 3-phosphoglycerate and one of phosphoglycollate (Fig. 1b). The phosphoglycollate can then be dephosphorylated to give glycollic acid which, after conversion to glyoxylate through the action of a specific transaminase, yields glycine.

## Glycine Formation through the Operation of the Glyoxylate Cycle

Various higher plants and many micro-organisms have a metabolic cycle called the glyoxylate cycle. In this series of reactions isocitrate is cleaved by the enzyme isocitrate lyase to give succinate and glyoxylate (Fig. 2). As

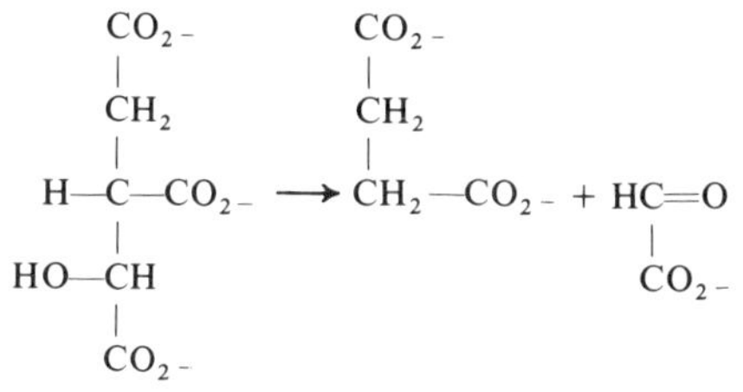

$$
\begin{array}{ccc}
CO_2{}^- & CO_2{}^- & \\
| & | & \\
CH_2 & CH_2 & \\
| & | & \\
H\!-\!C\!-\!CO_2{}^- \longrightarrow & CH_2\!-\!CO_2{}^- & +\ HC\!\!=\!\!O \\
| & & | \\
HO\!-\!CH & & CO_2{}^- \\
| & & \\
CO_2{}^- & &
\end{array}
$$

FIG. 2. The formation of glyoxylate in the glyoxylate cycle.

mentioned above, glyoxylate can then be transaminated to give glycine. The last reaction can occur in animals, but the formation of glyoxylate is quantitatively unimportant, and thus this reaction is unlikely to make a significant contribution to the total amount of glycine synthesized in mammals.

## The Reversible Conversion of Serine to Glycine

This conversion is catalysed by an enzyme called L-serine hydroxymethyl transferase, or L-serine tetrahydrofolate 5,10-methylene transferase. The enzyme is widely distributed in animals, plants and micro-organisms, where it is found both in the cytosol and mitochondria. The reaction is as follows:

Tetrahydrofolate + L-serine $\rightleftharpoons$
5,10-methylene-tetrahydrofolate + glycine

The enzyme catalyses the transfer of a $C_1$ moiety of the oxidation stage of formaldehyde from a tetrahydrofolate derivative to the $\alpha$-carbon atom of glycine. It is an enzyme which requires pyridoxal phosphate as a co-factor, and the mechanism of its action and the stereochemistry of the reaction have been fully investigated by Akhtar (Jordan and Akhtar, 1970; Akhtar *et al.*, 1975). The physiological importance of this enzyme is two-fold. On the one hand it is probably the most important reaction for the formation of glycine, but in addition, it also supplies 1-carbon units which are necessary for a very large number of 1-carbon biological reactions. It has become increasingly clear that quantitatively the main source of 1-carbon units is indeed the $\beta$-carbon atom of serine and the $\alpha$-carbon atom of glycine.

## The Glycine Cleavage Complex

Nearly 20 years ago Sagers and Gunsalus (Sagers and Gunsalus, 1961; Klein and Sagers, 1966*a,b*; 1967*a,b*) discovered a series of reactions in a micro-organism which degrades glycine by a complicated series of reactions as shown in Fig. 3. The enzyme complex is widely distributed in nature and in recent years Kikuchi (1973) has demonstrated the presence of a similar complex in mammalian liver. The reaction is catalysed by four different proteins named $E_1$ (a pyridoxal phosphate containing serine hydroxymethylase), $E_2$, $E_3$ and $E_4$. The co-factors involved are: pyridoxal phosphate, a compound containing a disulphide group, FAD, NAD, and tetrahydrofolate. The first step consists of a reaction of glycine with pyridoxal-5-phosphate leading to a Schiff's base, which is then decarboxylated to give bicarbonate and an active methylene group which is bound to $E_3$. This cleavage of the C–C bond is associated with a reduction of an S–S group to give vicinyl sulphydryl groups. The active methylene group becomes a hydroxymethyl group by the addition of water, and is then condensed with the second molecule of glycine to give serine. A flavoprotein $E_4$ is then responsible for the formation of an S–S group in $E_2$, and this results in the transfer of electrons to NAD. The oxidation of two molecules of glycine then leads to the formation of one molecule of serine, one of $CO_2$ and one of ammonia, as shown in Fig. 4. The system, working under

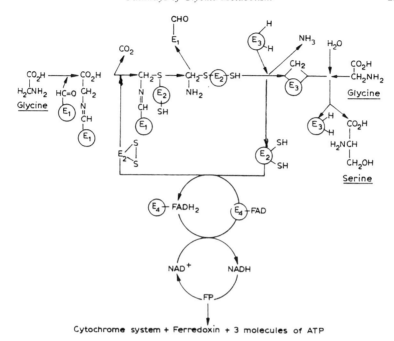

FIG. 3. The operation of the glycine cleavage complex. The conversion of two molecules of glycine to one molecule of serine is shown with the liberation of one molecule of $CO_2$ and one of ammonia. No free $C_1$ fragment appears in this conversion, the $C_1$ fragment originating from the α-carbon atom from one of the glycine molecules goes off and it appears as a complex of the $CH_2$ moiety and $E_3$, which directly combines with the second glycine molecule to give serine. The aldehydic function combines with $E_1$ and is the aldehyde group of pyridoxal phosphate.

oxidative conditions, has been calculated to produce three molecules of ATP for each molecule of serine formed. In the presence of oxygen the reaction will go largely in the direction of degradation of glycine and the formation of serine. On the other hand, by a combination of mitochondrial and cytosol reactions active formaldehyde will be produced as discussed previously:

(Mitochondrion) 2 Glycine $+ \frac{1}{2}O_2 \rightarrow 1$ Serine $+ CO_2 + NH_3$

(Cytosol) 2 Serine $\rightarrow 2$ Glycine $+ 2$ 'active' formaldehyde

Sum:

1 Serine $+ \frac{1}{2}O_2 \rightarrow 2$ 'active' formaldehyde $+ CO_2 + NH_3$

FIG. 4. The formation of active formaldehyde from glycine and serine.

In the absence of oxygen, the degradation of glycine would give, apart from serine, one molecule each of ammonia and $CO_2$, and also two electrons which might be used for a variety of other metabolic reactions.

In the so-called non-ketotic hyperglycinaemia, the plasma glycine is greatly increased and the metabolic block involved is almost certainly either the absence of the glycine cleavage system in the liver, or a great reduction in its activity. It is clear that this enzyme complex represents probably the most important mechanism for the degradation of glycine in animals.

## Conclusion

Glycine and serine are exceptional in that they are non-essential amino acids which are substrates for an unusually large number of anabolic reactions. As an example of this, glycine produces one out of the four nitrogen atoms of purines and two out of the five carbon atoms; two others are derived from 1-carbon units which again are formed from glycine or serine. Serine is also the precursor of the ethanolamine moiety of phospholipids, of the choline part of phosphatidyl chloline, of sphingosines and of related nitrogen-containing lipids. Serine again is important in the metabolism of cysteine and methionine through its incorporation into cystathionine. Glycine is also a component of glutathione. In animals and micro-organisms glycine is a precursor of aminolaevulinic acid (ALA), from which both porphyrins and chlorophyll are derived, whilst in green plants ALA appears to be mainly formed from glutamate. Finally, glycine is important in detoxication reactions and in the formation of creatine and creatinine. Thus, these two amino acids, which are so closely related metabolically, are involved in the biosynthesis of a very large number of nitrogen-containing components of living tissues.

However, although the metabolism of glycine and serine in mammals is known in its general outline, there is urgent need for further research on quantitative aspects. We have some knowledge of the amount of glycine which is required for porphyrin synthesis, but we know very little about the amounts of these amino acids used in the formation of 1-carbon units, or in the formation of nucleic acids. In fact, we lack any quantitative understanding of how far, for instance, methionine and choline are required for sources of methyl groups in man, and how much can be supplied by the two non-essential amino acids. Nor do we know exactly how ethanolamine and choline are formed, and we certainly have no exact quantitative information. There is thus a large field of further research which would involve a great variety of important biological compounds.

## Discussion

*Jackson* (*University of the West Indies, Kingston, Jamaica*): Do you think that glyoxylate has any significant role to play in glycine metabolism? The aminotransferase enzyme has been described for converting alanine and glyoxylate to pyruvate and glycine (Noguchi and Takada, 1979). This is one way in which glycine could potentially be synthesized without going through serine.

*Neuberger:* In C3 plants, when photorespiration occurs, glycollic acid is produced, which is easily oxidized to glyoxylic acid, and this is transformed to glycine. The loss of $CO_2$ in photorespiration is attributed nowadays to the working of the glycine cleavage complex in the mitochondria. I am sure that a large part of the glycine in plants is formed from glyoxylic acid. In micro-organisms, where the glyoxylic acid cycle operates, glycollic acid is again an important source. In animals the only sources of glyoxylic acid of which I am aware are the catabolism of ascorbic acid and of hydroxyproline. Otherwise I know of no obvious pathways by which glyoxylic acid can be formed, other than by the action of glycine oxidase on glycine.

*Munro* (*Tufts University, Boston, USA*): Glycine gives rise to glyoxylic acid and then to oxalate (Krebs, 1964), and I wonder if in cases of hyperglycinaemia there is any tendency to generate oxalic acid stones.

*Nyhan* (*University of California, San Diego, USA*): No, but it may be relevant that some years ago we studied some of these patients with labelled glycine and labelled glyoxylate (Ando *et al.*, 1968; Gerritsen, *et al.*, 1969). The upshot was that in man there is ready conversion of administered glyoxylate to glycine, but essentially no conversion of glycine to glyoxylate.

*Garlick* (*London School of Hygiene and Tropical Medicine, UK*): To emphasize the point Neuberger made about the interconversion of serine and glycine: some years ago Fern and I infused glycine in rats for 6 hours, and found that the serine in liver became at least as highly labelled as the glycine (Fern and Garlick, 1974); in fact, it has a slightly higher specific activity, reflecting the fact that the third carbon atom of serine is also derived from glycine. I think the conversion is mostly in the liver because in other tissues the specific activity of serine was much lower.

*Felig* (*Yale University School of Medicine, USA*): In prolonged fasting we found that glycine was the only amino acid whose concentration in plasma increased. This is very different from protein malnutrition, where the levels of the non-essentials go up as a group compared with the essentials. Why is there this unique response of glycine?

*Young (Massachusetts Institute of Technology, USA):* That is difficult to answer, because, as Neuberger indicated, there is very little information about the rate of glycine synthesis in man. We have recently measured the glycine flux with $^{15}$N-glycine in healthy young adults and find that the rate of *de novo* glycine synthesis is about 350 $\mu$mol/kg/h. This rate is markedly influenced by the adequacy of total protein intake (Gersovitz *et al.*, 1980).

# References

ANDO, T., NYHAN, W. L., GERRITSEN, T., GONG, L., HEINER, D. C. and BRAY, P. F. (1968). Metabolism of glycine in the nonketotic form of hyperglycinemia, *Pediat. Res.*, **2**, 254–63.

AKHTAR, M., EL-OBEID, H. A. and JORDAN, P. M. (1975). Mechanistic, inhibitory and stereochemical studies on cytoplasmic and mitochondrial serine transhydroxymethylases, *Biochem. J.*, **145**, 159–68.

ARNSTEIN, H. R. V. and NEUBERGER, A. (1953*a*). The effect of cobalamin on the quantitative utilization of serine, glycine and formate for the synthesis of choline and methyl groups of methionine, *Biochem. J.*, **55**, 259–71.

ARNSTEIN, H. R. V. and NEUBERGER, A. (1953*b*). The synthesis of glycine and serine by the rat, *Biochem. J.*, **55**, 271–80.

FERN, E. B. and GARLICK, P. J. (1974). The specific radioactivity of the tissue free amino acid pool as a basis for measuring the rate of protein synthesis in the rat *in vivo*, *Biochem. J.*, **142**, 413–19.

GERRITSEN, T., NYHAN, W. L., REHBERG, M. and ANDO, T., (1969). Metabolism of glyoxylate in nonketotic hyperglycinemia, *Pediat. Res.*, **3**, 269–74.

GERSOVITZ, M., BIER, D., MATTHEWS, D., UDALL, J., MUNRO, H. N. and YOUNG, V. R. (1980). Dynamic aspects of whole body glycine metabolism: influence of protein intake in young adults and elderly males, *Metabolism*, **29**, 1087–94.

JORDAN, P. M. and AKHTAR, M. (1970). The mechanism of action of serine transhydroxymethylase, *Biochem. J.*, **116**, 277–86.

KIKUCHI, G. (1973). The glycine cleavage system: composition, reaction mechanism, and physiological significance, *Mol. Cell Biochem.*, **1**, 169–89.

KLEIN, S. M. and SAGERS, R. D. (1966*a*). Glycine metabolism. I. Properties of the system catalyzing the exchange of bicarbonate with the carboxyl group of glycine in *Peptococcus glycinophilus*, *J. Biol. Chem.*, **241**, 197–205.

KLEIN, S. M. and SAGERS, R. D. (1966*b*). Glycine metabolism. II. Kinetic and optical studies on the glycine decarboxylase system from *Peptococcus glycinophilus*, *J. Biol. Chem.*, **241**, 206–9.

KLEIN, S. M. and SAGERS, R. D. (1967*a*). Glycine metabolism. III. A flavin-linked dehydrogenase associated with the glycine cleavage system in *Peptococcus glycinophilus*, *J. Biol. Chem.*, **242**, 297–300.

KLEIN, S. M. and SAGERS, R. D. (1967b). Glycine metabolism. IV. Effect of borohydride reduction on the pyridoxal phosphate-containing decarboxylase from *Peptococcus glycinophilus*, *J. Biol. Chem.*, **242**, 301–5.

KREBS, H. A. (1964). The metabolic fate of amino acids, in *Mammalian Protein Metabolism*, Vol. I (Ed. H. N. Munro and J. B. Allison), Academic Press, New York, pp. 125–76.

NOGUCHI, T. and TAKADA, Y. (1979). Peroxisomal localisation of alanine: glyoxylate aminotransferase in human liver, *Arch. Biochem. Biophys.*, **196**, 645–7.

SAGERS, R. D. and GUNSALUS, I. C. (1961). Intermediary metabolism of *Diplococcus glycinophilus*. I. Glycine cleavage and one-carbon interconversions, *J. Bacteriol.*, **81**, 541–9.

# 21

# Creatine and Creatine Metabolism

P. J. REEDS

*Protein Biochemistry Department,*
*Rowett Research Institute, Bucksburn, Aberdeen, UK*

## Introduction

The central role of creatine phosphate as a store of energy for muscle contraction has long been recognized. A very high proportion of body creatine is found in skeletal muscle and the excretion of the end product of creatine breakdown, creatinine, has been used extensively as an indirect index of the skeletal muscle mass (Cheek, 1968). This paper is concerned with the unusual nature of the catabolism of creatine and the impact that this has upon the metabolism and requirement of those amino acids, particularly glycine, which act as precursors for the synthesis of creatine.

## Synthesis and Incorporation of Creatine

The creatine pool of the body may be derived from two sources: dietary creatine and synthesis *de novo* from three amino acids—glycine, arginine and methionine (Bloch and Schoenheimer, 1941). It is clear that these sources of creatine interact, addition of creatine to the diet increasing the creatine pool of the body (Crim *et al.*, 1975) and suppressing the synthesis of creatine (Crim *et al.*, 1976).

The synthesis of creatine occurs at a number of sites which vary between species, synthesis in the liver being particularly important in man (Walker, 1980). Its synthesis does not occur in skeletal muscle, which possesses a specific transport system for creatine (Fitch *et al.*, 1968) and maintains a positive concentration gradient with respect to the blood. As pointed out below, the scope for control of creatine breakdown is limited and the ability of skeletal muscle to accumulate creatine may be an important factor in the control of the size of the creatine pool. In patients with those myopathies which are associated with a reduction in the creatine pool, accelerated loss of creatine appears to be associated with a defect in the ability of the muscle to accumulate creatine (Fitch and Sinton, 1964). In severe malnutrition the concentration of creatine in skeletal muscle is increased, suggesting an enhanced ability to maintain a positive concentration gradient (Reeds et al., 1978).

## Creatine Catabolism

In 1939 Bloch and Schoenheimer demonstrated that in normal individuals the sole breakdown product of creatine is creatinine. The physiological importance of this metabolic relationship is that the conversion of creatine and creatine phosphate to creatinine is non-enzymic (Borsook and Dubnoff, 1947). It is not surprising, therefore, that measurements in adult man of the fractional rate of breakdown of creatine have yielded a comparatively narrow range of values (Table 1) with an average of $1.57\%$/day $\pm 0.18$ (SD).

TABLE 1

THE RATE OF CONVERSION OF THE TOTAL CREATINE POOL TO CREATININE IN ADULT MAN. MEAN VALUES $\pm$ 1SD

| Reference | Creatine breakdown %/day | Number of subjects |
|---|---|---|
| Fitch and Sinton (1964) | $1.60 \pm 0.26$ | 3 |
| Fitch et al., (1968) | $1.60 \pm 0.21$ | 4 |
| Kreisberg et al., (1970) | $1.63 \pm 0.21$ | 4 |
| Crim et al., (1976) | $1.47 \pm 0.12$ | 5 |

The non-enzymic nature of creatinine synthesis means that continued loss of body creatine is inevitable and there is only a limited range over which the fractional rate of loss can vary. For example, attempts to demonstrate an effect of severe depletion of body protein upon the rate of creatine catabolism have failed (Waterlow et al., 1972; Reeds et al., 1978; Table 2). From a knowledge of the fractional rate of creatine breakdown, creatinine excretion and the total creatine concentration in muscle it is

## TABLE 2
THE EFFECT OF SEVERE PROTEIN DEPLETION IN RATS AND SEVERE PROTEIN ENERGY
MALNUTRITION IN INFANTS UPON BODY CREATINE TURNOVER

| Species | State | Creatine breakdown %/day | Reference |
|---------|-------|--------------------------|-----------|
| Rat | Malnourished | 2·28 ± 0·31 | Waterlow *et al.*, (1972) |
| | Normal | 2·15 ± 0·36 | |
| Infants | Malnourished | 2·08 ± 0·37 | Reeds *et al.*, (1978) |
| | Recovered | 2·15 ± 0·42 | |

possible to calculate the size of the creatine pool (Kreisberg *et al.*, 1970; Picou *et al.*, 1975). In adult man this amounts to some 100 g and the inevitable loss of about 1·6 % of this pool per day requires, in the absence of creatine in the diet, 14 mmol (1·1 g) of glycine per day for its replacement.

This represents approximately one quarter of the loss of glycine by all catabolic routes and 8 % of the total apparent irreversible loss (the flux) of glycine. In Table 3 are calculated (from the results for 3 infants given in Reeds *et al.*, 1978) the amounts of glycine which were deposited in muscle protein and muscle creatine during recovery from severe protein-energy malnutrition. If the amount of creatine which was lost irreversibly over the period of time required for recovery is taken into account, of the 16 g of glycine which could be required for muscle growth, 5 g were utilized for creatine synthesis.

## TABLE 3
THE POTENTIAL REQUIREMENT OF GLYCINE FOR CREATINE SYN-
THESIS DURING RECOVERY FROM SEVERE PROTEIN-ENERGY MAL-
NUTRITION (CALCULATED FROM REEDS *et al.*, 1978)

| | |
|---|---|
| Change in creatine pool | 2·3 g |
| Glycine required* | 1·3 g |
| Time for recovery | 60 days |
| Creatine catabolized* | 7·0 g |
| Glycine required | 4·0 g |
| Total glycine required for creatine | 5·3 g |
| Total glycine required for muscle protein | 11·3 g |

\* Assuming no creatine in the diet, and a rate of breakdown of
2 %/day.

# Conclusion

The metabolism of amino acids is generally interpreted largely in the light of the metabolism of protein. Glycine however acts as the precursor for a variety of physiologically important compounds which include nucleic acids and porphyrins. This paper is concerned with the synthesis of creatine, a pathway which can account for a significant proportion of the requirement of glycine for maintenance and growth.

# Discussion

*Rennie* (*University College Hospital Medical School, London, UK*): Are you suggesting that creatinine can be derived from creatine without going through creatine phosphate?

*Reeds:* Yes. Borsook and Dubnoff (1947) showed clearly that it can be derived from both. *In vitro*, at neutral pH, creatine forms creatinine at the rate of 1 % per day, whereas creatine phosphate, if it can be kept stable under those conditions, forms it at 2 % per day. In theory, therefore, if one knows the fractional rate of conversion of creatine to creatinine, one can calculate the relative steady state concentrations of creatine and creatine phosphate.

*Waterlow* (*London School of Hygiene and Tropical Medicine, UK*): Do conditions such as physical activity or bed rest affect the relative amounts of creatine and creatine phosphate in muscle? If they do, and if the rate of creatinine formation depends on the relative amounts of these two precursors, this will affect the interpretation of measurements of creatinine output as an index of muscle mass.

*Reeds:* It is difficult to find reliable information on the relative concentrations of creatine and creatine phosphate in muscle. A number of workers have examined the effect of exercise on creatinine excretion, but there seems to be no consensus about it.

*Rennie:* I think it is well recognized that in resting muscle two thirds of the total creatine should be bound as creatine phosphate. In exercise creatine phosphate drops to about 30 % of the resting level, at which point the ATP concentration starts to fall. There is the further complication during exercise of a fall in renal plasma flow, leading to a reduction in creatinine clearance and hence an apparent fall in the rate of creatinine excretion.

*Munro (Tufts University, Boston, USA):* Since creatinine formation from creatine is a non-enzymic reaction occurring at a constant rate, a question which has always intrigued me is the control of the creatine content of muscle. To what extent do sex and age influence the creatine content? Secondly, what happens in tissue cultures of muscle cells? Pardridge *et al.* (1980) have reported experiments which suggest that the biosynthesis of creatine may indeed take place in muscle itself, contrary to the traditional statements in textbooks.

*Reeds:* I would be interested to see those results. What I said was based on a recent extensive review by Walker (1980).

*Young (Massachusetts Institute of Technology, (USA):* I think you may have overestimated the drain which creatine synthesis makes on the total glycine flux. Glycine is not only entering the pool by protein breakdown; there is also substantial endogenous synthesis, equivalent to about 50 g glycine/day. Therefore the formation of 2 g creatinine is hardly significant in terms of an irreversible loss of glycine.

*Reeds:* I agree. However, as Neuberger pointed out, glycine moves into a variety of compounds, which represent a real loss of glycine to the body. My object was to quantify one route of loss. Its significance will depend on a number of factors, including the supply of creatine in the diet.

*Walser (John Hopkins University, Baltimore, USA):* I have been trying to find out how much the creatine pool falls on a creatine- and creatinine-free diet and I have been unable to do so. From your numbers the half-life should be about 46 days. Such a diet is scarcely tolerable, so the conduct of such an experiment for a period of 3 to 5 half-lives is obviously very difficult. Do you have any information on this point?

*Reeds:* Crim's work (Crim *et al.*, 1975) showed that creatinine excretion|fell by about 30% on a creatine-free diet, and because their data were based solely upon creatinine excretion they do not allow one to work out the fractional rate of turnover. Their longest period on the diet was 62 days.

*Walser:* Not nearly long enough.

*Reeds:* Nevertheless, they apparently reached a steady state of creatinine excretion, presumably because the rate of creatine synthesis increased and minimized the reduction in the creatine pool size and hence the absolute reduction in creatinine excretion.

268 *P. J. Reeds*

# References

BLOCH, K. and SCHOENHEIMER, R. (1939). Studies in protein metabolism. XI. The metabolic relation of creatine and creatinine studied with isotopic nitrogen, *J. Biol. Chem.*, **131**, 111–19.

BLOCH, K. and SCHOENHEIMER, R. (1941). The biological precursors of creatine, *J. Biol. Chem.*, **138**, 167–94.

BORSOOK, H. and DUBNOFF, J. W. (1947). The hydrolysis of phosphocreatine and the origin of urinary creatinine, *J. Biol. Chem.*, **168**, 493–510.

CHEEK, D. B. (1968). *Human Growth*, Lea & Febiger, Philadelphia.

CRIM, M. C., CALLOWAY, D. H. and MARGEN, S. (1975). Creatine metabolism in man. Urinary creatine and creatinine excretion with creatine feeding, *J. Nutr.*, **105**, 428–38.

CRIM, M. C., CALLOWAY, D. H. and MARGEN, S. (1976). Creatine metabolism in man. Creatine part size and turnover in relation to creatine intake, *J. Nutr.*, **106**, 371–81.

FITCH, C. D. and SINTON, D. W. (1964). A study of creatine metabolism in diseases causing muscle wasting, *J. Clin. Invest.*, **43**, 444–52.

FITCH, C. D., LUCY, D. D., BORNHOFEN, J. H. and DALRYMPLE, G. V. (1968). Creatine metabolism in skeletal muscle. II. Creatine kinetics in man, *Neurology*, **18**, 32–40.

KREISBERG, R. A., BOURDOIN, B. and MEADOR, C. K. (1970). Measurement of muscle mass in humans by isotopic dilution of creatine [14]C, *J. Appl. Physiol.*, **28**, 264–7.

PARDRIDGE, W. M., DUDUCGIAN-VARTAVARIAN, L., CASANELLO-ERTL, D., JONES, M. R. and KOPPLE, J. D. (1980). Amino acid and creatine metabolism in adult rat skeletal muscle cells in tissue culture, *Fed. Proc.*, **39**, 1179.

PICOU, D., REEDS, P. J., JACKSON, A. A. and POULTER, N. R. (1975). The measurement of muscle mass in children using [15]N creatine, *Pediat. Res.*, **10**, 184–8.

REEDS, P. J., JACKSON, A. A., PICOU, D. and POULTER, N. R. (1978). Muscle mass and composition in malnourished infants and children and changes seen after recovery, *Pediat. Res.*, **12**, 613–18.

WALKER, J. B. (1980). Creatine: Biosynthesis, regulation and function, *Adv. Enzym.*, **48**, 177–242.

WATERLOW, J. C., NEALE, R. J., ROWE, L. and PALIN, I. (1972). Effects of diet and infection on creatine turnover in the rat, *Am. J. Clin. Nutr.*, **25**, 371–5.

# 22

# Metabolism of $^{15}$N Nucleic Acids in Children

M. H. N. GOLDEN

*Tropical Metabolism Research Unit,*
*University of the West Indies, Kingston, Jamaica*

and

J. C. WATERLOW

*Department of Human Nutrition,*
*London School of Hygiene and Tropical Medicine,*
*London, UK*

and

D. PICOU

*Mount Hope Medical Complex Task Force,*
*Port of Spain, Trinidad*

Most interest in nucleic acid metabolism in man has been directed towards elucidating the control of purine turnover rates, stemming directly from the investigation of gout. This work has shown not only the precise steps in purine synthesis, but also the complicated system of control and feed-back inhibition that exists in purine biosynthesis and degradation (Wyngaarden and Kelly, 1972).

Our interest in nucleic acid metabolism arose from a totally different consideration. We were attempting to validate the methods for measuring protein synthesis rates *in vivo* by using $^{15}$N-labelled yeast, and at first found some gross anomalies.

In an initial experiment, yeast was grown in culture with $^{15}$NH$_4$Cl as the sole source of nitrogen. After extraction of fat, but not of nucleic acids, the dried yeast was given orally as a single dose to six children who had recovered from malnutrition. Rates of protein turnover were calculated from the excretion of $^{15}$N in urea and total urinary N (Waterlow *et al.*,

1978). The urea was unexpectedly highly labelled, giving anomalously low rates of turnover, compared to those previously obtained with [15]N-glycine as tracer. In this experiment the faecal loss of [15]N amounted to $18\cdot5 \pm 3\cdot2\%$ of the dose. This is much higher than the loss found with [15]N-glycine. Malabsorption of the labelled yeast would tend to give estimates of protein turnover which are too high rather than too low.

In a second study nucleic acids were removed from the yeast by extraction with trichloroacetic acid (TCA) at 80 °C. This preparation was given to five children, again as a single oral dose. The test was then repeated in the same children under identical conditions, with [15]N-glycine as tracer. The results are shown in Fig. 1. The turnover rates with yeast, calculated as before from

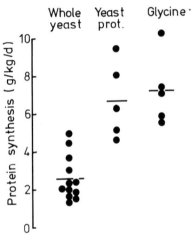

FIG. 1.   Derived values for protein synthesis rate determined from the cumulative excretion of [15]N-urea after a single dose of [15]N-whole yeast, [15]N-isolated yeast protein and [15]N-glycine. All the subjects, who had recently recovered from severe malnutrition, weighed 4–9 kg and were aged 6 months to 2 years.

the [15]N abundance in urea, were now only 7% lower than those obtained with glycine. This suggested that the underestimate of turnover rates in the first experiment arose from excretion of highly labelled degradation products of the nucleic acids.

In a third experiment nucleic acids were again extracted from labelled yeast with hot TCA. After removal of the TCA with ether the nucleic acids were given to three children as a single oral dose. High pressure liquid chromatography of the hydrolysed nucleotides gave the following base composition expressed as $\mu$mol/ml: adenine, 25·6; guanine, 15·2; uracil,

13·3; cytosine, 6·9; thymidine, 1·0; total purines, 40·8; total pyrimidines, 21·2. Urine was collected for 48 h after the dose and the enrichment measured in total-N, urea-N and ammonia-N. During this time the children excreted 45·2, 41·0 and 40·7 % of the administered [15]N in their urine. It is difficult to reach any definite conclusions concerning the metabolism of any specific nitrogen moiety in an experiment such as this where a heterogeneous dose is given; however, if one assumes that the dose of pyrimidine nitrogen was excreted as urea and ammonia and purine nitrogen as non-urea and non-ammonia-N, then we find that 96, 104 and 100 % of administered pyrimidine-N was excreted and 33, 26 and 27 % of administered purine-N.

It is fortuitous that the excretion of pyrimidine-N is virtually equal to the administered dose, because the assumption of a strict division of N is unwarranted for four reasons. First, one fifth of adenine-N and guanine-N will enter the free ammonia pool during conversion to inosine and hypoxanthine by the action of adenylate deaminase and guanase respectively. Secondly, uric acid undergoes an enterohepatic circulation (Sorensen and Levinson, 1975), so that on average, in adults, only 68 % of an injected dose is excreted in the urine. The remaining third is degraded in the bowel presumably to free ammonia (Geren *et al.*, 1950). Thirdly, urea itself is not completely excreted. About one third of the urea produced is cycled through the bowel, is degraded and returns to the liver as ammonia, about 10 % of the original urea produced is not resynthesized into urea (Picou and Phillips, 1972). Fourthly, it is unlikely that the administered nucleic acids were completely absorbed.

Nevertheless, these results show that contamination of biologically generated [15]N tracer with purines or pyrimidines can lead to grossly inaccurate results. Our findings in children are in complete agreement with those of Wilson *et al.* (1954), derived from two normal subjects and one gouty subject.

The relative non-availability of dietary nucleic acids is in marked contradistinction to protein. The various purines and pyrimidines are metabolized quite differently if given orally or parenterally; however, even with parenteral administration their incorporation into body nucleic acids is extremely modest (Roll *et al.*, 1949).

## *Acknowledgement*

M.H.N.G. thanks the Wellcome Trust for support.

## Discussion

*Munro (Tufts University, Boston, USA):* To what extent can you get information from your data about the salvage pathway for purines? Pyrimidine salvage has never been considered very active, but purine salvage is a very active pathway in many tissues. In the case of the gut it may in fact be the major or even exclusive mechanism within the mucosal cells (MacKinnon and Deller, 1973).

*Golden:* We cannot get any quantitative evidence on this from our results. They suggest that the administered pyrimidines in particular were very rapidly and extensively catabolized. There is evidence in the literature (Brown, 1950) of a large difference in the metabolism of the bases and nucleotides given orally and parenterally. If adenine is given parenterally, about 20 % is excreted in the urine and 80 % retained; if it is given orally, the reverse is found (Savaiano et al., 1980). This is difficult to understand. The rapid catabolism of orally administered nucleotides suggests that there must be extensive *de novo* synthesis and/or recycling *within* the cell.

*Nyhan (University of California, San Diego, USA):* I think there is much evidence that reutilization is very active *in vivo*. An example is xanthinuria, in which there is a complete block in xanthine oxidase. In these patients there must be virtually 100 % reutilization of hypoxanthine.

*Neuberger (Charing Cross Hospital Medical School, London, UK):* Has anyone got any accurate information on how far the salvage reactions operate in various tissues and how far a complete synthesis of nucleic acids occurs? The literature is very confused. It is an important problem, in the interpretation of isotope data, to know the relative quantitative importance of these two reaction sequences.

*Millward (London School of Hygiene and Tropical Medicine, UK):* There is some evidence that following the degradation of RNA the nucleotides are not as extensively reutilized for further RNA synthesis as one might expect. We have been interested in the turnover of ribosomal RNA and to begin with we assumed that the methods based on the measurements of decay rates after giving labelled orotic acid would be erroneous, because of reutilization of nucleotides in the same way as occurs with the recycling of labelled amino acids. We measured the methylation rates of nucleosides by constant infusion of labelled methionine. From the labelling of S-adenosyl methionine we calculated the rate of incorporation into the ribose methylated nucleosides (Grimble and Millward, 1977). The turnover rates of RNA obtained by this method are more or less the same as the decay rates found with labelled orotic acid. This was surprising. If true, it indicates that there is not extensive reutilization of the nucleotides which are labelled from orotic acid, and that there must be a large energy expenditure for the *de novo* synthesis of the nucleotides.

*Cohen (University of Wisconsin-Madison, USA):* On the other hand, if you think of the CPS2 activity in the liver cells as the initiating reaction for biosynthesis of pyrimidines, the level of activity is between 1/500 and 1/1000 of that of CPS1 in the mitochondria. I realize that kinetically this does not tell one how much is formed, but it suggests that the rate of pyrimidine synthesis must be some two orders of magnitude slower than the rate of urea synthesis. It must be recognized that the initial step—the glutamine-dependent carbamyl phosphate synthetase—is an enzyme of relatively low activity. Therefore the rate of *de novo* synthesis must be very slow, and one would expect there to be a salvage mechanism to meet at least minimal needs for pyrimidine synthesis.

# References

BROWN, G. B. (1950). Biosynthesis of nucleic acids in the mammal, *Fed. Proc.*, **9**, 517–23.
GEREN, W., BENDICH, A., BODANSKY, O. and BROWN, G. B. (1950). The fate of uric acid in man, *J. Biol. Chem.*, **183**, 21–31.
GRIMBLE, G. K. and MILLWARD, D. J. (1977). Measurement of ribosomal RNA synthesis in rat liver and muscle *in vivo*, *Biochem. Soc. Trans.*, **5**, 913–16.
MACKINNON, A. M. and DELLER, D. J. (1973). Purine nucleotide biosynthesis in gastrointestinal mucosa, *Biochim. Biophys. Acta*, **319**, 1–4.
PICOU, D. and PHILLIPS, M. (1972). Urea metabolism in malnourished and recovered children receiving a high or low protein diet, *Am. J. Clin. Nutr.*, **25**, 1261–6.
ROLL, P. M., BROWN, G. B., DI CARLO, F. J. and SCHULTZ, A. S. (1949). The metabolism of yeast nucleic acid in the rat, *J. Biol. Chem.*, **180**, 333–40.
SAVAIANO, D. A., HO, C. Y., CHU, V. and CLIFFORD, A. J. (1980). Metabolism of orally and intravenously administered purines in rats, *J. Nutr.*, **110**, 1793–804.
SORENSEN, L. B. and LEVINSON, D. J. (1975). Origin and extrarenal elimination of uric acid in man, *Nephron*, **14**, 7–20.
WATERLOW, J. C., GOLDEN, M. H. N. and GARLICK, P. J. (1978). Protein turnover in man measured with $^{15}N$: comparison of end products and dose regimes, *Am. J. Physiol.*, **235**(2), E165–E174.
WILSON, D., BEYER, A., BISHOP, C. and TALBOTT, J. H. (1954). Urinary uric acid excretion after the ingestion of isotopic yeast nucleic acid in the normal and gouty human, *J. Biol. Chem.*, **209**, 227–32.
WYNGAARDEN, J. B. and KELLY, W. N. (1972). Gout, in *The Metabolic Basis of Inherited Disease*, 3rd Edition (Eds. Stanbury, Wyngaarden and Fredrickson), pp. 889–968, McGraw-Hill, New York.

# Summing Up: Free Amino Acid Metabolism and its End Products

H. N. MUNRO

*USDA Human Nutrition Research Center on Aging,*
*Tufts University, Boston, Massachusetts, USA*

I shall try to evaluate this session of the proceedings of the workshop in terms of the needs for future research and their directions. The comments I shall make reflect the banquet of free amino acids at which we have sat down today.

First, it is clear that we have reached the point in nitrogen metabolism studies where we need to pay close attention to quantitation of the dynamics of amino acid metabolism. This need permeates many of the presentations today. Thus inter-organ exchange of amino acids and their metabolites demands adequate quantitation. While we can accept the data accumulated for the magnitude of the flux of alanine between muscle, gut and liver, and less satisfyingly for glutamine because exchange across the kidney provides a less clear contribution to glutamine balance, we have to recognize that other body sources of these amino acids complicate the quantitation of their fluxes between organs. For example, it will be important in future studies to evaluate the contribution of adipose tissue to body alanine and glutamine, especially since release of these amino acids from the adipocyte and the muscle cell responds differently to insulin and to starvation (Tischler and Goldberg, 1980). Consequently, an adequate and comprehensive description of the dynamics of glutamine and alanine will be required in order to provide a picture of alanine and glutamine metabolism in response to diet, hormones, exercise and other factors causing perturbations in the steady state.

No sooner have we accepted this proposition than we have to acknowledge that any such quantitative description must specify the species of animal, notably recognizing the differences between rat and man. This emerges in connection with glutamine metabolism in the two species, and in terms of branched chain intermediary metabolism as distributed between muscle and liver. But one must also know the fate of the branched chain keto acids made in the kidney of rat and man, and the quantitative products of branched chain amino acid metabolism in the brain and elsewhere.

Again, it has only been obliquely alluded to here that experimental species of different body size have intensities of metabolism that decrease as the body weight of the species increases (Munro, 1969). This not only applies to the well-known effect of body size on energy metabolism but also to various aspects of nitrogen metabolism. Furthermore, the relative ratios of organ and tissue sizes, such as liver and muscle, change systematically with body size. This must inevitably alter the balance between inter-organ metabolism as we have shown for muscle versus visceral protein synthesis in man versus the rat (Munro, 1969).

We come now to the relevance of studies of amino acid metabolism to needs for essential amino acids and indeed for dietary protein. The regulation of free amino acid pools requires adaptive enzyme changes for the catabolic pathways of the essential amino acids. The study of the responses of such pathways to the level of intake provides a first step towards a better basis for measuring dietary requirements of essential amino acids. When $CO_2$ production from essential amino acids fed at various levels is coupled to estimates of tissue protein synthesis, then we can begin to provide a more secure basis for understanding the metabolic needs for essential amino acids. The study of cases with inborn errors of amino acid metabolism suggest additional opportunities for measuring needs of individual essential amino acids.

It is clear that more research on the metabolism of branched chain amino acids promises to yield dividends. We can ask what is the evolutionary advantage of excluding these amino acids from major catabolism in the liver, and transferring the incoming load to the peripheral tissues? Speculation on this question suggests special functions of the branched chain amino acids. These amino acids affect not only muscle metabolism, but also entry of other neutral amino acids across the blood–brain barrier, notably through competition between them and tryptophan. For example, Anderson (1981) has found in the rat that appetite for dietary protein is controlled by dietary tryptophan intake. Once within the brain, the concentration of free tryptophan is rate-limiting for the synthesis of the neurotransmitter serotonin. Is it possible that, with increasing dietary levels of protein, tryptophan entry into the brain is diminished by the concomitant rise in free branched chain amino acids in the blood, thus complicating the effect of tryptophan intake on the cerebral control of appetite for protein? The disposal of branched chain amino acids in muscle and its response to insulin stimulation have relevance to disease processes, such as diabetes and notably liver cirrhosis in which the excessive lowering of branched chain plasma levels should promote tryptophan entry into the brain and will promote serotonin synthesis, which favours drowsiness, a feature of hepatic incompetence (Munro *et al.*, 1975).

The study of amino acid metabolism has progressed from identification of pathways and their details to quantitation and distribution between organs. Nevertheless, there are still some surprises such as the very recent description of a new leucine pathway leading to transfer of the amino group to the $\beta$-carbon atom and eventual resynthesis of leucine (Poston, 1980).

Finally, study of the end products of amino acid metabolism, principally urea and ammonia, is important, not only in its own right, but as a basis for measuring rates of protein turnover. Elucidation of the mechanism by which the rate of urea biosynthesis is controlled represents a major advance in our understanding of biological regulations. Nevertheless, as the discussions have shown, there is much still to be learnt about the precursors of these end products and the way in which the pattern of precursors changes under different conditions. The recycling of urea through the colon is well established, but there appears still to be no consensus about its physiological importance.

We therefore end this part of the symposium with a great many unsolved questions, but it is precisely the purpose of such a symposium to define the areas of ignorance and the growth points for the future.

# *References*

ANDERSON, G. H. (1981). Diet, neurotransmitters and brain function. *Brit. med. Bull.*, **37**, 95–100.

MUNRO, H. N. (1969). Evolution of protein metabolism in mammals, in *Mammalian Protein Metabolism* (Ed. H. N. Munro), Vol. III, pp. 133–82. Academic Press, New York.

MUNRO, H. N., FERNSTROM, J. D. and WURTMAN, R. J. (1975). Insulin, plasma amino acid imbalance and hepatic coma, *Lancet*, **i**, 722.

POSTON, J. M. (1980). Cobalamin-dependent formation of leucine and $\beta$-leucine by rat and human tissues, *J. Biol. Chem.*, **255**, 10067–72.

TISCHLER, M. E. and GOLDBERG, A. L. (1980). Leucine degradation and release of glutamine and alanine by adipose tissue, *J. Biol. Chem.*, **255**, 8074–81.

# PART III

# Turnover Methods

# 23

## Methods of Measuring Protein Turnover

J. C. WATERLOW

*Department of Human Nutrition,
London School of Hygiene and Tropical Medicine, London, UK*

This session is concerned with methods of measuring protein turnover in the living organism. With good tools an enormous range of enquiries becomes possible. In the field of general biology and physiology we have subjects such as the relationship of protein turnover to energy metabolism, to growth and senescence, and to processes of physiological adaptation, such as temperature acclimatization (Haschemeyer, 1978). The applications to pathological states are obvious: malnutrition, hormonal imbalance, effects of infection and trauma. The central mystery is the mechanism by which rates of protein synthesis and degradation are matched, so that normally a steady state is maintained. It is obvious that the first step in solving this fundamental problem is to be able to measure these rates. I think, therefore, that no excuse is needed for continuing to concentrate on methods, even though that part of the subject may now seem rather uninspiring.

I shall confine myself to the crudest approach, the measurement of total turnover in the whole body, because this is of greatest interest to me as a clinician. The starting point of our work in Jamaica in the 1950s was that we were confronted with malnourished children who died, and we did not understand why. The hypothesis put forward was that perhaps as a result of protein depletion the essential machinery for protein synthesis was destroyed, so that restoration of the tissues lost would be impossible. Therefore we looked for means of measuring protein synthesis in children.

The problems in such an enterprise fall into three groups: (1) theoretical, by which I mean the definition of appropriate models and acceptable methods of kinetic analysis; (2) metabolic, by which I mean the validity of assumptions about aspects of protein metabolism on which our calculations of turnover rate depend, whatever method of kinetic analysis is used and (3) technical. In the early days the limitations on the last front were very severe. I might just mention that in our first measurements with U-$^{14}$C-lysine in man (Waterlow, 1967), lysine in plasma was measured enzymatically in the Cartesian diver (Waterlow and Borrow, 1949), and $^{14}CO_2$, liberated by lysine decarboxylase, in a Conway unit, was measured in a low background gas-flow counter with a background of 1 cpm and a total count of 2 cpm—a very tedious process. We have come a long way since then, as later speakers will show.

I shall confine myself to the first set of problems, leaving the metabolic and technical ones to those who follow. I hope I may be forgiven if I deal with the subject from the rather personal point of view of someone trying to get to grips with it without being a mathematician.

The starting point for us was the paper of Sprinson and Rittenberg (1949). They injected a single dose of $^{15}$N-glycine and calculated the turnover rate of the metabolic pool from an exponential expression for the cumulative abundance of $^{15}$N in the urine over a period of about 30 h. This, of course, is a very simple example of compartmental analysis based on a 2-pool model. I think we can ignore the method proposed by San Pietro and Rittenberg (1953) which, although popular for a time, turned out to be a red herring for reasons discussed elsewhere (Waterlow *et al.*, 1978). The next paper which made a great impression on me was that by Olsen *et al.* (1954) in Denmark. They also gave a single dose of $^{15}$N-glycine, followed the excretion of isotope for two weeks and derived three exponentials from the excretion curve. From these they calculated the parameters of a model with a metabolic pool, a rapid protein pool and a slower protein pool which was 'blind'. This work clearly showed that the excretion of isotope after a single dose is not really linear when plotted semi-logarithmically, although it may be made to appear so over short periods.

This kind of compartmental approach clearly had serious limitations. If we are approximating a curve by one or more straight lines, even when it is done by computer, small errors in measurement can have a disproportionate effect on the values of the parameters derived. In the early stages particularly, the accuracy of time-points is crucial, and when we are working on urine, how can we know the time lag between the production of an end product and its excretion? Finally, in the study of Olesen *et al.* (1954) it was necessary to continue the measurements for 14 days in order to obtain the parameters of the 4-pool model. By comparing models with one versus two protein pools I calculated that in order to demonstrate the existence of a second protein pool, the measurements have to be continued for 10 days (Waterlow *et al.*, 1978). This is not practical for application to patients.

It seemed to me that many of these difficulties could be avoided if one could produce an isotopic steady state by continuous infusion of tracer. Theoretically, this should be possible, however complex the system of precursor and product pools, provided that re-entry of tracer from protein could be neglected (Aub and Waterlow, 1970). Experiments on rats by Gan and Jeffay (1967) and by us (Waterlow and Stephen, 1968) suggested that such a steady state (pseudoplateau) was in fact produced. This has great practical advantages. To define a straight line with reasonable precision requires fewer measurements than to define the shape of a curve. Secondly, when we are working on end products the exact timing of their production does not matter, once an isotopic steady state has been achieved.

Only later did I realize that this simple and obvious approach is called 'stochastic', and I still do not know very clearly what the word means, except that it comes from the Greek 'to guess'. However, from that point it was a short step to seeing that, in theory, exactly the same result can be achieved with a single dose of tracer if, instead of analysing the slope of the abundance–time curve by the compartmental method we determine the area under it. This is what we are now doing in many of our studies, since it is a method which is very easy to use in clinical situations. The drawbacks of this simple approach depend on the metabolic assumptions rather than on the kinetic ones.

Nevertheless, I think that some discussion would be appropriate about the merits of the stochastic versus the compartmental approach for studies of nitrogen metabolism *in vivo*. The stochastic method is at present in the ascendant; Young drew my attention to a paper which compares stochastic and deterministic methods (Matis and Tolley, 1980) and concludes that 'the real world of tracer kinetics is stochastic'. All the same, the compartmental method is *potentially* able to give us more information. For example, it would be extremely useful if we could, *in vivo*, distinguish separately turnover rates in a fast and slow protein pool, even if we could not delimit these pools precisely.

My hesitations about compartmental analysis are practical and biological rather than mathematical. First, it is necessary to make far more measurements, and in some situations this is a serious disadvantage. Secondly, the models are often indeterminate (Shipley and Clark, 1972) and no amount of computer fitting of curves will solve this problem. Even though it may be possible to get extra information, e.g. to derive sizes and turnover rates for several different pools, the physiological meaning of such results needs to be established by independent evidence. For example, Long *et al.* (1977), from a multi-exponential analysis of a 4-pool model, concluded that in septic patients the 'active metabolic pool' (which is of the same order of magnitude as the free amino-N pool) increases in size by a factor of 3. This is an interesting observation, but surely it needs to be confirmed by direct measurements.

Thirdly, the conventional compartmental analysis relies on the basic assumption that exchanges occur at constant fractional rates. We have suggested that this assumption is probably not valid for many aspects of nitrogen metabolism and protein turnover (Waterlow *et al.*, 1978). For example, it is almost certainly wrong to suppose that a constant fraction of the precursor amino acid pool is necessarily taken up into protein in unit time. Synthesis can be *described* as a constant fractional process, and in the steady state the description will fit the facts, but it will not do so if the precursor pool size changes. Similarly, I find it difficult to believe that the process of amino acid oxidation is adequately described by a model in which a constant fraction of the free pool is being oxidized. There is indeed evidence (e.g. Brookes *et al.*, 1972) of a more or less linear relationship between amino acid supply and oxidation rate, which would fit in with a constant fractional process. However, the situation alters the moment that enzyme adaptations occur and it is likely that these adaptations take place quite rapidly (Potter *et al.*, 1968). In other words, although compartmental analysis can in theory give us more information about the detailed structure and working of the system, it is at the expense of over-simplifying it in other ways and destroying its flexibility. I put forward these views with diffidence since, as I have said, I am no mathematician.

Before leaving this subject I would like to give one example of the kind of problem that we encounter. When we give $^{15}$N-glycine in hourly doses, we get what looks like the beginning of a plateau of labelling of urinary $NH_4^+$ after 6–8 h. At any rate, there is at this time a sharp inflection in the labelling curve. Putting together the results of fifteen such studies, I find that the so-called plateau has an upward slope which averages 4 %/h. This is by no means a negligible amount. There are three possible explanations: (1) recycling from labelled protein; (2) slowness of equilibration in the large free glycine pool (I am indebted to Dr P. Garlick for this suggestion); (3) the influence of a slowly turning over protein pool, which delays equilibration of the precursor pool. Which is the correct explanation?

With 1-$^{14}$C-leucine Golden showed that a continuous intragastric infusion produced a plateau which was constant for 30 h, with no detectable slope (Golden and Waterlow, 1977). This seems to rule out explanations 1 and 3, since recycling or a slow protein pool should make themselves felt just as much with leucine as with glycine. On the other hand, if we assume from the data of Bergström *et al.* (1974) on human muscle that the free glycine pool is about 1 mmol/kg (a value, incidentally, which fits in well with the estimate of Watts and Crawhall (1959) of the 'first glycine pool' in man); if we assume also that glycine constitutes 5 % of the protein flux (taken as 4 g protein/kg/day); then we can calculate the theoretical time-course of labelling of free glycine produced by a constant infusion (Fig. 1). The large

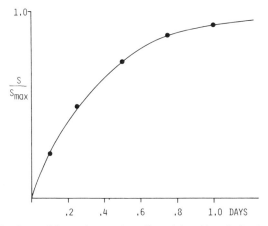

FIG. 1.     Calculated rate of rise to plateau of specific activity of free glycine during continuous infusion. Pool size = 1 mmol/kg, flux = 2·67 mmol/kg/day.

size of the free glycine pool may be expected to produce a significant delay in equilibration, which is by no means complete after 8 h. Further mathematical analysis will not help us. What has to be done is to measure by actual experiment the time course of the labelling of free glycine in plasma. Technically this now presents no difficulty. It may be noted that Lapidot and Nissim (1980) found a very rapid exponential disappearance of labelled glycine from plasma after a single injection of tracer. They attributed this rapid disappearance to mixing in the hepatic pool, but their observations were continued for only 1 h. The studies of Neuberger and co-workers many years ago (Henriques *et al.*, 1955) suggest that full equilibration takes much longer. This seems to be confirmed by the recent report of Gersovitz *et al.* (1980). These workers determined glycine flux by measurements of [15]N-glycine in plasma, the tracer being given orally every 3 h. The blood samples from which the flux was calculated were taken 36–60 h after the first dose of tracer. This implies, although it is not explicitly stated, that plateau labelling was not reached for 30 h or so.

Now that accurate methods are available for measuring the abundance of stable isotopes in quite small blood samples (see Bier *et al.*, p./289), might it not be worthwhile, in spite of what I have said earlier, to go back to compartmental analysis, either of a decay curve or of the curve of rise to plateau, to see whether it is possible to distinguish fast and slowly turning over protein pools? If this could be done, it would be very valuable. In the steady state, if the bicarbonate pool remains constant and if $CO_2$ production does not change, the same kind of information could presumably be obtained from measurement of respiratory $^{13}CO_2$ if a $^{13}C$-amino acid is given.

At present, as a result of the development of better instruments, measurement of $^{13}C$ abundance in plasma amino acids is becoming for many of us the method of choice for studies of protein turnover in man. This is clearly a powerful tool. However, in ending I would like to put in a word for the end product methods based on the measurement of $^{15}N$ in urinary urea and ammonia. These methods have the merit of great practical simplicity, so that they can be used in a wide range of clinical circumstances, and the analysis does not need a very complicated instrument. The main problem with these methods is the metabolic assumptions on which they depend. One benefit which has come from this approach is that it has stimulated us to explore in more detail important aspects of nitrogen metabolism in man, as later papers in this symposium will show.

## Discussion

*Bier (Washington University School of Medicine, St. Louis, USA):* Compartmental analysis by computer has one advantage—it can tell which models are incompatible with the data. However, once one gets into 4- or 5-pool models interpretation becomes very difficult.

*Pratt (Institute of Psychiatry, London, UK):* I agree. The physical reality is always much more complicated than the most intricate compartmental model, and whatever one may be able to do on the computer, in practice it is difficult to deal effectively with a model with more than three compartments. Another problem with multi-compartmental models is that two or three models may give identical results, and you cannot distinguish between them unless it is possible to sample several compartments. My feeling is to stick to what works in practice, and to use other measures to find out what is happening in physical reality.

*Waterlow:* You are referring to the indeterminacy of the models. I personally would very much like to be able to distinguish between a fast and a slowly turning over protein pool. Garlick and Swick (1977) distinguished three protein pools in a study on the breakdown rate of rat liver protein, but we have not yet found any way of doing this in the whole body except by the very long drawn out method of Olesen *et al.* (1954). If we have a compartmental analysis which identifies a physical entity, that is fine, but very often the pools described do not seem to have any real existence.

*Klein (Baylor College of Medicine, Houston, USA):* I agree. However, in line with what Dr Bier has said, the use of the computer can be specified in several ways; with the techniques that Dr Mones Berman at NIH has developed, which are now generally available as computer programmes (Berman and Weiss, 1977), you can determine the best estimate from the parameters that are known. Then you can interrogate the model and ask what would be the effect on any measurement of changes in blood flow, for example. This is very important, because then you can determine which parameters are insensitive to the factors you are not interested in studying. I think this is an approach which will give us a great deal of power in the future. There is a very good article called *Good Manners in Modelling* (Yates, 1978), which deals with the question of how many compartments you should propose and what the criteria are for introducing the compartments.

*Waterlow:* I believe you have used this approach to analyse Garlick's equations (Garlick *et al.*, 1973) and showed that the estimate of protein synthesis rate is very insensitive to changes in the slope of the precursor specific activity.

*Klein:* That is correct.

*Munro (Tufts University, Boston, USA):* In the stochastic model the chasm into which the isotope disappears is assumed to be protein. Are there any dangers that with specific precursors some of the label will be lost by going into other compounds, thus giving falsely high rates of synthesis?

*Waterlow:* In our work we have assumed that from a quantitative point of view the only important routes are synthesis and excretion, but with particular amino acids there may be problems. With glycine, as we saw earlier (Reeds, p. 263) there may be some question about the quantitative importance of creatine formation. I believe that the error will not be more than some 5 % if we take account of only those two paths of exit for the precursor.

*Golden* (*University of the West Indies, Kingston, Jamaica*): In measurements with $^{15}$N-glycine, the loss of label into other compounds such as purines and creatine will affect the estimate of total flux, but not necessarily the estimate of synthesis rate. If the formation of these compounds is balanced by their excretion, one can to some extent correct for the loss of label by calculating synthesis as flux minus total N rather than flux minus urea and ammonia-N (see Waterlow *et al.*, 1978, p. 268). However, this correction does not allow for the fact that compounds such as pyrimidines may be degraded to urea with a proportionate dilution, reducing the enrichment attributable to protein and increasing the estimated synthesis rate.

*Neuberger* (*Charing Cross Hospital Medical School, London, UK*): One could compare an amino acid like glycine, which has anabolic reactions of considerable importance, with, for example lysine, for which the only outlets are oxidation and protein synthesis, and see whether the synthesis rates obtained are very dissimilar. That would give some idea of how far the neglect of other reactions is important.

*Waterlow:* Garlick did precisely that. We had measurements for lysine flux in the rat (Waterlow and Stephen, 1967) and he then measured glycine flux (Garlick, 1969). However, the problem with glycine is the recycling through serine, as well as the channelling off into other pathways.

*Young* (*Massachusetts Institute of Technology, USA*): You have data which compare estimates of whole body protein synthesis based on leucine and glycine simultaneously, and as I recall, the estimates were comparable (Golden and Waterlow, 1977; Garlick *et al.*, 1980). From our data, other pathways associated with glycine metabolism do not make an important contribution to the overall utilization of the amino acid in relation to the magnitude of total body protein turnover.

*Waterlow:* But when we use $^{15}$N-glycine and estimate protein turnover from the excretion of isotope in an end product, we are not measuring glycine flux; we are estimating total N flux. The estimate, of course, depends on a mass of assumptions, which we have discussed in detail elsewhere (Waterlow *et al.*, 1978).

*Golden:* The assumptions and the errors seem to balance. Just because we get the same numbers it does not mean that we are measuring the same thing.

*Klein:* The elegant studies which Bier and Matthews have been doing by constant infusion are very precise, but when you are dealing with free-living populations a bolus-type study has advantages in terms of simplicity and acceptability, and it doesn't involve admission to hospital or cannulation of veins, etc. Therefore from a practical standpoint you have access to a much wider population by the bolus technique.

*Waterlow:* I would not argue the pros and cons of constant infusion versus single dose, because both are useful. The single dose method has indeed the merit of simplicity—for example, one of my colleagues has used it in a bush hospital in Northern Nigeria, where any other method would be very difficult (see p. 543). However, from the theoretical point of view, the interesting question still is whether the stochastic approach best represents the 'real world', or whether there is a place for compartmental analysis.

# *References*

AUB, M. R. and WATERLOW, J. C. (1970). Analysis of a five-compartment system with continuous infusion and its application to the study of amino acid turnover, *J. theoret. Biol.*, **26**, 243–50.

BERGSTRÖM, J., FÜRST, P., NORÉE, L-O. and VINNARS, E. (1974). Intracellular free amino acid concentration in human muscle tissue, *J. applied Physiol.*, **36**, 693–7.

BERMAN, M. and WEISS, M. F. (1977). *Users Manual for SAAM-27*, US Department of Health, Education and Welfare, Public Health Service, National Institute of Health.

BROOKES, I. M., OWENS, F. N. and GARRIGUS, U. S. (1972). Influence of amino acid level in the diet upon amino acid oxidation by the rat, *J. Nutr.*, **102**, 27–36.

GAN, J. C. and JEFFAY, H. (1967). Origins and metabolism of the intracellular amino acid pools in rat liver and muscle, *Biochim. Biophys. Acta*, **148**, 448–59.

GARLICK, P. J. (1969). Turnover rate of muscle protein measured by constant intravenous infusion of $^{14}$C-glycine, *Nature*, **223**, 61–2.

GARLICK, P. J. and SWICK, R. W. (1977). Determination of the average degradation rate of mixtures of protein, in *Intracellular Protein Catabolism II* (Eds. V. Turk and N. Marks), Plenum, New York and London.

GARLICK, P. J., MILLWARD, D. J. and JAMES, W. P. T. (1973). The diurnal response of muscle and liver protein synthesis *in vivo* in meal-fed rats, *Biochem. J.*, **136**, 935–45.

GARLICK, P. J., CLUGSTON, G. A. and WATERLOW, J. C. (1980). Influence of low energy diets on whole body protein turnover in obese subjects, *Am. J. Physiol.*, **238**, E235–E244.

GERSOVITZ, M., BIER, D., MATTHEWS, D., UDALL, J., MUNRO, H. N. and YOUNG, V. R. (1980). Dynamic aspects of whole body glycine metabolism: influence of protein intake in young adult and elderly males, *Metabolism*, **29**, 1087–94.

GOLDEN, M. H. N. and WATERLOW, J. C. (1977). Total protein synthesis in elderly people: a comparison of results with $^{15}$N glycine and $^{14}$C leucine, *Clin. Sci. Mol. Med.*, **53**, 277–88.

HASCHEMEYER, A. V. (1978). Protein metabolism and its role in temperature acclimation, in *Biochemical and Biophysical Perspectives in Marine Biology* (Eds D. C. Malins and J. R. Sargent), Vol. 4, |Academic Press, New York.

HENRIQUES, O. B., HENRIQUES, S. B. and NEUBERGER, A. (1955). Quantitative aspects of glycine metabolism in the rabbit, *Biochem. J.*, **60**, 409–23.

LAPIDOT, A. and NISSIM, I. (1980). Regulation of pool sizes and turnover rates of amino acids in humans: $^{15}$N-glycine and $^{15}$N-alanine single-dose experiments using gas-chromatography–mass spectrometry analysis, *Metabolism*, **29**, 230–9.

LONG, C. L., JEEVANANDUM, M., KIM, B. M. and KINNEY, J. M. (1977). Whole body protein synthesis and catabolism in septic man, *Am. J. clin. Nutr.*, **30**, 1340–4.

MATIS, J. H. and TOLLEY, H. D. (1980). On the stochastic modelling of tracer kinetics, *Fed. Proc.*, **39**, 104–9.

OLESEN, K., HEILSKOV, N. C. S. and SCHØNHEYDER, F. (1954). The excretion of $^{15}$N in urine after administration of $^{15}$N glycine, *Biochim. Biophys. Acta*, **15**, 95–107.

POTTER, V. R., BARIL, E. F., WATANABE, M. and WHITTLE, E. D. (1968). Systematic oscillations in metabolic functions in liver from rats adapted to controlled feeding schedules, *Fed. Proc.*, **27**, 1238–45.

SAN PIETRO, A. and RITTENBERG, D. (1953). A study of the rate of protein synthesis in humans. II. Measurement of the metabolic pool and the rate of protein synthesis, *J. biol. Chem.*, **201**, 457–73.

SHIPLEY, R. A. and CLARK, R. E. (1972). *Tracer Methods for In Vivo Kinetics*, Academic Press, New York and London.

SPRINSON, D. B. and RITTENBERG, D. (1949). The rate of interaction of the amino acids of the diet with the tissue proteins, *J. biol. Chem.*, **180**, 715–26.

WATERLOW, J. C. (1967). Lysine turnover in man measured by intravenous infusion of L-[U-$^{14}$C]lysine, *Clin. Sci.*, **33**, 507–15.

WATERLOW, J. C. and BORROW, A. (1949). Experimental observations on the Cartesian diver technique, *Comptes rendus trav. Lab. Carlsberg, Sér. Chim.*, **27**, 93–123.

WATERLOW, J. C., GARLICK, P. J. and MILLWARD, D. J. (1978). *Protein Turnover in Mammalian Tissues and in the Whole Body*, Elsevier North Holland, Amsterdam.

WATERLOW, J. C. and STEPHEN, J. M. L. (1967). The measurement of total lysine turnover in the rat by intravenous infusion of L-[U-$^{14}$C]-lysine, *Clin. Sci.*, **33**, 489–506.

WATERLOW, J. C. and STEPHEN, J. M. L. (1968). The effect of low protein diets on the turnover rates of serum, liver and muscle proteins in the rat, measured by continuous infusion of L-[$^{14}$C]-lysine, *Clin. Sci.*, **35**, 287–305.

WATTS, R. W. E. and CRAWHALL, J. C. (1959). The first glycine metabolic pool in man, *Biochem. J.*, **73**, 277–86.

YATES, E. F. (1978). Good manners in modelling: mathematical models and computer simulation of physiological systems, *Am. J. Physiol.*, **234**, R159–R160.

# 24

# Practical Advantages of Gas Chromatography–Mass Spectrometry for Stable Isotope Measurement in Biological Samples

D. M. BIER and D. E. MATTHEWS

*Department of Internal Medicine,*
*Washington University School of Medicine, St. Louis, Missouri, USA*

and

V. R. YOUNG

*Department of Nutrition and Food Sciences,*
*Massachusetts Institute of Technology, Cambridge, Massachusetts, USA*

Stable isotopes had, until recently, enjoyed only limited use in clinical investigation, despite their obvious advantages for research in human subjects and their absolute necessity for tracing certain elements (nitrogen for instance) that lack a convenient radioisotope. In large part this was the result of a short supply of enriched materials for chemical synthesis and relatively difficult analytical methods compared with those available for radiotracers. These problems have now been largely solved. Production of highly enriched $^{13}C$, $^{15}N$ and oxygen isotopes at the National Stable Isotope Resource, Los Alamos Scientific Laboratory, exceeds current demands and many practical computer systems for gas chromatography–mass spectrometry (GCMS) are commercially available. We will focus on the advantages of this analytical technique. Several recent monographs have reviewed these developments (Watson, 1976; Baillie, 1978; Matwiyoff and Ott, 1973; Ligon, 1979; Caprioli and Bier, 1980).

Compared with classical dual-inlet, dual-collector isotope ratio mass spectrometry, GCMS techniques offer several distinct advantages. The most obvious are those related to sample size and ease of sample preparation for analysis. Thus, while microgram amounts of carbon (as $CO_2$), or nitrogen (as $N_2$) are required for isotope ratio mass spectrometry, GCMS ion monitoring techniques can measure isotope enrichments in

289

picogram samples of complex molecules. Since nitrogen constitutes less than 20 % of the mass of amino acids, this difference becomes magnified in circumstances where, for example, one wishes to trace nitrogen flow in plasma amino acids. Furthermore, since a sample for isotope ratio mass spectrometry must be introduced as a low molecular weight gas generated from a pure sample of the material of interest, sample preparation can be tedious and difficult. While preparation of urinary urea and ammonia for subsequent isotope ratio analysis is relatively simple, the procedures necessary for similar isotopic measurements in a single plasma amino acid preclude application of isotope ratio mass spectrometry to routine clinical studies which generate large numbers of samples.

GCMS techniques, however, have additional advantages related to the confidence one can place in the analytical result (Table 1).

TABLE 1

LEVELS OF SPECIFICITY IN GCMS ANALYSIS OF BIOLOGICAL MATERIALS

1. Conventional sample 'clean up'
2. Derivative selectivity
3. Gas chromatographic resolution
4. Choice of ionization mode
5. Ions selected for monitoring
6. Resolving power of the mass spectrometer

Routine sample preparatory methods offer the first level of selectivity. For example, in analysis of plasma amino acids a simple cation exchange 'clean up' step removes potential neutral and anionic contaminants (Bier *et al.*, 1977; Matthews *et al.*, 1979; Bier and Christopherson, 1979). Choice of the appropriate derivative presents the second level of specificity. In the simplest case, one can pick the appropriate derivative to tailor the analysis with regard to GC characteristics, constraints of analysis time, and the like. In addition, if one selects a reaction that will only take place with certain classes of chemical compounds (including the compound of interest), one effectively removes contaminants of different chemical classes from subsequent analytical steps. Thus, for example, we have employed the diboronate derivative for GCMS analysis of plasma glucose since this derivative is relatively specific for the glucose contained in a neutral fraction obtained from plasma (Bier *et al.*, 1977; Wiecko and Sherman, 1976), whereas other less selective reagents such as the trimethylsilyl (TMS) would form products with almost all compounds in that fraction. Similar considerations hold for certain amino acid derivatives as well (Bier and Christopherson, 1979).

Choice of the appropriate GC stationary phase offers the next level of specificity since, with the almost infinite array of phases now available, it is relatively easy to pick the phase which will permit complete separation of the compound of interest from its neighbours. Thus, an essentially pure material elutes from the GC and is directly introduced into the mass spectrometer for subsequent analysis.

In the ionization chamber, use of electron impact (EI) or chemical ionization (CI) further enhances selectivity and specificity. The difference between these two modes of ionization is shown in Table 2. Under EI

TABLE 2

THE MECHANISMS OF ELECTRON IMPACT IONIZATION (EI) AND CHEMICAL IONIZA-TION (CI) OF A NEUTRAL MOLECULE (M)

1. *Electron impact ionization:*

$$A + e^- \rightarrow M^+ + 2e^-$$
$$\rightarrow m_1^+$$
$$\rightarrow m_2^+$$
$$\rightarrow m_3^+$$
$$\rightarrow \text{etc.}$$

2. *Chemical ionization:*

$$CH_4 + e^- \rightarrow CH_4^+ + 2e^-$$
$$CH_4^+ + CH_4 \rightarrow CH_5^+ + CH_3$$
$$M + CH_5^+ \rightarrow MH^+ + CH_4$$

conditions, a high energy incident electron ejects an outer shell electron from a neutral molecule (M) to produce a positively charged ion. However, since the ion usually contains excess energy transferred from the incident electron, the molecular ion breaks up into many smaller charged fragments ($m^+$) of the initial molecule. This can be both an advantage and a disadvantage. In the former case, choice of the appropriate fragment allows one to obtain specific localization of the label in the molecule under study. However, small fragments of similar mass are likely to be produced by compounds in the same (or different) chemical class and thus there is increased potential for interference from unwanted substances.

In chemical ionization, on the other hand, one first ionizes a reagent gas (methane in Table 2) which subsequently transfers its charge in a gentle fashion to the molecule of interest. This low energy process produces little if any fragmentation of the molecule. Therefore, while the structural information found in EI fragments is lost, the problem of potential contaminating ions is dramatically reduced (Matthews *et al.*, 1979).

Under many circumstances the ionization pattern (mass spectrum) of the material of interest will provide several ions potentially useful for measurement both under EI and CI conditions. Choice of the appropriate ones, taking into account constraints of related and unrelated contaminants yet remaining, further increases the specificity of analysis.

At this point, the odds of unwanted substances still interfering with measurement of the compound under study are extraordinarily small. However, if such interference exists, the resolving power of the mass spectrometer offers a final degree of specificity. Thus, while low resolution analysis is generally sufficient at this point, one can resort to various degrees of high resolution measurement if necessary. In the latter case, one can achieve separation of ions which are only minutely different in mass. This mass difference is the result of the small differences in chemical composition between the ions in question and permits virtually specific elimination of contaminants.

In summary, GCMS ion monitoring analytical techniques permit the measurement of stable isotope enrichment in specific chemical compounds present in complex biological materials with a high degree of specificity and a minimal degree of preparation. Although GCMS is viewed as a complicated analytical approach by many investigators who use radio-tracers, stable isotope methodology is, in fact, simpler to use for a wide variety of experimental circumstances. Furthermore, because of the considerations outlined above, the confidence one obtains in the measured result far exceeds that usually possible by conventional 'black box' liquid scintillation methods.

## Discussion

*Munro* (*Tufts University, Boston, USA*): Is the derivatization step subject to incomplete recovery? Are you liable to have an isotope dilution effect that discriminates against the labelled compound?

*Bier:* At the level of isotope that we administer for conventional GCMS measurements, and with the chemical conditions used for the derivatization processes, isotopic fractionation is not a problem.

*Kerr (Case Western Reserve University, Cleveland, USA):* Would you comment on the measurement of the doubly labelled $^{13}C$-$^{15}N$-leucine which has been used by Young?

*Matthews:* This illustrates the great advantage of being able to choose between two different methods of ionization, as described by Bier. For di-labelled leucine the peak of the leucine derivative by chemical ionization would be at mass 218. Molecules that have either a single $^{13}C$ or a single $^{15}N$ atom would be located at mass 217. Further selectivity is obtained by using electron impact ionization which produces a fragment not containing the carboxyl-carbon, so that only the $^{15}N$ is measured. This is an excellent example of how, with a single instrument and with two modes of ionization, one can separate two single labels in the same molecule.

*Walser (Johns Hopkins University, Baltimore, USA):* What degree of enrichment do you need to get reasonable precision?

*Bier:* Most of our studies are done at a plasma substrate isotopic enrich-ment of the order of 1–2 atom % excess. The relative measurement precision at this enrichment is 0·5–1 %. The absolute minimal detectable enrichment is about 0·1–0·2 atom % excess.

*Klein (Baylor College of Medicine, Houston, USA):* In similar studies that we have done with $^{13}C$-labelled bile acids, the tracer is usually measurable from 2 to about 0·02 atom % excess.

*Bier:* When I took up this line of work I was not a mass spectrometrist, and I was told by several very good mass spectrometrists that one had to introduce multiple labels into a molecule, because one could not measure precisely a few tenths of a percent enrichment on top of a large natural isotopic background in an intact derivatized molecule. For example, the natural $m + 1/m$ ratio for the molecular ion of *N*-acetyl-n-propyl leucine is approximately 12·5 %. This opinion has proved to be totally wrong, as shown by the work of several people here.

*References*

BAILLIE, T. A. (1978). *Stable Isotopes: Applications in Pharmacology, Toxicology and Clinical Research*, University Park Press, Baltimore.

BIER, D. M., ARNOLD, K. J., SHERMAN, W. R., HOLLAND, W. H., HOLMES, W. F. and KIPNIS, D. M. (1977). *In vivo* measurement of glucose and alanine metabolism with stable isotopic tracers, *Diabetes*, **26**, 1005–15.

BIER, D. M. and CHRISTOPHERSON, H. L. (1979). Rapid micromethod for determination of $^{15}$N enrichment in plasma lysine: application to measurement of whole body protein turnover, *Anal. Biochem.*, **94**, 242–8.

CAPRIOLI, R. M. and BIER, D. M. (1980). Use of stable isotopes, in *Biochemical Applications of Mass Spectrometry, First Supplementary Volume* (Eds G. R. Waller and O. C. Dermer), pp. 895–925, Wiley, New York.

LIGON, W. V. (1979). Molecular analysis by mass spectrometry, *Science*, **205**, 151–519.

MATTHEWS, D. E., BEN-GALIM, E. and BIER, D. M. (1979). Determination of stable isotopic enrichment in individual plasma amino acids by chemical ionization mass spectrometry, *Anal. Chem.*, **51**, 80–4.

MATWIYOFF, N. A. and OTT, D. G. (1973). Stable isotope tracers in the life sciences and medicine, *Science*, **181**, 1125–33.

WATSON, J. T. (1976). *Introduction to Mass Spectrometry: Biomedical, Environmental, and Forensic Applications*, Raven, New York.

WIECKO, J. and SHERMAN, W. R. (1976). Boroacetylation of carbohydrates, correlations between structure and mass spectral behaviour in monoacetyl-hexose cyclic boronic esters, *J. Am. Chem. Soc.*, **98**, 7631–7.

# 25

# Advances in the Measurement of Stable Isotopes

D. HALLIDAY

*Clinical Research Centre, Harrow, UK*

The accurate and precise measurement of isotopic enrichment or depletion of a given atom demands the use of mass spectrometry. Isotope ratio mass spectrometers of varying degrees of sophistication have been used for measurements of $^{15}N$ abundance in studies of amino acid and protein metabolism in man since the pioneer work of David Rittenberg. It is therefore appropriate in this symposium to investigate the present 'state of the art' of isotope ratio mass spectrometers and the preparation of samples for isotopic analysis together with their general areas of involvement in studies of protein metabolism.

## *The Isotope Ratio Mass Spectrometer*

Isotope ratio mass spectrometers are low resolution, high sensitivity instruments. Improvements in vacuum technology and the use of polyphenyl ether diffusion pump oils provide a vacuum of $10^{-4}$ torr in the gas inlet region and $10^{-9}$ torr in the analyser region. All gas inlet and analyser components can be baked to $300\,^\circ C$ to reduce the background spectrum.

The gas inlet system is of a dual nature such that reference and sample gases may be handled independently. Variable bellows on each inlet permit balancing of the two major ion currents—an essential procedure, as the measured enrichment is pressure-dependent to a marked degree. The gases pass through 'crimped' capillary leaks into a common changeover valve that facilitates alternate entry of reference and sample gas into the source region of the analyser tube on a time basis. The 'crimped' region of the capillary converts the gas flow from viscous to molecular, so that on entering the ion source the gas has an identical isotopic composition to that in its gas reservoir (Halsted and Nier, 1950). The mass spectrometer proper consists of the analyser tube that contains an ion source, ion collector and deflection region all within a vacuum envelope. For convenience and to reduce the volume, the vacuum envelope is crescent-shaped with the source and collector housed at opposite ends.

Ionization is caused by the bombardment of the gas sample in the source region and the required ions so generated are then extracted from the source region by the creation of an electrically non-desirable environment. The ions, on leaving the source, are subjected to an accelerating voltage and a magnetic field which cause the masses to separate. Their radius of curvature depends upon their mass, their velocity and the magnetic field. By adjusting the relationship between the magnetic field and the accelerating voltage the required ions are focussed into the collector region. The collector is either a Faraday plate or bucket, more usually the latter, which can be of either single or multiple bucket configuration depending on the isotopic measurement required. Major and minor resolving slits are positioned before the buckets and electron suppressor electrodes minimize ion drift between the collectors and reject secondary electrons formed by ion collisions. The improvement of ion optics within the analyser tube has been dramatic over the last decade and results primarily from the use of electrochemically etched slit and focussing plates and the electropolishing of stainless steel surfaces to preclude rapid contamination.

The ion current charge (major or minor) imparted to its respective collector bucket, resulting from the ion beam impact, is converted into a usable voltage. These voltages are measured in a ratio, so that the ion beam attributable to the isotope is given as a ratio of the major component. Advances in electronic technology have meant that the actual ratio measurement has developed from the bridge-type systems, employing a recorder as one leg of the bridge, to more sophisticated pulse-counting techniques which are directly proportional to the voltage input.

It is difficult to state the absolute accuracy of isotopic measurement obtainable from this type of instrument because gaseous sample preparation, introduction to the inlet system and instrumental errors are additive. Recently we prepared three separate samples of $^{15}$N-labelled molecular nitrogen by the action of LiOBr on ($^{15}$NH$_4$)$_2$SO$_4$. These samples were analysed in the mass spectrometer by three operators and each sample was run six times against the reference gas without retuning the instrument. The results demonstrating instrumental performance are given in Table 1.

TABLE 1

INSTRUMENTAL ACCURACY AND PRECISION OF ISOTOPIC
RATIO MEASUREMENT

| Mass spectrometer run | $^{15}N$ atom % excess[a] | | |
|---|---|---|---|
| | Sample 1 | Sample 2 | Sample 3 |
| 1 | 0·025 038 | 0·024 902 | 0·025 119 |
| 2 | 0·025 031 | 0·024 917 | 0·025 131 |
| 3 | 0·025 041 | 0·024 923 | 0·025 106 |
| 4 | 0·025 044 | 0·024 889 | 0·025 119 |
| 5 | 0·025 016 | 0·024 906 | 0·025 115 |
| 6 | 0·025 015 | 0·024 917 | 0·025 118 |
| mean | 0·025 031 | 0·024 909 | 0·025 118 |
| SD | 0·000 012 | 0·000 012 | 0·000 008 |
| CV | 0·05 | 0·05 | 0·03 |

[a] $0·5$ mg $(^{15}NH_4)_2SO_4$—reference gas BOC Research grade $N_2$.

## Gas Sample Preparation for Isotopic Analysis

Part of any investigation with stable isotopes is the preparation of a pure gaseous sample, be it $H_2/HD$, $CO_2$ or $N_2$. The preparation should be conducted in such a manner that isotopic fractionation is precluded at every stage.

The analysis of the deuterium content of a biological fluid or the water of combustion derived from a specific organic species is performed on hydrogen gas. The preparation is ideally carried out in a uranium reduction furnace at 800 °C (Friedman, 1953) and the $H_2/HD$ gas so formed is transferred to the mass spectrometer. Methods of preparing $CO_2$ for isotopic analysis depend on the nature of the starting material. For example, following 1-$^{13}C$-glucose administration to measure turnover, plasma and glucose may be enzymatically decarboxylated (Kalhan *et al.*, 1977). Alternatively, if an estimate of glucose oxidation is required, the subject has already performed the oxidative combustion and thus only collection and separation of $CO_2$ from expired air is required. This may be performed either by direct cryogenic trapping of $CO_2$ in liquid nitrogen (Lacroix, 1972) or by the precipitation of $CO_2$ as carbonate (Schoeller *et al.*, 1975) and the liberation of the $CO_2$ by the action of 100 % phosphoric acid.

Despite the continual use of $^{15}N$, not only in clinical investigations but also in animal and agricultural studies, analytical techniques have changed little in the last few decades. Essentially the $^{15}N_2$ gas required for isotopic analysis is obtained either by Dumas combustion or by the Kjeldahl–Rittenberg technique (Rittenberg *et al.* 1948). The ammonium salt resulting from the Kjeldahl digestion is recovered and oxidized to molecular nitrogen by the action of sodium or lithium hypobromite. This final oxidation stage, originally performed in Rittenberg tubes, has been modified to the stage of semi-automation (Ross and Martin, 1970; Porter and O'Deen, 1977).

Several of the preparation systems outlined above are commercially available in an automated form under computer control. An alternative system for gas preparation is that of a direct link from a combustion furnace to the mass spectrometer. Selective trapping of $H_2O$, $CO_2$ or $N_2$ permits entry of only the gas of interest into the mass spectrometer. The incorporation of a detector in the line allows quantitation of the gas before isotopic assessment. Sano *et al.* (1976) used the resolving power of a gas chromatograph in series with a combustion furnace for initial component separation. Matthews and Hayes (1978) have used this system to determine $^{15}N$-labelling in individual amino acids obtained from the hydrolysis of human serum albumin following the ingestion of a quantity of $^{15}N$-glycine. The intervention of a combustion furnace will obviously reduce the measured enrichment relative to the number of unlabelled like atoms.

A range of sample preparation configurations are shown diagrammatically in Fig. 1 and the expected precision of isotopic measurement for a given sample size is presented in Table 2.

The majority of systems for sample preparation cited above require previous isolation and/or purification of the isotopically labelled compound of interest before production of gas. Investigation of nitrogen

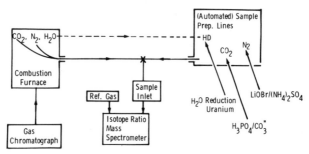

FIG. 1.    Possible configurations for gas sample preparation and introduction into an isotope mass spectrometer.

## TABLE 2

SAMPLE SIZE AND PRECISION OF MEASUREMENT OF THE ISOTOPE CONTENT OF A PURE GAS
SAMPLE

| Isotope | Gas analysed | Sample size— standard inlet ($cm^3$ at NTP) | Precision of measurement | | Gas transfer into minimum sample inlet |
|---------|--------------|------------|----------|----------|------------|
| | | | ($\delta‰$) | (atom % excess) | |
| D | $H_2$/HD | 0·2 | ±0·5 | ±0·000 5 | uranium 80° act. charcoal |
| $^{13}C$ | $CO_2$ | 0·1 | ±0·1 | ±0·000 1 | liq. $N_2$ |
| $^{15}N$ | $N_2$ | 0·1 | ±0·1 | ±0·000 1 | liq. He, mol. sieve |
| $^{18}O$ | $CO_2$ | 0·1 | ±0·1 | ±0·000 1 | liq. $N_2$ |

metabolism in man with $^{15}$N-labelled compounds necessarily results in the final estimation of the isotopic content of urinary ammonia, urea or α-amino nitrogen and often the corresponding plasma components. Classical methods for the isolation of urinary ammonia and urea are at best laborious and at worst may well result in a degree of isotopic fractionation. In our hands the aeration method for extracting ammonia from urine requires some two-and-a-half to three hours to displace 90% of the ammonia while the often used xanthydrol derivative of urea is notoriously difficult to recrystallize. The use of urease to produce ammonia from urea has to be followed by an aeration or a distillation process which may not result in full recovery of the product.

Ion exchange resins have been used to remove ammonia from urine, but a resin which specifically combines with ammonia has not been used in isotope work. Hutchinson and Labby (1962) have described the use of the $Na^+$/$K^+$ form of Dowex 50 to extract ammonia from whole blood and thence the direct reaction with Nesslers reagent to measure the blood ammonia. We have developed and extended this principle to bind ammonia to the $Na^+$/$K^+$ form of Bio-Rad AG50-X8. The complex so formed liberates molecular nitrogen on adding alkaline LiOBr without further processing. Ammonia liberated from urea by the action of urease or from α-amino acids following treatment with ninhydrin may be handled in a similar manner by a modification of a method described by Kennedy (1965). The appeal of this resin-based method relates to its specificity, stability of the ammonia complex and the ease of processing samples in batches. Only methylamine, rarely present in urine, has been found to bind to the resin and thus interfere with isotopic measurements. The general areas of application of this resin system are schematically shown in Fig. 2.

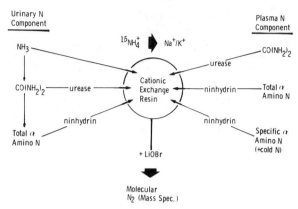

FIG. 2. Cationic resin–ammonia complex for the batch preparation of selected plasma and urinary nitrogenous components.

# Isotope Mass Spectrometry used in Studies of Protein Metabolism

Human tissues will in general reflect the isotopic ratios of their major food sources (Gaffney *et al.*, 1978). This in itself suggests that isotopic fractionation resulting from differences in rates of reaction and diffusion or equilibrium constants are minimal within the body. However, these effects require evaluation when measuring small enrichments in human tissues or metabolic end products.

Whole body protein turnover is commonly estimated from the isotopic analysis of a urinary end product following a pulse dose or constant infusion of $^{15}$N-amino acid. Whether these estimates are calculated from the change in end product labelling with time (compartmental analysis) or from the cumulative excretion of isotope (stochastic methods), the enrichment measured will be low due to vast dilution within the metabolic pool. Similar quantitative considerations apply when measuring expired $^{13}CO_2$ to calculate the oxidation rate and indirectly the whole body protein synthesis rate during the constant infusion of a $^{13}$C-amino acid. The use of an isotope ratio mass spectrometer in this area of protein metabolism is obligatory where sample size is not a limiting factor but low isotopic enrichments require high accuracy of measurement.

Constant infusion of a single $^{15}$N-labelled amino acid where estimates of protein flux are calculated from plasma isotopic equilibrium has been attempted. Halliday and McKeran (1975) used $^{15}$N-lysine in this manner to calculate both whole body protein turnover and the fractional synthesis rate of muscle protein from serial muscle biopsies obtained when plasma isotopic plateau of $^{15}$N-lysine had been achieved. The analytical protocol required large blood samples (30 ml), the isolation of plasma lysine and a protracted sample work-up before the final isotopic analysis. This analytical procedure has been superseded by the development of sophisticated gas chromatography–mass spectrometry (GC/MS) techniques capable of estimating $^{13}$C or $^{15}$N enrichment at the 2–5 atom % level in individual plasma amino acids from less than 0·5 ml blood (Bier and Christopherson, 1979; Matthews *et al.*, 1980; see also p. 289).

Protein flux is calculated from the known rate of infusion of labelled material divided by the isotopic enrichment at plateau. This holds whether an individual amino acid or whole protein is infused. We have infused $^{15}$N-yeast protein nasogastrically into two subjects on repeated occasions for 22 h. Protein flux was calculated from the plateau labelling of total $\alpha$-amino-$^{15}$N which occurred after approximately 16 h into the infusion. $^{13}$C-leucine was added to the infusate on one occasion for each subject and the synthesis rate was calculated from expired $^{13}$CO$_2$ by the method of Golden and Waterlow (1977). A summary of the results obtained are given in Table 3.

Attention has been directed towards the capabilities and relevant applications of isotope ratio mass spectrometry to studies of nitrogen metabolism in man. The complementary nature of GC/MS techniques has been discussed elsewhere in this symposium (Paper 24).

TABLE 3

PROTEIN TURNOVER CALCULATED FROM TOTAL PLASMA $\alpha$-AMINO-$^{15}$N PLATEAU DURING A NASOGASTRIC INFUSION OF $^{15}$N-YEAST PROTEIN IN TWO SUBJECTS

| Subject | Protein intake (g/day) | Flux (g/day) | | Protein turnover Synthesis (g/kg/day) | Catabolism (g/kg/day) |
|---|---|---|---|---|---|
| 1. Normal | 72 | 398 | | 5·35 ± 0·23 | 5·22 ± 0·24 |
| | 72 | 411 | (5·71)[a] | 5·46 ± 0·29 | 5·43 ± 0·32 |
| | 30 | 350 | | 4·74 ± 0·23 | 5·12 ± 0·27 |
| 2. Brainstem | 40 | 405 | | 6·74 ± 0·22 | 6·73 ± 0·29 |
| CVA | 40 | 394 | (6·30)[a] | 6·56 ± 0·25 | 6·30 ± 0·24 |

[a] Synthesis calculated from $^{13}$C-leucine oxidation rate (Golden and Waterlow, 1977) when $^{13}$C-leucine was infused nasogastrically with $^{15}$N-yeast protein for 22 h.

# References

BIER, D. M. and CHRISTOPHERSON, H. L. (1979). Rapid micromethod for determination of $^{15}N$ enrichment in plasma lysine: application to measurement of protein turnover, *Anal. Biochem.*, **94**, 242–8.

FRIEDMAN, I. (1953). Deuterium content of natural water and other substances, *Geochim. Cosmochim. Acta.*, **4**, 89–103.

GAFFNEY, J. S., IRSA, A. P., FRIEDMAN, L. and SLATKIN, D. N. (1978). Natural $^{13}C/^{12}C$ ratio variations in human populations, *Biomed. Mass Spectr.*, **5**, 495–7.

GOLDEN, M. H. N. and WATERLOW, J. C. (1977). Total protein synthesis in elderly people: a comparison of results with $^{15}N$-glycine and $^{14}C$-leucine, *Clin. Sci. & Mol. Med.*, **53**, 277–88.

HALLIDAY, D. and MCKERAN, R. O. (1975). Measurement of muscle protein synthetic rate from serial muscle biopsies and total body protein turnover in man by continuous intravenous infusion of L-[$\alpha$-$^{15}N$] lysine, *Clin. Sci. & Mol. Med.*, **49**, 581–90.

HALSTED, R. E. and NIER, A. O. (1950). Gas flow through the mass spectrometer viscous leak, *Rev. Sci. Instr.*, **21**, 1019–21.

HUTCHINSON, J. H. and LABBY, D. H. (1962). New method for micro-determination of blood ammonia by use of a cation exchange resin, *J. Lab. Clin. Med.*, **60**, 170–8.

KALHAN, S. C., SAVIN, S. M. and ADAM, P. A. J. (1977). Estimation of glucose turnover with stable tracer glucose -1-$^{13}C$, *J. Lab. Clin. Med.*, **89**, 285–94.

KENNEDY, I. R. (1965). Release of nitrogen from amino acids with ninhydrin for $^{15}N$ analysis, *Anal. Biochem.*, **11**, 105–10.

LACROIX, M. (1972). Thesis, University of Liege.

MATTHEWS, D. E. and HAYES, J. M. (1978). Isotope-ratio-monitoring gas chromatography–mass spectrometry, *Anal. Chem.*, **50**, 1465–73.

MATTHEWS, D. E., MOTIL, K. J., ROHRBAUGH, D. K., BURKE, J. F., YOUNG, V. R. and BIER, D. M. (1980). Measurement of leucine metabolism in man from a primed, continuous infusion of L-[-$^{13}C$] leucine, *Am. J. Physiol.*, **238**, E473–9.

PORTER, L. K. and O'DEEN, W. A. (1977). Apparatus for preparing nitrogen from ammonium chloride for nitrogen-15 determinations, *Anal. Chem.*, **49**, 514–616.

RITTENBERG, D., NIER, A. O. C. and REIMANN, S. (1948). In *Preparation and Measurement of Isotopic Tracers* (Ed. J. W. Edwards), University of Michigan Press, Ann Arbor, pp. 31–42.

ROSS, P. J. and MARTIN, A. E. (1970). A rapid procedure for preparing gas samples for nitrogen -15 determination, *Analyst*, **93**, 817–22.

SANO, M., YOTSUI, Y., ABE, H. and SASAKI, S. (1976). A new technique for the detection of metabolites by the isotope $^{13}C$ using mass fragmentography, *Biomed. Mass Spectr.*, **3**, 1–3.

SCHOELLER, D. A., KLEIN, P. D. and SCHNEIDER, J. F. (1975). In *Proceedings of the Second International Conference on Stable Isotopes.* (Oak Brook, Illinois) (Ed. E. R. Klein and P. D. Klein) pp. 246–51.

# 26

# Measurement of Whole Body Protein Turnover by Constant Infusion of Carboxyl-Labelled Leucine

P. J. Garlick and G. A. Clugston

*Clinical Nutrition and Metabolism Unit,
London School of Hygiene and Tropical Medicine, London, UK*

## Introduction

Carbon-labelled amino acids were first used to measure the rate of whole body protein turnover in man by Waterlow (1967), who gave $^{14}$C-lysine by constant infusion. After 24 h of infusion the specific radioactivity of the free lysine in plasma had reached a constant value (plateau), from which the turnover rate of the free lysine pool of the body (the lysine flux) could be calculated. Subsequent refinements enabled the flux to be compartmented between protein synthesis and protein oxidation by collection of respiratory $^{14}CO_2$. Although this was initially achieved during infusion of $^{14}$C-tyrosine (James *et al.*, 1974, 1976), the preferred labelled material is now generally recognized to be leucine labelled in the carboxyl group with $^{14}$C or $^{13}$C. The advantage of this amino acid is that the metabolism of the labelled material is restricted to protein synthesis and to oxidation, with subsequent excretion of labelled $CO_2$ in the breath. Rates of protein metabolism can then be calculated with the aid of the model shown in Fig. 1.

At plateau specific activity ($S$) the amount of isotope leaving the free leucine pool is equal to the rate of entry of isotope from the infusion ($i$). Hence

$$i = Q.S \tag{1}$$

where $Q$ is the rate of leucine flux. Similarly, if the size of the free leucine pool remains constant, the rates of entry and exit of leucine must be the same. Hence

$$Q = E + Z = B + I \tag{2}$$

These equations then enable the rates of protein synthesis, oxidation and

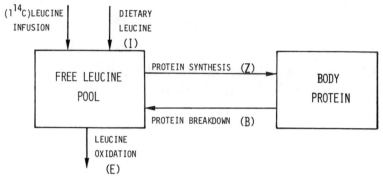

FIG. 1.    Simple 2-pool model for leucine metabolism in the whole body, used for calculating rates of protein turnover from information obtained during constant infusion of $1\text{-}^{14}C$ leucine.

breakdown to be calculated, as described in more detail by Waterlow *et al.* (1978).

This technique has a number of advantages which have been discussed previously (Waterlow *et al.*, 1978). It is particularly convenient for observing rapid changes in protein metabolism, since the plateau takes 5 h

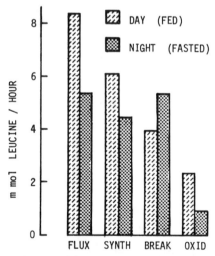

FIG. 2.    Diurnal changes in rates of flux, synthesis, degradation and oxidation of protein measured in 10 obese subjects by constant intravenous infusion of $1\text{-}^{14}C$-leucine for 24 h. Rates were calculated from eqns (1) and (2) and expressed as mmol leucine/h. During the day subjects received 12 hourly feeds containing approximately 75 g protein and 1600 kcals. During the night there was no food intake.

or less to be achieved, or to be re-established after a change in experimental conditions during the infusion. For example, a change of food intake results in a rapid readjustment of the plateaux for leucine specific activity and $^{14}CO_2$ excretion, which is complete within 5 h (Garlick *et al.*, 1980). Rates of protein synthesis and oxidation were shown to be higher during the daytime, when food was taken regularly, than at night, when the subjects fasted (Fig. 2). By contrast, protein breakdown was highest during fasting and larger than originally thought (Clugston and Garlick, unpublished data). However, if we are to have confidence that the observed changes in the metabolism of labelled leucine represent real changes in the metabolism and turnover of body protein, it is important that we examine the sources of error in this approach. We have therefore attempted to assess the validity of some of the assumptions that are necessary to enable rates of protein metabolism to be derived from information obtained during constant infusion of $1$-$^{14}C$-leucine.

## Changes in Size of the Free Leucine Pool

It was stated above that eqn (2) is only correct if the body pool of free leucine does not change in size during the period of measurement. Clearly this is a reasonable assumption if the subject has been in a steady state for some time before measurement, but when acute changes are being observed, it may not hold. For example, in the study shown in Fig. 2, it was deduced from the relative rates of oxidation of leucine during feeding and fasting that a proportion of the leucine intake during the day was stored, and then withdrawn from store and oxidized at night. The question is, was this leucine stored in the form of body protein, or was it merely an expansion of the free leucine pool?

Table 1 shows the concentration of leucine in the plasma of three subjects during successive periods of feeding and fasting. In the period when hourly feeds were given the concentrations were higher than the post-absorptive values by about 40 $\mu$mol/litre. If, for the sake of argument, leucine was assumed to be uniformly distributed in a body water volume of 40 litres, this change would amount to 1·6 mmol leucine. Obviously this is a very rough estimate, since the concentration in the tissues may be higher than in the plasma (in muscle about 40% higher, Möller *et al.*, 1979) and may not change in the same way. However, when this amount of leucine is compared with the estimate of that stored while feeding, about 11 mmol in 12 h, it is clear that a change in the size of the free leucine pool is very unlikely to

TABLE 1

LEUCINE CONCENTRATION IN PLASMA OF
SUBJECTS INFUSED WITH 1-$^{14}$C-LEUCINE
FOR 24 H

| Subject | Leucine concentration ($\mu mol/litre$) | |
| --- | --- | --- |
| | Fed | Fasted |
| EK | 149 ± 22 | 109 ± 9 |
| RO | 148 ± 19 | 100 ± 4 |
| KG | 102 ± 10 | 71 ± 6 |

The infusions were divided into 12-h periods of feeding and fasting as in Fig. 2. Measurements (means ± SD) on several samples taken during the last 6 h of each period (Clugston and Garlick, unpublished).

account for the storage. In addition, it was observed that the change in concentration in the blood took place within the first few hours after the change in food intake, before the plateaux, used to calculate the values shown in Table 1, were reached.

A second piece of evidence also leads us to think that the accumulation of amino acids during feeding occurs by storage as tissue protein. Table 2

TABLE 2

LEUCINE OXIDATION AND NITROGEN EXCRETION IN FED AND FASTED NORMAL SUBJECTS

| | Day (fed) | Night (fasted) | Day/night |
| --- | --- | --- | --- |
| Leucine oxidation (mmol/h) | 2·19 ± 0·17 (5) | 0·96 ± 0·08 (5) | 2·3 |
| N excretion (g/9 h) | 5·84 ± 0·13 (20) | 2·54 ± 0·16 (8) | 2·3 |

Rates of leucine oxidation were measured as in Fig. 2. Nitrogen excretion, corrected for changes in the size of the body urea pool, was measured in a different group of subjects over a 9-h period during which they received either a regular intake of protein and energy (daytime) or were fasted (night). Mean rates ± SEM; number of observations in parentheses. Data of Fern et al. unpublished.

shows the relative rates of leucine oxidation during feeding and fasting compared with the rates of urinary nitrogen excretion, which have been corrected for changes in the body urea pool. The ratio of feeding/fasting rates is the same for leucine oxidation and nitrogen excretion. The most likely explanation for this parallelism between leucine and total nitrogen is that all amino acids are being stored together as protein, rather than independently as free amino acids whose concentrations and distribution in the body vary considerably.

## *Specific Activity of the Free Amino Acid at the Site of Protein Synthesis*

In calculating the rate of flux from eqn (1) it must be assumed that the free leucine pool is homogeneous and that its specific activity is represented by that in the plasma. It is known from studies on animals, however, that the specific activity of free leucine in the tissues is lower than in plasma, and is different in each separate tissue. The problem is aggravated by the fact that even in a single tissue the free amino acid pool is compartmented, so that it is not certain whether the specific activity at the site of protein synthesis better approximates that in the tissue or that in the plasma (for review, see Waterlow *et al.*, 1978). An illustration of this problem during constant infusion of labelled amino acids is shown in Table 3. This shows that when $^{14}$C-glycine was infused into rats the ratio of specific activity of free serine to glycine was higher in the liver than in the plasma because of intracellular conversion of glycine to serine. However, the ratio in the proteins synthesized in the liver was the same as in the intracellular pool, suggesting that intracellular amino acids were used for protein synthesis. By contrast, however, when a mixture of U-$^{14}$C- and 2,3-$^{3}$H-tyrosine was infused into

TABLE 3

SPECIFIC ACTIVITY RATIO AFTER A 6-H CONSTANT INFUSION OF $^{14}$C-GLYCINE OR A MIXTURE OF U-$^{14}$C-TYROSINE AND 2,3-$^{3}$H-TYROSINE

|               | *Serine/glycine* | $^{3}H$-/$^{14}C$-*tyrosine* |
|---------------|------------------|------------------------------|
| Plasma        | $0.31 \pm 0.01$  | $1.02 \pm 0.001$             |
| Liver, free   | $0.96 \pm 0.03$  | $0.78 \pm 0.04$              |
| Liver protein | $0.90 \pm 0.02$  | $1.05 \pm 0.02$              |

Data of Fern and Garlick (1976) and Nicholas *et al.* (1977).

rabbits, the difference between the ratio of $^3H$ to $^{14}C$ in tyrosine in the plasma and that in the tissue was a result of the intracellular removal of $^3H$ attached to C-2 during transamination. The labelling of protein in this case more closely resembled that in the plasma. This suggests that there is no easy solution to the problem of which compartment should be used for measurement of the precursor specific activity, because this may vary depending on the conditions of the experiment.

It can be seen, therefore, that the flux rate estimated from measurements on the plasma is likely to underestimate the true flux by an amount which will vary depending on the amino acid used and probably on other conditions, such as the nutritional state. Table 4 demonstrates the magnitude of this problem during constant infusion of labelled leucine and lysine into young pigs. The specific activity of free leucine in a range of tissues was between 22–100 % of that in the plasma. With lysine, however, the intracellular specific activity was even lower, between 11–56 % of that in the plasma. Not surprisingly, the flux rate calculated in terms of g protein/day was far lower with lysine than with leucine. This probably explains why flux rates determined in a number of other laboratories by constant infusion of labelled lysine appear to be somewhat lower than those obtained by other

TABLE 4

THE RATIO OF SPECIFIC ACTIVITY OF FREE LEUCINE AND FREE LYSINE IN TISSUES TO THAT IN PLASMA MEASURED AT THE END OF 6-H INFUSIONS OF U-$^{14}$C-LEUCINE AND U-$^{14}$C-LYSINE INTO 30 KG PIGS

| Tissue | Intracellular/plasma specific activity | |
|---|---|---|
| | Leucine | Lysine |
| Liver | 0·22 | 0·11 |
| Jejunum | 0·57 | 0·18 |
| Muscle | 0·58 | 0·25 |
| Heart | 1·03 | 0·56 |
| Skin | 0·43 | 0·23 |
| Flux ($\mu$mol/min) | 192 | 69 |
| Flux (g protein/kg/day) | 16·4 | 8·6 |

The flux was calculated from the specific activity of the amino acid in plasma and converted from $\mu$mol/min to g protein/kg/day by assuming that leucine comprises 7·4 g/100 g protein and lysine 5·6 g/100 g protein. (Data of Simon et al., 1978.)

techniques, both in man (e.g. Waterlow, 1967; Halliday and McKeran, 1975) and in animals (e.g. Buttery *et al.*, 1975).

No values are currently available in the literature for the specific activity in tissues of man during constant infusion of labelled leucine. A possible way round this problem, however, is to assume that the precursor of protein synthesis is the same as that for leucine oxidation. The rate of leucine oxidation can be calculated independently from the rate of nitrogen excretion in the urine, on the assumption that leucine oxidation will represent the same proportion of nitrogen excretion as dietary leucine intake does of nitrogen intake. This is shown in Table 5. The rate of leucine

TABLE 5

COMPARISON OF MEASURED LEUCINE OXIDATION WITH THAT CALCULATED FROM URINARY NITROGEN EXCRETION IN OBESE AND NORMAL SUBJECTS UNDERGOING 24-H CONSTANT INFUSION WITH 1-$^{14}$C-LEUCINE, WITH 12 H PERIODS OF FEEDING AND FASTING AS IN FIG. 2

| *Subjects* | *Leucine oxidation (mmol leucine measured)* $E_l$ | 24-h excretion | | $E_l/E_u$ (%) |
|---|---|---|---|---|
| | | *(g nitrogen)* | *(mmol leucine calculated)* $E_u$ | |
| Obese (10) | $40.4 \pm 4.8$ | $13.14 \pm 1.10$ | $59.7 \pm 4.1$ | $67.7 \pm 6.7$ |
| Normal (5) | $37.8 \pm 5.9$ | $13.50 \pm 2.14$ | $54.4 \pm 7.9$ | $65.0 \pm 15.1$ |

The total oxidation over 24 h ($E_l$) was compared with the apparent rate of leucine oxidation ($E_u$) calculated from the 24-h urinary nitrogen excretion in the same subjects. Means $\pm$ SD (Clugston and Garlick, unpublished).

oxidation measured from the production of $^{14}CO_2$ in the breath was about 67% of the value estimated from nitrogen excretion, indicating that the specific activity of leucine at the site of oxidation was 67% of that in the plasma. A slightly higher value of 80% was reported by Golden and Waterlow (1977), perhaps reflecting variations due to different dietary intakes.

The apparent underestimate of oxidation, and by inference, of flux, could influence the validity of results in two ways. First, it is possible that the relationship shown above could vary under different experimental conditions. In the study of feeding and fasting shown in Fig. 2 we do not believe that this had occurred, because there were parallel changes in leucine oxidation (from $^{14}CO_2$ excretion) and nitrogen excretion (Table 2). Secondly, values for protein synthesis and leucine oxidation adjusted for this factor (Fig. 3) are higher than the original values in Fig. 2, but the changes brought about by feeding and fasting are proportionally the same. The effect on protein breakdown is, however, quite different, because this

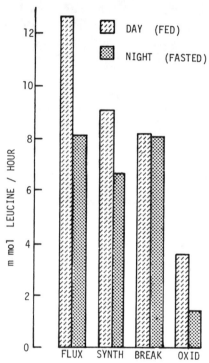

FIG. 3.    Diurnal changes in rates of protein turnover after correction for an apparent error resulting from the inhomogeneity of the free leucine pool of the body. Data were obtained in the same experiment as described in Fig. 2, but it was assumed that the specific activity of free leucine at the sites of protein synthesis and leucine oxidation was 67 % of that in plasma, as described in more detail in the text.

was calculated from the flux, which is now higher, and the dietary leucine intake, which is unaltered (eqn 2). The adjusted values indicate no change in protein breakdown when the subjects are fed or fasted. This contrasts with the values shown in Fig. 2 where breakdown appeared to rise with fasting. It should be emphasized that we do not know whether Fig. 3 is a better representation of the changes in protein metabolism on feeding than Fig. 2, because the assumption that the specific activity at the sites of leucine oxidation and protein synthesis were the same, may not be valid. What is more important however is that, in spite of the problems, we can have some confidence that the acute changes in protein synthesis and leucine oxidation brought about by feeding are real, but that the effect on protein breakdown is less certain.

## *Collection of Labelled $CO_2$*

Determination of the rate of leucine oxidation requires an estimate of the rate of production of respiratory $^{14}CO_2$. It has been shown, however, that recovery of infused $H^{14}CO_3^-$ in the breath is not complete and that about 20 % cannot be accounted for (James *et al.*, 1976). A correction factor has usually been applied on the assumption that the same recovery of $^{14}CO_2$ will be appropriate during an infusion of 1-$^{14}$C- leucine. Since we do not know why this loss of $^{14}CO_2$ occurs, it is important to consider the possibility that it may vary under different experimental conditions. In particular, with infusions lasting 24–36 h, such as those of Fig. 2, it is possible that the recovery of $^{14}CO_2$ could gradually increase if the compartment into which it was lost should become saturated with labelled material. However, we have found the recovery to be constant both with time and with dietary state (Table 6), suggesting that errors in rates of protein oxidation from this source are unlikely.

An important source of error can, however, arise from the use of a valve and Douglas bag to collect $CO_2$. We have found that subjects frequently do not breathe naturally under these circumstances. Even after a period of

TABLE 6

PERCENTAGE RECOVERY OF $^{14}CO_2$ AFTER INFUSION OF $H^{14}CO_3^-$ OVER PERIODS OF 36 H

| Feeding<br>*0–12 h* | Fasting<br>*12–24 h* | Refeeding<br>*24–36 h* |
|---|---|---|
| $88\cdot2 \pm 4\cdot9$ | $92\cdot6 \pm 6\cdot1$ | $91\cdot3 \pm 2\cdot9$ |

Means $\pm$ SD in 10 feeding and fasting subjects and in 6 refed subjects calculated from the excretion of $^{14}CO_2$ in the later part of each time period (Clugston and Garlick, unpublished).

adjustment they can give erroneously high rates of $^{14}CO_2$ production. We find it is more reliable to measure the specific radioactivity of $CO_2$ by the technique of Kaihara and Wagner (1968) and to multiply this by the rate of $CO_2$ production measured with a ventilated hood. In practice we use a large oxygen tent of approximately 1000 litres volume. This allows reasonable freedom to move, eat and sleep during infusions lasting up to 36 h, coupled with almost continuous monitoring of $CO_2$ production and $O_2$ consumption (Clugston and Garlick, unpublished data).

## Conclusion

The above examples illustrate some potential and real sources of error in the measurement of whole body protein turnover by constant intravenous infusion of labelled amino acids, in particular $1$-$^{14}C$-leucine. The advantage of this amino acid, by comparison with $^{15}N$-glycine (which is both easier to use and disturbs the subject less) is that the metabolism of the labelled material is better understood. This makes it possible to predict potential errors and to quantitate their influence on the final result. The inability to measure the specific activity of the amino acid at the site of protein synthesis remains the major difficulty, perhaps resulting in an underestimate of true turnover rates by as much as one third. This may have serious consequences for the measurement of protein breakdown, but despite this there is evidence that measured changes in protein synthesis and oxidation can be accepted with some confidence.

## Discussion

*Walser (Johns Hopkins University, Baltimore, USA):* Did you correct the $CO_2$ output for the missing 11 % in your calculation of oxidation?

*Garlick:* Yes. All the values are corrected on the basis that the $CO_2$ produced by the oxidation of leucine is handled in the same way as infused bicarbonate.

*Walser:* Is there any problem in that assumption, since the $CO_2$ derived from oxidation is not simply injected into the antecubital vein?

*Garlick:* I think the assumption is probably reasonable, because the bicarbonate pool mixes rapidly. However, there are some inconsistencies.

We do not know where the 'missing' $CO_2$ goes to, but there is quite a lot of it. On the basis of available data one would expect the bound carbonate pool to have a half-life of about 12 h, in which case the labelled material would recycle and the recovery should increase appreciably towards 100 % later on. This does not appear to happen.

*Bier (Washington University School of Medicine, St. Louis, USA):* Flux is calculated from the dilution of tracer in the sampling pool. For some amino acids the choice of sampling pool is very important; for example, the dilution of tracer in the venous pool is quite inappropriate for estimating alanine flux; for leucine turnover measurements this problem is somewhat less since venous tracer dilution is closer to that found in the arterial pool. Expired air is yet another sampling pool. The relative confidence we can place in results, subsequently calculated from the dilution of labelled material measured in this pool, depends upon how representative we consider the sampling pool to be.

*Neuberger (Charing Cross Hospital Medical School, London, UK):* In an extreme case there might be a part of the body where the free amino acid pool communicates with the tissue at an infinitely slow rate. If protein turnover is taking place in that organ you would not be able to measure it at all. In reality there may be certain parts of the body or certain tissues which come close to this situation.

*Garlick:* I agree. This is the process which we call internal recycling, and if it is totally internal we cannot measure it. In the study of Simon *et al.* (1978) on pigs, lysine in liver had a specific activity only 11 % of that in plasma. If this is correct, we are likely to be underestimating the rate of protein synthesis in liver by a factor of nine.

*Pratt (Institute of Psychiatry, London, UK):* The movement of glycine across the blood–brain barrier is much slower than that of essential amino acids (Pratt, 1976). There is a large amount of glycine in the brain, as there is of all the non-essentials, and there is evidence of compartmentation for some of these non-essentials. Have you thought about the brain as a possible place where there may be problems of glycine not exchanging rapidly?

*Garlick:* Compartmentation is a very difficult problem, particularly for glycine. In the studies I described we were using leucine but I am sure some of the same arguments apply.

*Cohen (University of Wisconsin-Madison, USA):* I would like to make a speculation which may possibly provide a new approach to this problem. According to the signal hypothesis, for which there is now much evidence (Blobel, 1980), proteins are synthesized with 'signal' peptides attached. In the co- or post-translational processing of many proteins, the attached peptides have molecular weights of the order of 3000–5000 da. Many of them are very rich in hydrophobic amino acids, because the purpose of the signal peptide is to provide a code for transport into and through membranes. This suggests that in an organ with a rapid rate of protein turnover, there must be a significant sequestered pool of amino acids formed as a result of cleavage of the signal peptide from the pre-protein when the mature protein is finally formed. The peptides that result from this cleavage may represent a separate pool, or they may even have a feedback effect in regulating the level of certain enzymes by affecting transcription. I have not calculated the amounts, but those of you who know more than I do about actual rates of protein turnover might give some thought to this possibility of a hidden pool of amino acids in the oligo-peptide form, which may be contributing to the difficulty we have been discussing.

*Millward* (*London School of Hygiene and Tropical Medicine, UK*): There is another approach to the problem of determining whether the rates of protein turnover which we measure by the usual methods are valid. In animal experiments we can back up the kinetic measurements of rates of protein synthesis with indirect measurements of tissue RNA concentrations. These give us approximate values, and then we can see whether the methods match. This is very relevant to the question of the brain, because we have consistently found in the rat that the rate of protein synthesis per unit RNA, measured by constant infusion of tyrosine, is lower in the brain than in any other tissue. This would suggest that either there is something fundamentally different in the translational mechanism in the brain, or there is a special methodological problem, perhaps reflecting the precursor pool and the rate of penetration of amino acids into it.

*Waterlow* (*London School of Hygiene and Tropical Medicine, UK*): If we come back to the practical problem of measuring protein turnover in man, the question is: how important is it that the specific activity in plasma, which we sample, is not the same as in the true precursor pool? What is the size of the error? Golden and I (1977) tried to answer this question by comparing the flux calculated in two ways: direct, from the specific activity of leucine at plateau during a constant infusion of [14]C-leucine; and indirect, from the fraction of dose excreted and the total urinary nitrogen output. The assumptions underlying this calculation are discussed in more detail elsewhere (Waterlow *et al.*, 1978, p. 311 *et seq.*). In the direct method the plasma is, of course, taken as the precursor pool, whereas in the indirect method we are sampling the site of oxidation, which is presumably closer to the true precursor site. We came up with the answer that the plasma measurements underestimated the flux by about 20 % (see p. 309).

*Reeds (Rowett Research Institute, Aberdeen, UK):* We have made similar measurements on pigs, which I will be presenting in more detail later. I calculated urea + ammonia excretion from the flux of leucine (multiplied by the conversion factor 6·7 g leucine/100 g protein) and the proportion of dose oxidized, and compared this indirect estimate with the directly measured excretion of urea + ammonia. Figure 4 shows that there is

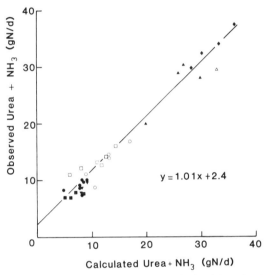

FIG. 4.    Prediction of urea and ammonia excretion calculated from leucine catabolism.

excellent agreement between the two estimates ($r = 0.98$), with a slope of unity. The indirect calculation depends on a number of assumptions, particularly that the dietary protein is similar in composition to body protein (Waterlow *et al.*, 1978) and that the 'conversion factor' for amino acid to body protein is correct. It is possible that errors in these assumptions cancel out.

*Garlick:* On the face of it, it would seem better to use, as you have done, urea + ammonia for this calculation rather than total urinary nitrogen.

However, you cannot then make use of the following assumption:

$$\frac{\text{leucine-N in diet}}{\text{total-N in diet}} = \frac{\text{leucine-N excreted}}{\text{total-N excreted}}$$

an assumption which must be true for a subject in balance, since the body cannot accumulate individual amino acids. How do you work out the contribution of leucine to urea + ammonia, as opposed to other end products?

*Golden* (*University of the West Indies, Kingston, Jamaica*)*:* With some amino acids, such as glycine, there is indeed diversion into other end products. In using excretion of total-N to calculate synthesis by the end product method, we correct for these other pathways (see page 334).

*James* (*Dunn Clinical Nutrition Centre, Cambridge, UK*)*:* That is why I have doubts about the use of glycine.

*Waterlow:* A further difficulty arises: in order to compare the two estimates of flux, one of which is in fact leucine flux and the other total-N flux, it is necessary to convert one to the other by a factor for the leucine content of whole body protein. We have used the factor 8 %, but it may not be correct.

*Reeds:* From our results we conclude that a more appropriate value would be about 6·7 %. This was arrived at by two different methods: direct measurement of the leucine content of whole body protein; and estimates of the contribution of leucine turnover in different organs to whole body protein turnover. I will give the details later (page 394).

*Millward:* When you calculated the leucine content of muscle, did you do it on the non-collagen protein or on the whole muscle? In the adult animal collagen will not participate substantially in turnover. In studies that we did on chickens (Laurent *et al.*, 1978) in the steady state collagen synthesis was a very small proportion of the total.

*Reeds:* We did our measurements on the whole protein. I think there is very little information on the relation between collagen synthesis and total protein synthesis. Our comparison does nothing but check a package of assumptions. One is the precursor–product relationship; another is the relationship between the amino acid mixture in the diet and that in body protein, which is very critical (Waterlow *et al.*, 1978, page 312). If we had used milk protein, containing about 11 % leucine, we might have got a completely different answer by the indirect method. As it was, we balanced the dietary protein to make the leucine content as close as possible to that of body protein. The comparison I presented tests a prediction. All it tells one is that the various assumptions lead to an answer which seems to be consistent between diets and animals.

*Clague (Royal Victoria Infirmary, Newcastle upon Tyne, UK):* In fact one can get the same figure for flux with many different combinations of the two values we have been discussing, e.g.: (i) intracellular specific activity 80 % of that in plasma and leucine content 8 % of whole body protein, or (ii) intracellular specific activity 100 % of that in plasma and leucine 6·7 % of whole body protein; both give the same answer.

I would like to suggest an alternative approach which, unlike that of Waterlow and Reeds, gives a separate and independent estimate of the error produced by calculating flux from plasma specific activity.

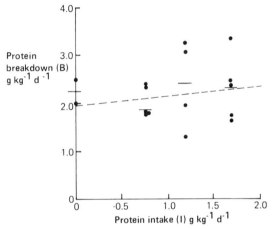

FIG. 5.    Relation between protein breakdown and protein intake in pre-operative patients. Breakdown rates measured with 1-$^{14}$C-leucine and calculated on the assumption that the intracellular specific activity is 80 % of that in plasma. There is no significant relationship.

Our results show that protein degradation remains unchanged in pre-operative non-stressed patients fed different dietary intakes of protein (Fig. 5). Plasma leucine levels, however, rise with increasing intake of protein, and hence of leucine, according to the relation:

$$y = 0\cdot462x + 99$$

where $y$ = plasma leucine concentration ($\mu$mol/litre) and $x$ = leucine intake (mg/kg/day, Fig. 6). From this relationship the fasting leucine level of

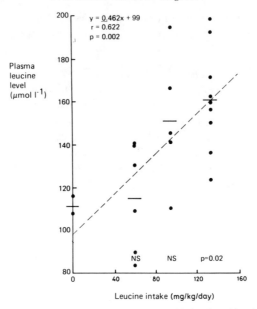

FIG. 6.   Correlation of plasma leucine concentration with intake of leucine derived from
protein.

99 $\mu$mol/litre must represent an input of 214 mg leucine/kg/day from the
degradation of protein. Our calculations of the leucine flux from the plasma
specific activity gave an average value for leucine flux of only 168 mg/kg/
day. This suggests that the intracellular specific activity is 168/214 = 79 %
of that in plasma.

This approach gives an independent estimate of the specific activity ratio,
but it does not give any information about the most appropriate value for
the leucine content of whole body protein.

I think the data support a composite picture of a leucine content of whole
body protein of 8 %, with a specific radioactivity at the site of synthesis of
whole body protein of only 80 % of the plasma value.

*Garlick:* You made the assumption that protein breakdown is constant
with dietary intake, whereas the data I showed suggest that an error in the
estimate of precursor specific activity would alter the conclusion regarding
protein breakdown; it would no longer be constant. Therefore, your
argument is a circular one and I cannot accept the conclusion.

*Clague:* If the ratio of specific activity at the site of synthesis to that in plasma is altered downwards to 65 % or upwards to 100 %, there is still no significant correlation between intake and breakdown rate.

*Bier:* The starting point for your calculations was a figure (Fig. 6) showing an intercept on the vertical axis of about 90 $\mu$mol/litre. Yet, from the data shown, the standard error of the regression estimate of that value must be close to 50 $\mu$mol/litre. This means that you can subsequently calculate almost any value you like and have it agree with this regression data.

*Clague:* I accept that. However, I should like to put forward another point which I think may be relevant. In our infusions of $^{14}$C-leucine 6·7 $\pm$ 0·5 % (mean $\pm$ SD, $n = 16$) of the infused labelled material becomes incorporated into plasma proteins. When an allowance is made for extravascular exchange and for the high leucine content of plasma proteins (11·5 %), the synthesis rate of plasma proteins, based on the specific activity of leucine in plasma, comes to 105 mg/kg/day. This is only a quarter of the expected rate of total plasma protein synthesis—350–470 mg/kg/day—quoted by Waterlow *et al.* (1978). This suggests that the specific activity of leucine at the site of synthesis of these proteins must be only about 25 % of that in plasma. To counteract this extremely low value in liver, and to produce a mean value of 80 % in the body as a whole, the specific activity of free leucine in the remaining tissues, of which muscle is a large component, must be about 90 % of that in plasma. This agrees with a value of 85 % found in human muscle *in vitro* (Lundholm and Schersten, 1975).

*Waterlow:* We estimated that albumin synthesis might account for about 7·5 % of whole body protein synthesis (Waterlow *et al.*, 1978, Table 14.10), and the synthesis of other plasma proteins might be as much again (Waterlow *et al.*, 1978, Table 15.1), so Clague's estimate is not all that far out.

*James:* From what has been said, perhaps we have been worrying too much about the discrepancy between plasma and intracellular specific radio-activities. However, this is a key question, and we do need independent methods to check its importance, for example by measuring the specific radioactivity of the keto acids, which should reflect that of the intracellular pool.

*Garlick:* I agree that it is a key question. For one thing, it makes comparisons between different tracer amino acids difficult because the relationship between the plasma specific activity and that of the true precursor might differ. Secondly, when comparisons are being made by the same method, these sources of error might not affect the relative rates of synthesis, but they can seriously affect the calculated rates of breakdown, for the reasons given in my paper.

*Cohen:* After listening to this discussion I would like to make a challenge. As someone who has been at the simple end of the spectrum, concerned only with what an enzyme does rather than with what a whole organism does, I am willing to admit that studying an isolated enzyme only gives you the information that an isolated enzyme can give, and extrapolating to the whole organism has serious limitations. I think a sophisticated enzymologist would agree that this is where he should stop. He may project what he thinks might happen, but he will not go as far as he might have gone 20 years ago, and say with certainty that things go on at this particular rate. Now we hear from people who are trying to measure events in the whole body, who have made the jump from one end of the spectrum to the other, and they are not sure what they are doing. Whereas perhaps 15 years ago you might have pounded on the table and argued with each other that such and such is the correct final answer, it is now clear that you are less sure of it. Should you continue to do this if the correct methodology is not available? The techniques are very complicated, time-consuming and expensive, and you admit that 'fudge-factors' may be necessary in trying to arrive at smooth curves. What is the point of doing it? I am reminded of an aphorism by Dr Salvador Luria in a discussion of a research proposal. Someone commented 'The idea is not very promising, but he will do it well, thoroughly and carefully,' to which Dr Luria replied, 'If something is not worth doing, it is not worth doing well.'

*Garlick:* If we had not thought it was worth doing, we would never have progressed at all, and if we continue to think in that way we will never learn anything about human metabolism. We should not let our awareness of potential errors and limitations prevent us from ever trying. We do have other approaches with which we can check the validity of our results. The $^{15}N$ end product methods which we shall be discussing later have entirely different sources of error. We hope that if we make measurements by two different ways and come to the same conclusion, it is more likely to be correct.

*Klein (Baylor College of Medicine, Houston, USA):* One answer to Cohen is that one can develop operational criteria which enable one to make decisions. The detailed mechanism may not be known, but if you can separate patients into two categories as a result of the information obtained, you have a basis for treating them.

*Waterlow:* I want to reply to Cohen—a scientist whom I greatly admire. I really do not think you should throw stones at us, living as you do in a glasshouse. I have always taken as my motto in this work that we advance by a process of successive approximations—you call it fudging—but what did we hear this morning? The enzymologists are saying that measurements of $K_m$ are valueless for telling one what is going on *in vivo*. So what is the point of them?

*Harper* (*University of Wisconsin-Madison, USA*): I want to follow up that point, and comment on the arrogance of molecular biologists and sometimes enzymologists. Do you remember how the dicarboxylic- and tricarboxylic-acid cycles were taught before Krebs put them together? And then for a considerable period after that, how they were taught before Ogston showed how the tricarboxylic acids fitted together? I think a little caution might be in order.

*Cohen:* Playing the role of devil's advocate has had a very salutary effect. I find the response to my remarks encouraging.

# References

BLOBEL, G. (1980). Intracellular protein topogenesis, *Proc. Natl. Acad. Sci. USA*, **77**, 1496–500.

BUTTERY, P. J., BECKERTON, A., MITCHELL, R. M., DAVIES, K. and ANNISON, E. F. (1975). Turnover rate of muscle and liver protein in sheep, *Proc. Nutr. Soc.*, **34**, 91A.

FERN, E. B. and GARLICK, P. J. (1976). Compartmentation of albumin and ferritin synthesis in rat liver *in vivo*, *Biochem. J.*, **156**, 189–92.

GARLICK, P. J., CLUGSTON, G. A., SWICK, R. W. and WATERLOW, J. C. (1980). Diurnal pattern of protein and energy metabolism in man, *Am. J. Clin. Nutr.*, **33**, 1983–6.

GOLDEN, M. H. N. and WATERLOW, J. C. (1977). Total protein synthesis in elderly people: a comparison of results with ($^{15}$N) glycine and ($^{14}$C) leucine, *Clin. Sci. Mol. Med.*, **53**, 277–88.

HALLIDAY, D. and McKERAN, R. O. (1975). Measurement of muscle protein synthetic rate from serial muscle biopsies and total body protein turnover in man by continuous intravenous infusion of L-($\alpha$-$^{15}$N) lysine, *Clin. Sci. Mol. Med.*, **49**, 581–90.

JAMES, W. P. T., SENDER, P. M., GARLICK, P. J. and WATERLOW, J. C. (1974). Choice of label and measurement technique in tracer studies of protein metabolism in man, in *Dynamic Studies with Radioisotopes in Medicine*, International Atomic Energy Agency, Vienna, p. 461–72.

JAMES, W. P. T., GARLICK, P. J., SENDER, P. M. and WATERLOW, J. C. (1976). Studies of amino acid and protein metabolism in normal man with (U-$^{14}$C)tyrosine, *Clin. Sci. Mol. Med.*, **50**, 525–32.

KAIHARA, S. and WAGNER, H. N. (1968). Measurement of intestinal fat absorption with carbon-14 labelled tracers, *J. Lab. Clin. Med.*, **71**, 400–11.

LAURENT, G. J., BATES, P. C., SPARROW, M. P. and MILLWARD, D. J. (1978). Protein turnover in the adult fowl. III. Collagen content and turnover in cardiac and skeletal muscles in the adult fowl and the changes during stretch-induced growth, *Biochem. J.*, **176**, 393–405.

LUNDHOLM, K. and SCHERSTEN, T. (1975). Incorporation of leucine into human skeletal muscle protein. A study of tissue amino acid pools and their role in protein biosynthesis, *Acta Physiol. Scand.*, **93**, 433–41.

MÖLLER, P., BERGSTROM, J., ERIKSSON, S., FURST, P. and HELLSTROM, K. (1979). Effect of ageing on free amino acids and electrolytes in leg skeletal muscle, *Clin. Sci.*, **56**, 427–32.

NICHOLAS, G. A., LOBLEY, G. E. and HARRIS, C. I. (1977). Use of the constant infusion technique for measuring protein synthesis in the New Zealand White rabbit, *Br. J. Nutr.*, **38**, 1–17.

PRATT, O. E. (1976). Transport of metabolizable substances into the living brain, in *Transport Phenomena in the Nervous System: Physiological and Pathological Aspects* (Eds G. Levi, L. Battistin and A. Lajtha), Plenum Press, New York, pp. 55–75.

SIMON, O., MUNCHMEYER, R., BERGNER, H., ZEBROWSKA, T. and BURACZEWSKA, L. (1978). Estimation of rate of protein synthesis by constant infusion of labelled amino acids in pigs, *Br. J. Nutr.*, **40**, 243–52.

WATERLOW, J. C. (1967). Lysine turnover in man measured by intravenous infusion of L-(U-$^{14}$C) lysine, *Clin. Sci.*, **33**, 507–15.

WATERLOW, J. C., GARLICK, P. J. and MILLWARD, D. J. (1978). *Protein Turnover in Mammalian Tissues and in the Whole Body*, Elsevier North Holland, Amsterdam.

# 27

# Assumptions and Errors in the Use of $^{15}$N-excretion Data to Estimate Whole Body Protein Turnover

M. H. N. GOLDEN and A. A. JACKSON

*Tropical Metabolism Research Unit,*
*University of the West Indies, Kingston, Jamaica*

## Introduction

The end product methods, with one exception (Golden and Waterlow, 1977), have used a $^{15}$N-labelled amino acid as the tracer and urinary ammonia or, urea as the end product. The general equations for measurement of turnover by this method have been developed by Waterlow *et al.* (1978). It is the purpose of this paper to examine the validity of the assumptions and consider how an alteration in metabolic state or experimental technique may affect the assumptions. As in most studies $^{15}$N-glycine has been used as the tracer, this amino acid will be taken to illustrate the problems that arise.

The basic assumptions are:

1. $^{15}$N is treated in the same way as $^{14}$N.
2. Dietary total nitrogen and glycine-N are treated in the same way as nitrogen derived from tissue breakdown.
3. Glycine-N and amino acids deriving $^{15}$N from glycine are distributed between protein and excretory products in the same proportion as total amino-N.
4. The dose of glycine-N is a tracer dose and does not itself affect turnover.
5. $^{15}$N-glycine, and any amino acid deriving $^{15}$N from glycine, equilibrate throughout their respective pools so that their enrichment is the same at all sites of protein and end product synthesis.
6. Glycine-N does not give rise to quantitatively significant amounts of products other than protein and excretory products and the end products do not derive nitrogen from non-protein sources.

## Treatment of $^{15}N$ and $^{14}N$

There is biological fractionation of $^{15}N$ and $^{14}N$. Gaebler *et al.* (1966) maintained rats for 8 weeks on diets with corn or casein as the source of protein and then measured the $^{15}N$ abundance in the individual amino acids of the diet and tissue protein. Each amino acid was fractionated to a different extent. The highest abundance of $^{15}N$, in proline, serine, glutamic acid and valine, was about 0·0033 atoms % excess. The $^{15}N$ species was preferentially incorporated into protein by about 0·9%. This is the magnitude of error likely to arise from biological fractionation.

Of more importance is the assessment of baseline enrichment when low doses of isotope are used or the product is not highly enriched. A change in the precursor mix for end product synthesis could alter the enrichment independently of the administered isotope. We have observed a smooth diurnal variation in enrichment of plasma glutamic acid, with an amplitude of 0·002 atoms % excess. The timing of the baseline sample may thus be important.

## Treatment of Dietary Nitrogen and Nitrogen Released from Protein Breakdown

Results of turnover measurements with $^{15}N$-glycine when the isotope is given by the oral and intravenous route do not show any consistent difference with urea as the end product (Picou and Taylor-Roberts, 1969; Steffee *et al.*, 1976). A similar comparison of the effect of the route of administration of food has not been made.

## Distribution of Glycine-N into Protein and Excretory Products in the same Proportion as Total Amino-N

This condition will obviously be met if there is rapid and free exchange of nitrogen between glycine and all the other amino acids so that all amino-N achieves the same enrichment. This does not happen with any amino acid (Aqvist, 1951*a*). Some amino acids receive and donate nitrogen more readily than others. There are at least six lines of evidence to show that glycine-N is particularly poorly distributed throughout the α-amino-N pool.

(1) When large doses of $^{15}$N-glycine are given to rats, the amino acids isolated from tissue proteins are all lowly enriched except glycine and serine (Ratner *et al.*, 1940; Shemin and Rittenberg, 1944; Aqvist, 1951*a,b*; Vitti and Gaebler, 1963).

(2) When large doses of $^{15}$N-glycine are given to man the enrichment of the various nitrogen atoms in isolated uric acid reflects the relative enrichments of their precursors: N7 reflects the enrichment of glycine, N1 that of aspartate, N3 that of glutamine, and N9 that of glutamine or ammonia. Three studies have reported these measurements. Howell *et al.* (1961) found that glutamine and aspartic acid reached about 10 % of the enrichment of glycine after a dose of 1·5 g $^{15}$N-glycine. Even after 24 days the relative enrichment of these three amino acids had not changed, showing that after the initial redistribution of $^{15}$N there was little further donation of $^{15}$N from glycine to glutamine or aspartate. Gutman *et al.* (1962) and Sperling *et al.* (1973), who both gave about 4·5 g of $^{15}$N-glycine, found aspartate and glutamine to have about 14 % of the enrichment of glycine.

(3) When low doses of $^{15}$N-glycine are infused into normal man, no label can be found in glutamine or alanine (Jackson and Golden, 1980). The amide-N of glutamine does become slightly enriched.

(4) When $^{14}$C-$^{15}$N-glycine is given to man there is no difference in the turnover of the two labels, measured in hippuric acid and uric acid (N7 and C4). This steady relationship held for 24 days (Howell *et al.*, 1961). If there was significant recycling of nitrogen within the free amino-N pool one would not expect identical behaviour of the nitrogen and carbon moieties.

(5) When bile salts labelled with both $^{14}$C- and $^{15}$N-glycine are given to patients with blind loop syndrome, leading to fluctuating rates of deconjugation of the bile salts and hence varying rates of entry of $^{15}$N-glycine into the metabolic pool, the time course and magnitude of enrichment of urinary ammonia and breath $^{14}$CO$_2$ are congruous (Taruvinga, *et al.*, unpublished data). This indicates that the kidney is a major organ of glycine metabolism and that the label entering urinary ammonia is derived directly from glycine without passing through other metabolites.

(6) After administration of oral $^{15}$N-glycine, there is an apparent precursor–product relationship between the enrichment of urinary ammonia and urea. The peak labelling of urea occurs at the point in time when enrichment of urea equals that of ammonia (Wu, 1951).

Zilversmit *et al.* (1943) say of this relationship 'the isotope concentration of each successive product of an unbranched pathway will achieve a maximal value when its enrichment equals that of its *immediate precursor*'. Because urinary ammonia and urea are known to have different immediate precursors, Wu and Sendroy (1959) interpreted this relationship as showing 'a single pool of amino acids, a single urea pool and an equality of isotope concentration of urinary ammonia and of the amino acid mixture'. This is not the case; even urinary ammonia and free $\alpha$-amino-N do not have the same enrichment (Golden and Jackson, 1977). Even $^{15}$N-alanine, which distributes $^{15}$N throughout $\alpha$-amino-N more than most amino acids (Aqvist, 1951*a*), fails to follow this precursor–product relationship (Jeevanandam *et al.*, 1979). No clear precursor–product relationship is shown with $^{15}$N-labelled yeast protein (Crane and Neuberger, 1960). If there is general admixture of $^{15}$N, so that many compounds are becoming labelled at different rates and to different extents, this precursor–product relationship should not hold. It is evidence for the non-distribution of glycine-N. With our present state of knowledge the only explanation for this relationship which comes to mind seems unlikely to be correct: that the kidney is the major site for glycine deamination and conversion to serine, and either the excess free ammonia or the serine leaving the kidney is the major source of isotope for incorporation into urea. It is of great interest that the same precursor–product relationship seems to hold between urinary ammonia and N7 of uric acid (Sperling *et al.*, 1973). At present we cannot adequately explain the relationships except to say that it is difficult to be precise in timing the peak of both N7 uric acid and of urea.

If glycine does not mix with the $\alpha$-amino-N pool, neither does serine, because serine rapidly becomes highly enriched after giving $^{15}$N-glycine. Gersovitz *et al.* (1980*a*) have shown in man that with a constant infusion of $^{15}$N-glycine, enrichment of serine at plateau is $57 \pm 5\%$ of the enrichment of plasma glycine. Because of the phenomenon of internal recycling the enrichment of plasma glycine will be higher than the intracellular enrichment and serine will have a higher enrichment within the cell than in the plasma. From the data of Fern and Garlick (1974) it is likely that the intrahepatic enrichment of serine and glycine are similar and that from an isotopic point of view the two amino acids can be considered a single species.

Assumption (3), that glycine and serine contribute their nitrogen to protein and excretory products in the same proportion as total amino-N, now hinges upon the nature of the precursors for protein and end product synthesis.

We must not think of a precursor in classical terms, as a single species with a single product,. The amount of isotope incorporated into the product, i.e. the enrichment, is determined both by the individual enrichments of each precursor and by the relative contribution each makes to the product (not the relative contribution each makes to the precursor pool). The precursor pool will thus have an actual α-amino-N enrichment and a functional amino-N enrichment (Golden and Jackson, 1977). These two are not the same. The functional enrichment is the enrichment of importance. It cannot be measured in the free pool. The same consideration holds for any tracer amino acid and any product.

Any amino acid which is conserved more than average, for instance lysine (Read *et al.*, 1971), will overestimate flux; whereas, if another amino acid contributes more than its fair share to an excretory product it will underestimate flux. Underestimation is to be expected with those amino acids which lie directly on the excretory pathway for nitrogen. Thus, aspartic acid, a direct precursor of both urea and ammonia through the adenine nucleotide cycle, gives lower values for turnover than the branched chain amino acids or glycine (Taruvinga *et al.*, 1979). Glutamine contributes excessively to urinary ammonia. When we used this tracer and ammonia as the end product we derived a very low value for flux, equivalent to 2 g protein/kg/day. There are insufficient data to assess what happens with most other amino acids.

Glycine contributes directly to urinary ammonia. Indeed, shortly after a $^{15}$N-glycine load urinary ammonia is fifty times as enriched as glutamine (Sperling *et al.*, 1973). In chronically acidotic dogs about 4% of urinary ammonia is derived from glycine (Pitts *et al.*, 1965). In normal man up to 10% of ammonia may be derived from glycine (Jackson and Golden, 1980 and unpublished data).

What conditions affect the contribution of glycine and serine-N to protein and excretory products? To attempt to answer this problem we have constructed a model and observed the effects of alteration in each variable in turn, keeping the other variables constant, upon the resultant enrichment in the excretory products. The basic characteristics of the model are shown in Fig. 1 and the Appendix (p. 336).

The effect of altering the proportion of dietary glycine and serine-N to total-N is illustrated in Fig. 2. As glycine plus serine-N is increased or reduced the end product enrichment rises or falls, so that, for instance, a

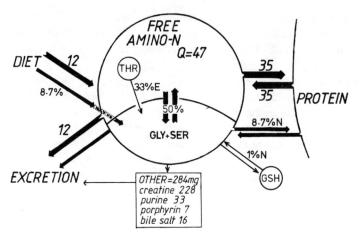

Fig. 1. Diagram showing the parameters of the model (see Appendix A).

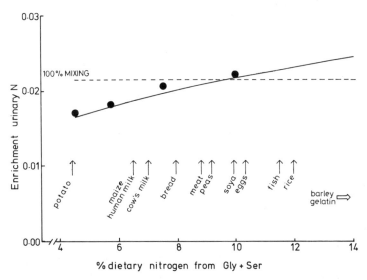

Fig. 2. Prediction of the effect of altering the proportion of dietary glycine and serine-N to total-N on the enrichment of urinary N. The dashed line represents the enrichment if complete mixing occurred in the precursor-N pool; the solid line is derived from the model, assuming 50 % mixing. The points marked have been derived from the data of Choitz et al. (1963).

potato diet would give 70 % of the enrichment of a rice diet at the same level of total-N intake (12 g N/day). The model also predicts that as cold glycine or serine is added to the diet the effect of dilution of the $^{15}N$ is far outweighed by the effect of an increased proportion of the excreted product coming from glycine. Toxic levels of glycine or serine (100 g N) would need to be taken before the enrichment fell back to standard conditions. Only one study, in dogs, has looked at the effect of adding graded doses of glycine to the diet (Choitz *et al.*, 1963). We have calculated the end product enrichment from the dietary composition, urinary nitrogen and % dose excreted (see Appendix, p. 336). There is close agreement between the slopes of the lines predicted from the model and the data of Choitz *et al.* (1963). This indicates that with the high doses of glycine used in this experiment about half behaved as if it was distributed evenly throughout the amino-N pool, while the other half behaved as if it was not distributed at all. It is likely that the actual proportion distributed will, under any given physiological circumstances, itself be a function of the dietary glycine intake and the balance between glycine and other amino acids in the diet. However, we have chosen the value which is in agreement with the data of Choitz for the other figures and calculations.

It is perhaps fortuitous that most foods contain about the same proportion of glycine plus serine-N to total-N and will thus give a reasonable measure for turnover. However, when gelatin is given, high enrichments are to be expected (Garlick *et al.*, 1980), and with milk-based diets low enrichments (Pencharz *et al.*, 1977). Many parenteral feeding solutions and formula diets contain very unbalanced non-essential amino acid mixtures which could affect enrichment dramatically.

The effect of both breaking down and synthesizing protein which contains more or less glycine plus serine-N is shown in Fig. 3. The effect of alteration in this ratio is enormous. Structural proteins and glycoproteins contain up to 30 % of their nitrogen as glycine or serine respectively. A small excess or deficit in the breakdown or synthesis of these proteins will give large changes in end product enrichment. These conditions may be expected to occur in skin disease, trauma, bone growth, rapid lean tissue regeneration, cachexia, gout, albuminuria, malabsorption syndrome, neoplasia, haemolytic states, etc. Indeed it is probable that all pathological states affect structural and non-structural proteins differently. This will be a major source of error when interpreting $^{15}N$ turnover data. It is probable that the extremely low enrichments of excretory products observed by Stein *et al.* (1976) in rats on a protein-free diet are to be explained in this way.

When the diet contains either 5 % or 15 % N as glycine plus serine (but the

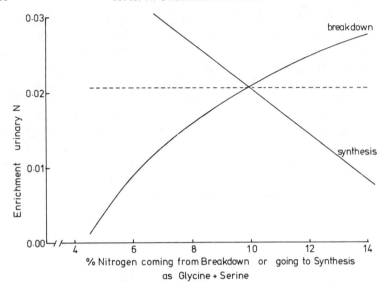

FIG. 3. Prediction of the effect of an alteration in the ratio of glycine and serine-N to total-N, either coming from protein breakdown or going to protein synthesis, upon the enrichment of urinary N. The symbols are the same as in Fig. 2.

body proteins 8·7%) and there is alteration in the dietary intake, breakdown rate or synthesis rate, equally dramatic changes occur in end product enrichment (Figs. 4(a), (b) and (c). The effects are exaggerated when nitrogen is being used efficiently at low-N intakes and when synthesis exceeds breakdown, Even quite small deviations from the point of balance with a generous nitrogen intake may give large erroneous changes in enrichment. It is noteworthy that with the particular circumstances chosen there may be anomalous changes in enrichment, so that as flux rises so does enrichment. Thus the results under particular circumstances may be qualitatively, as well as quantitatively, wrong.

Altering any of the numbers for the characteristics of the model (see Appendix, p. 336) will have effects upon the enrichment of the end product. Recalculation of the curves without allowing for the minor pathways of glycine metabolism (Fig. 1 and Appendix A.4) showed very minor alterations in the shape of the enrichment curves. The extent of mixing with the pool is however a major variable which changes the curves substantially.

To summarize, a number of factors can produce important changes in the enrichment of urinary nitrogen and hence lead to errors in the estimation of flux. These factors are: a change in the proportion of glycine

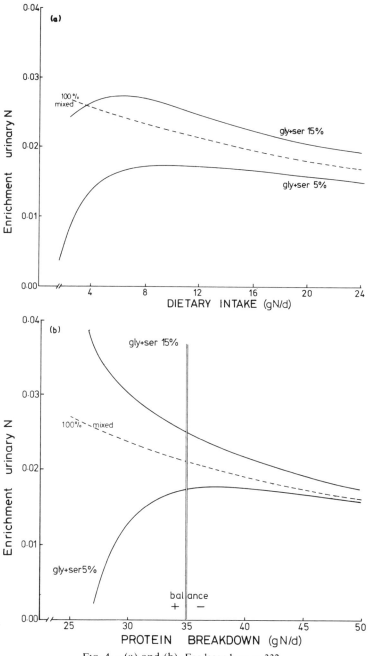

FIG. 4.—(a) and (b). For legend see p. 332.

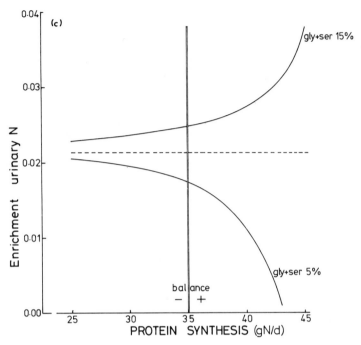

FIG. 4. Prediction of the effect of altering (a) dietary intake, (b) protein breakdown rate, (c) protein synthesis rate on the enrichment of excreted N, when diets containing either 5 % or 15 % glycine plus serine are ingested. The meaning of the different lines are as in Fig. 2. All the variables were held constant (diet = 12, breakdown = 35, synthesis = 35) except the one under study.

plus serine-N in the diet or in the protein being synthesized or degraded; the degree of mixing of glycine-N in the amino-N pool; extremes of imbalance between synthesis and breakdown; decreased urinary nitrogen excretion through dietary nitrogen restriction. These constraints should be recognized in the design of turnover experiments in which the end product method is used.

Nevertheless, when turnover values obtained with [15]N-glycine are compared with those found with other isotopes, values are similar. Thus, compared with 1-[14]C-leucine in the fed elderly (Golden and Waterlow, 1977) or with [15]N-labelled egg protein (Picou and Taylor-Roberts, 1969) and with [15]N-labelled yeast protein (Golden, unpublished) in normally growing, fed children, [15]N-glycine gives similar values. Comparison of label has not been made under the conditions of stress suggested by the model.

Reeds and Lobley (1980) have shown that with different species ranging from rat to cow, when in energy balance and when the same isotope is used in the same way, similar values are obtained. However, they have suggested that abnormal results are obtained with some amino acids by the precursor method, because of the unbalanced nature of the amino acid mixture absorbed by conventionally fed ruminants. Agreement between different labels does not necessarily show absence of systematic error.

## *The Dose of Glycine-N as a Tracer Dose*

Figure 5 shows the effect predicted by the model of altering the dose of glycine (of any enrichment) when there is a zero protein intake, breakdown is 35 g N/day and synthesis 30 g N/day. Under these circumstances even

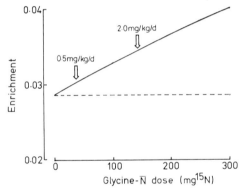

FIG. 5. The possible dose effects from giving [15]N-glycine according to the model (see Part C of the Appendix). Characteristics used: dietary non-glycine N intake = zero, protein breakdown = 35 g N/day, protein synthesis = 30 g N/day, degraded and synthesized protein contains 8·7% glycine plus serine. The positions of a dose of glycine of 0·5 mg/kg and 2·0 mg/kg for a 70 kg adult are shown. The tracer [15]N is assumed to contribute an infinitesimal amount of nitrogen.

0·5 mg glycine-N/kg/day would lead to a 7% increase in enrichment and 2 mg glycine-N/kg/day to a 23% increase in enrichment (with unit dose [15]N in both cases). We cannot assume that doses of this magnitude are negligible. The effect of dose is much less with high dietary intakes and with constant infusion rather than single shot (or loading dose) methods. Dose effects of this order of magnitude have been observed (Waterlow *et al.*, 1978).

## Equilibration of [15]N-glycine and Serine throughout their own Pools and Attainment of the same Enrichment at all Sites of Protein Synthesis and End Product Synthesis

This assumption is incorrect. The pools fail to equilibrate for two reasons: first, because of various degrees of internal recycling within different tissues. The magnitude of this effect is determined by the ratio of the rate of exchange of intracellular and extracellular amino acids and the rate of protein synthesis and breakdown. Fern and Garlick (1974) have shown that in the rat the concentration of the glycine isotope in the liver is half that in the plasma. This could lead to a serious error, but the difference is not likely to be as great in man. The error introduced may well be less than that seen with the plasma precursor method (Golden and Waterlow, 1977) because both the end products are synthesized from the intracellular pool. However, there is evidence for functional compartmentation of glycine in the liver as sarcosine and hippurate seem to be synthesized from different pools (Benedict et al., 1956).

The second potential cause of non-equilibration within the course of an experiment is the very slow exchange with the large muscle pool of free glycine. In rats at 6 h the muscle pool is far from reaching isotopic equilibrium (Fern and Garlick, 1974). In man, because of his much slower turnover, a commensurately longer period will presumably be required. The muscle free pool may act as an isotope sink in experiments of short duration. The error will be much greater with single dose than with continuous infusion experiments.

After the muscle pool has become labelled it will presumably lose its label equally slowly. This could be the reason for the great prolongation of [15]N-ammonia and urea synthesis after a single dose of [15]N-glycine, and the explanation for the reversal of enrichments after 5 days, so that ammonia becomes more highly enriched than urea (Benedict et al., 1952).

## Products other than Protein and Excretory Products from the Glycine-N Dose and the Derivation of Nitrogen by the End Product from Non-protein Sources

In the calculation of flux the loss of label to other pathways such as creatine, porphyrin and purines will increase the flux, even though this effect is marginal. However, if the synthesis of these compounds equals the excretion of their end products, creatinine, bilirubin and uric acid respectively, there will be no error in the computation of synthesis rate because *total* nitrogen is taken from flux and this will balance the losses of nitrogen into other pathways.

Of more importance is the second half of this assumption. Pyrimidines are broken down to form urea. This influx of unlabelled urea derived from pyrimidine will specifically dilute the urea pool and thus give a lower enrichment of urea than of ammonia. If pyrimidine breakdown is similar to purine excretion, on a normal diet, there will be a 7–10 % dilution of the urea pool. On low protein diets, high nucleic acid diets or in net catabolic states, this contribution may be more marked. This flux of pyrimidine urea may explain the differential enrichment of urea and ammonia often observed. The precursor–product relationship for these two excretory products after a single dose of $^{15}$N-glycine indicates that they should both reach the same enrichment when at isotopic equilibrium from a constant infusion.

## Steady State

It is a further assumption that the metabolic state does not alter during the course of a measurement. This assumption is frequently violated.

## Conclusion

The measurement of protein turnover with $^{15}$N-glycine is analogous to the use of Newtonian physics. Under normal circumstances, with a normal diet and turnover rate when normal tissues are being synthesized and broken down, the method gives reasonable answers. However, at the extremes of dietary intake, synthesis and breakdown, or when unusual mixtures of proteins are made or disposed of, the results will be in error, just as Newtonian physics breaks down near the speed of light. Careful consideration has to be given to the design of experiments to minimize the errors which we are not in a position to quantify or to allow for. Perhaps in the future we will be able to assess such things as the proportion of glycine-N coming from breakdown by looking at tissue markers of specific tissue breakdown such as hydroxyproline.

## Acknowledgements

M.H.N.G. thanks the Wellcome Trust for support. We owe a great debt to the pioneers who preceded us, particularly J. C. Waterlow and D. I. M. Picou.

## Appendix

**A. Characteristics of a Model to Simulate Glycine/Serine Flux**

1.  It is assumed that glycine and serine enter the pool from both diet and protein breakdown.

2.  On a standard diet of 12 g N/day, with protein synthesis and breakdown rates of 35 g N/day and a urinary output of 12 g N/day, there is no *de novo* glycine or serine synthesis.

3.  8·74 % of nitrogen incorporated into tissues, released from tissues and in the diet is in glycine plus serine (value derived from standard food tables).

4.  Other pathways of glycine and serine are estimated as follows:

    (i)   Purine excretion of about 400 mg/day is replaced by *de novo* synthesis utilizing 33 mg glycine-N/day.

    (ii)  35 mg glycine is used for haem synthesis which removes 7 mg glycine-N/day.

    (iii) 18 % of the bile salt pool is degraded daily and replaced, with the consumption of 16 mg glycine-N/day.

    (iv)  Creatine (52 mg/day) and creatinine (1800 mg/day) lost in the urine are replaced, utilizing 228 mg glycine-N/day.

    (v)   No allowance has been made for phosphoserine synthesis.

    (vi)  Threonine is broken down via glycine at the rate of 3·3 % of excreted nitrogen, and increases glycine flux by this amount.

    (vii) Glutathione synthesis and degradation increase glycine flux by 1 % of total-N coming from breakdown or going to synthesis. If there is net breakdown or synthesis glutathione breakdown or synthesis is in proportion to that of total protein. Glutathione does not form a sink for label. (These estimates are likely to be in error. Glutathione turnover values in the literature vary by up to two orders of magnitude.)

5. A unit dose of $^{15}N$ is given which does not contribute to the glycine plus serine flux.

6. It is assumed that 50 % of $^{15}N$ from glycine plus serine is evenly distributed among the other amino acids, and that this is replaced by *de novo* synthesis of glycine plus serine with $^{14}N$.

## B. Calculations

The general formulae for the measurement of whole body protein turnover with $^{15}N$ are:

1. $e/d = E/Q$ (where $e$ = the amount of isotope excreted, $d$ = the dose of isotope, $E$ = total urinary N excretion, and $Q$ = nitrogen flux).

2. $Q = Z + E = B + I$ (where $Z$ = nitrogen incorporated into protein, $B$ = nitrogen coming from protein breakdown, and $I$ = dietary N intake).

3. The enrichment of excreted nitrogen is the ratio of excreted isotope to that of total nitrogen $(^{14}N + {}^{15}N) = e/E$.

   With even distribution of isotope throughout the total amino-N pool the equations all refer to total-N and the amount of isotope excreted ($e$) is derived from eqn (1).

   If there is no distribution of isotope and no *de novo* synthesis of glycine an exactly similar series of equations may be written with reference to glycine-N alone. Under these circumstances the amount of isotope excreted ($e_{gly}$) will be:

4. $$e_{gly} = d \cdot E_{gly}/Q_{gly}$$

   If a certain proportion ($p$) of the glycine-N exchanges with the rest of the amino-N pool, so that that proportion of the dose of isotope will behave as if it is evenly distributed throughout the amino-N pool, the isotope excretion from this distribution nitrogen will be:

5. $$e = pd \cdot E/Q$$

   The amount of isotope coming from the remaining fraction of the glycine pool which does not distribute its nitrogen will then be:

6. $$e_{gly} = (1 - p)d \cdot E_{gly}/Q_{gly}$$

   The total amount of isotope excreted ($e_t$) will be the sum of that coming from the glycine-N which exchanges and from that which does not.

7. $$e_t = pd \cdot E/Q + (1 - p)d \cdot E_{gly}/Q_{gly}$$

8.  Contributions to the glycine flux will be derived from the breakdown of glutathione and of threonine, and part of the flux will be distributed to pathways other than synthesis and excretion (see Appendix A).

## C. Illustrative Calculation

The calculation is designed to show the effect on the isotope enrichment of total urinary N ($e/E$) of changes: (a) in the intake of nitrogen from food; (b) in the proportion of glycine-N in food nitrogen.

*Assumptions*
- (i)    Protein synthesis and breakdown are constant and equal.
- (ii)   Glycine-N represents 8·74% of nitrogen in protein being synthesized or broken down.
- (iii)  Other pathways of glycine metabolism are as given in Appendix A.
- (iv)   Unit dose of $^{15}$N as glycine-N.
- (v)    50% of $^{15}$N is distributed from glycine to other amino acids ($p = 0·5$, Appendix B, eqn (7)).

*Examples*
- (i)    Nitrogen intake of food ($I$) = 4 g
         Glycine-N = 5% of food nitrogen.
- (ii)   Nitrogen from food ($I$) = 12 g
         Glycine-N = 15% of food nitrogen.

| Variable | Derivation | Example (i) | Example (ii) |
|---|---|---|---|
| *Total N* (g/day) | | | |
| 1. $I = E$ | — | 4 | 12 |
| 2. $Z = B$ | — | 35 | 35 |
| 3. $Q$ | — | 39 | 47 |
| *Glycine-N* (g/day) | | | |
| 4. $I_{gly}$ | — | 0·20 | 1·80 |
| | | (5% of $I$) | (15% of $I$) |
| 5. $B_{gly}$ | 8·74% of $B$ | 3·06 | 3·06 |
| 6. From GSH | 1% of $Q*$ | 0·39 | 0·47 |
| 7. From threonine | 3·3% of $E*$ | 0·13 | 0·40 |
| 8. $Q_{gly}$ | (4 + 5 + 6 + 7) | 3·78 | 5·73 |
| 9. $Z_{gly}$ | 8·74% of $Z$ | 3·06 | 3·06 |

| Variable | Derivation | Example (i) | Example (ii) |
|---|---|---|---|
| *Glycine-N* (g/day) | | | |
| 10. To GSH | 1 % of $Q$* | 0·39 | 0·39 |
| 11. To other pathways* | | 0·28 | 0·28 |
| 12. $E_{gly}$ | (8 + 9 + 10 + 11) | 0·05 | 2·00 |
| 13. $e_t$ = isotope excreted | from eqn (7), Appendix B | 0·058 | 0·302 |
| 14. $e_t/E$ = enrichment | 13/1 | 0·0145 | 0·0252 |
| 15. Calculated $Q$ | from eqn (1), Appendix B | 69·0 | 39·7 |

\* See Appendix A.

These calculations represent extreme cases, but illustrate the errors which may arise from variations in glycine content of the components of flux.

## Discussion

*Munro (Tufts University, Boston, USA):* I congratulate you on pointing out both the complexity of the problems and the analytical methods for looking into them. One particular problem which interests us (Gersovitz *et al.*, 1980*b*) is the study of albumin synthesis by continuous feeding of labelled glycine. Glycine is a precursor of arginine, which is used for the synthesis of both urea and plasma albumin. It took 36 h to reach plateau labelling of plasma urea, whereas albumin started to be labelled almost from zero time. The explanation must be that the arginine, and hence the urea, in the liver reached plateau labelling very quickly, whereas the body pool of urea is so large that it takes a long time to reach an isotopic steady state. This is a situation in which critical inspection of the data tells one something about the kinetics of the system.

*Golden:* What you say is fascinating. It seems that the specific activity of both free glycine and arginine in the liver during your experiments must reach plateaux extremely quickly. This supports my thesis of non-exchange of glycine-N. If there were free exchange one would expect a delay in arginine reaching a plateau caused by the time taken for the other amino acids to receive glycine-N, reach isotopic steady state and then pass it on to arginine.

*Young (Massachusetts Institute of Technology, USA):* We were greatly intrigued by these observations. It is clear that the urea pool in the liver is rapidly labelled following a continuous oral administration of [15]N-glycine at the level of tracer administration we used.

*Golden:* One of the main assumptions of the end product method is that the specific activity ([15]N enrichment) of glycine will be the same at all sites of protein synthesis and at the site of synthesis of the end product. It seems to me that in many situations the muscle free amino acid pool acts as a kind of isotopic sink, particularly when the isotope is given as a single dose. Benedict *et al.* (1952) showed with large single doses of glycine that after 5 days urinary $NH_3$ again becomes more highly enriched than urea, and this remains so for at least 26 days. I take it that this labelling represents [15]N coming back out of protein and out of muscle. Why it should label ammonia more highly than urea I do not know.

*Garlick (London School of Hygiene and Tropical Medicine, UK):* You mentioned a precursor–product kind of relationship between ammonia and urea. As is very well known, the specific activities of precursor and product cross when that of the product is at its peak. Secondly, the areas under the two curves, taken to infinite time, should be the same. They do not appear to me to be so. Another point is that, as Walser and others have shown, when one compares the labelling of urea in plasma and urine, there is a delay of 1–2 hours in the excretion of labelled urea. This would alter the timing of the urea peak and alter the so-called precursor–product relationship, so that it may not be as good as it appears at first sight.

*Golden:* There may be some controversy about how long the delay is—it may be shorter than you say. Walser and Bodenlos (1959) in their classic paper found that in 19 periods in three subjects the delay ranged from 0–1·6 hours (mean 1·0 ± 0·4). It is by no means easy to define the time at which the labelling of urea reaches its peak, and the same goes for the N7 atom of uric acid.

*Walser (Johns Hopkins University, Baltimore, USA):* With regard to the urinary delay time, I thought that the findings we reported (Walser and Bodenlos, 1959) had been generally confirmed. In chronic renal failure there is scarcely any delay time because there is no appreciable renal pool of urea (Walser, 1974). Perhaps that is what you were referring to.

*James (Dunn Clinical Nutrition Centre, Cambridge, UK):* You seem to be worried about explaining the 'crossover' in labelling of ammonia and urea.

As I see it, if glycine is to some extent contributing directly to the formation of ammonia, the enrichment of ammonia will fall very rapidly as labelled glycine is diluted in the body pool after a single dose. In contrast, the labelling of urea will depend not only on dilution but on the contribution of nitrogen from many other metabolic pathways.

*Golden:* That kind of general statement does not really explain the relationship that we find, which suggests that urea and ammonia must be synthesized from the same precursor pool of serine plus glycine, the difference being in their rates of turnover. Wu and Sendroy (1959) found the same thing when phenylalanine was used as a tracer. I think this is a remarkable phenomenon that has to be explained.

*Alleyne* (*University of the West Indies, Kingston, Jamaica*): You are concerned that the enrichment of ammonia should be higher than that of urea?

*Golden:* No, I am concerned that they have the same enrichment at the urea peak, as if ammonia, or some metabolite rapidly equilibrating with ammonia, is the precursor of urea.

*Alleyne:* Theoretically, could not this be explained on the basis of two facts which perhaps are not sufficiently appreciated: first, a proportion of urinary $NH_3$ is formed directly from glycine; in Pitts' experiments in acidotic dogs this amounted to 3–4 % of urinary $NH_3$ (Pitt *et al.*, 1965). Secondly, there may be a considerable quantity of ammonia in the renal vein, which will form urea when it reaches the liver.

*Golden:* Yes: the system behaves as if renal ammonia is the precursor for urea or perhaps serine derived from glycine in the kidney, although this is unlikely. However, this explanation cannot be right because we know that glycine is extensively metabolized in the liver and will thus contribute nitrogen to urea without renal ammonia as an intermediate. Our data on the relatively small proportion of glutamine flux disposed of in the kidney would suggest that a figure of 20 % of the ammonia reaching the liver to be derived from kidney is too high.

*Garlick:* You mentioned the problem that may arise when the diet contains different amounts of glycine. In some fluids used for parenteral feeding a large part of the non-essential nitrogen is made up of glycine. Dr J. Powell-Tuck from the Clinical Nutrition Metabolism Unit has been trying to find out whether the amount of glycine in the diet or the infusion makes any difference to estimates of protein turnover with labelled glycine. So far no effect has been found.

*Golden:* I should just like to make a point about the effects of a meal which relates also to Garlick's paper (p. 303). In the course of the studies we did with $^{14}$C-leucine in the elderly some years ago (Golden and Waterlow, 1977), in one experiment I looked at the effect of stopping the protein intake but keeping on the tracer. There was an immediate fall in $^{14}CO_2$ excretion within minutes, suggesting that the body adapts very rapidly to the reduced inflow of leucine.

*Walser:* You imply that this rapid fall indicates some 'wisdom of the body'. However, plasma isoleucine concentration falls very rapidly following an isoleucine-free meal as we reported at another meeting held last week (Walser *et al.*, 1981). Thus the metabolic clearance in the experiment you described may not have changed, which would indicate 'stupidity' of the body rather than wisdom. The fall in plasma leucine or isoleucine of course reflects continuing protein synthesis from the infused amino acids and from circulating leucine, thus rapidly depleting the leucine or isoleucine pool.

*Golden:* You will reach a different plasma concentration after a time, but the response in oxidation occurs before that. My point is that the system is exquisitely sensitive; the time interval between perturbing it and seeing a response is very small—a few minutes only.

*Walser:* Did plasma leucine concentration fall as fast as the labelling of expired $CO_2$?

*Golden:* No: according to my measurements the plasma leucine concentration did not change within 20 min of altering the dietary input, but subsequently it did change.

## References

AQVIST, S. (1951*a*). Metabolic interrelationships among amino acids studied with isotopic nitrogen, *Acta Chem. Scand.*, **5**, 1046–64.
AQVIST, S. (1951*b*). Amino acid interrelationship during growth, studied with $^{15}$N-labelled glycine in regenerating rat liver, *Acta Chem. Scand.*, **5**, 1065–73.
BENEDICT, J. D., KALINSKY, H. J., SCARRONE, L. A., WERTHEIM, R. and SWETTEN, D. (1956). The origin of urinary creatine in progressive muscular dystrophy, *J. Clin. Invest.*, **34**, 141–5.
BENEDICT, J. D., ROCHE, M., YU, T. F., BIEN, E. J., GUTMAN, A. B. and SWETTEN, D. (1952). Incorporation of glycine nitrogen into uric acid in normal and gouty man, *Metabolism*, **1**, 3–12.
CHOITZ, H. C., KURRIE, D. and GAEBLER, O. H. (1963). Utilisation of glycine nitrogen at various levels of glycine intake, *J. Nutr.*, **80**, 365–9.

CRANE, C. W. and NEUBERGER, A. (1960). The digestion and absorption of protein by normal man, *Biochem. J.*, **74**, 313–23.

FERN, E. B. and GARLICK, P. J. (1974). The specific radioactivity of the tissue free amino acid pool as a basis for measuring the rate of protein synthesis in the rat *in vivo*, *Biochem. J.*, **142**, 413–19.

GAEBLER, O. H., VITTI, T. G. and VUKMIROVICH, R. (1966). Isotope effects in metabolism of $^{14}$N and $^{15}$N from unlabelled dietary proteins, *Canad. J. Biochem.*, **44**, 1249–57.

GARLICK, P. J., CLUGSTON, G. A. and WATERLOW, J. C. (1980). Influence of low-energy diets on whole body protein turnover in obese subjects, *Amer. J. Physiol.*, **238**, E235–44.

GERSOVITZ, M., BIER, D., MATTHEWS, D., VDALL, J., MUNRO, H. N. and YOUNG, V. R. (1980*a*). Dynamic aspects of wholebody glycine metabolism: influence of protein intake in young adult and elderly males, *Metabolism*, **29**, 1087–94.

GERSOVITZ, M., MUNRO, H. N., UDALL, J. and YOUNG, V. R. (1980*b*). Albumin synthesis in young and elderly subjects using a new stable isotope methodology: response to level of protein intake, *Metabolism*, **29**, 1075–86.

GOLDEN, M. H. N. and JACKSON, A. A. (1977). Tissue enrichments and protein turnover measured with $^{15}$N-glycine, *Nature*, **265**, 563–4.

GOLDEN, M. H. N. and WATERLOW, J. C. (1977). Total protein synthesis in elderly people: a comparison of results with [$^{15}$N] glycine and [$^{14}$C] leucine, *Clin. Sci. Mol. Med.*, **53**, 277–88.

GUTMAN, A. B., YU, T. F., ADLER, M. and JAVITT, N. B. (1962). Intramolecular distribution of uric acid-$N^{15}$ after administration of glycine-$N^{15}$ and ammonium-$N^{15}$ chloride to gouty and nongouty subjects, *J. Clin. Invest.*, **41**, 623–36.

HOWELL, R. R., SPEAS, M. and WYNGAARDEN, J. B. (1961). A quantitative study of recycling of isotope from glycine-1-$C^{14}$, $\alpha$ $N^{15}$ into various subunits of the uric acid molecule in a normal subject, *J. Clin. Invest.*, **40**, 2076–82.

JACKSON, A. A. and GOLDEN, M. H. N. (1980). ($^{15}$N) Glycine metabolism in normal man: the metabolic $\alpha$-amino-nitrogen pool, *Clin. Sci.*, **58**, 517–22.

JEEVANANDAM, M., LONG, C. L. and KINNEY, J. M. (1979). Kinetics of intravenously administered $^{15}$N-L-alanine in the evaluation of protein turnover, *Amer. J. Clin. Nutr.*, **32**, 975–80.

PENCHARZ, P. B., STEFFEE, W. P., COCHRAN, W., SCRIMSHAW, N. S., RAND, W. M. and YOUNG, V. R. (1977). Protein metabolism in human neonates: nitrogen-balance studies, estimated obligatory losses of nitrogen and whole-body turnover of nitrogen, *Clin. Sci. Mol. Med.*, **52**, 485–98.

PICOU, D. and TAYLOR-ROBERTS, T. (1969). The measurement of total protein synthesis and catabolism and nitrogen turnover in infants in different nutritional states and receiving different amounts of dietary protein, *Clin. Sci.*, **36**, 283–96.

PITTS, R. F., PILKINGTON, L. A. and DE HAAS, J. C. M. (1965). $N^{15}$ tracer studies on the origin of urinary ammonia in the acidotic dog, with notes on the enzymatic synthesis of labelled glutamic acid and glutamines, *J. Clin. Invest.*, **44**, 731–45.

RATNER, S., RITTENBERG, D., KESTON, A. S. and SCHOENHEIMER, R. (1940). Studies in protein metabolism. XIV. The chemical interaction of dietary glycine and body proteins in rats, *J. Biol. Chem.*, **134**, 665–76.

READ, W. W. C., MCLAVEN, D. S. and TCHALIAN, M. (1971). Urinary excretion of nitrogen from ¹⁵N-labelled amino acids in the malnourished and recovered child. I glycine and lysine, *Clin. Sci.*, **40**, 375–80.

REEDS, P. J. and LOBLEY, G. E. (1980). Protein synthesis: are there real species differences?, *Proc. Nutr. Soc.*, **39**, 43–52.

SHEMIN, D. and RITTENBERG, D. (1944). Some interrelationships in general nitrogen metabolism, *J. Biol. Chem.*, **153**, 401–21.

SPERLING, O., WYNGAARDEN, J. B. and STARMER, C. F. (1973). The kinetics of intramolecular distribution of ¹⁵N in uric acid after administration of (¹⁵N) glycine, *J. Clin. Invest.*, **52**, 2468–85.

STEFFEE, W. P., GOLDSMITH, R. S., PENCHARZ, P. B., SCRIMSHAW, N. S. and YOUNG, V. R. (1976). Dietary protein intake and dynamic aspects of whole body nitrogen metabolism in adult humans, *Metabolism*, **25**, 281–97.

STEIN, T. P., ORAM-SMITH, J. C., LESKIW, J., WALLACE, H. W., LONG, L. C. and LEONARD, J. M. (1976). Effect of nitrogen and calorie restriction on protein synthesis in the rat, *Amer. J. Physiol.*, **230**, 1321–5.

TARUVINGA, M., JACKSON, A. A. and GOLDEN, M. H. N. (1979). Comparison of ¹⁵N labelled glycine, aspartate, valine and leucine for measurement of whole body protein turnover, *Clin. Sci.*, **57**, 281–3.

VITTI, T. G. and GAEBLER, O. H. (1963). Effects of growth hormone on metabolism of nitrogen from several amino acids and ammonia, *Arch. Biochem. Biophys.*, **101**, 292–8.

WALSER, M. (1974). Urea metabolism in chronic renal failure, *J. Clin. Invest.*, **53**, 1385–92.

WALSER, M. and BODENLOS, L. J. (1959). Urea metabolism in man, *J. Clin. Invest.*, **38**, 1617–26.

WALSER, M., SAPIR, D. G., MITCH, W. E. and CHAN, W. (1981). Effects of branched-chain ketoacids in normal subjects and patients. In *Metabolism and Clinical Implications of Branched Chain Amino and Keto Acids* (Eds. Walser, M. and Williamson, J. R.), Elsevier North-Holland, Amsterdam.

WATERLOW, J. C., GOLDEN, M. H. N. and GARLICK, P. J. (1978). Protein turnover in man measured with ¹⁵N: comparison of end products and dose regimes, *Amer. J. Physiol.*, **235**, E165–74.

WU, H. (1951). Relative concentrations of N¹⁵ in urinary ammonia N and urea N after feeding N¹⁵ labelled compounds, *J. Gen. Physiol.*, **34**, 403–9.

WU, H. and SENDROY, J. (1959). Pattern of ¹⁵N-excretion in man following administration of ¹⁵N-labelled L-phenylalanine, *J. Appl. Physiol.*, **14**, 6–10.

ZILVERSMIT, D. B., ENTENMAN, C. and FISHLER, M. C. (1943). On the calculation of 'turnover time' and 'turnover rate' from experiments involving the use of labelling agents, *J. Gen. Physiol.*, **26**, 325–31.

# 28

# $^{15}$N-Glycine as a Tracer to Study Protein Metabolism *in vivo*

T. P. STEIN

*Department of Surgery,*
*University of Pennsylvania School of Medicine,*
*Philadelphia, Pennsylvania, USA*

## Introduction

Glycine was the first amino acid labelled with $^{15}$N to be used for measuring protein synthesis rates. It was selected because $^{15}$N-glycine is the easiest amino acid to synthesize and has no optical enantiomers. These reasons are as valid now as they were 30 years ago and are why most investigators persist in using $^{15}$N-glycine. This paper is concerned with one question: given adequate financing and availability of other $^{15}$N-labelled amino acids, is $^{15}$N-glycine still the preferred tracer?

During the past 30 years there have been a number of 'definitive' studies 'proving' that $^{15}$N-glycine was unsuitable for use as a tracer for studying protein metabolism (Wu and Bishop, 1959; Wu *et al.*, 1959; Tschudy *et al.*, 1959). In fact, what was disproved in these rather pessimistic studies was the use of a particular approach to obtain absolute synthesis values from urinary isotope excretion patterns following a single pulse of a $^{15}$N-labelled amino acid. If the aim is to obtain reliable data for comparative purposes, then decay curve analysis is satisfactory (Long *et al.*, 1977). The values and trends found in various pathological states are the same as those obtained with other methods based on the Rittenberg–San Pietro model.

The basic requirement for using $^{15}$N-glycine is that its behaviour can be taken as representative of the typical amino acid, i.e. that its amino-N is partitioned between synthesis and excretion in the same proportions as amino-N in general. When the correspondence is not perfect, then the divergence from the true average ratio must be invariate.

In theory, the question could be answered by examining all the detailed steps involved in glycine metabolism. This classic biochemical approach has not been seriously pursued because of the enormous complexity of the problem. Consequently, the approach in recent years has been to compare one method against another, with either the same or different tracers, the argument being that the assumptions behind each procedure are a little different, so that if two methods give the same answer, either the non-common assumptions are both correct or both wrong. Implicit in this approach is that one of the test methods is more likely to be correct than the other. Usually the glycine method is assumed to be the more suspect of the methods. This is the legacy of the papers by Wu and Tschudy as well as the suggestion that $^{15}$N-glycine may possibly disproportionately reflect liver metabolism (James et al., 1974).

Measurements in vivo of protein synthesis rate with a single isotopically labelled amino acid can be classified in two groups ('direct' and 'indirect') each with a number of subgroups. A direct method is one in which the amount of isotope incorporated into protein is actually measured. Indirect methods depend on measuring the rate of isotope excretion, or the isotopic flux in the plasma, urinary urea, urinary ammonia or expired air. The approach is termed indirect because the amount of isotope incorporated into protein is calculated and not directly measured. The methodological comparisons that can be made are (i) indirect versus indirect, (ii) direct versus direct and (iii) indirect versus direct.

## Indirect versus Indirect

There are two types of indirect method. In one, the calculation is based on the specific activity of the precursor pool, in the other, on the isotopic enrichment of what is excreted. Part of the theory behind the two approaches is different. The precursor pool method assumes that the pool samples are part of the body's free amino acid pool and that an isotopic steady state has been achieved between labelled amino acid and the amino acid originating from diet and endogenous sources (Waterlow and Stephen, 1966). The end product methods assume that labelled nitrogen is partitioned between protein synthesis and excretion in the same proportion as the total amino acid nitrogen from the diet, de novo synthesis and protein breakdown. It does not matter how the various components of the protein synthesis precursor amino acid pool get there, the crucial factor is that they are partitioned equally. In other words, the precursor pool method depends on the free amino acid pool per se and the excretion method on what happens to it.

Indirect methods have been frequently used during the past decade for a variety of studies, and it is now apparent that, irrespective of the methodological details, the labelled amino acid used or the way in which the experimental data are handled, all the results are fairly close (i.e. within a range of 2). The differences between the variants are not due to random experimental factors.

Golden and Waterlow (1977) compared flux methods (with [14]C-leucine) against end product methods (with [15]N-glycine) in a series of elderly patients (Table 1). Although the means for the four calculations are

TABLE 1

COMPARISON OF FLUX AND END PRODUCT METHODS FOR MEASUREMENT OF SYNTHESIS RATE IN SIX ELDERLY PATIENTS. EACH PATIENT WAS GIVEN [14]C-LEUCINE AND [15]N-GLYCINE SIMULTANEOUSLY

| Labelled amino acid | Method of measurement | Measurement made in: | Synthesis rate g protein/kg/day |
|---|---|---|---|
| [14]C-leucine | (A) Flux | Plasma leucine | $2.67 \pm 0.29$ |
| [14]C-leucine | (B) Flux | Expired $CO_2$ | $3.43 \pm 0.37$ |
| [15]N-glycine | (C) End product | Urea | $2.34 \pm 0.26$ |
| [15]N-glycine | (D) End product | Ammonia | $3.3 \pm 0.11$ |

Data of Golden and Waterlow (1977).

different, methods A, B and D ranked the individual patients in almost identical order. Golden and Waterlow argued that the [14]C-leucine-$CO_2$ method (B) was probably the most sound theoretically, but concluded that either of the other methods could be used for comparative studies. The difference in the two [15]N determinations is due to glycine contributing a little more to urinary ammonia than it does to liver urea, although in the case of the data in Table 1, the difference is not statistically significant.

Likewise, similar studies in rats revealed no differences between synthesis rates calculated from ammonia and urea (Golden and Jackson, 1977; Taruvinga et al., 1979, Stein et al., 1980a), although different labelled amino acids gave different results (Table 2). Other methods (plasma flux or $CO_2$ excretion) gave values in the range 35–45 g protein/kg body weight/day for male rats (Waterlow et al., 1978). Although the results obtained with [15]N-labelled aspartate are low compared to those with [15]N-labelled glycine or with [14]C-labelled amino acids, the values based on urea and ammonia are reasonably close to each other (Table 2), However, this is

TABLE 2

WHOLE BODY PROTEIN TURNOVER RATES (g PROTEIN/kg/DAY) DERIVED FROM
DIFFERENT END PRODUCTS BY THE USE OF VARIOUS [15]N-LABELLED AMINO ACIDS

| Labelled amino acid | End product | | | |
|---|---|---|---|---|
| | Urea | | Ammonia | |
| | Liver | Kidney | Liver | Kidney |
| [15]N-glycine | $37.7 \pm 1.6$ | $35.0 \pm 1.6$ | $35.6 \pm 1.6$ | $36.5 \pm 1.7$ |
| [15]N-aspartate | $29.6 \pm 3.8$ | $29.3 \pm 2.7$ | $26.4 \pm 5.0$ | $21.3 \pm 2.1$ |
| [15]N-valine | $74.0 \pm 7.8$ | $70.5 \pm 5.4$ | $73.1 \pm 4.5$ | $54.7 \pm 3.1$ |
| [15]N-leucine | $60.8 \pm 5.6$ | $78.7 \pm 10.1$ | $70.3 \pm 6.2$ | $36.0 \pm 3.4$ |

Data of Taruvinga et al. (1979).
Values are for male rats in the 250–400 g weight range.

not the case with the labelled branched chain amino acids, because they are metabolized preferentially by muscle and a variable fraction is subsequently re-exported to the liver.

## Direct versus Direct

[15]N-glycine can be used for measuring tissue protein synthesis rates in exactly the same way as any other isotopically labelled amino acid. Synthesis rates are calculated from the ratio of $S_B/S_I$, where $S_B$ is the amount of isotope incorporated into protein and $S_I$ the isotopic enrichment of the precursor pool. There are two benefits from using [15]N-glycine: (i) not being radioactive, it is suitable for human use and (ii) where the objective is to obtain data for comparative purposes, the enrichments of total tissue protein-N and of total free amino acid-N can be used for $S_B$ and $S_I$, i.e. it is unnecessary to isolate the free glycine for [15]N analysis, a cumbersome procedure requiring an amino acid analyser.

The use of the mixed amino acid N to estimate $S_B$ and $S_I$ assumes that the [15]N enrichment of the mixed amino acids is directly proportional to the amount of [15]N-glycine present. Experimentally, this has been shown to be so for rat liver (Stein et al., 1980a). It is also likely to be true for other tissues, as most of the [15]N is in glycine and serine with much less in alanine, glutamine, etc. (Jackson and Golden, 1980; Stein et al., 1980a; Matthews et *al.*, 1981; Gersovitz *et al.*, 1980). As long as the ratio $^{15}N$-glycine: $^{15}N$-serine is constant, the relationship should be valid. However, as shown in Table 3, fractional synthesis rates found with the mixed amino acids for liver are lower than those found where the glycine-$^{15}N$ enrichment is used for $S_B$ and $S_I$ because there is proportionately more glycine in the tissue free amino acids than in the tissue protein. If the relative abundance of $^{15}N$ in

### TABLE 3

FRACTIONAL SYNTHESIS RATES OF LIVER PROTEIN IN PARENTERALLY NOURISHED 180 g FEMALE RATS AS MEASURED WITH $^{15}N$-GLYCINE AND CALCULATED BY TWO DIFFERENT METHODS

| Time of infusion (h) | $^{15}N$ enrichment measurement | Method of calculation | Fractional synthesis rate (%/day) |
|---|---|---|---|
| 2 and 9 | Mixed amino acids | Initial rate[a] | $157 \pm 41$ |
| 3 | Mixed amino acids | Initial rate | $100 \pm 18$ |
| 3 | Glycine | Initial rate | $144 \pm 32$ |
| 8 | Mixed amino acids | Steady state[b] | $48.3 \pm 5.1$ |
| 8 | Glycine | Steady state | $85.7 \pm 41.3$ |
| 12 | Mixed amino acids | Steady state | $41.2 \pm 6.6$ |
| 12 | Glycine | Steady state | $98.2 \pm 14.7$ |
| 18 | Mixed amino acids | Steady state | $28.2 \pm 5.5$ |
| 24 | Mixed amino acids | Steady state | $30.9 \pm 6.1$ |
| 24 | Glycine | Steady state | $53.5 \pm 10.2$ |

[a] Stein *et al.* (1980a).
[b] Garlick *et al.* (1973).

the other amino acids (mainly serine) were known, the molar fractions of glycine and serine in the tissue free amino acids and protein would in theory permit the calculation of $S_B^{gly}$ and $S_I^{gly}$ from $S_B^{mixed}$ and $S_I^{mixed}$. In reality, the exercise is pointless because (a) the data are for comparative use only, and (b) there are more serious unresolved problems if it is desirable to obtain the absolute synthesis rate, such as the location of the amino acid precursor pool used for tRNA synthesis and whether its relationship to the tissue free and plasma amino acids pools is always the same.

It does not follow, however, that in liver the tissue free amino acid enrichment is equal to the liver urea enrichment. Urea-N is not derived equally from all amino acid-N and the relationship between the enrichments of liver urea and liver mixed amino acids will vary with the mixture of amino acids present in the liver. Thus, if $^{15}$N-glycine is given in the absence of other amino acids the ratio of $^{15}$N enrichments urea:mixed amino acids is 0·5 (Golden and Jackson, 1977; Stein *et al.*, 1976), but if amino acids are given concomitantly the ratio is 1 (Stein *et al.*, 1976). The $^{15}$N enrichment of mixed amino acids cannot be used for calculating protein synthesis rates either by flux or end product methods, but only where the ratio $S_B/S_I$ is used in the calculation.

One problem, common to both the mixed amino acids and glycine approaches is that long infusions ($>$4h) can underestimate the true fractional synthesis rate if there is a sizable amount of proteins turning over fast as in liver (Stein *et al.*, 1980a). The reason is that if an approximate isotopic steady state is attained in the precursor pool after 3 h, then the enrichment of glycine in a protein with a half-life of say 30 min will be about the same at 4, 6 or 10 h because the protein will have turned over often enough to have equilibrated with the precursor pool. Experimentally, this can be shown to be the case for rat liver by plotting the ratio $S_B:S_I^{-t}$ for various infusion times against time. $S_I^{-t}$ is the mean value of $S_I$ for time $t$ (Fig. 1). If there were no proteins turning over fast, the lines should go through the origin and the fractional synthesis rate ($k_s$) should be

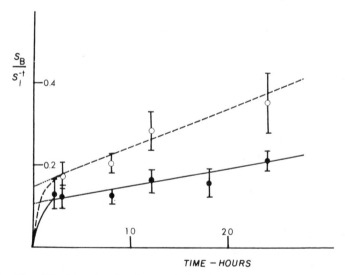

TIME – HOURS

FIG. 1.    Plot of $S_B/S_I^{-t}$ against time for the mixed amino acids, ●, and the glycine data, ○. Means ± 1 SD. The best fit lines (by linear regression) are the solid lines. Since at $t = 0$, $S_B/S_I^{-t}$ must also $= 0$, the dashed and solid lines indicate the actual curves and the dotted lines the extrapolated linear regression lines.

independent of the infusion time. In fact, $k_s$ decreases with longer infusion times because there is an initial burst of isotope uptake into protein followed by a slower incorporation rate (Fig. 1). The same phenomenon is found if the mixed amino acids are used for the $^{15}N$ analysis. The close correspondence is further confirmatory evidence that the $^{15}N$ enrichment of mixed amino acids can be used in determinations of synthesis rate dependent on the ratio $S_B:S_I$.

A second type of direct versus direct comparison is when $^{15}N$-glycine is used to measure synthesis rates of plasma protein. Two such studies have been reported. In one, by Gersovitz *et al.* (1980), $^{15}N$-glycine was given to adults at a constant rate and the synthesis rate estimated from the ratio of the $^{15}N$ enrichment in the arginine of albumin ($S_B$) and in urea ($S_I$), as originally described by McFarlane (1963) and Reeve *et al.* (1963). The problem with albumin is that it is compartmentalized, with the extravascular–intravascular pools equilibrating with a half-life of 12 h. Thus, unless the infusion is allowed to go on for 45 + h, as in the studies of Gersovitz *et al.*, the $S_B:S_I$ ratio cannot be translated into a true synthesis. Shorter infusion times (12–16 h) reflect mainly the plasma compartment. For comparative purposes, comparison of the ratio $S_B:S_I$ at any time is satisfactory, provided that the infusion times are equal. With this approach we were able to show that total parenteral nutrition improved albumin synthesis in depleted cancer patients (Stein *et al.*, 1980*b*).

One of the goals in the development of $^{15}N$ methodology is, as Waterlow *et al.* stated (1978), to develop methods suitable for use in 'the field' on patients rather than research volunteers. For such studies, very long periods of $^{15}N$ administration are impractical. For this reason we used a shorter, non-isotopic steady state method to measure the synthesis rate of fibrinogen. $^{15}N$-glycine or $^{15}N$-ammonium chloride was given at a constant rate for 4 h concomitantly with sodium benzoate for 4 h. The $^{15}N$ enrichment of the protein-bound glycine and of the urinary hippuric acid were used for $S_B$ and $S_I$, respectively. The values for the half-life of fibrinogen for the two $^{15}N$ carriers were different (Table 4), again indicating lack of homogeneity in the amino acid precursor, although when the period of the infusion was extended to 10–12 h, the difference became less. We do not know which source of glycine, exogenous or *de novo* (synthesis from $^{15}N$-$NH_4Cl$), is the better approximation to the precursor pool. The mean $t_{1/2}$ with $^{15}N$-glycine, 2·9 days, agrees well with literature data obtained by other methods. The initial rate (non-steady state) method is also suitable for albumin, but in our particular case we lacked access to a sufficiently sensitive mass spectrometer for the $^{15}N$ assay.

### TABLE 4

COMPARISON OF THE HALF-LIFE OF FIBRINOGEN AFTER THE ADMINISTRATION
OF $^{15}$N-GLYCINE OR $^{15}$NH$_4$Cl

| Time of infusion (h) | $^{15}N$ carrier | Half-life of fibrinogen | Method of calculation |
|---|---|---|---|
| 3–4 | Glycine | $3 \cdot 4 \pm 1 \cdot 5$ | Non-isotopic |
|  | NH$_4$Cl | $1 \cdot 4 \pm 0 \cdot 6$ | steady state[a] |
| 10–12 | Glycine | $2 \cdot 4 \pm 0 \cdot 6$ | Isotopic |
|  | NH$_4$Cl | $1 \cdot 6 \pm 0 \cdot 4$ | steady state[b] |

[a] Stein et al. (1980a).
[b] Garlick (1973).

## Direct versus Indirect

One such study has been reported, a comparison of the method of Picou and Taylor-Roberts (1969) and the Garlick method (Garlick et al., 1973) using a whole rat homogenate for the tissue analysis (Stein et al., 1980a). Estimation of $S_B$ and $S_I$ from the enrichment of glycine gives synthesis rates a little higher than those from the enrichment of the mixed amino acids because of the greater proportion of glycine in the tissue free amino acids (Table 5). The significant feature of the data in Table 5 is that the values are sufficiently similar to be accounted for by the different moieties analysed for $^{15}$N. There is no evidence that there may be something amiss with either approach. Probably the best value is given by the $^{15}$N-glycine method of Garlick because it is subject to one less assumption than the other methods.

### TABLE 5

WHOLE BODY PROTEIN SYNTHESIS RATES IN HOMOGENATES FROM 170–190 g
FEMALE RATS AS DETERMINED BY DIRECT AND INDIRECT METHODS

| Product analysed for $^{15}N$ | Method of calculation | Protein synthesis rate (g protein/kg/day) |
|---|---|---|
| Urine | Indirect[a] | $20 \cdot 0 \pm 3 \cdot 0$ |
| Urea | Indirect | $22 \cdot 1 \pm 4 \cdot 0$ |
| Ammonia | Indirect | $17 \cdot 8 \pm 3 \cdot 2$ |
| Glycine | Direct (steady state)[b] | $47 \cdot 2 \pm 15 \cdot 7$ |
| Mixed amino acids | Direct (steady state) | $21 \cdot 5 \pm 6 \cdot 3$ |

[a] Picou and Taylor-Roberts (1969).
[b] Garlick et al. (1973).
Data of Stein et al. (1980a).

Glycine is only one of twenty amino acids, and in view of the original reasons for using glycine, it is surprising that it works so well. Because no other $^{15}$N-labelled amino acid has been shown to be better than glycine, it is likely to remain the $^{15}$N-labelled amino acid of choice. There is, however, one caveat: in indirect methods where amino acids are severely limiting and energy is excessive, such as in parenterally nourished rats given only hypertonic glucose, urea as the end product in the Picou–Taylor-Roberts method gives anomalously high synthesis rates. The reason is that the excessive energy permits maximal reutilization and compartmentation within the tissues, so 'cold' nitrogen transferred to the liver for excretion mixes incompletely with the free amino acid pools in the liver and contributes disproportionately to urea synthesis. Obviously, in the rats the result was obtained under highly unnatural conditions (Stein *et al.*, 1978). But we have recently noted the possibility of a similar phenomenon in cancer patients when a variety of end product methods were used with $^{15}$N-glycine. Of a total of about 100 patients studied with a method in which the cumulative excretion of $^{15}$N in urinary ammonia was measured, about 10 % had synthesis rates much higher ( > 2) than normal healthy subjects. In a mixed sample of 40 normal subjects, we have seen no similarly 'anomalous' values. At present we are uncertain as to whether the discordant values are due to a systemic effect of the tumour, the nutritional status of the patients (most of the cancer patients were chronically malnourished and the synthesis rates become more normal after refeeding), or a methodological problem in using glycine in certain nutritionally depleted states. However, we found an expected decrease with acutely starved adults. In malnourished children, $^{15}$N-glycine end product methods do show an expected decrease in protein synthesis, which also argues against a methodological error as the explanation of the anomaly.

In summary, $^{15}$N-glycine is an acceptable amino acid for use as a tracer in all types of determinations of protein synthesis, with two reservations, both applying only to indirect end product methods: (1) in extreme nutritional states great care should be taken in interpreting the results and (2) to have confidence in interpreting the data, it is best to compare closely related groups or to use each subject as his own control.

## Discussion

*Millward (London School of Hygiene and Tropical Medicine, UK):* There is a problem in comparing the labelling of the tissue α-amino N pool with that of the protein. For this to be valid one has to assume that the distribution of label between individual amino acids in these two pools is the same. Since the glycine content of the free amino acid pool is different from that of protein it is likely to introduce an error. Do you know how big this error is?

*Stein:* We have determined this error by measuring synthesis rates of rat liver by both the method of mixed amino acids and that of the isolation of glycine (Table 3; Stein, 1980a). The values for the fractional synthesis rate of liver based on isolating the glycine are double those found with the mixed amino acids, because the proportion of glycine in the tissue free amino acids is about double that in liver protein. As long as the ratio is constant, either approach can be used in experiments where the aim is to compare one group against another.

*Golden (University of the West Indies, Kingston, Jamaica):* In general, the results we get with the Picou–Taylor-Roberts model are valid, I think, maybe not in terms of absolute numbers, but certainly in terms of comparative numbers. It is only under particularly stressful or difficult circumstances that the Picou–Taylor-Roberts method breaks down. Rats in your experiments (Stein *et al.*, 1978) had been force fed very high levels of glucose until nigh unto death. I think this is a very stressful situation.

*Stein:* I agree with you that it is a stressful situation, but it is one of great clinical interest because total parenteral nutrition involves giving excessively high levels of glucose to patients, and some patients are unable to handle the glucose load. In this situation the Picou–Taylor-Roberts model is no longer valid if urea is used as the end product. However, ammonia seems satisfactory.

# References

GARLICK, P. J., MILLWARD, D. J. and JAMES, W. P. T. (1973). The diurnal response of muscle and liver protein synthesis *in vivo* in meal-fed rats, *Biochem. J.*, **136**, 935–45.

GARLICK, P. J., CLUGSTON, G. A. and WATERLOW, J. C. (1980). Influence of low-energy diets on whole-body protein turnover in obese subjects. *Am. J. Physiol.*, **238**, E235–244.

GERSOVITZ, M., MUNRO, H. N., UDALL, J. and YOUNG, V. R. (1980). Albumin synthesis in young and elderly subjects using a new stable isotope methodology: response to level of protein intake, *Metabolism*, **29**, 1075–86.

GOLDEN, M. H. N. and JACKSON, A. A. (1977). Tissue enrichments and protein turnover measured with [15]N-glycine, *Nature*, **265**, 563–4.

GOLDEN, M. H. N. and WATERLOW, J. C. (1977). Total protein synthesis in elderly people: a comparison of results with $^{15}$N-glycine and $^{14}$C-leucine, Clin. Sci. and Mol. Med., **53**, 277–88.

JACKSON, A. A. and GOLDEN, M. H. N. (1980). $^{15}$N-glycine metabolism in normal man: the metabolic $\alpha$-amino nitrogen pool, Clin. Sci., **58**, 517–22.

JAMES, W. P. T., GARLICK, P. J., SENDER, P. M. and WATERLOW, J. C. (1974). The choice of label and measurement technique in tracer studies of body protein metabolism in man, in Dynamic Studies with Radioisotopes in Medicine, pp. 461–72, International Atomic Energy Agency, Vienna.

LONG, C. L., JEEVANANDAM, M., KIM, B. M. and KINNEY, J. M. (1977). Whole body protein synthesis and catabolism in septic patients, Am. J. Clin. Nutr. **30**, 1340–8.

MATTHEWS, D. E., CONWAY, J. M., YOUNG, V. R. and BIER, D. M. (1981). Glycine metabolism in man, Metabolism, in press.

MCFARLANE, A. S. (1963). Measurement of synthesis rates of liver produced plasma proteins, Biochem. J., **89**, 277–90.

PICOU, D. and TAYLOR-ROBERTS, T. (1969). The measurement of total protein synthesis and catabolism and nitrogen turnover in infants in different nutritional states and receiving different amounts of dietary protein, Clin. Sci., **36**, 283–96.

REEVE, E. B., PEARSON, J. B. and MARTZ, D. C. (1963). Plasma protein synthesis in the liver. Method for measurement of albumin formation in vivo, Science, **139**, 914–16.

STEIN, T. P., LESKIW, M. J. and WALLACE, H. W. (1978). Measurement of half-life of human plasma fibrinogen, Am. J. Physiol., **234** (Endocrinol. Metab. Gastrointest. Physiol.|3), E504–E510.

STEIN, T. P., ORAM-SMITH, J. C., LESKIW, M. J., WALLACE, H. W., LONG, L. C. and LEONARD, J. M. (1976). The effect of nitrogen and calorie restriction on protein synthesis in the rat, Am. J. Physiol., **230**: 1321–5.

STEIN, T. P., LESKIW, M. J., BUZBY, G. P., GIANDOMENICO, A. R., WALLACE, H. W. and MULLEN, J. L. (1980a). Measurement of protein synthesis rates with $^{15}$N-glycine, Am. J. Physiol., **239**, E294–E300.

STEIN, T. P., BUZBY, G. P., GERTNER, M. H., HARGROVE, W. C., LESKIW, M. J. and MULLEN, J. L. (1980b). Effect of parenteral nutrition on protein synthesis and liver fat metabolism in man, Am. J. Physiol., **239**, G280–G287.

TARUVINGA, M., JACKSON, A. A. and GOLDEN, M. H. N. (1979). Comparison of $^{15}$N-labelled glycine, aspartate, valine and leucine for measurement of whole body protein turnover, Clin. Sci.,**57**, 281–3.

TSCHUDY, D. P., BACCHUS, H., WEISSMAN, S., WATKIN, D. M., EUBANKS, M. and WHITE, J. (1959). Studies of the effect of dietary protein and calorie levels on the kinetics of nitrogen metabolism using $^{15}$N-L-aspartic acid, J. Clin. Invest., **38**, 892–901.

WATERLOW, J. C., GOLDEN, M. H. N. and GARLICK, P. J. (1978). Protein turnover in man measured with $^{15}$N-glycine. Comparison of end products and dose regimes, Am. J. Physiol., **235**, E165–E174.

WATERLOW, J. C. and STEPHEN, J. M. L. (1966). Adaptation of the rat to a low protein diet: the effect of a reduced intake on the pattern of incorporation of L-$^{14}$C-lysine. Br. J. Nutr., **20**, 461–84.

*T. P. Stein*

WU, H. and BISHOP, C. W. (1959). Pattern of $^{15}$N-excretion in man following the administration of $^{15}$N-labelled glycine, *J. Appl. Physiol.*, **14**, 1–5.

WU, J., SENDROY, J. JR. and BISHOP, C. W. (1959). Interpretation of urinary $^{15}$N excretion data following the administration of $^{15}$N-labelled amino acids, *J. Appl. Physiol.*, **14**, 11–15.

# 29

# A Comparison of Urinary Ammonia and Urea as End Products in the Measurement of Protein Synthesis with $^{15}$N-Glycine

P. J. GARLICK

*Clinical Nutrition and Metabolism Unit,*
*London School of Hygiene and Tropical Medicine,*
*London, UK*

In spite of what was said earlier about the better accuracy of $^{14}$C (precursor) methods, I am still interested in end product methods because I think it is important for the purposes of comparison to have results obtained by independent techniques. Methods involving end products in the urine also involve less disturbance to the patient.

Figure 1 shows apparent rates of protein synthesis obtained by giving single doses of 200 mg $^{15}$N-glycine to four volunteers under a number of different conditions (Fern et al., 1981). Measurements were made in the fasted state and when the subjects were receiving meals at intervals of 2 h. In each of the fed and fasted states the dose of isotope was given orally and intravenously and the rate of protein synthesis was determined separately from the excretion of $^{15}$N in ammonia and in urea during the 9 h following the dose. For ammonia, the flux ($Q_{NH_3}$) was calculated from eqn (1), as described previously (Waterlow et al. 1978):

$$Q_{NH_3} = E_a \frac{d}{e_a} \tag{1}$$

where $E_a$ and $e_a$ are the excretion of unlabelled and labelled ammonia respectively during the 9 h, and $d$ is the dose of isotope given. Protein synthesis ($Z_{NH_3}$) was then calculated from eqn (2):

$$Z_{NH_3} = Q_{NH_3} - E_T \tag{2}$$

where $E_T$ is the excretion of total nitrogen during the 9 h. The value of $E_T$ was adjusted for changes in the size of the body urea pool, estimated from

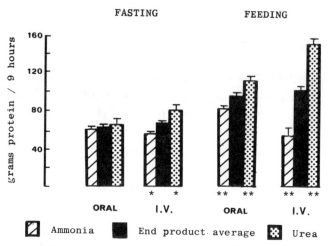

FIG. 1. Rates of whole-body protein synthesis in four normal subjects measured by single dose of $^{15}$N-glycine. Each subject was studied four times, with oral and intravenous dosage and while feeding and fasting. The histograms show rates of synthesis calculated from the excretion of $^{15}$N in ammonia and urea and, in addition, the average of these two rates (end product average). Differences between rates from ammonia and urea by paired $t$-test: $*p < 0.05$, $**p < 0.01$.

measurements of the urea concentration in plasma. For urea the flux ($Q_{urea}$) was calculated from a modified formula (eqn (3)) to take account of labelled urea retained within the body urea pool at 9 h, i.e.

$$Q_{urea} = \frac{(E_u + \Delta R_u)d}{(e_u + r_u)} \tag{3}$$

where $E_u$ and $e_u$ are the excretion of unlabelled and labelled urea during the 9 h, $\Delta R_u$ is the change in the size of the body pool of unlabelled urea during this period, and $r_u$ is the size of the body pool of labelled urea at the end of 9 h. The rate of synthesis was calculated from eqn (2) as with values derived from ammonia.

Figure 1 shows the mean values of synthesis from fed and fasted subjects, with intravenous and oral dosage and calculated from ammonia and from urea. In addition, in each case the average of the rates given by urea and ammonia was calculated (end product average). In fasted subjects the rates given by urea and ammonia were very similar with the oral dose but not with the intravenous dose. When the subjects were fed, the differences between rates given by urea and ammonia were greater, particularly when the isotope was given intravenously.

The other comparison that can be made is between fed and fasted states. With constant infusion of $1\text{-}^{14}\text{C}$-leucine, synthesis in fed subjects was about 50 % higher than in fasted subjects (Garlick and Clugston, p. 303). With urea as the end product, it can be seen from Fig. 1 that rates of synthesis in the fed state are nearly twice those in the fasted state. This is true whether comparisons are made between studies where the dose was given intravenously or orally. However, when the same comparison is made for ammonia, there is relatively little difference between fed and fasted states.

The reason for calculating the end product average is that there appears to be an inverse relationship between the rates given by urea and those from ammonia. This is particularly apparent when comparisons are made between intravenous and oral dosage (Fig. 1). A possible explanation for this lies in the physical separation within the body of the sites for synthesis of urea and ammonia, the liver and the kidney. Not only will the liver take preference when the dose is given orally and the kidney when it is given intravenously, but this relationship may be expected to be influenced in fed subjects by dilution of the isotope in the liver by unlabelled amino acids from the gut. The end product average is relatively little affected by the route of administration of isotope. Also, the difference in the end product average between fed and fasted subjects is about 50 % with both oral and intravenous dosage. This is very similar to the difference estimated from $1\text{-}^{14}\text{C}$-leucine infusion.

We can conclude from these results that we cannot automatically accept the result from either end product. Differences between urea and ammonia may vary depending on a number of factors, among them the route of administration of the isotope and the dietary intake. The end product average may be one way of minimizing this difficulty. It is worth continuing to study ways of improving the end product method because its convenience for use in sick patients is of such great potential value.

## *References*

FERN, E. B., GARLICK, P. J., McNURLAN, M. A. and WATERLOW, J. C. (1981). The excretion of isotope in urea and ammonia for estimating protein turnover in man with $^{15}\text{N}$-glycine, *Clin. Sci.*, **61**, 217–28.

WATERLOW, J. C., GOLDEN, M. H. N. and GARLICK, P. J. (1978). Protein turnover in man measured with $^{15}\text{N}$: comparison of end product and dose regimes, *Amer. J. Physiol.*, **235**, E165–74.

# 30

# Interrelationships of Amino Acid Pools and Protein Turnover

A. A. JACKSON and M. H. N. GOLDEN

*Tropical Metabolism Research Unit,*
*University of the West Indies,*
*Kingston, Jamaica*

'A cardinal assumption of both this and previous methods of measuring nitrogen turnover and total protein synthesis after administration of a single amino acid labelled with $^{15}N$ is that urinary data reflect the behaviour of the total amino-N mixture of the metabolic pool' (Picou and Taylor-Roberts, 1969).

It is clear from theoretical considerations (p. 323) that there must be limitations to the cardinal assumption that underlies the measurement of whole body protein turnover by the end product method. Our concern has been to identify the extent to which these limitations affect the results obtained, and the degree of confidence one can use in applying the results obtained to clinical situations. One of the criticisms of the end product method was that as urea is synthesized in the liver the enrichment of the precursor nitrogen pool for protein synthesis in the liver may determine the enrichment of urea.

Thus we infused a series of rats with $^{15}N$-glycine and measured the enrichment in urea, ammonia and total $\alpha$-amino-N in the liver and kidneys after 6 h (Golden and Jackson, 1977). The results produced three useful pieces of information. Firstly the flux calculated from urea and ammonia enrichment in both kidney and liver gave the same results (36·4 g/kg/day). The value was equivalent to that obtained by other methods where $^{14}C$-tyrosine or $^{14}C$-lysine had been used. Golden and Waterlow (1977) have shown that using $^{14}C$-leucine and $^{15}N$-glycine in man, comparable results can be derived with both tracers. Therefore a certain confidence is obtained in the validity of the results. More surprisingly, in both liver and kidney the enrichment of the $\alpha$-amino-N was completely different from that of both urea and ammonia. Therefore, the end product does not reflect the enrichment in the total α-amino-N pool of the organ in which it is formed. What then are the precursors for the end product and what relationship do these bear to the overall enrichment in the α-amino-N pool, the precursor pool for protein synthesis, and the end product? In all the human data in the literature there is a consistent difference in the enrichment of urinary ammonia and urinary urea following a continuous infusion of [15]N-glycine. The plateau enrichment of ammonia is higher than that of urea and hence ammonia gives a lower value for flux than urea (Waterlow *et al.*, 1978).

There are two immediately obvious reasons why urea and ammonia should give different results. The precursor mix for the synthesis of the two end products may be different in the liver and kidney in man, whereas they are similar in the rat. The difference may be methodological in origin, in that either urea has not reached plateau at the time of sampling, or that ammonia has gone beyond plateau and is showing the effects of recycling of label. Picou and Phillips (1972) have shown that with an infusion of [15]N-urea plateau is not achieved in the urea pool until 32–36 h in infants. In normal adults it takes about the same time (Jackson and Landman, unpublished). Therefore, as [15]N-glycine has to equilibrate first in the glycine pool and then the resultant labelled urea equilibrates in the urea pool, it is unlikely that plateau in urea could be achieved in less than 36 h.

We followed the enrichment of urea and ammonia in urine in a normal adult who was receiving a continuous intake of feed containing a tracer dose of [15]N-glycine. We also looked at the enrichment of the plasma alanine, glutamate and glutamine amide-N, three likely precursors for urea and ammonia (Jackson and Golden, 1980). There was little or no label in these amino acids. Ammonia enrichment rises to plateau over a period of 16 to 20 h. This plateau is maintained with virtually no perturbation up to 36 h, and gives a value for flux of 2·34 g protein/kg/day. The urea enrichment rises at a slower rate, and shows perturbations which may be due to diurnal variations, but essentially the line is still rising and shows no evidence of plateau at 36 h, when it is still significantly below the ammonia plateau. However, the rate of rise is sufficiently slow so that if a few samples had been taken over any 6 h period they might reasonably be considered to represent a satisfactory plateau. Flux calculated from the ammonia plateau would at all times be less than that calculated from the urea values. The difference between the two results would be critically dependent upon the time at which the samples were collected. A similar study was carried out in two children who had recovered from severe protein-energy malnutrition, for a longer duration (72 h) (Driver, Golden, Jackson and Picou, unpublished). Ammonia achieved plateau by 12–15 h and this was maintained for 72 h

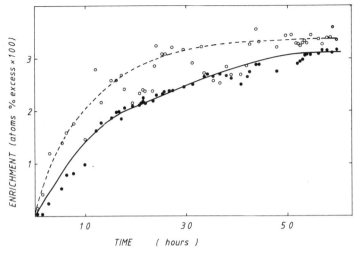

FIG. 1.    Enrichment in urinary ammonia (○) and urea (●) in an 8 month-old child, weight 5·25 kg, following the continuous infusion of $^{15}$N-glycine (Driver *et al.* unpublished data).

(results for one child are shown in Fig. 1). The plateau enrichment in urea was essentially the same as the ammonia plateau, but was not reached until 40 h. Therefore flux calculated from the two plateaux gives results that agree within 10 % (Table 1). There was some variability in the enrichment of different urine specimens, which was less marked for urea than ammonia. Most of this is technical in origin, due to failure in the infusion pump or power cuts. This emphasizes the sensitivity, particularly of ammonia, to a non-constant input of isotope. Improvements in the technique reduced this source of variability in subsequent studies.

TABLE 1

PROTEIN TURNOVER CALCULATED FROM THE PLATEAU ENRICHMENT OF URINARY AMMONIA AND UREA IN TWO CHILDREN RECOVERED FROM MALNUTRITION

| Subject | Calculated flux (g protein/kg/day) | | Ammonia/ Urea (%) |
|---|---|---|---|
| | From urea | From ammonia | |
| AG | 9·55 | 8·78 | 92 |
| KK | 9·46 | 9·39 | 99 |

These studies suggest that turnover calculated from the ammonia plateau may reasonably be used as a reference in this situation, and that any difference between urea and ammonia may be due to a failure of urea to reach an isotopic steady state.

In a completely different study (Taruvinga, Golden and Jackson, unpublished) we have had cause to give $^{15}$N- and $^{14}$C-labelled glycine as a conjugated bile salt in a single dose to normal people and patients suspected of having excessive bile salt deconjugation. The appearance of $^{14}$C in breath as $^{14}CO_2$ and of $^{15}$N in urine as $^{15}$N-ammonia were followed over time. Both $^{14}CO_2$ and $^{15}$N-ammonia appeared at the same time and in equivalent amounts, suggesting that they arise from the same source, namely the kidney. We interpret this to mean that the net renal uptake of glycine (Squires *et al.*, 1976) represents a major breakdown pathway for this amino acid. As glycine-N hardly distributes to amino acids other than serine, the appearance of labelled ammonia in urine represents the dilution of labelled glycine in the glycine pool of the body. The rise of urea to plateau has, in addition, a component representing equilibration of labelled urea in the urea pool. This is dependent on the turnover of the urea pool, which may vary widely with metabolic state. For example, in a patient with renal disease, who had a urea pool that was expanded to twice normal and received a low protein diet, labelled urea had not equilibrated in the urea pool by 72 h.

These considerations gave a reasonable explanation for the data to hand at that time. However, the next series of experiments demanded a reassessment of the situation. Turnover was being measured in preterm infants with $^{15}$N-glycine. They were fed donor human breast milk in balances lasting 72 h. We wanted to do longitudinal measurements so as to be able to assess changes with time (Jackson *et al.*, 1981). An infant born at 32 weeks gestation weighing 1000 g was studied first on the eighth day of life, and again one week later. In the first study ammonia reached plateau at 12 to 18 h and urea in 40 h, and both gave identical values for flux of 12·75 g protein/kg/day. In the second study ammonia followed a similar rise to plateau and gave a flux of 7·56 g protein/kg/day. Urea, however, failed to become enriched at all over 72 h and gave a derived value for flux of infinity. Clearly $^{15}$N from glycine was not reaching urea, and this situation was obtained in six out of eight studies. In one study the urea plateau was 66 % of the ammonia plateau, giving a derived flux 150 % of that of ammonia. We have argued that these results may be explained by the fact that the metabolic demands of the premature infant for glycine are very high; because the dietary intake of preformed glycine is relatively low, an *a priori* need for endogenous glycine synthesis is created. The actual endogenous synthesis, *de novo*, of glycine is insufficient to satisfy the demand, probably because the enzyme systems involved are immature. Thus glycine becomes a semi-essential amino acid. Therefore, there are situations in which the body is able to limit the flow of nitrogen from glycine to urea, and glycine fails to label the urea precursor pool adequately. There is no reason to suppose that the precursor pool for protein is unlabelled; therefore, depending upon the extent to which the urea precursor pool fails to become labelled, protein turnover will be overestimated to a varying extent. Again we have used the ammonia value as a reference, and conclude that the underlying assumption of the end product method cannot be substantiated for urea in this situation.

This relationship between the flux calculated from the two end products brings us back to the difference found in children recovering from malnutrition between the values for flux calculated from the ammonia and urea data. In individual cases flux calculated from urea data may be more than twice that derived from ammonia data (Waterlow *et al.*, 1978).

A series of studies was conducted in children who were recovering from malnutrition, at different planes of recovery (Golden, Golden and Jackson, to be published). Each child was studied as he started to gain weight, at 50 % repletion of weight deficit, and as he approached complete repletion of weight deficit. The feed was given by continuous infusion for 72 h, and contained a tracer dose of [15]N-glycine. A consistent pattern was seen in all the children and is represented in Fig. 2. Ammonia reached plateau in 12 to 20 h, whereas urea took 40 h, on occasion longer. In the first study the urea plateau was invariably significantly lower than the ammonia plateau, hence giving relatively high values for flux. As recovery progressed there was a diminution in the difference between the two plateaux with successive studies. By the time the child was approaching complete repletion of his weight deficit both ammonia and urea gave values for flux that differed by less than 10 %. At each stage the child was gaining weight at a rate greater than normal. The difference between the studies may be due to change in the glycine composition of the balance of tissue being synthesized as recovery progresses, so that the demands for glycine become less with time. Alternatively, the mechanism for synthesizing glycine *de novo* may have been sacrificed, in part or in whole, as part of the adaptation to malnutrition, and these results represent the time course of its recovery.

Whatever the explanation, it would seem that we are looking at the same situation in the recovering children as we have seen in the preterm infants,

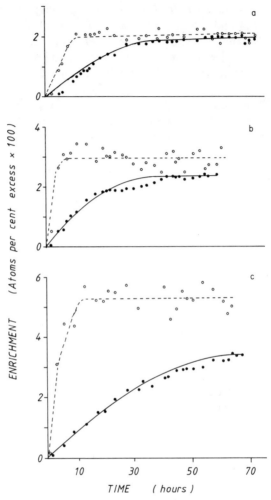

Fig. 2. Enrichment in urinary ammonia (○) and urea (●) in a child at late (a), middle (b) and early (c) recovery from severe protein-energy malnutrition (Golden *et al.* (to be published)).

but in a milder form. In the face of the metabolic demands for new tissue synthesis, glycine availability is marginal, and the extent to which glycine-N contributes to the urea precursor pool is limited. If this is so, then one might reasonably presume that provision of adequate exogenous glycine in the diet would make good the short-fall, and allow for the appropriate amount of label to appear in urea.

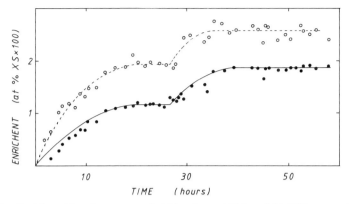

FIG. 3.   Enrichment in urinary ammonia (○) and urea (●) in a child 50 % recovered from
severe malnutrition. At 26 h unlabelled glycine was added to the feed.

A child recovering from malnutrition, who had repleted half his weight
deficit, was studied by the same technique (Royes, Golden, Jackson,
unpublished) (Fig. 3). We would expect to find a difference in the plateaux
of urea and ammonia. By 30 h both urea and ammonia had reached a
significantly different plateau, labelling of urea being 66 % of that of
ammonia. Additional unlabelled glycine was then added to the feed, to give
an extra 150 mg/h of N. This introduction of additional cold glycine caused
the urea plateau to rise to a new steady state within 12 h, to a level that was
identical to the earlier ammonia plateau. However, the ammonia plateau
itself changed, increasing by 25 %, so that it was still higher than the urea
plateau, although the relative difference between the two was less than
before (73 %) (Table 2). Which is the correct value for flux ? At first sight one

TABLE 2

PROTEIN TURNOVER DERIVED FROM THE PLATEAU ENRICHMENT
OF URINARY AMMONIA AND UREA IN A CHILD RECOVERING FROM
MALNUTRITION, BEFORE AND AFTER A DIETARY GLYCINE LOAD

|  | Time | Calculated flux (g protein/kg/day) | |
| --- | --- | --- | --- |
|  |  | From urea | From ammonia |
| Before load | 0–30 h | 10·31 | 6·31 |
| After load | 30–60 h | 6·66 | 4·79 |

may reasonably consider that the values for ammonia and urea which correspond reflect the correct result in different situations. Thus during the first 30 h ammonia reflects the activity in the glycine pool and gives a reference value for turnover on a diet in which the balance of amino acids may be considered normal. At this time the demands for glycine in metabolic pathways are sufficiently high to limit the extent to which glycine-N gives rise to urea, and distributes its N to other amino acids. During the second 30 h excessive glycine in the diet allows glycine-N to pass to urea in the same proportion as it passes to protein synthesis. However, a glycine load disturbs the glycine pool, possibly leading to excess ammonia formation (Kamin and Handler, 1951). This is reflected in the perturbation of the ammonia plateau, which represents an increased flux through an expanded glycine pool, so that glycine contributes proportionately more to the precursor mix for ammonia synthesis (Pitts and Pilkington, 1966). There is no evidence to suggest that with a glycine load the N from glycine is distributed more evenly among other amino acids (Jackson and Golden, 1980). However, the equivalence of the urea and ammonia plateaux may be fortuitous, the rise in both being accounted for by the properties of the model presented earlier (Golden, p. 323). The rapidity of the response and the short time required to reach a new steady state would support either interpretation.

   The balance of precursor amino acids for ammonia formation in the kidney depends among other things on the acid–base status of the body (Owen and Robinson, 1963). Therefore we would expect chronic acidosis or starvation to exert an effect upon the ammonia plateau enrichment. A normal adult was fasted for 96 h and then received a continuous infusion of $^{15}$N-glycine in water for 24 h. The enrichment in ammonia reached a low plateau after 4 h. The urea enrichment rose at the same rate as it had when the isotope had been given in the fed state and exceeded the ammonia enrichment towards the end of the 24 h. At this point food was introduced as an infusion providing adequate energy, essential amino acids to recommended requirements and non-essential N as glycine. Glycine provided about 60 % of the total-N provided. The infusion of $^{15}$N-glycine was continued at the same rate. There was an immediate and dramatic increase in the ammonia enrichment to a new steady state after 6 h. Flux calculated from this plateau was the same as the value derived in the same individual in the fed state. There was no obvious perturbation in the slope of the urea line. During the first 30 h $^{15}$N-glycine had equilibrated in the body glycine pool, but because the precursor mix for renal ammonia synthesis had a proportionately greater contribution from amino acids containing little or no isotope the enrichment in urinary ammonia was decreased. Refeeding with a glycine-rich diet resulted in glycine having a greater preponderance in the precursor mix for ammoniagenesis.

The effect that extreme changes in dietary intake, protein synthesis or protein breakdown could have on the enrichment of the end product was described in the model presented on p. 336. The model allows us to predict the direction and extent to which the enrichment would be expected to change, with no change in protein turnover itself, for the different situations. The enrichment of two end products, ammonia and urea, is affected differently by different situations, and hence the estimate of flux derived from their plateau enrichments during a constant infusion differs. The enrichment in ammonia reflects the enrichment in the glycine pool, provided that the proportion of glycine-N going to total ammonia-N is that seen in the normal, probably about 5–10 %. If the proportion of glycine-N in arterial blood increases compared with other precursors of ammonia, the plateau will be elevated and flux underestimated. If the proportion of glycine-N for ammoniagenesis falls, e.g. in acidosis, flux will be overestimated. For urea, the tendency is to overestimate flux. We have demonstrated situations where the body limits the flow of glycine-N as a proportion of total-N to urea. This is most likely to occur in situations where the metabolic requirements for glycine significantly exceed the total available from the diet and from endogenous synthesis. We have not definitely been able to describe a situation where glycine-N is represented in urea in excess of the proportion going to protein synthesis, although this may have been the reason for the rise in the urea plateau in the child given a glycine load. However, there is a limit to the extent to which glycine-N can enter urea. Glycine loads predispose to hyperammonaemia, and maybe the kidney has an important role to play in disposing of excess glycine-N as urinary ammonia (Kamin and Handler, 1951).

We have concentrated on the use of $^{15}$N-glycine for the continuous infusion end product method for the calculation of protein turnover because this is the system with which we are most familiar. Glycine is generally considered to be a relatively simple amino acid with straightforward metabolic relations. We have demonstrated limitations in the model using both urea and ammonia in certain metabolic states. However, the important question is not whether these limitations exist, but to what extent do they limit our ability to make useful measurements and draw useful conclusions. Consideration of these variables should enable us to design studies in such a way that useful measurements can be made.

## Acknowledgements

This work was supported by The Wellcome Trust, MRC Great Britain and DANIDA through FAO Rome.

## Discussion

*Stein (University of Pennsylvania School of Medicine, Philadelphia, USA):* By how much did you increase the dosage of glycine in the experiment you showed in Fig. 3, in which you found that the plateau changed by 50 %?

*Jackson:* Glycine accounted for 50 % of the total-N intake, so it was an enormous increase in glycine as a proportion of total-N.

*Bier (Washington University School of Medicine, St. Louis, USA):* Presumably if you give so much glycine the flux of nitrogen through the nitrogen pool has to increase. Thus, the isotopic plateau should go down not up.

*Jackson:* The fact that the plateau went up was the main reason why we had to develop the new model described by Golden (p. 336). If you take that model and apply the change in the glycine intake to it, you can predict that the plateau will go up in all but the most extreme situations (where glycine-N approaches total-N intake). If the proportion of glycine going to the end product increases, then the enrichment in the end product will always increase.

*Matthews (Washington University School of Medicine, St. Louis, USA):* When we measured the $^{15}$N enrichment of plasma glycine in man during repeated oral $^{15}$N-glycine administration, we found that the plasma $^{15}$N-glycine rose to a plateau with a half-life of about 9 h. It took about 30 h to approach plateau in plasma glycine, so I think you are right in saying that much of the delay in the $^{15}$N-labelling of urinary urea arises from slow equilibration of the free glycine pool.

*Waterlow (London School of Hygiene and Tropical Medicine, UK):* That is what I predicted in my paper (p. 279).

*Bier:* We have recently measured plasma leucine and lysine flux in newborn infants with infusions of L-1-$^{13}$C-leucine and L-$\alpha$-$^{15}$N-lysine (Frazer and Bier, 1980). If we use the kind of assumptions that we have heard about in the last few days we obtain calculated values for whole body protein turnover which are similar to those which you found, but far different from those reported previously (Pencharz *et al.*, 1977).

*Young (Massachusetts Institute of Technology, USA):* As a co-author of the paper referred to (Pencharz *et al.*, 1977), in which we attempted to quantify protein turnover in prematures, I accept the criticism that our estimates were probably too high. This was probably due to the fact that the [15]N-urea plateau had not been achieved. Indeed, a computer fit of the time-course of the urea-[15]N enrichment curves gives an estimate that approximates to 15 g/kg/day, or more in line with your estimates and those reported by Nicholson some years ago (Nicholson, 1970).

*Betton (University of the West Indies, Kingston, Jamaica):* Jackson made the point that preterm infants present special metabolic problems. He described measurements of protein turnover in these infants with a constant infusion of [15]N-glycine, and found that in most of the studies urinary urea failed to become enriched. He postulated that the metabolic demands for glycine during rapid growth exceeded the amount available from the diet plus endogenous synthesis, so that glycine became in effect a semi-essential amino acid.

We have tried to test this hypothesis by measuring protein turnover in preterm infants by the single dose–end product method, using [15]N-glycine, before and after supplementation with dietary glycine. Nine preterm infants with a mean gestational age of 33 weeks were studied at various times after birth, when they were judged clinically to be in a stable metabolic state. On average they weighed about 1·7 kg. The diet provided about 2·84 g protein/kg/day. A single dose of [15]N-glycine (0·94 mgN/kg) was given and urine collected for 12 h. The diet was then supplemented with unenriched glycine (88·5 mg/kg/day) and 4 days later the measurements of turnover were repeated.

The results are shown in Table 3. The values for flux obtained with urea as end product are very high, presumably as a result of a limited flow of glycine-N to urea, as suggested by Jackson, whereas the values obtained with ammonia fall within the range expected.

In the group as a whole there was no difference in calculated flux before and after glycine supplementation, but when we looked at individual cases in relation to growth rate a pattern emerged. In infants gaining weight at more than 10 g/kg/day the enrichment of urinary urea tended to decrease, whereas in those growing at a slower rate the enrichment of urea tended to increase. Rapid deposition of new tissue will increase the demand for glycine, making less available for urea synthesis. The approximate need for

TABLE 3

THE ENRICHMENT OF URINARY UREA AND AMMONIA IN 12 h URINE
SAMPLES FROM NINE PRETERM INFANTS FOLLOWING A SINGLE DOSE
OF [15]N-GLYCINE, BEFORE AND AFTER DIETARY SUPPLEMENTATION
WITH UNENRICHED GLYCINE

| | Enrichment ($\times 10^3$) | | Flux (g N/kg/day) | |
| | Ammonia | Urea | Ammonia | Urea |
|---|---|---|---|---|
| Day 1 | $62 \pm 15$ | $27 \pm 22$ | $3\cdot24 \pm 0\cdot88$ | $10\cdot42 \pm 6\cdot9$ |
| Day 4 | $72 \pm 32$ | $22 \pm 12$ | $2\cdot72 \pm 1\cdot19$ | $10\cdot58 \pm 5\cdot7$ |

glycine in an infant of this gestational age is about 30 mg/kg/day; a normal daily intake of breast milk or humanized cow's milk formula will provide less than one-third of this requirement. A dietary glycine supplement of 100 mg/kg/day may fail to satisfy the demands, and hence the rate of growth would critically determine the amount of glycine-N available for urea synthesis. Supplementation at a higher level may lead to increased label appearing in urea.

*Munro* (*Tufts University, Boston, USA*): From what has been said, it looks as if there may be two possible explanations for the low labelling of urea in these studies. The one proposed by Jackson and Betton is that all glycine is diverted into protein synthesis. That suggested by Matthews is that the equilibration of the glycine pool is slow, so that a 12 h period of urine collection would be too short.

*Waterlow:* Since we cannot resolve this issue, I would like to turn to another question. Jackson referred to renal ammonia production. We have heard that by far the greater part of urinary ammonia comes from glutamine (p. 172), and this glutamine comes mainly from muscle. Am I wrong in thinking that the labelling of the ammonia basically reflects what is happening in the muscle pool and not in the kidney pool?

*Jackson:* I think you are quite right. We are discussing two end products and the nitrogen for both of them has to come from the balance of amino acids in the body. The concepts which we have been presented with earlier show that nitrogen reaches the viscera from the periphery mainly as alanine and glutamine. Both the kidney and the liver use alanine and glutamine for the production of their end products. The proportion of glycine which is used must relate to the proportion of these other nitrogen sources which is used.

# References

FRAZER, T. E. and BIER, D. M. (1980). Essential amino acid turnover in the human newborn, *Pediat. Res.*, **14**, 571A.

GOLDEN, M. H. N. and JACKSON, A. A. (1977). Tissue enrichments and protein turnover measured with $^{15}$N-glycine, *Nature*, **265**, 563–4.

GOLDEN, M. H. N. and WATERLOW, J. C. (1977). Total protein synthesis in elderly people: a comparison of results with $^{15}$N glycine and $^{14}$C leucine, *Clin. Sci. Mol. Med.*, **53**, 277–88.

JACKSON, A. A. and GOLDEN, M. H. N. (1980). $^{15}$N glycine metabolism in normal man: the metabolic α-amino-nitrogen pool, *Clin. Sci.*, **58**, 517–22.

JACKSON, A. A., SHAW, J. C. L., BARBER, A. and GOLDEN, M. H. N. (1981). Nitrogen metabolism in preterm infants fed human donor breast milk: the possible essentiality of glycine, *Pediat. Res.*, in press.

KAMIN, H. and HANDLER, P. (1951). The metabolism of parenterally administered amino acids. III. Ammonia formation, *J. Biol. Chem.*, **193**, 873.

NICHOLSON, J. F. (1970). Rate of protein synthesis in premature infants. *Pediat. Res.*, **4**, 389–97.

OWEN, E. E. and ROBINSON, R. R. (1963). Amino acid extraction and ammonia metabolism by the human kidney during the prolonged administration of ammonium chloride, *J. Clin. Invest.*, **42**, 263–76.

PENCHARZ, P. B., STEFFEE, W. P., COCHRAN, W., SCRIMSHAW, N. S., RAND, W. M. and YOUNG, V. R. (1977). Protein metabolism in human neonates: nitrogen-balance studies, estimated obligatory losses of nitrogen, *Clin. Sci. Mol. Med.*, **52**, 485–98.

PICOU, D. and TAYLOR-ROBERTS, T. (1969). The measurement of total protein synthesis and catabolism and nitrogen turnover in infants in different nutritional states and receiving different amounts of dietary protein, *Clin. Sci.*, **36**, 283–96.

PICOU, D. and PHILLIPS, M. (1972). Urea metabolism in malnourished and recovered children receiving a high or low protein diet, *Amer. J. Clin. Nutr.*, **25**, 1261–6.

PITTS, R. F. and PILKINGTON, L. A. (1966). The relation between plasma concentrations of glutamine and glycine and utilization of their nitrogens as sources of urinary ammonia, *J. Clin. Invest.*, **45**, 86–93.

SQUIRES, E. J., HALL, D. E. and BROSNAN, J. T. (1976). Arteriovenous differences for amino acids and lactate across kidneys of normal and acidotic rats, *Biochem. J.*, **160**, 125–8.

WATERLOW, J. C., GOLDEN, M. H. N. and GARLICK, P. J. (1978). Protein turnover in man measured with $^{15}$N: comparison of end products and dose regimes, *Amer. J. Physiol.*, **235**, E165–74.

# 31

# Autontic Measurements† of Protein Turnover

R. W. Swick

*Department of Nutritional Sciences,*
*University of Wisconsin-Madison,*
*Madison, Wisconsin, USA*

Because of the uncertainties in estimating turnover rates, particularly from the disappearance of tracer accompanied by an unknown probability of reutilization, Schimke (1970) urged that whenever possible more than one method should be used. Waterlow *et al.* (1978) suggested that a comparison of methods in the same animals would be a crucial test. The criterion usually applied is that the method that gives the fastest renewal rate is probably the most accurate. The renewal rates of a number of enzymes have been estimated by a variety of methods and, in many cases, rates which are similar have been obtained (Schimke, 1970; Waterlow *et al.*, 1978). Rarely, however, have two methods been compared in the same animals (Swick, 1958).

In the first experiment the kinetics of the loss of radioactivity was compared to the kinetics of enzyme activity change (Chee and Swick, 1976). Ornithine aminotransferase, an adaptive enzyme found in liver mitochondria, was chosen because earlier experiments (Swick *et al.*, 1968) had shown that its renewal rate, estimated from the kinetics of enzyme activity change (Schimke, 1970), was much more rapid (about 90 %/day) than that of the bulk of mitochondrial proteins (15 %/day), calculated from tracer incorporation. A later experiment, in which arginine in ornithine aminotransferase was labelled by the administration of $^{14}$C-bicarbonate, still indicated that the rate, although different from that obtained earlier, was more rapid (about 40 %/day) than those of the mitochondrial proteins (Ip *et al.*, 1974). To resolve the question of the different renewal rates of ornithine aminotransferase we repeated the two experiments; however, in this instance, in the same animals.

† Autontic measurements are defined as 'measurements made in the same individuals'.

In this experiment the fractional rate of degradation was calculated from the decline in total radioactivity derived from $^{14}$C-bicarbonate, a method suggested by Millward (1970) and developed by us (Swick and Ip, 1974). Because it gives the fastest rates of renewal for hepatic proteins, it would seem to be the method of choice. Rats were fed diets containing either 12 % or 60 % casein for 5–6 days. The latter diet produces an increase in the activity of ornithine aminotransferase of about three-fold. The animals were injected with NaH$^{14}$CO$_3$ and groups were killed daily for 5 days. Ornithine aminotransferase was purified to homogeneity and its radio-activity measured. Other animals were fed one of the diets, injected with tracer, and switched to the other diet. The total radioactivity remaining in ornithine aminotransferase was measured daily until the enzyme activity had achieved a new equilibrium.

When the enzyme activity was at equilibrium, either high or low, the fractional degradation rate was 37–40 % per day (Table 1). During transition from a low to a high protein diet, the enzyme activity increased, the fractional rate of breakdown measured isotopically decreased transiently, and the calculated rate of synthesis increased (data not shown). When the shift was from high protein to low protein, the enzyme activity fell, the fractional rate of degradation doubled briefly, and the calculated rate of synthesis decreased. If the intake of tryptophan was continued as it was when 60 % casein was fed, the enzyme activity fell more slowly and the fractional degradation rate remained constant (Table 1).

TABLE 1

FRACTIONAL DEGRADATION RATES OF ORNITHINE AMINOTRANSFERASE DETERMINED FROM DECLINE IN RADIOACTIVITY AND KINETICS OF ENZYME ACTIVITY CHANGE

| Method | Fractional degradation rate, $k_d$, (%/day) | | | |
|---|---|---|---|---|
| | Dietary protein | | | |
| | 12% or 60% | 12%→60% | 60%→12% | 60%→12% + tryptophan |
| Kinetics of total isotope decay | 37–40 | 26[a] | 90[b] | 39[c] |
| Kinetics of enzyme activity change | | 83 | 162 | 46 |

[a] Mean of days 1, 2 and 3.
[b] Mean of days 1 and 2.
[c] Mean of days 1–5.
Data from Chee, P. Y. and Swick, R. W., 1976.

The conclusions are different, however, when degradation was calculated from the kinetics of enzyme activity change. During the shift from low to high protein the fractional rate appeared to be increased, a change which is counterproductive to the achievement of an increase in enzyme activity. When the shift in dietary protein was downward, the rate also appeared to be higher, almost double that given by the tracer portion of the experiment, but at least in the right direction. Two basic assumptions of the enzyme perturbation method are that the rate of activity change is exponential and that the rate of synthesis changes instantaneously. Apparently the mathematical model does not fit in this instance where both synthesis and degradation rates are changing over time (Fig. 1). As originally proposed by Berlin and Schimke (1965), the fractional degradation rate can be obtained from the time taken to achieve one half of the total excursion, a value apparently more readily discerned by inspection than by calculation.

When tryptophan was included in the diet, there was no transient increase in the breakdown rate and the mathematical model gave a value for degradation similar to that given by the tracer experiment. Therefore, the

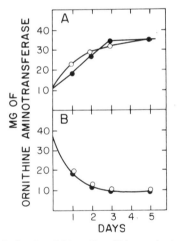

FIG. 1. Observed and calculated activities of ornithine aminotransferase during dietary transition. A, the rise in ornithine aminotransferase activity after a switch from a 12% to a 60% casein diet. When the observed values (●) were fitted to the equation $(E_t - E_0) = (E_e - E_0)(1 - e^{-kt})$, a value of 0·83/day was calculated for $k$. This value was used to predict the activity of ornithine aminotransferase at various times (○). The differences in the observed and calculated values are large. B, the fall in ornithine aminotransferase activity after a switch from a 60% to a 12% casein diet. The observed values (●) were fitted to the equation $(E_t - E_0) = (E_e - E_0)e^{-kt}$ and $k$ was 1·62/day. The values for ornithine aminotransferase activity (○) agree better with the observed values (Chee and Swick, 1976).

method based on the perturbation of enzyme activity may be useful only in obtaining a first approximation of the fractional breakdown rate.

The second experiment (MacDonald et al., 1979) was undertaken to compare two different tracer methods, one measuring the fractional rate of synthesis from the rate of incorporation of $^3$H-tyrosine, the other measuring the fractional rate of degradation from the decline in radioactivity derived from $^{14}$C-bicarbonate. The first is unaffected by the problems of tracer reutilization; we assume the second is not significantly compromised by tracer recycling.

In the first part, rats were fed a diet containing 60 % casein. After 5 days NaH$^{14}$CO$_3$ was injected. One, three and five days later groups of rats were continuously infused intravenously with L-2,6-$^3$H-tyrosine for 6 h. Ornithine aminotransferase was prepared from the livers and counted. The fractional rate of synthesis ($k_s$) was calculated from the incorporation of tyrosine as described by Garlick et al. (1973) and the fractional rate of breakdown ($k_d$) was calculated from the loss in specific $^{14}$C-radioactivity or total radioactivity since the rats were growing slowly. The two methods gave similar values (Table 2); this suggests that they are valid methods for measuring the turnover rates of hepatic proteins. The rate derived from the incorporation of tyrosine, however, requires comment. Ward and Mortimore (1978) showed that half of the free tyrosine in the liver may be sequestered in a compartment that does not readily mix with tyrosine in the precursor pool. Because the calculation of the synthesis rate is dependent on the specific activity of the free tyrosine in the homogenate, this could lead to a serious overestimation of the fractional synthesis rate. Indeed, there

TABLE 2

FRACTIONAL RATES OF SYNTHESIS AND OF DEGRADATION OF HEPATIC ORNITHINE AMINOTRANSFERASE AND SKELETAL MUSCLE ALDOLASE

| Method | Fractional turnover rate (%/day) | | | |
| | Liver ornithine aminotransferase | | Skeletal muscle aldolase | |
| | $k_s$ | $k_d$ | $k_s$ | $k_d$ |
| --- | --- | --- | --- | --- |
| Kinetics of isotope incorporation | 42 | — | 5·1 | — |
| Kinetics of specific activity decay | — | 45 | — | 2·2 |
| Kinetics of total isotope decay | — | 40 | — | ND |

Data from MacDonald, M. L. et al. 1979.

have been occasions when the use of labelled tyrosine has seemed to give unexpected results (Garlick, pers. comm.). We must conclude, however, that under the conditions of the experiment described here, because they gave similar rates, both methods appear appropriate.

In the second phase of this experiment, rats were fed an amino acid diet in agar-gel form (Harney *et al.*, 1976) for 5 days and injected with $NaH^{14}CO_3$. Four, eleven, and twenty days later groups of animals were fed a meal of the agar-gel diet containing L-2,6-$^3$H-tyrosine and killed 6 h later. Aldolase was prepared from the muscles (Ting *et al.*, 1971) and a portion counted. Glutamate and aspartate were isolated (Partridge, 1949) from another portion and counted.

In contrast to the results in liver, the rate of degradation of muscle aldolase, calculated from the decline in specific radioactivity derived from $^{14}$C-bicarbonate, appeared to be less than half as rapid as the rate of synthesis, although the animals in this study were only growing very slowly (Table 2). The decay in $^{14}$C-glutamate activity from days 11 to 20 gave a degradation rate of 4·2 %/day, which is closer to the rate of synthesis. However, there was no significant decline from day 4 to day 11, perhaps because of the additional error introduced by the isolation and quantification of the amino acids. Nor were the values obtained from aspartate useful. These results and those of Swick and Song (1974) demonstrate that the decline in total radioactivity derived from $^{14}$C-bicarbonate is unsuitable for the measurement of muscle protein breakdown. Although labelled glutamate and aspartate may give reasonable rates of degradation in muscle under basal conditions, the long decay periods required are unacceptable for the study of most treatments of interest. Furthermore, these amino acids may also be reutilized in muscle, especially during dietary deprivation and refeeding (Young *et al.*, 1971; Nettleton and Hegsted, 1973).

In summary, both the measurement of protein degradation via labelling with $^{14}$C-bicarbonate and protein synthesis via continuous exposure to tracer appear to be valid methods for the measurement of the turnover of hepatic proteins; the kinetics of enzyme activity change and the labelling of skeletal muscle proteins via $^{14}$C-bicarbonate may not provide accurate estimates of these parameters.

## *Acknowledgement*

Work supported in part by University of Wisconsin-Madison, College of Agriculture and Life Sciences and USPHS Grant No. 14704.

# Discussion

*Millward (London School of Hygiene and Tropical Medicine, UK):* Would you agree that there is a fundamental problem in trying to obtain the same rates of protein synthesis and degradation in muscle, because even in non-growing animals there may be small diurnal changes in the mass of protein in the muscle following a meal. This means that at certain times of the day the actual synthesis rate may be different from the breakdown rate. Since the turnover rate in muscle is very slow these differences may well be quite marked, so that if one were to measure the synthesis rate in the fed state one might obtain rates of synthesis that were a little higher than the fractional breakdown rate which was actually occurring at that particular time. But in the case of the liver where turnover is so rapid and where at any particular time the discrepancy between synthesis and breakdown is proportionately much less, then perhaps we can get closer to similar measured rates of synthesis and breakdown.

*Swick:* I agree completely. But the agreement isn't good in the liver either. The turnover rate of 40 % that we measured here is an average for the 24 h. We have made measurements at six-hourly intervals. Synthesis increases by about 20 % after a meal and decreases when the animal is fasted. We find that immediately after eating breakdown in the liver is practically zero, but 12 h later, when the animal is hungry, the fractional breakdown rate has gone up to over 100 %/day. It is very difficult to make meaningful decay measurements in a period as short as 6 h.

*Neuberger (Charing Cross Hospital Medical School, London, UK):* Does this apply to all proteins?

*Swick:* These results refer to total protein.

*Neuberger:* Do all the proteins stop turning over or is that an average?

*Swick:* We cannot measure the really rapidly turning over proteins in these experiments, so they may or may not change. But the bulk protein of liver just stops breaking down after a meal.

*Munro (Tufts University, Boston, USA):* When arginine is labelled with $^{14}CO_2$, are there quantitative estimates of its uptake into slowly turning over muscle proteins? Featherston et al. (1973) showed that citrulline was made into arginine in the kidney and went to muscle and other peripheral tissues, whereas liver arginine seldom got out into the periphery. His experiments suggested that liver arginine might not be a major contributor to other tissues, whereas other evidence suggests that liver arginine does appear in labelled form in muscle proteins.

*Swick:* If you give labelled arginine muscle is much more heavily labelled than liver and there is a feedback of label from muscle to liver. Scornik (1972) did partial hepatectomies, and finding no decay in arginine counts said that there was no breakdown. He was right for the wrong reasons.

## *References*

BERLIN, C. M. and SCHIMKE, R. T. (1965). Influence of turnover rates on the response of enzymes to cortisone, *Mol. Pharmacol.*, **1**, 149–52.

CHEE, P. M. and SWICK, R. W. (1976). Effect of dietary protein and tryptophan on the turnover of rat liver ornithine aminotransferase, *J. Biol. Chem.*, **251**, 1029–34.

FEATHERSTON, W. R., ROGERS, Q. R. and FREEDLAND, R. A. (1973). Relative importance of kidney and liver in synthesis of arginine by the rat, *Am. J. Physiol.*, **224**, 127–9.

GARLICK, P. J., MILLWARD, D. J. and JAMES, W. P. T. (1973). The diurnal response of muscle and liver protein synthesis *in vivo* in meal-fed rats, *Biochem. J.*, **136**, 935–45.

HARNEY, M. E., SWICK, R. W. and BENEVENGA, N. J. (1976). Estimation of tissue protein synthesis in rats fed diets labelled with [U-$^{14}$C]tyrosine, *Am. J. Physiol.*, **231**, 1018–23.

IP, M. M., CHEE, P. Y. and SWICK, R. W. (1974). Turnover of hepatic mitochondrial ornithine aminotransferase and cytochrome oxidase using [$^{14}$C]carbomate as tracer, *Biochim. Biophys. Acta*, **354**, 29–38.

MACDONALD, M. L., AUGUSTINE, S. A., BURK, T. L. and SWICK, R. W. (1979). A comparison of methods for the measurement of protein turnover *in vivo*, *Biochem. J.*, **184**, 473–6.

MILLWARD, D. J. (1970). Protein turnover in skeletal muscle. I. The measurement of rates of synthesis and catabolism of skeletal muscle protein using Na$_2$$^{14}$CO$_3$ to label protein, *Clin. Sci.*, **39**, 577–90.

NETTLETON, J. A. and HEGSTED, D. M. (1973). Effect of level of protein and calories on tissue protein catabolism, *Fed. Proc.*, **32**, 922.

PARTRIDGE, S. M. (1949), Displacement chromatography on synthetic ion-exchange resins. 3. Fractionation of a protein hydrolysate, *Biochem. J.*, **44**, 521–7.

SCHIMKE, R. T. (1970). Regulation of protein degradation, in *Mammalian Protein Metabolism* (Ed. Munro, H. N.), Vol. 4, Academic Press, New York, pp. 177–228.

SCORNIK, O. E. (1972). Decreased in vivo disappearance of labelled liver protein after partial hepatectomy, *Biochem. Biophys. Res. Comm.*, **47**, 1063–6.

SWICK, R. W. (1958). Measurement of protein turnover in rat liver, *J. Biol. Chem.*, **231**, 751–64.

SWICK, R. W. and IP, M. M. (1974). Measurement of protein turnover in rat liver with $^{14}$C-carbonate. Protein turnover during liver regeneration, *J. Biol. Chem.*, **249**, 6836–41.

SWICK, R. W., REXROTH, A. K. and STANGE, J. L. (1968). The metabolism of mitochondrial proteins. III. The dynamic state of rat liver mitochondria, *J. Biol. Chem.*, **243**, 3581–7.

SWICK, R. W. and SONG, H. (1974). Turnover rates of various muscle proteins, *J. An. Sci.*, **38**, 1150–7.

TING, S. M., SIA, C. L., LAI, C. Y. and HORECKER, B. L. (1971). Frog muscle aldolase: purification of the enzyme and structure of the active site, *Arch. Biochem. Biophys.*, **144**, 485–90.

WARD, W. F. and MORTIMORE, G. E. (1978). Compartmentation of intracellular amino acids in rat liver. Evidence for an intralysosomal pool derived from protein degradation, *J. Biol. Chem.*, **253**, 3581–7.

WATERLOW, J. C., GARLICK, P. J. and MILLWARD, D. J. (1978). *Protein Turnover in Mammalian Tissues and in the Whole Body*, Elsevier North Holland, Amsterdam.

YOUNG, V. R., STOTHERS, S. C. and VILAIRE, G. (1971). Synthesis and degradation of mixed proteins and composition changes in skeletal muscle of malnourished and refed rats, *J. Nutr.*, **101**, 1379–90.

# 32

# $^{13}$C Breath Tests as Potential Probes of Protein Nutritional Status

P. D. Klein, C. S. Irving and W. W. L. Wong

*Children's Nutrition Research Center,*
*Baylor College of Medicine and Texas Children's Hospital,*
*Houston, Texas, USA*

It has been said that the United Kingdom and the United States are two countries separated by a common language, and by analogy, $^{15}$N and $^{13}$C are two stable isotopes separated by a common methodology. Although the discovery and initial applications of $^{13}$C in modern biochemistry predate those of $^{15}$N, the exclusive availability of a single stable isotope of nitrogen for tracer studies has to date predicated its central role in protein metabolism techniques. By the time interest in the metabolism of amino acids had broadened to include the fate of the carbon skeleton, the availability of $^{14}$C and the rapid convenient methods for its determination channelled most investigators to the use of the radioisotopic as opposed to the stable isotopic form of carbon in their studies. $^{14}$C was and is still the ideal tool for animal studies but its use has become limited or prohibited in human studies. As we search for quantitative measurements in human subjects in a wide variety of nutritional and clinical conditions, the need to use $^{13}$C as a metabolic tracer has become more and more apparent.

Within recent years, the development of gas chromatography–mass spectrometry has enabled the determination of isotopic abundances in intact organic molecules; these techniques are equally applicable to compounds labelled with $^{2}$H, $^{13}$C, $^{15}$N, and $^{18}$O, as we have seen in previous presentations. Two considerations limit the extension of these techniques: (1) the labour intensive aspects of sample preparation, involving isolation and derivatization of the molecules of interest, and (2) the detection sensitivity which is usually between $10^{-3}$ and $10^{-4}$. On the other hand, isotope ratio measurements on gases such as $CO_2$ or $N_2$ can determine changes in isotopic concentration as small as 5 parts in $10^6$. Such determinations are accomplished by the use of a dual inlet gas isotope ratio mass spectrometer, in which the isotopic abundance of the sample is compared to that of a standard. In the case of $^{15}N$, the results often are expressed as atoms percent excess, whereas $^{13}C$ measurements are expressed as $\delta\%_{00}$ (delta per mil, or parts per thousand) differences from the standard, PDB, which is a limestone carbonate from the Pee Dee region of North Carolina, defined as having $1\cdot111\%$ $^{13}C$.

Conversion of $^{15}N$-containing metabolites to $N_2$ gas usually requires Kjeldahl digestion and conversion to $(NH_4)_2SO_4$, which is treated with sodium hypobromite to produce the end sample. Carbon dioxide, however, once produced by metabolism or combustion requires only cryogenic purification before introduction into the mass spectrometer. The inherent simplicity of $^{13}C$ isotopic studies, in which the end product is $CO_2$, prompted the development in our laboratory of simple, inexpensive, and disposable storage containers that could be used for breath samples (Schoeller and Klein, 1978). The containers were chosen for their adaptability to an automated system for sample purification and introduction into the mass spectrometer. Such an automated inlet system subsequently was designed and constructed in our laboratory, and can carry out 25–30 analyses of $^{13}C$ isotopic abundance, including all calculations without operator intervention, in an 8 h day (Schoeller and Klein, 1979).

The difference between $^{13}C$ and $^{15}N$ isotopic abundance measurements is that the proportion of $^{13}C/^{12}C$ in nature is not constant, but varies according to the photosynthetic mechanism by which carbon dioxide is fixed into carbohydrate. During subsequent transformations into protein and fatty acids, these substrates become depleted further in $^{13}C$. Since the gas isotope ratio mass spectrometer has a precision of $\pm0\cdot3\%_{00}$, and under fasting conditions, the variation in respiratory $^{13}C$ abundance is $\pm0\cdot7\%_{00}$ (Klein and Schoeller, 1976), it is possible to detect the $7–11\%_{00}$ change that results from switching from fatty acids to carbohydrate as an endogenous fuel source. Thus we may obtain information on substrate utilization by the simple process of exploiting the natural differences in $^{13}C$ enrichments between soybean oil ($-30\delta\%_{00}$ vs. PDB) and cornstarch ($-10\delta\%_{00}$ vs. PDB).

For most metabolic studies, however, it is desirable to have a larger signal, which can be generated by the use of compounds labelled in specific carbon atoms with highly enriched (90–99 %) $^{13}C$. The release of excess $^{13}C$ from the metabolism of this carbon in the molecule has been employed in a variety of studies collectively designated as 'breath tests'. In these tests, the rate and extent of $^{13}CO_2$ production has been used to infer the presence (or absence) of enzymes specifically responsible for cleaving off the labelled atom or group, and to obtain information about the rate and extent of metabolism. The studies include functional assessments of pancreatic lipase (Watkins *et al.*, 1977), cholestatic and mucosal disorders of fat absorption (Watkins *et al.*, 1980), carbohydrate absorption (Schoeller *et al.*, 1980), bacterial overgrowth (Solomons *et al.*, 1977), and hepatic microsomal mass (Schneider *et al.*, 1978). Development of the breath tests has been directed specifically toward the diagnostic information to be obtained when these tests are applied to infants, children, and women of child-bearing age.

How might such techniques be applied to the study of human protein metabolism? The oxidation of a $^{13}C$-labelled amino acid to $CO_2$ cannot be related directly to the turnover or flux of that amino acid in a subject. However, ample evidence exists in animal studies that oxidation of labelled amino acids is responsive to a number of nutritional factors, including protein adequacy of the diet, levels of the amino acid in the diet, and vitamin status. Some selected examples of the use of $^{14}C$-labelled amino acids to demonstrate this are shown in Table 1. These findings indicate that when the limiting nutrient is below the animal's requirement for maintenance, oxidation of the amino acid is low, but when the requirement has been fulfilled, the rate of oxidation begins to rise rapidly. Thus, a $^{13}C$-amino acid breath test might be capable of establishing the adequacy of dietary

TABLE 1

THE RESPONSE OF AMINO ACID OXIDATION TO NUTRITIONAL FACTORS OR STATUS

| Amino acid | Factors influencing % dose as *$CO_2$ | References |
|---|---|---|
| L-His-2R-$^{14}C$ | $B_{12}$ and folic acid-deficient megaloblastic anaemias in man | Fish *et al.* (1963) |
| DL-Try-2-$^{14}C$ | L-Tryptophan loads in man | Hankes *et al.* (1967) |
| L-Ser-3-$^{14}C$ | Simultaneous injection of unlabelled formate in man | Kretchmar and Price (1969) |
| L-Phe-UL-$^{14}C$ ⎫ DL-Leu-2-$^{14}C$ ⎬ | Protein-calorie deficiency in rats | McFarlane and von Holt (1969) |
| L-Ser-3-$^{14}C$ | $B_{12}$ and folate deficiency in man | DeGrazia *et al.* (1972) |
| L-Leu-UL-$^{14}C$ | Fasting in rats | Meikle and Klain (1972) |
| L-Lys-UL-$^{14}C$ | Dietary level of amino acid in rats | Brookes *et al.* (1972) |
| L-His-UL-$^{14}C$ ⎫ L-Thr-UL-$^{14}C$ ⎬ | Dietary level of amino acid in rats | Kang-Lee and Harper (1977, 1978) |
| L-Met-Me-$^{14}C$ | Dietary level of amino acid in rats | Aguilar *et al.* (1974) |
| L-Met-1-$^{14}C$ | Glycine and serine in rats | Benevenga and Harper (1970) |
| | $B_6$ deficiency in rats | Everett *et al.* (1979) |

protein intake in a given individual, in a given circumstance. Although such a test would bypass the questions of actual protein turnover, synthesis, and breakdown, it might provide the more specific information required to assess the protein and nutritional status of the individual.

When one embarks on the development of $^{13}$C breath tests in this area, the choice of candidate amino acids is limited, as shown in Table 2. At

### TABLE 2
SOME CANDIDATE $^{13}$C AMINO ACID OXIDATION PROBES

| Amino acid | Approximate price (US $/g) | Cost ($) for 70 kg individual | |
|---|---|---|---|
| | | 2 mg/kg | 5 mg/kg |
| L-Met-Me-$^{13}$C | 400 | 56 | 140 |
| L-Leu-1-$^{13}$C | 800 | 112 | 280 |
| L-Lys-1-$^{13}$C | 975 | 136 | 341 |
| L-Phe-1-$^{13}$C | 1 300 | 182 | 455 |
| L-Phe-ring-$^{13}$C | 1 500 | 210 | 525 |
| L-His-2 ring-$^{13}$C | 4 000 | 560 | 1 400 |

present there are five commercially available amino acids that could be utilized; the range in cost per gram is tenfold: from US $400 to $4000. Two levels of dosage are computed (2 and 5 mg/kg) and the costs of substrate for an individual study in a 70 kg individual are presented. Although stable isotopes are ideally suited for studies in infants from the standpoint of safety, the reciprocal advantage of infants in stable isotope studies is the lower substrate cost! Nevertheless, with the exception of histidine, initial trials of such tests appear feasible, and currently we are developing protocols for this purpose in the Children's Nutrition Research Center.

There are overall requirements for $^{13}$C breath test protocols arising, in part, from the differences between $^{13}$C and $^{14}$C in their natural abundances. The large background abundance of $^{13}$C in $CO_2$, which in $^{14}$C measurements would be the equivalent of $10^9$ dpm, requires that the baseline isotopic abundance be maintained within as narrow limits as possible, usually by conducting the test under fasting conditions. Moreover, the substrate oxidation rates must be in excess of the natural fluctuations in order to be apparent during the test (Schoeller et al., 1977). As in any test into a labelled amino acid, the absorption of the substrate must be complete for the yield of labelled $CO_2$ to be meaningful; fortunately, the measurement of faecal losses by combustion to $CO_2$ can be determined with a high degree of accuracy (Schoeller et al., 1980). Because the yield of respiratory $^{13}CO_2$ can be altered in an individual by changes in the bicarbonate kinetics, it is important to know either that these are within normal limits, or to establish them for the particular population under study. Finally, one must be certain that other factors affecting intermediate pools in the oxidative pathway, e.g., a $C_1$ pool if present, are controlled adequately to avoid complicated assessments.

Despite the host of initial boundary conditions that must be met if amino acids are to be used as probes of protein nutritional status, the non-invasive nature of such tests and their potential for application outside the metabolic ward suggests that the investment of effort is indeed worthwhile.

## Acknowledgements

This work is a publication of the USDA/SEA, Children's Nutrition Research Center at Baylor College of Medicine and Texas Children's Hospital and is supported by NIH AM 28129.

## Discussion

*Waterlow* (*London School of Hygiene and Tropical Medicine, UK*): It seems that in theory you ought to be able to do everything with $CO_2$ in the breath as an end product that you can do with nitrogen in the urine as an end product. The cumulative excretion of the label after a single dose of $^{13}C$-labelled amino acid ought to give us a lot of information about where the label has gone to.

*Klein:* I think that's true. These types of measurement do not attempt the complete accounting of the synthesis, degradation and turnover rates obtained by constant infusion studies, which can only be done on a very limited number of individuals. The question we set out to answer, perhaps in a pragmatic sense, is what is the nutritional status of this person right here in front of us with regard to his protein intake?

*Neuberger* (*Charing Cross Hospital Medical School, London, UK*): Has there been any investigation carried out in which $^{13}C$- and $^{15}N$-labelled amino acid was given at the same time and values for $^{13}CO_2$ excretion were compared with urea-$^{15}N$ excreted?

*Waterlow:* I expect that this experiment will almost certainly have been done in a year from now.

*Millward* (*London School of Hygiene and Tropical Medicine, UK*): There seems to be one very specific advantage over using carbon as an end product rather than nitrogen: that the sensitivity enables measurements to be made of acute changes in oxidation, which simply cannot be detected with nitrogen as an end product. Since we are unable to use $^{14}C$ in man, it seems that $^{13}C$ will become the isotope of choice.

*Neuberger:* You are not up against the problem of transamination, for instance.

*Young* (*Massachusetts Institute of Technology, USA*): Do you think the $^{13}C$ breath test might lend itself to the determination of dietary amino acid availability?

*Klein:* I would imagine so, though I am a little diffident about what we are going to be able to do in terms of nutritional state.

*Kerr* (*Case Western Reserve University, Cleveland, USA*): Do you know how complete the recovery of $CO_2$ entering the bicarbonate pool is?

*Klein:* It has been our experience that it ranges fairly widely. Winchell's data (Winchell *et al.*, 1970) on this show three pools in equilibrium with the final $CO_2$. We are very concerned as to whether or not the bicarbonate kinetics are constant in the same individual and whether you can make assumptions about this for a group of individuals. In some instances, by following the $^{13}CO_2$ output, we have recovered as little as 30 % of the administered label as $CO_2$. I am not confident that 90 % of the $CO_2$ appears, and that is why I think knowledge of the bicarbonate pool is a very great necessity.

*Cohen* (*University of Wisconsin-Madison, USA*): I am curious as to whether the library of $^{13}C$ compounds is increasing rapidly enough and whether they are getting cheaper. Some non-nitrogenous compounds such as some of the carbohydrates and their derivatives are relatively cheap. I think they now have at least two towers operating at Los Alamos. What is the present situation?

*Klein:* The cost of the $^{13}C$ is a fairly small consideration in the final cost of the compound. The real problem is that if you ask an organic chemist to synthesize a compound from scratch you are paying for the development and labour of very small scale synthesis. In the initial stages individual custom syntheses are always expensive. The price reductions do not begin to accrue until you have identified substrates which are consumed on a large scale. It looks, for instance, as if the use of trioctanoene to measure pancreatic insufficiency can replace a number of faecal fat measurements so that it is unnecessary to hospitalize a patient for 3 days in order to collect a 72 h stool.

*Bier (Washington University School of Medicine, St. Louis, USA):* The cost of $^{13}C$ is approximately 70 dollars per gram as CO or $CO_2$. The nitrogen and carbon columns at Los Alamos are operating at a fraction of their true capacity because isotope production is linked to (and limited by) isotope sales via a Government inventory account. At present, more isotopic raw material is available than the biomedical community wants to use.

*Klein:* The cheapest organic labour that we can find is probably micro-organisms. We are very keenly interested in producing a di-$^{15}N$-labelled lysine by bacterial fermentation at very low cost, starting with ammonia sulphate. I believe that for the future one has to think in terms of microbial mutants producing these amino acids for this sort of study.

## *References*

AGUILAR, T. S., BENEVENGA, N. J. and HARPER, A. E. (1974). Effect of dietary methionine level on its metabolites in rats, *J. Nutr.*, **104**, 761–71.

BENEVENGA, N. J. and HARPER, A. E. (1970). Effect of glycine and serine on methionine in rats fed diets high in methionine, *J. Nutr.*, **100**, 1205–14.

BROOKES, I. M., OWENS, F. N. and GARRIGUS, U. S. (1972). Influence of amino acid level in the diet upon amino acid oxidation by the rat, *J. Nutr.*, **102**, 27–36.

DEGRAZIA, J. A., FISH, M. B., POLLYCOVE, M., WALLERSTEIN, R. O. and HOLLANDER, L. (1972). The oxidation of the beta carbon of serine in human folate and vitamin $B_{12}$ deficiency, *J. Lab. Clin. Med.*, **80**, 395–404.

EVERETT, G. B., MITCHELL, A. D. and BENEVENGA, N. J. (1979). Methionine transamination and catabolism in vitamin B-6 deficient rats, *J. Nutr.*, **109**, 597–605.

FISH, M. B., POLLYCOVE, M. and FLEICHTMER, T. V. (1963). Differentiation between vitamin $B_{12}$-deficient and folic acid-deficient megaloblastic anemias with $^{14}C$-histidine, *Blood*, **21**, 447–61.

HANKES, L. V., BROWN, R. R., LIPPINCOTT, S. and SCHMAELLER, M. (1967). Effects of L-tryptophan load on the metabolism of tryptophan-2-$^{14}C$ in man, *J. Lab. Clin. Med.*, **69**, 313–24.

KANG-LEE, Y. A. and HARPER, A. E. (1977). Effect of histidine intake and hepatic histidase activity on the metabolism of histidine *in vivo*, *J. Nutr.*, **107**, 1427–43.

KANG-LEE, Y. A. and HARPER, A. E. (1978). Threonine metabolism *in vivo*: Effect of threonine intake and prior induction of threonine dehydratase in rats, *J. Nutr.*, **108**, 163–75.

KLEIN, P. D. and SCHOELLER, D. A. (1976). Sources of variability in the use of $^{13}C$ labelled substrates as 'breath tests' in clinical research and diagnosis, *Z. Anal. Chem.*, **279**, 134.

KRETCHMAR, A. L. and PRICE, E. J. (1969). Use of respiration pattern analysis for study of serine metabolism *in vivo*, *Metabolism*, **18**, 684–91.

MCFARLANE, I. G. and VON HOLT, C. (1969). Metabolism of amino acids in protein-calorie deficient rats, *Biochem. J.*, 111, 557–63.

MEIKLE, A. W. and KLAIN, G. J. (1972). Effect of fasting and fasting–refeeding on conversion of leucine into $CO_2$ and lipids in rats, *Am. J. Physiol.*, **222**, 1246–50.

SCHNEIDER, J. F., SCHOELLER, D. A., NEMCHAUSKY, B., BOYER, J. L. and KLEIN, P. D. (1978). Validation of $^{13}CO_2$ breath analysis as a measurement of demethylation of stable isotope labelled aminopyrine in man, *Clin. Chem. Acta.*, **84**, 153–62.

SCHOELLER, D. A., SCHNEIDER, J. F., SOLOMONS, N., WATKINS, J. B. and KLEIN, P. D. (1977). Clinical diagnosis with the stable isotope $^{13}C$ in $CO_2$ breath tests: Methodology and fundamental considerations, *J. Lab. Clin. Med.*, **90**, 412–21.

SCHOELLER, D. A. and KLEIN, P. D. (1978). A simplified technique for collecting breath $CO_2$ for isotope ratio mass spectrometry, *Biomed. Mass Spectrom.*, **5**, 29–31.

SCHOELLER, D. A. and KLEIN, P. D. (1979). A microprocessor controlled mass spectrometer for the fully automated purification and isotopic analysis of breath $CO_2$, *Biomed. Mass Spectrom.*, **6**, 350–5.

SCHOELLER, D. A., KLEIN, P. D., WATKINS, J. B., HEIM, T. and MACLEAN, W. C. JR. (1980). $^{13}C$ abundances of nutrients and the effect of variations in $^{13}C$ isotopic abundances of test meals formulated for $^{13}CO_2$ breath tests, *Am. J. Clin. Nutr.*, in press.

SCHOELLER, D. A., KLEIN, P. D., MACLEAN, W. C., JR., WATKINS, J. B. and VAN SANTEN, E. (1981). Fecal $^{13}C$ analysis for the detection and quantitation of intestinal malabsorption: Limits of detection and application to disorders of intestinal cholylglycine metabolism, *J. Lab. Clin. Med.*, in press.

SOLOMONS, N., SCHOELLER, D. A., WAGONFELD, J., OTT, D. G., ROSENBERG, I. H. and KLEIN, P. D. (1977). Application of a stable isotope ($^{13}C$) labelled glycocholate breath test to diagnosis of bacterial overgrowth and ileal dysfunction, *J. Lab. Clin. Med.*, **90**, 431–9.

WATKINS, J. B., SCHOELLER, D. A., KLEIN, P, D., OTT, D. G., NEWCOMER, A. D. and HOFMANN, A. F. (1977). $^{13}C$ trioctanoin: A non-radioactive test to detect fat malabsorption, *J. Lab. Clin. Med.*, **90**, 421–30.

WATKINS, J. B., PARK, R., PERMAN, J., SCHOELLER, D. A. and KLEIN, P. D. (1980). Detection and identification of fat malabsorption with non-radioactive $^{13}CO_2$ breath test, *Gastroenterology*, **78**, 1288.

WINCHELL, H. S., STAHELIN, H., KUSUBOVM, N., SLANGER, B., FISH, M., POLLYCOVE, M. and LAWRENCE, J. H. (1970). Kinetics of $CO_2$-$HCO_3^-$ in normal adult males. *J. Nuclear Med.*, **11**, 711–15.

**PART IV**

# Turnover in Various States

# 33

# Protein Turnover in Animals: Man in his Context

P. J. REEDS and C. I. HARRIS

*Rowett Research Institute, Aberdeen, UK*

## Introduction

In an attempt to compare the metabolism of different species it is important to emphasize the truism that it is only on the basis of similar techniques that a valid comparison can be made. If such a comparison includes man there are further limitations as, almost invariably, the measurements on man will be indirect. In the context of measurements of protein turnover this constraint limits comparisons of man with other mammals to measurements of protein synthesis in the whole body, since with the exception of estimates of the rate of degradation of actin and myosin (see Munro, p. 495), of plasma protein turnover (see James and Coward, p. 457) and a single report on muscle protein synthesis (Halliday and McKeran, 1975) there seems to be no published information on the rate of synthesis or degradation of the proteins of a single organ or tissue in human beings.

To some extent the emphasis of studies in man is upon pathological or quasi-pathological states and comparison of results are made with measurements in animals in which metabolic derangements have been invoked which are believed to be analogous to those in man. If the aim is to infer, from experiments in animals, the metabolic basis of pathological changes in man it is important to establish how similar are 'normal' animals and 'normal' human beings.

Much of the information on protein synthesis in laboratory animals is related to individual tissues (see Waterlow *et al.*, 1978a) and there is less information about protein synthesis in the whole body (Lobley *et al.*, 1977; Albertse *et al.*, 1979; McNurlan and Garlick, 1980). In larger animals there has been a somewhat greater emphasis upon measurements of whole body protein synthesis and these measurements are more readily compared with those in man.

## Methodological Considerations

Comparisons between species can readily be made on an essentially qualitative basis but, as the title of this paper states, we wish to place man in his context: to investigate whether the amounts of protein synthesized by man and the changes which occur in response to, say, changing nutrition are quantitatively similar. The immediate problem in carrying out such an exercise is one of methodology.

The majority of reports upon human beings have used $^{15}$N-glycine as the tracer amino acid, basing calculations either on the end product method of Picou and Taylor-Roberts (1969) (see Golden, p. 323), or a minor modification of this method (Pencharz *et al.*, 1977; Waterlow *et al.*, 1978*b*), or on methods employing a single dose of $^{15}$N-glycine (Golden *et al.*, 1977*a*; Waterlow *et al.*, 1978*b*; Garlick *et al.*, 1980*a,b*). On the other hand, virtually all the estimates of whole body protein synthesis in animals have been based on precursor methods (see Garlick, p. 303) employing leucine or tyrosine. If we wish to make quantitative comparisons, either the estimates must be based on the same method or we have to determine whether the two commonly used approaches give similar estimates under similar circumstances, ideally in all the species concerned.

It is generally assumed that the flux of an amino acid is underestimated from measurements of the specific radioactivity of the free amino acid in blood (Golden and Waterlow, 1977; Waterlow *et al.*, 1978*a*). In a report on elderly humans, Golden and Waterlow (1977) showed that whole body protein synthesis calculated from the flux of leucine was some 15 % less than that estimated from the isotopic abundance of $^{15}$N-urea during a constant infusion of $^{15}$N-glycine. By comparing an estimate of leucine excretion derived from urinary-N with the rate of leucine catabolism determined from measurements of the leucine flux, they concluded that the specific radioactivity of blood leucine underestimated the true flux by some 20 %. A comparison of the measured and calculated excretion of urinary nitrogen in growing pigs (Fig. 1, Reeds *et al.*, 1980*a*) also implies a similar underestimate but does not indicate a close and linear relationship between the two values over a wide range of nitrogen and energy intakes. These results appear to support the general validity of the technique, but it should be emphasized that they apply only to one component of the flux of amino-N, namely amino acid catabolism, and test not one assumption but their totality.

One of these assumptions is that it is possible, by the use of a constant, to convert the flux of the tracer amino acid to that of total amino-N. It is not

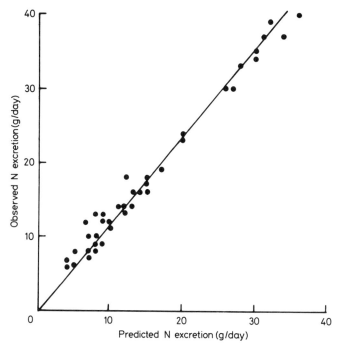

FIG. 1. A comparison in pigs, of daily nitrogen excretion with that calculated from the apparent rate of catabolism of leucine. $y = 1\cdot17\ (\pm0\cdot02)x$; $n = 47$; $r = 0\cdot975$.

absolutely necessary to perform such a calculation (see, for example, Garlick *et al.*, 1980*a*), but as the basic information derived from experiments with [15]N-glycine is in terms of nitrogen, data derived from experiments using leucine or tyrosine must be expressed in similar terms if we wish to compare the results obtained with these amino acids and those obtained with [15]N-glycine. The derivation of the appropriate constant with which to convert the turnover of a single amino acid to that of protein is difficult. Theoretically it is the average contribution of the tracer amino acid to body protein *synthesis*. Although it is possible to estimate an average value for this contribution from a calculation based upon the leucine content of the mixed proteins of various tissues and the contribution of their synthesis to that of whole body protein (Table 1), it is difficult to take account of changes which occur in the amino acid composition of some proteins, (for example, collagen) between translation and assembly within the tissue protein mass. A value based upon the average composition of body protein is often used (Golden and Waterlow, 1977; Reeds and Lobley,

## TABLE 1

THE AVERAGE CONTRIBUTION OF LEUCINE TO BODY PROTEIN AND AN ESTIMATE
OF ITS CONTRIBUTION TO BODY PROTEIN SYNTHESIS

| Species | Leucine content (g/100 g protein) | | Reference |
| | Body protein | Body protein synthesis | |
| --- | --- | --- | --- |
| Rat | 6·9 | 7·0 | 1 |
| Rabbit | 6·7 | 6·6 | 2 |
| Cattle | 6·3 | 6·5 | 3 |

1. Composition (Lobley and Milne, unpublished); distribution of protein synthesis from McNurlan *et al.* (1980).
2. Composition and synthesis, Lobley, Milne and Reeds, unpublished.
3. Composition and synthesis, Lobley *et al.* (1980).

1980), but opinions differ as to the correct value. Golden and Waterlow (1977) and O'Keefe *et al.* (1974) have used, for leucine, a value of 8 g leucine/100 g protein whilst we (Lobley *et al.*, 1980; Reeds *et al.*, 1980*a*) have proposed that a value of between 6·5 and 7 g/100 g protein is more appropriate (Table 1).

Table 2 compares the estimates of daily protein synthesis made with either glycine or leucine and tyrosine in young and middle-aged adults. If the estimate with leucine and tyrosine is calculated with our preferred constants, there appears to be a statistically significant difference of some 30 % between this value and that derived with glycine, but if it is calculated using the constants of O'Keefe *et al.* (1974) and James *et al.* (1976) there is

## TABLE 2

BODY PROTEIN SYNTHESIS IN ADULT MAN MEASURED EITHER BY THE METHOD OF PICOU
AND TAYLOR-ROBERTS (1969) OR WITH LEUCINE OR TYROSINE (MEAN VALUES ± 1 SD)

| Method | [15]N-glycine[a] | Leucine or tyrosine[b,c] | Leucine or tyrosine[b,d] |
| --- | --- | --- | --- |
| n | 21 | 18 | — |
| Body weight (kg) | 68 | 71 | — |
| Protein synthesis (g/day) | 242 ± 35 | 328 ± 38 | 276 ± 30 |

[a] Steffee *et al.* (1976); Crane *et al.* (1977); Sim *et al.* (1979).
[b] O'Keefe *et al.* (1974); James *et al.* (1976); Motil *et al.* (1979); Garlick (pers. comm.).
[c] Assuming that 6·7 % body protein is leucine and 3 % is tyrosine.
[d] Assuming that 8 % body protein is leucine and 4 % is tyrosine.

little difference between the estimates. It is still not certain, of course, which value is more correct but, as the quantitative comparisons which follow will involve the comparison of measurements made with both methods, this possible source of differences should be borne in mind.

## Protein Synthesis in Adults and at Energy Equilibrium

In making quantitative comparisons of the metabolic activities of different species account must be taken of differences in body weight. In the study of the energy exchanges of animals this problem has received particular attention and Blaxter (1972), in an extensive review of this subject, concluded that between species of the same order of mammals the fasting heat production of adults was directly proportional to body weight raised to the power 0·75, although between orders heat production (kJ/kg$^{0·75}$) varies. This also appears to apply, in adults, to daily heat production at energy equilibrium and it has become a common practice to make interspecific comparisons of other processes on a similar basis.

A general parallelism between whole body protein synthesis and metabolic rate has long been recognized (Waterlow, 1968). Table 3 contains

TABLE 3

BODY PROTEIN SYNTHESIS ESTIMATED WITH LEUCINE OR TYROSINE IN ADULT MAMMALS AND IN IMMATURE INDIVIDUALS RESTRICTED TO A STATE OF ENERGY EQUILIBRIUM

| Species | Body weight (kg) | Protein synthesis g/day | g/kg/day | g/kg$^{0·75}$/day | Reference |
|---|---|---|---|---|---|
| Rat | 0·2 | 5·6 | 28·1 | 18·8 | Reeds and Lobley, unpublished |
| | 0·35 | 7·7 | 22·0 | 16·9 | Reeds and Lobley, unpublished |
| | 0·82 | 11·1$^a$ | 13·5 | 12·8 | Millward (quoted by Waterlow, 1980) |
| Rabbit | 3·6 | 33 | 9·2 | 12·6 | Lobley, Milne and Reeds, unpublished |
| Pig | 32 | 268 | 8·1 | 18·9 | Reeds et al. (1980a) |
| Sheep | 63 | 351 | 5·6 | 15·7 | Reeds and Chalmers, unpublished |
| Man | 71 | 328 | 4·6 | 13·4 | See Table 2 |
| Cattle | 575 | 1 740 | 3·0 | 14·8 | Reeds et al. (1981) |

$^a$ Measured with tyrosine.

a comparison of measurements, made with leucine or tyrosine, of the amounts of protein synthesized by adult and immature animals restricted to energy equilibrium. These measurements demonstrate that daily protein synthesis at energy equilibrium also appears to be proportional to so-called metabolic body weight ($W^{0.75}$), and whilst this implies that protein synthesis makes a nearly constant contribution to daily heat production, differences in heat production per $kg^{0.75}$ between species must be borne in mind. In addition it is to be noted that, as with the basal metabolic rate, the highest values for protein synthesis /$kg^{0.75}$/day are obtained in immature animals and the lowest for those animals (man, rabbit and rat) that are the oldest in a developmental sense.

The physiological basis of the observation that fasting heat production in adults is proportional to body weight ($kg^{0.75}$) has never been satisfactorily explained (Blaxter, 1972), but it appears that when protein synthesis is measured with similar techniques adult man synthesizes protein in a quantity which is to be expected for an adult mammal of 70 kg body weight.

## *Protein Synthesis, Growth and Development*

Despite the general similarity of adults of different species, immature human beings differ, particularly with respect to their protein requirement, from other immature animals. The questions which we wish to attempt to answer in this section are (a) are there also quantitative differences between the rate of protein synthesis in young human beings and the young of other mammals, and (b) if so, can we identify the factors responsible for these differences?

Figure 2 shows that, as has been recognized for a long time, young mammals synthesize protein at a higher rate per unit of metabolic body weight ($W^{0.75}$) than adults. What is equally clear is that, with the exception of premature infants, immature human beings synthesize protein less rapidly than other young mammals. Although in Fig. 2 the estimates of protein synthesis in the children were all made with $^{15}$N-glycine and those in other animals with $^{14}$C-tyrosine and $^{14}$C-leucine the difference between children and animals seems too great to be due to a methodological difference alone, as for example at 25 % of mature body weight the rate in children is only half that of any other species.

It is generally assumed that the reduction with age in the rate of protein synthesis per unit of metabolic body weight is due to a decline in the fraction associated with growth, although a real change with development in 'basal'

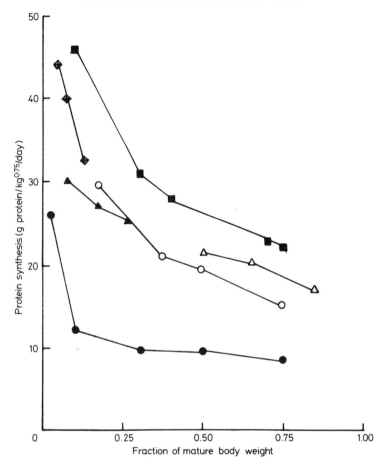

FIG. 2. Body protein synthesis per unit metabolic body weight and its relationship to age. (■) rat: Millward (quoted by Waterlow, 1980); Reeds and Lobley, unpublished, (◇) lambs: Soltesz *et al.* (1973), (▲) pigs: Reeds *et al.* (1980*a*), (○) rabbits: Lobley, Reeds and Milne (unpublished), (△) cattle: Lobley, pers. comm. and Lobley *et al.* (1980), (●) man: Pencharz *et al.* (1977); Golden *et al.* (1977*b*); Picou and Taylor-Roberts (1969); Kien *et al.* (1978*a,b*).

protein synthesis cannot be ruled out (Waterlow *et al.*, 1978*a*). The growth rate and *ad libitum* intake of children are, under normal circumstances, very much less than those of animals at a similar proportion of mature body weight (see Blaxter, 1978 for comparisons) and this is the obvious explanation of the difference that is apparent in Fig. 2. Certainly the results in Table 4 in which the relationship between the flux of amino-N and

## TABLE 4

THE RELATIONSHIP BETWEEN PROTEIN INTAKE AND AMINO-N FLUX IN IMMATURE
MAMMALS (MEAN VALUES $\pm$ 1 SEM)

| Species | Weight or age | $\left(\dfrac{N\ intake}{Amino\text{-}N\ flux}\right)$ | Reference |
|---------|---------------|------------------------|-----------|
| Rat | 200 g | $0\cdot30 \pm 0\cdot02$ | Reeds and Lobley (unpublished) |
|     | 350 g | $0\cdot33 \pm 0\cdot01$ | |
| Pig | 30 kg | $0\cdot36 \pm 0\cdot02$ | Reeds et al. (1980a) |
|     | 60 kg | $0\cdot36 \pm 0\cdot03$ | |
|     | 90 kg | $0\cdot38 \pm 0\cdot02$ | |
| Man | Premature | $0\cdot14 \pm 0\cdot01$ | Pencharz et al. (1977) |
|     | 1–2 years | $0\cdot44 \pm 0\cdot02$ | Golden et al. (1977a) and Golden (pers. comm.); recovered malnourished children |
|     | 4–6 years | $0\cdot41 \pm 0\cdot06$ | Kien et al. (1978a, 1978b); subjects before reconstructive surgery of the skin |
|     | 10–15 years | $0\cdot38 \pm 0\cdot01$ | Kien et al. (1978a, 1978b); subjects before reconstructive surgery of the skin |
|     | 11 years | $0\cdot44 \pm 0\cdot04$ | Adeniyi-Jones et al. (1979); subjects recovering from cystic fibrosis |

protein intake are compared in growing rats, pigs and children suggests a generally similar relationship between total protein intake and body protein turnover in these species.

Total protein synthesis in growing individuals can be divided into two components: one related to the turnover of body protein which occurs at nitrogen equilibrium, the other to the additional protein synthesis that is required to support growth. As regards the growth component, Waterlow et al. (1978a) have made the important point that if one gram of protein accretion requires only one gram increase in protein synthesis, the changes in protein synthesis with development in man are too great to be due to a change in growth alone. Whilst there appears to be a change with age in protein synthesis at energy equilibrium (Table 3), this observation of Waterlow's group also implies that an increase of one gram in protein deposition requires an increase in protein synthesis in the whole body of more than one gram, similar in kind to that which occurs in skeletal muscle during accelerated growth (Millward et al., 1975; Laurent et al., 1978).

Comparisons of protein synthesis and nitrogen retention in children during catch-up growth (Golden et al., 1977a and M. H. N. Golden, pers. comm.) and in growing pigs (Reeds et al., 1980a) (Fig. 3) suggest that when body nitrogen retention is accelerated by an increase in food intake it does

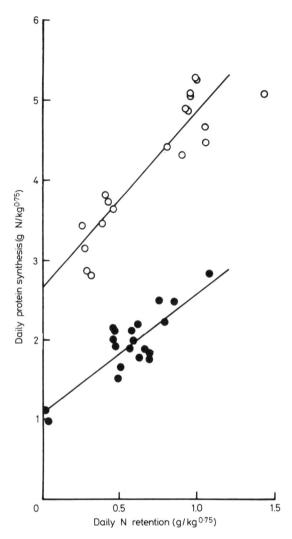

FIG. 3.  The relationship between nitrogen retention ($x$) and body protein synthesis ($y$) in growing malnourished children ($\bullet$) and growing pigs ($\bigcirc$) of 30 kg body weight. For pigs: $y = 2\cdot21(\pm0\cdot29)x + 2\cdot65(\pm0\cdot23)$; $n = 21$; $r = 0\cdot898$. For children: $y = 1\cdot48(\pm0\cdot25)x + 1\cdot08(\pm0\cdot08)$; $n = 20$; $r = 0\cdot839$.

indeed require a disproportionate increase in body protein synthesis. There is, however, a marked and statistically significant interspecific difference in both the rate of protein synthesis at nitrogen equilibrium and in the increment of protein synthesis per unit increment in nitrogen retention. It is possible that the differences in both the intercepts and gradients could be ascribed in part to the techniques (single dose [15]N-glycine versus continuous infusion of [14]C-leucine), but both seem too great to arise only from the differences in methods.

With the information that is available at present it is not possible to draw any final conclusions on the physiological basis of the difference in the slopes of the two lines in Fig. 3. Apart from the fact that the results are from different species there are other highly pertinent differences between the experiments. Not the least of these concerns the possible influence of the preceding nutritional state upon the control of protein deposition and this consideration may be particularly important in interpreting the difference between the rates of protein synthesis at nitrogen equilibrium (Golden *et al.*, 1977*b*). There is, however, additional information in children and in pigs (Table 5) which suggests one explanation for the difference in the slopes of the lines in Fig. 3.

In the experiments summarized in Fig. 3 the change in protein synthesis which is associated with a change in growth can also be regarded as a response to an alteration in the intake of dietary protein and energy. Both Picou and Taylor-Roberts (1969) and Golden *et al.* (1977*b*) have emphasized that in children receiving a constant intake of non-protein energy, an increase in protein intake increases the rate of nitrogen retention but appears to have little effect upon protein synthesis, so that the rate of body protein breakdown must be reduced. We (Reeds *et al.*, 1980*b*) have found that in pigs a higher rate of nitrogen retention associated with increased protein intake is accompanied by higher rates of both protein synthesis and protein breakdown and in this circumstance the increment in protein synthesis per unit of increment of protein deposition is even higher than that in Fig. 3. On the other hand the response of protein synthesis in the pig to an increase in the intake of non-protein energy is apparently quantitatively similar to that in children undergoing catch-up growth during recovery from cystic fibrosis (Adeniyi-Jones *et al.*, 1979).

It appears, then, that in children the increased rate of protein synthesis accompanying accelerated growth due to an increase in food intake contains a component due to a response to dietary energy alone, with a small or negligible component due to protein intake, while in the pig the protein synthetic response contains components due both to increased

## TABLE 5

### THE RESPONSE, IN CHILDREN AND PIGS, OF NITROGEN RETENTION AND BODY PROTEIN SYNTHESIS TO CHANGES IN DIETARY PROTEIN OR ENERGY INTAKE

| Species | N intake (mg/kg/day) | Energy intake (kJ/kg/day) | N balance (mg N/kg/day) | Protein synthesis (g N/kg/day) | Reference |
|---|---|---|---|---|---|
| Man | 192 | 502 | 110 | 1·02 | Golden et al. (1977b) |
|  | 832 | 502 | 352 | 1·06 |  |
| Pig | 930 | 502 | 376 | 2·26 | Reeds et al. (1980b); recalculated for comparison with data from children |
| Man | 1 808 | 523 | 473 | 3·20 | Adeniyi-Jones et al. (1979) |
|  | 480 | 365 | 102 | 0·64 |  |
|  | 480 | 550 | 160 | 0·83[a] |  |
| Pig | 896 | 480 | 395 | 2·24 | Reeds et al. (1980b); recalculated for comparison with data from children |
|  | 882 | 747 | 650 | 2·59[a] |  |

[a] The increase in daily protein synthesis per kJ increase in energy. Intake was 1·03 g N for children and 1·31 g N for pigs.

protein intake and to increased energy intake. While the explanation agrees with the observations it must be remembered that there are differences of technique, the influence of which can be assessed only when both methods are applied simultaneously in one or other species. However, one of the advantages of pursuing comparisons of different species, and one which might well have been emphasized in the introduction to this paper, is that, paradoxically, the identification of differences may point to factors which are common to the control of metabolic processes.

## Conclusion

As we have pointed out previously (Reeds and Lobley, 1980), difficulties of technique continue to overshadow attempts to place man in his metabolic context. On the basis of a number of measurements, which include the rate of protein synthesis in the body, adult human beings appear to be little different from other mature mammals. Immature human beings, on the other hand, are characterized by a slow rate of both growth and protein synthesis and increases in nitrogen retention appear to incur a lower cost in terms of protein synthesis. It is possible that this is related to an interspecific difference in the response of protein turnover to a change in dietary protein intake.

## Acknowledgements

We are grateful to Drs M. Golden, V. Young and G. Lobley for giving us access to unpublished results. The many interesting discussions with Dr M. Fuller are also acknowledged.

## Discussion

*Neuberger (Charing Cross Hospital Medical School, London, UK):* During growth and development the biological species differences in the time scale are much greater than the differences at the adult stage, for example in the rate of maturation of the central nervous system. I wonder whether you are correct in assuming there could ever be equality between man and animals.

*Reeds:* Heat production in growing animals and in growing humans per $kg^{0.75}$ is higher than in adults. What is not really clear is how the underlying maintenance component changes with development. What I was trying to say was that the principal quantitative difference between protein synthesis in man and in growing animals can be attributed to just these major differences in the pattern and the time scale of development in man on the one hand and animals on the other.

*Jackson (University of the West Indies, Kingston, Jamaica):* If you expressed the rate of protein synthesis in relation to rate of weight change per unit weight (fractional weight gain) you would expect to eliminate the problem between the different rates of growth in relation to absolute weight.

*Reeds:* I have done such a calculation. The problem is to get good growth data on the subjects in whom protein synthesis is being measured. The calculation does not normalize the data completely; it still suggests that man tends to synthesize less protein per unit growth than animals.

*Bier (Washington University School of Medicine, St. Louis, USA):* In Table 3 where data were normalized to $kg^{0.75}$ you had a rat which weighed 820 g. I don't know whether the rats in Britain are bigger than ours, but that is a fairly hefty rat by US standards.

*Millward (London School of Hygiene and Tropical Medicine, UK):* Yes, it was a big rat.

*Reeds:* You get virtually the same answer if you use data, again from Millward, from 500 g rats.

*Pratt (Institute of Psychiatry, London, UK):* It seems to me that to compare the maximum possible growth or synthesis rate in different species one wants to have a situation where there is deprivation for a period and then a maximal catch-up growth rate. A useful model might be partial hepatectomy. Some of my colleagues have been working on this and we are struck by the speed with which the rat can replace its liver—in a matter of only a few days.

*Millward:* In the child during catch-up growth after malnutrition, the rate of degradation per unit increase in growth does not seem to be as great as that which occurs in the pig. One of the big differences between the malnourished child and the pig at this time is a very marked difference in body composition because the malnourished child starts with a very reduced lean body mass. There might be important differences in the individual constituents of the lean body mass and in particular muscle mass, which, during the rehabilitation stage, might be a smaller proportion of the whole body mass than in the growing pig. There is a lot of evidence to suggest that the increased protein turnover associated with growth may be in muscle. Therefore, perhaps during catch-up growth in the malnourished child muscle contributes a smaller proportion to the total body, so that the influence of the increased turnover in muscle might not be as apparent as you see in the pig.

*Reeds:* That is a very good point. Unfortunately there are very few data for the child and its composition during growth. In fact, according to our data, muscle makes quite a substantial contribution to growth during recovery. However, I agree with your point.

*Waterlow (London School of Hygiene and Tropical Medicine, UK):* The data from Jamaica on malnourished children growing at different rates, admittedly with different food intakes, show that for every gram of protein laid down, $1.4$ g have to be synthesized. Therefore the process was $67\%$ efficient. Those children would be growing at 10 or 15 times the normal rate, that is, at rates comparable to that of a pig in terms of fraction of body weight gained per day. I should like to know what would be the comparable figure for the growing pig.

*Reeds:* In our original experiments we varied the intake of the whole diet and obtained the line in Fig. 3. This suggests that for each unit increase in nitrogen retention, two units of additional protein synthesis are required. We have since done experiments in which we have manipulated either the protein or the energy intake to supply either the amount of protein that animals on the high intake were receiving with the lower energy or the high energy intake with the lower protein intake. Our general conclusion was that the increases in nitrogen retention associated with increased energy intake incur lower costs in terms of synthesis, and in our animals increases in nitrogen retention associated with higher protein intake incur higher costs. One can alter the slope of the line in Fig. 3 by manipulating the composition of the diet.

*Waterlow:* But in general, if the composition of the diet remains more or less constant but the amount varies in relation to the rate of growth, then your value in the pig, compared with ours in children, would be about 2 compared with $1.4$?

*Reeds:* Yes.

*Kerr (Case Western Reserve University, Cleveland, USA):* Are you comparing states of either inadequate nitrogen intake or inadequate energy intake with states where there is adequate intake of one or the other?

*Reeds:* In these experiments I had always been concerned that in the children there was just not enough energy to allow a response in protein synthesis. But when I calculated the energy intakes of the children and the pigs, I found them broadly similar in terms of $kJ/kg^{0.75}$. In that respect, if energy is limiting, both situations are limited equally, yet we still get an apparent increase in synthesis with a higher protein intake and Golden *et al.* (1977*b*) do not.

*Stein (University of Pennsylvania School of Medicine, Philadelphia, USA):* In considering growth, a distinction should be made between normal growth and catch-up growth, because the latter involves a repletion component as well and that may proceed by a different mechanism from normal growth.

*Reeds:* I take your point. The problem in trying to look at the relation between total body protein synthesis and total nitrogen retention in animals and man is that the normal 2-year-old child is growing at a rate which if observed in a 30 kg pig would lead one to say that the animal was virtually at maintenance. Unfortunately the only circumstance where the same nitrogen retention in relation to body weight is observed in the human is during catch-up growth.

*Stein:* When elderly depleted patients are repleted parenterally, it seems that as much as 40 % or more of the increased nitrogen retention is due to decreased breakdown.

*Munro (Tufts University, Boston, USA):* The plasma proteins provide an accessible protein series which can be studied in this way and which might supplement your information. The allometric equation applies to them equally well (Munro, 1969) and has been validated extensively for albumin, ceruloplasmin and very recently for transferrin (Recoeczi and Hatton, 1980). It would be interesting to know if the turnover of these is also affected by the various dietary and growth changes.

*Young (Massachusetts Institute of Technology, USA):* Is there any information about differences between breeds of pigs with respect to the protein metabolic responses you have described?

*Reeds:* No. Simon in his experiments used the same breed of pig as we did

406     *P. J. Reeds and C. I. Harris*

because we were both working in the field of animal production and studied strains which are commonly encountered in commercial practice (Simon *et al.*, 1978).

## References

ADENIYI-JONES, S., SUSKIND, R., POLOMBO, J., PENACRUZ, V. and KHAW, K. T. (1979). Effect of increased energy intake alone on whole body protein turnover during catch-up growth in cystic fibrosis, *Fed. Proc.*, **38**, 708.

ALBERTSE, E. C., GARLICK, P. J. and PAIN, V. M. (1979). Whole-body protein synthesis and oxidation rates in streptozotocin diabetic rats, *Proc. Nutr. Soc.*, **38**, 125A.

BLAXTER, K. L. (1972). Fasting metabolism and the energy required by animals for maintenance, in *Festskrift til Knut Breirein* (Eds L. S. Spildo, T. Homb and H. Hvidsten) Mariendals Bortrykkeri, A. S. Gjørik, pp. 19–36.

BLAXTER, K. L. (1978). Comparative aspects of nutrition, in *Diet of Man: Needs and Wants* (Ed. J. Yudkin), Applied Science Publishers, London, pp. 145–58.

CRANE, C. W., PICOU, D., SMITH, R. and WATERLOW, J. C. (1977). Protein turnover in patients before and after elective orthopaedic operations, *Brit. J. Surg.*, **64**, 129–33.

GARLICK, P. J., BURK, T. L. and SWICK, R. W. (1976). Protein synthesis and RNA in tissues of the pig, *Am. J. Physiol.*, **230**, 1108–16.

GARLICK, P. J., CLUGSTON, G. A. and WATERLOW, J. C. (1980*a*). Influence of low-energy diets on whole-body protein turnover in obese subjects, *Am. J. Physiol.*, **238**, E235–44.

GARLICK, P. J., MCNURLAN, M. A., FERN, E. B., TOMKINS, A. M. and WATERLOW, J. C. (1980*b*). Stimulation of protein synthesis and breakdown by vaccination, *Brit. Med. J.*, **ii**, 263–5.

GOLDEN, M. H. N. and WATERLOW, J. C. (1977). Total protein synthesis in elderly people: a comparison of results with [$^{15}$N]-glycine and [$^{14}$C]-leucine, *Clin. Sci. Mol. Med.* **53**, 277–88.

GOLDEN, M. H. N., WATERLOW, J. C. and PICOU, D. (1977*a*). The relationship between dietary intake, weight change, nitrogen balance and protein turnover in man, *Am. J. Clin. Nutr.*, **30**, 1345–8.

GOLDEN, M. H. N., WATERLOW, J. C. and PICOU, D. (1977*b*). Protein turnover, synthesis and breakdown before and after recovery from protein-energy malnutrition, *Clin. Sci. Mol. Med.*, **53**, 473–7.

HALLIDAY, D. and MCKERAN, R. O. (1975). Measurement of muscle protein synthetic rate from serial muscle biopsies and total body protein turnover in man by continuous intravenous infusion of L-[$\alpha$-$^{15}$N] lysine, *Clin. Sci. Mol. Med.*, **49**, 581–90.

JAMES, W. P. T., GARLICK, P. J., SENDER, P. M. and WATERLOW, J. C. (1976). Studies of amino acid and protein metabolism in normal man with L-(U-$^{14}$C) tyrosine, *Clin. Sci. Mol. Med.*, **50**, 525–32.

KIEN, C. L., ROHRBAUGH, D. K., BURKE, J. F. and YOUNG, V. R. (1978a). Whole body protein synthesis in relation to basal energy expenditure in healthy children and in children recovering from burn injury, *Pediat. Res.*, **12**, 211–16.

KIEN, C. L., YOUNG, V. R., ROHRBAUGH, D. K. and BURKE, J. F. (1978b). Whole-body protein synthesis and breakdown rates in children before and after reconstructive surgery of the skin, *Metabolism*, **27**, 27–34.

LAURENT, G. L., SPARROW, M. P., BATES, P. C. and MILLWARD, D. J. (1978). Turnover of muscle protein in the fowl. Collagen content and turnover in cardiac and skeletal muscles of the adult fowl and the changes during stretch induced growth, *Biochem. J.*, **176**, 419–27.

LOBLEY, G. E., WEBSTER, A. J. F. and REEDS, P. J. (1977). Protein synthesis in lean and obese Zucker rats, *Proc. Nutr. Soc.*, **37**, 20A.

LOBLEY, G. E., MILNE, V., LOVIE, J. M., REEDS, P. J. and PENNIE, K. (1980). Whole body and tissue protein synthesis in cattle, *Brit. J. Nutr.*, **43**, 491–501.

MATTHEWS, D. E., MOTIL, K. J., ROHRBAUGH, D. K., BURKE, J. F., YOUNG, V. R. and BIER, D. M. (1980). Measurement of leucine metabolism in man from a primed, continuous infusion of L-[1-$^{13}$C] leucine, *Am. J. Physiol.*, **238**, E473–9.

MCNURLAN, M. A. and GARLICK, P. J. (1980). Contribution of rat liver and gastrointestinal tract to whole-body protein synthesis in the rat, *Biochem. J.*, **186**, 381–3.

MCNURLAN, M. A., PAIN, V. M. and GARLICK, P. J. (1980). Conditions that alter rates of tissue protein synthesis *in vivo*, *Biochem. Soc. Trans.*, **8**, 283–385.

MILLWARD, D. J., GARLICK, P. J., STEWART, R. J. C., NNANYELUGO, D. O. and WATERLOW, J. C. (1975). Skeletal-muscle growth and protein turnover, *Biochem. J.*, **150**, 235–43.

MILLWARD, D. J., GARLICK, P. J. and REEDS, P. J. (1977). The energy cost of growth, *Proc. Nutr. Soc.*, **35**, 339–49.

MOTIL, K. J., MATTHEWS, D., ROHRBAUGH, D., BIER, D., BURKE, J. F. and YOUNG, V. R. (1979). Simultaneous estimates of whole body leucine and lysine flux in young men: effect of reduced protein intake, *Fed. Proc.*, **38**, Abs. 2533.

MUNRO, H. N. (1969). Evolution of protein metabolism in mammals. In *Mammalian Protein Metabolism* Vol. 3 (Ed. H. N. Munro), Academic Press, New York, pp. 133–82.

O'KEEFE, S. J. D., SENDER, P. M. and JAMES, W. P. T. (1974). 'Catabolic' loss of body nitrogen in response to surgery, *Lancet*, **ii**, 1035–8.

PENCHARZ, P. B., STEFFEE, W. P., COCHRAN, W., SCRIMSHAW, N. S., RAND, W. M. and YOUNG, V. R. (1977). Protein metabolism in human neonates: nitrogen balance studies, estimated obligatory losses of nitrogen and whole-body turnover of nitrogen, *Clin. Sci. Mol. Med.*, **52**, 485–98.

PICOU, D. and TAYLOR-ROBERTS, T. (1969). The measurement of total protein synthesis and catabolism and nitrogen turnover in infants in different nutritional states and receiving different amounts of dietary protein, *Clin. Sci. Mol. Med.*, **36**, 283–96.

RECOECZI, E. and HATTON, M. W. C. (1980). Transferrin catabolism in mammalian species of different body sizes, *Am. J. Physiol.*, **28**, R306–10.

REEDS, P. J., LOBLEY, G. E., HARRIS, C. I. and FULLER, M. F. (1979). Methods of measuring protein metabolism in large animals. In *Protein Transactions in the Ruminant* (Ed. P. J. Buttery), ARC, London, 8·1–8·9.

REEDS, P. J. and LOBLEY, G. E. (1980). Protein synthesis: are there real species differences?, *Proc. Nutr. Soc.*, **39**, 43–52.

REEDS, P. J., CADENHEAD, A., FULLER, M. F., LOBLEY, G. E. and MCDONALD, J. D. (1980*a*). Protein turnover in growing pigs. Effects of age and food intake, *Brit. J. Nutr.*, **43**, 445–55.

REEDS, P. J., FULLER, M. F., CADENHEAD, A. and LOBLEY, G. E. (1980*b*). The effects of dietary energy and protein on protein turnover and nitrogen balance in growing pigs, *Proc. 3rd EAAP Symposium on Protein Metabolism and Nutrition*, Braunsweig.

REEDS, P. J., ØRSKOV, E. R. and MACLEOD, N. A. (1981). Whole body protein synthesis in cattle sustained by infusion of volatile fatty acids and casein, *Proc. Nutr. Soc.*, **40**, 50A.

SIM, A. J. W., WOLFE, B. M., YOUNG, V. R., CLARKE, D. and MOORE, F. D. (1979). Glucose promotes whole body protein synthesis from infused amino acids in fasting man, *Lancet* **i**, 68–72.

SIMON, O., MUNCHMEYER, R., BERGNER, H., ZEBROWSKA, T. and BURACZEWSKA, L. (1978). Estimation of rate of protein synthesis by constant infusion of labelled amino acids in pigs, *Brit. J. Nutr.*, **40**, 243–52.

SOLTESZ, G., JOYCE, J. and YOUNG, M. (1973). Protein synthesis in the newborn lamb, *Biol. Neonate*, **23**, 139–48.

STEFFEE, W. P., GOLDSMITH, R. S., PENCHARZ, P. B., SCRIMSHAW, N. S. and YOUNG, V. R. (1976). Dietary protein intake and dynamic aspects of whole body nitrogen metabolism in adult humans, *Metabolism*, **25**, 281–97.

WATERLOW, J. C. (1968). Observation on the mechanism of adaptation to low protein intakes, *Lancet*, ii, 1091–7.

WATERLOW, J. C. (1980). Protein turnover in the whole animal, *Invest. Cell Pathol.*, **3**, 107–19.

WATERLOW, J. C., GARLICK, P. J. and MILLWARD, D. J. (1978*a*). *Protein Turnover in Mammalian Tissues and in the Whole Body*, Elsevier North Holland, Amsterdam.

WATERLOW, J. C., GOLDEN, M. H. N. and GARLICK, P. J. (1978*b*). Protein turnover in man measured with $^{15}N$: comparison of end products and dose regimes, *Am. J. Physiol.*, **235**, E165–74.

YOUNG, V. R., STEFFEE, W. P., PENCHARZ, P. B., WINTERER, J. C. and SCRIMSHAW, N. S. (1975). Total human body protein synthesis in relation to protein requirements at various ages, *Nature (Lond)*, **253**, 192–3.

# 34

# Protein Turnover and the Regulation of Growth

D. J. Millward, P. C. Bates, J. G. Brown and M. Cox
*Clinical Nutrition and Metabolism Unit,*
*London School of Hygiene and Tropical Medicine,*
*London, UK*

and

M. J. Rennie
*Department of Human Metabolism,*
*University College Hospital Medical School,*
*London, UK*

## Introduction

Tissue growth is in one sense a complex process in which the size and number of functional units of the tissue increase with often complex morphological or biochemical changes (see Goss, 1964), while in another sense it is the simple result of the rate of protein synthesis exceeding that of protein degradation. The latter interpretation of growth is valid in the strict sense of the process, but if the dynamics of growth are ever to be understood, then the changes in tissue structure and function associated with the increasing protein mass should also be considered alongside the changes in protein turnover. In other words, we should not expect to find that the changes in protein synthesis and degradation which induce the logarithmic growth of a bacterial culture are necessarily the same as those which occur during the growth of the different tissues and organs of animals and man. We do not have detailed information about changes in protein turnover during growth for many tissues, but what information there is indicates that the changes are complex and variable. The last point is particularly important because variability in the mechanism of growth is observed in the same tissue in different situations. This paper examines the changes in

409

protein turnover associated with the changes in growth rate in different circumstances in one tissue—skeletal muscle.

## Growth Associated with Increased Protein Turnover—the Anabolic Increase in Degradation

A surprising feature of growth in muscle is that in some cases growth is associated with elevated rates of degradation which require at the same time very high rates of protein synthesis. We have called this an anabolic increase in degradation to distinguish it from the catabolic increase observed in starvation (Millward et al., 1980). For example, during the rapid growth of very immature animals degradation is more rapid than in adults in all species examined (see Millward, 1980a,b). In this case it is perhaps not strictly true to describe the high rate of degradation as an increase, since as far as we know the rate is high from the earliest stages of muscle differentiation. However, an actual increase in degradation is observed during nutritional rehabilitation (Millward et al., 1975) and particularly during muscular hypertrophy, for example, in the diaphragm after unilateral phrenicectomy (Turner and Garlick, 1974), or during the stretch-induced growth of a chicken wing muscle (Laurent et al., 1978b). An anabolic increase in degradation may also occur in man judging by increased $N^r$-methylhistidine excretion in gonad-deficient boys treated with testosterone, and in a patient recovering from polymyositis in marked positive nitrogen balance (results of M. J. Rennie and R. H. T. Edwards quoted in Millward et al., 1980).

One suggestion for the nature of this increased degradation is that it is an accompaniment to myofibrillar remodelling, enlargement and pro-liferation (Millward et al., 1975) and as such has been termed 'wastage' (Laurent and Millward, 1980). This is to liken the apparent over-production of protein during muscle growth to the over-production of RNA during the formation of the ribosome (Cooper, 1972). If this explanation is correct, then when growth stops the degradation rate should fall. This does occur after induced growth of muscle has ceased, for example, in the stretched chicken wing muscle (Laurent et al., 1978b), but does not occur in very young chick muscle when growth is suppressed. The high rate of degradation observed in muscle in 1-week-old birds falls after 2 weeks, even though growth is accelerated, and is maintained in 2-week-old malnour-ished birds, even though growth has stopped (Maruyama et al., 1978). Thus, it would appear that the high rate of degradation in immature muscle reflects some feature of the structure and function which is not related to the fact that it grows rapidly. What these features might be is not really understood (see Millward, 1980*b*). This observation is important because in adult muscle, when growth is induced during work-induced hypertrophy, there are a small number of new fibres with the histological appearance of immature fibres (Sola *et al.*, 1973), and the increased degradation in these fibres must be one part of the overall increase in degradation, the other part being the changes occurring in the hypertrophying mature fibres.

In addition, in some diseases of muscle, such as muscular dystrophy, rates of degradation are high judging by the elevated rates of $N^r$-methylhistidine excretion (McKeran *et al.*, 1977; Ballard *et al.*, 1979). One feature of diseased muscle is the simultaneous presence of regenerating immature fibres and necrotic fibres. Thus the increased degradation may include that associated with repair as well as that due to the disease process and the atrophy.

In summary, the following points can be made about elevated degradation in growing muscle.

1.  The increased degradation in immature muscle may not be associated with the growth process.
2.  The increased degradation during induced muscle hypertrophy may be a necessary part of tissue remodelling.
3.  Because increased degradation in diseased muscles may be associated with repair as well as atrophy, it should not necessarily be assumed that the increased degradation is responsible for the reduced growth, and consequently attempts to suppress it with proteinase inhibitors may be inadvisable.

## Growth Associated with Decreased Protein Degradation—the Anabolic Decrease in Degradation

Since the anabolic decrease in degradation is a mechanism of growth in cell cultures (e.g. Hershko *et al.*, 1971), it was thought to be a common mechanism in many situations. In fact there are few examples obtained by reliable methods (see Waterlow *et al.*, 1978). One example is to be found in a rapid-growing strain of rat (CFY). In these rats the rate of degradation at weaning is lower than that in a slow-growing strain and furthermore the rate falls to the low adult value at a younger age (Bates and Millward, 1978) (Table 1). As a result more rapid growth can be achieved at a lower rate of

412    D. J. Millward, P. C. Bates, J. G. Brown, M. Cox and M. J. Rennie

## TABLE 1
DEVELOPMENTAL CHANGES IN PROTEIN TURNOVER IN MUSCLE OF TWO STRAINS OF RATS

| Age (days) | | Synthesis (%/day) | | Degradation (%/day) | | Growth (%/day) | |
|---|---|---|---|---|---|---|---|
| Hooded | CFY | Hooded | CFY | Hooded | CFY | Hooded | CFY |
| 25 | 23 | 28·6 | 15·6 | 22·5 | 9·8 | 6·1 | 5·8 |
| 32 | 46 | 16·1 | 15·2 | 13·1 | 9·5 | 3·0 | 5·7 |
| 52 | 65 | 11·5 | 7·3 | 9·8 | 4·4 | 1·7 | 2·9 |
| 101 | 130 | 5·3 | 5·2 | 4·6 | 4·1 | 0·7 | 1·1 |
| 320 | 330 | 4·9 | 4·5 | 4·9 | 4·5 | 0 | 0 |

Results of Millward et al., (1975) and Bates and Millward (1978; 1981).
Protein synthesis was measured by a 6-hour constant infusion of $^{14}$C-tyrosine.

protein synthesis and furthermore a greater proportion of protein synthesis results in net protein synthesis (Bates and Millward, 1981).

Another example is observed in female rats following growth acceleration caused by the anabolic steroid trienbolone acetate. In this case, as originally reported by Vernon and Buttery (1976) and recently confirmed by us (Table 2), the increased growth resulted from a decrease in degradation, which was only 69 % of that of the controls.

Why either of these changes should occur is not known. However we do know that the rate of protein degradation varies not only with the stage of development and with the type of muscle (Laurent et al., 1978a,b), but also with the thyroid status (Brown and Millward, 1980 and see p. 475). The latter factor is important because by manipulating thyroid status (for example, through $T_3$ administration by an implanted Alza minipump) growth can be increased at low levels of $T_3$ dosage when degradation has

## TABLE 2
EFFECT OF ANABOLIC STEROIDS ON MUSCLE PROTEIN TURNOVER
(TRIENBOLONE ACETATE 80 μg/day)

| | Synthesis (%/day) | Degradation (%/day) | Growth (%/day) |
|---|---|---|---|
| Control | 7·0 ± 0·8 | 4·82 | 2·18 ± 1·5 |
| Treated | 7·2 ± 0·6 | 3·33 | 3·9 ± 0·7 |

Brown and Millward, unpublished results. 100 g female rats were treated for 14 days. Protein synthesis was measured by a 6-hour constant infusion with $^{14}$C-tyrosine (means ± SD).

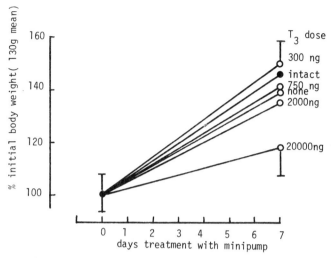

FIG. 1. The effect of $T_3$ administration on the growth of thyroidectomized rats. $T_3$ was administered for 7 days by an implanted osmotic minipump starting 3 days after thyroidectomy.

fallen (Fig. 1 and see p. 483), although it is suppressed when high doses are given.

We do not yet know whether the lower degradation in the rapid-growing rat strain is a result of a low thyroid status but we have examined this in the rats treated with anabolic steroids. Table 3 shows that, in one experiment, free $T_3$ levels did appear lower following treatment, but when we repeated this with more animals we could not observe any difference. Thus a reduced thyroid status cannot be the main mechanism. Another theory is that the anabolic agents are antagonistic to glucocorticoid action on muscle, thus

TABLE 3

THYROID STATUS OF RATS TREATED WITH ANABOLIC STEROIDS (TRIENBOLONE ACETATE 80 $\mu$g/day)

|  | Free $T_3$ (pg/ml) | |
| --- | --- | --- |
|  | Control | Treated |
| Experiment 1 ($n = 5$) | $5.84 \pm 0.76$ | $4.49 \pm 0.87$ |
| Experiment 2 ($n = 12$) | $5.55 \pm 1.00$ | $5.30 \pm 0.69$ |

Cox, Brown and Millward, unpublished results. 100 g female rats were treated for 14 days (means $\pm$ 1 SD).

desuppressing growth (Mayer and Rosen, 1975). However, we have observed that anabolic agents induce growth in adrenalectomized rats as effectively as in intact rats (Millward, Cox and Brown, unpublished). This precludes any mechanism involving an interaction with glucocorticoids since these hormones are absent in the adrenalectomized rat. Therefore the way in which this anabolic agent works remains to be determined.

## Anabolic Changes in Protein Degradation in Non-muscle Tissues

The changes in protein degradation described here have been limited to those occurring in muscle tissue. The developmental changes in degradation have not been described to any great extent in non-muscle tissues (see Waterlow et al., 1978). Furthermore the anabolic increase in degradation has not to our knowledge been described in non-muscle tissues, so this might be a specific characteristic of muscle. The anabolic decrease in degradation certainly occurs in non-muscle tissues and is particularly well documented in the liver (Waterlow et al., 1978; Mortimore and Schworer, 1980). However, whether the biochemical changes associated with these anabolic decreases in muscle are the same as those associated with the anabolic decrease in liver remains to be determined.

## Acknowledgements

This work was generously supported by the Muscular Dystrophy Group of Great Britain and the Medical Research Council.

## Discussion

*Garlick* (*London School of Hygiene and Tropical Medicine, UK*): Millward described some situations in which there is a decrease in protein breakdown during rapid growth. Another example that we have studied is in rats which had been diabetic for a few days and then treated with insulin. The muscles, gastrocnemius, diaphragm and heart, all grew at about twice the normal rate after this, at 14–15 %/day, presumably a form of catch-up growth. This was achieved both by an increase in the rate of protein synthesis and by a decrease in the rate of protein breakdown to very low levels (Albertse et al., 1980).

*Young (Massachusetts Institute of Technology, USA):* I wonder whether oestrogens have an overall catabolic effect on muscle protein metabolism.

*Millward:* The rôle of androgens and oestrogens in general in the regulation of muscle growth is a very poorly explored area and the question of why the female rat, for instance, grows at a markedly different rate from the male has hardly been attacked at all.

*Coore (University of the West Indies, Kingston, Jamaica):* Perhaps when a hormone increases synthesis and degradation at the same time it could be because of an increase in the amount of intracellular proteases. If one could measure changes in the amounts of these proteases, it might give a clue to what is happening in these different situations.

*Millward:* This is known to occur in response to thyroid hormones. As I indicate in my other paper (p. 475), $T_3$ appears to regulate rates of protein synthesis and degradation in muscle by changing the amounts of RNA and the lysosomal proteinases.

*Harper (University of Wisconsin-Madison, USA):* I was not quite clear about how you were using the word 'wastage' in relation to the change in degradation rate.

*Millward:* Some people don't like the use of this term. I was using it in the same way as we use it in relation to ribosomal RNA synthesis, where a large chunk is made and some of it is degraded in the course of maturation of the ribosome. That is usually termed wastage in the sense that it is synthesis that does not contribute to the anabolic process. It should not be seen as a process which is of no importance; it may well be very necessary and important in growth.

*Harper:* It seems to me to be important to distinguish between wastage of energy and wastage of building materials, for example, amino acids. To what extent do you imply wastage of amino acids?

*Millward:* None at all. The term is very strictly applied to the partition of protein synthesis that is occurring in relation to net accretion of protein and replacement of that lost by protein breakdown (see Reeds, p. 398). Subsequent changes in amino acid oxidation rates and the energetic implications of wastage are separate questions.

*Harper:* It is interesting that the young animal at the stage of most rapid growth has the highest rate of degradation, but also has extremely high efficiency of utilization of amino acids for growth. So the recapture of those amino acids must also be very efficient. Also during this time the animal has a high food intake. How much is the influx of food and the production of insulin responsible for some of those depressions in degradation rate during this period of rapid growth?

*Millward:* As far as the first of your points is concerned—the efficiency of amino acid oxidation—it seems to me that while amino acid oxidation rates and rates of net accretion of protein are obviously interdependent, they are separately regulated. I certainly do not think there is any evidence to suggest that rates of oxidation are necessarily correlated with rates of protein turnover in the sense that an amino acid is allowed to go in and out of a protein so many times before it is oxidized. This must be precluded as a basis for a regulatory mechanism simply because of the fact that the oxidation of individual amino acids occurs in specific, often different tissues. Thus leucine might be much more reutilized in the liver than, say, phenylalanine simply because of the low level of leucine transaminase activity. As for your second point, the interesting fact for me is that even with high intakes of amino acids and energy in young rats muscle protein degradation is still elevated. This must mean that the increased degradation in muscle at this age is in some way a necessary characteristic which is not subject to suppression by high concentrations of amino acids or insulin.

*Pratt (Institute of Psychiatry, London, UK):* You suggest that both the myofibrils formed during early growth and those formed by hypertrophy suffer more rapid degradation. Is anything known about how these differ from other myofibrils?

*Millward:* Not much. During development the muscle fibre can be shown to change quite considerably. According to Young (p. 505), the methylation of the histidine of the myosin heavy chain changes during the maturation of the myofibre. There are functional changes, for example, in the contraction time, which probably reflect alterations in the myosin light chains as well. The difficulty is to decide which of the changes has any bearing on the change in the rate of breakdown.

# References

ALBERTSE, E. C., PAIN, V. M. and GARLICK, P. J. (1980). Protein synthesis and breakdown in muscle and kidney of diabetic and insulin-treated rats, *Proc. Nutr. Soc.*, **39**, 19A.

BALLARD, F. J., TOMES, F. M. and STERN, L. M. (1979). Increased turnover of muscle contractile proteins in Duchenne muscular dystrophy as assessed by 3-methylhistidine and creatinine excretion, *Clin Sci.*, **56**, 347–52.

BATES, P. C. and MILLWARD, D. J. (1978). Muscle growth and protein turnover in a fast growing rat strain, *Proc. Nutr. Soc.*, **37**, 19A.

BATES, P. C. and MILLWARD, D. J. (1981). Characteristics of skeletal muscle growth and protein turnover in a fast growing rat strain, *Brit. J. Nutr.*, **46**, 7–14.

BROWN, J. G. and MILLWARD, D. J. (1980). Thyroid hormones and muscle protein turnover. *Biochem. Soc. Trans.*, **8**, 366.

COOPER, H. L. (1972). The induction of ribosomal RNA synthesis in lymphocytes, *Transplant. Rev.*, **11**, 3–38.

GOSS, R. J. (1964). *Adaptive Growth*, Academic Press, New York.

HERSHKO, A., MAMONT, P., SHIELDS, R. and TOMKINS, G. M. (1971). Pleiotypic response, *Nature (New Biol)*, **232**, 206–11.

LAURENT, G. J. and MILLWARD, D. J. (1980). Protein turnover during skeletal muscle hypertrophy, *Fed. Proc.*, **39**, 42–7.

LAURENT, G. J., SPARROW, M. P., BATES, P. C. and MILLWARD, D. J. (1978a). Muscle protein turnover in the adult fowl. 1. Rates of protein synthesis in fast and slow skeletal cardiac and smooth muscle of the adult fowl (*Gallus domesticus*), *Biochem. J.*, **176**, 393–405.

LAURENT, G. J., SPARROW, M. P. and MILLWARD, D. J. (1978b). Turnover of muscle protein in the fowl. Changes in rates of protein synthesis and breakdown during hypertrophy of the anterior and posterior *latissimus dorsi* muscles, *Biochem. J.*, **176**, 407–17.

MARUYAMA, K., SUNDE, M. L. and SWICK, R. W. (1978). Growth and muscle protein turnover in the chick, *Biochem. J.*, **176**, 573–82.

MAYER, M. and ROSEN, F. (1975). Interaction of anabolic steroids with glucocorticoid receptor sites in rat muscle cytosol, *Am. J. Physiol.*, **229**, 1381–6.

MCKERAN, R. O., HALLIDAY, D. and PURKISS, P. (1977). Increased myofibrillar protein catabolism in Duchenne muscular dystrophy measured by 3-methylhistidine excretion in the urine, *J. Neurol. Neurosurg. Psychiatr.*, **40**, 979–81.

MILLWARD, D. J. (1980a). Protein turnover in cardiac and skeletal muscle during normal growth and hypertrophy, in *Degradative Processes in Heart and Skeletal Muscle* (Ed. Wildenthal, K.), Elsevier North-Holland, Amsterdam, pp. 161–99.

MILLWARD, D. J. (1980b). Protein degradation in muscle and liver, in: *Comprehensive Biochemistry* (Eds Florkin, M., Neuberger, A. and van Deenan, L. L. M.), Elsevier North-Holland, Amsterdam, Vol. 19B, Part 2, 153–232.

MILLWARD, D. J., BATES, P. C., BROWN, J. G., ROSOCHACKI, S. R. and RENNIE, M. J. (1980). Protein degradation and the regulation of protein balance in muscle, in *Protein Degradation in Health and Disease*, Ciba Symposium No. 75, Excerpta Medica, Amsterdam, pp. 307–29.

MILLWARD, D. J., GARLICK, P. J., STEWART, R. J. C., NNANYELUGO, D. O. and WATERLOW, J. C. (1975). Skeletal muscle growth and protein turnover, *Biochem. J.*, **150**, 235–43.

MORTIMORE, G. E. and SCHWORER, C. M. (1980). Application of liver perfusion as an *in vitro* model in studies of intracellular protein degradation, in *Protein Degradation in Health and Disease*, Ciba Symposium No. 75, Excerpta Medica Amsterdam, pp. 281–98.

SOLA, O. M., CHRISTENSEN, D. L. and MARTIN, A. W. (1973). Hypertrophy and hyperplasia of adult chicken anterior *latissimus dorsi* muscles following stretch with and without denervation, *Exp. Neurol.*, **41**, 76–100.

TURNER, L. V. and GARLICK, P. J. (1974). The effect of unilateral phrenicectomy on the rate of protein synthesis in rat diaphragm *in vivo*, *Biochim. Biophys. Acta*, **349**, 109–113.

VERNON, B. G. and BUTTERY, P. J. (1976). Protein turnover in rats treated with trienbolone acetate, *Brit. J. Nutr.*, **36**, 575–9.

WATERLOW, J. C., GARLICK, P. J. and MILLWARD, D. J. (1978). *Protein Turnover in Mammalian Tissues and in the Whole Body*, Elsevier North-Holland, Amsterdam.

YOUNG, V. R. and MUNRO, H. N. (1980). Muscle protein turnover in human beings in health and disease, in *Degradative Processes in Heart and Skeletal Muscle* (Ed. Wildenthal, K.), Elsevier North-Holland, Amsterdam.

# 35

# Protein and Energy Intake in Relation to Protein Turnover in Man

V. R. Young, J. J. Robert and Kathleen J. Motil
*Department of Nutrition and Food Science,*
*Massachusetts Institute of Technology,*
*Cambridge, Massachusetts, USA*

and

D. E. Matthews and D. M. Bier
*Department of Internal Medicine,*
*Washington University School of Medicine,*
*St. Louis, Missouri, USA*

## Introduction

Protein synthesis and breakdown are regulated by multiple factors. These include the supply of amino acids that serve as substrates for formation of polypeptides, the availability of high energy intermediates (ATP and GTP), required for formation of the initiation complex, amino acyl-tRNAs, peptide bond formation, and, to some extent, for subsequent breakdown of completed proteins and release of their amino acids into the tissue free amino acid pools (Dean, 1980). In addition, protein and energy intakes influence endocrine function and balance (e.g. Crim and Munro, 1979), and this, in turn, will determine the status of body nitrogen metabolism. It is not surprising, therefore, that human nitrogen metabolism is highly sensitive to altered dietary intakes of protein and energy.

Much of our current knowledge concerning the effects of these dietary factors on protein metabolism in man has been derived from measurements of body nitrogen balance. Furthermore, the response of whole body nitrogen metabolism to one of these two major dietary components is modulated by the level of the other component (Munro, 1964), thereby

419

increasing the difficulty of arriving at a quantitative definition of protein-energy interrelationships in human protein metabolism. Therefore, a more precise description of the effects of protein and energy intakes on nitrogen metabolism in the human subject is of importance in resolving a number of problems concerned with estimation of protein and energy needs and nutritional status in health and disease.

Waterlow *et al.* (1978*b*) have extensively reviewed this field, and we will consider here only selected recent observations, giving particular emphasis to studies concerned with dynamic aspects of whole body nitrogen and amino acid metabolism in adult human subjects, including results obtained recently in our laboratories. To begin with, we review briefly the effects of changes in nitrogen and energy intake on body nitrogen balance. Studies of human protein and amino acid turnover will then be examined in detail. Because the responses of body nitrogen metabolism to dietary protein and energy intakes are determined also by the nutritional status of the individual we shall consider both the acute and chronic effects of dietary changes on whole body protein turnover.

## *Response of Nitrogen Balance to Protein and Energy Intakes*

Body N balance is sensitive to changes in the dietary level and source of the protein and major energy-yielding components (carbohydrate, lipid) in the diet. Although body N equilibrium can be maintained in the adult over a wide range of protein intake, N balance becomes progressively less positive or more negative when protein intakes approach and fall below a requirement level. However, the precise nature of the relationship between N intake and balance is still uncertain, because it is still debated whether the response of N balance to changes in N intake within the entire submaintenance range is best characterized by a rectilinear (Kishi *et al.*, 1978) or curvilinear (Young *et al.*, 1973) relationship (Fig. 1). Nevertheless, it is clear that the efficiency of dietary N retention is high at very low intakes of protein, even of poor quality proteins (Young and Scrimshaw, 1978; Inoue *et al.*, 1973, 1974) and this efficiency declines as N intake approaches requirement levels of N (Calloway and Margen, 1971). As pointed out by Waterlow in 1968, the metabolic basis for changes in N balance should be explored if improvements are to be made in the assessment of human protein requirements.

For a given intake of dietary protein, the actual N balance depends also

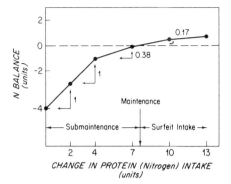

FIG. 1.    Schematic representation of the change in efficiency of dietary N retention with changes in the sub-maintenance and surfeit ranges of N (protein intake).

upon the amount and source of the dietary energy. The available data, previously reviewed by Munro (1951, 1964) and Calloway and Spector (1954), show that N balance is highly sensitive to altered energy intakes and that the degree of change is determined by the levels of both protein and energy in the diet (Munro, 1964). Indeed, changes that occur in N balance with alterations in energy intake may complicate estimation of requirements for dietary protein when these are derived from N balance criteria (e.g. Garza *et al.*, 1976, 1977, 1978). This problem also applies to estimates of requirements of individual essential amino acids (see Young, p. 137). A review of more recent data obtained in healthy young men and in depleted patients maintained with intravenous nutrition (Elwyn *et al.*, 1979) confirms that energy intake has a profound effect on N balance, especially when the diet supplies generous amounts of protein of high quality and where the initial energy intake is low. A summary of these findings is presented in Table 1.

    In addition to the influence of level of energy intake, the source of the energy-yielding nutrients may also affect dietary N retention. Munro (1964) concluded that carbohydrate exerts a specific action on the utilization of amino acids during their absorption from protein-containing meals, in addition to an effect shared by fat and carbohydrate on the metabolism of amino acids during the post-absorptive period.

    The practical nutritional significance of differences in energy source has not been determined, because most previous studies have involved intakes of protein that greatly exceed those considered to be just sufficient for long-term maintenance of N balance. Therefore, we (Richardson *et al.*, 1979)

TABLE 1

A SELECTED SURVEY OF SOME RECENT STUDIES ON THE RESPONSES OF N BALANCE IN
ADULTS TO CHANGES IN ENERGY INTAKE (E)

| Subjects | Dietary conditions | N balance response | Reference |
|---|---|---|---|
| Young men | Protein intake varied 0·28–0·76 g/ kg/day | ~2 mg N/kcal | Inoue *et al.* (1973) |
| Six young men | Protein intake 5–7 % of dietary energy, 12 day periods, at low E and excess E | 1·74 mg N/kcal at low E; 1·12 mg N/kcal at high E | Calloway (1975) |
| Four young men | Protein intake 0·6 g/kg, 3–4 week diet periods | 1·74 mg N/kcal | Garza *et al.* (1976) |
| Four young men | Protein intake 0·6 g/kg plus dispensable amino acids (≡0·23 g protein/kg) | ~3 mg N/kcal | Garza *et al.* (1978) |
| Young men (46 total) | Variable energy and protein intake among groups | ~3 mg N/kcal | Kishi *et al.* (1978) |
| Depleted patients | Intravenous glucose, N intake 173 mg/kg | 1·7 mg N/kcal | Elwyn *et al.* (1979) |

examined the relationship between N balance and energy intake in young men given diets high or low in carbohydrate and a protein intake that approximated their mean requirements. As shown in Fig. 2, we observed a strong association between N balance and energy intake; for the diet low in carbohydrate, 2·6 mg N and for the high carbohydrate diet, 1·3 mg N were retained for each additional kcal of total energy intake. Our results also indicated that the N-sparing effects of carbohydrate, relative to fat, were more pronounced for those subjects whose energy intakes were lowest. In summary, therefore, N balance in adults is affected by the relative proportion of non-protein energy sources when the protein intake is at the level recommended by the FAO/WHO (1973) for young adults and when changes in the proportion of dietary carbohydrate and fat are in the range found in usual diets.

Precisely how dietary carbohydrate exerts this specific effect on the sparing of body protein is not fully understood. Munro (1964) proposed a scheme for the protein-sparing action of dietary carbohydrate that is dependent on the secretion of insulin and he cites evidence to support this.

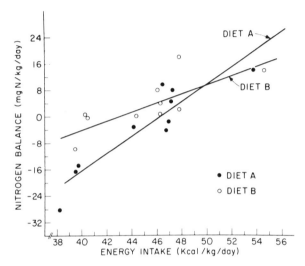

FIG. 2. Relation between N balance and energy intake in young men receiving a 0·6 g protein/kg/day diet for 21 days which supplied a carbohydrate:fat ratio of either 1:1 (diet A) or 2:1 (diet B). Taken from Richardson *et al.* (1979).

To examine this hypothesis Fuller *et al.* 1977) used a continuous infusion of physiological amounts of exogenous insulin in well-nourished pigs together with glucose infused at a rate sufficient to maintain plasma glucose concentrations within the normal physiological range. The response to this treatment over a period of three to seven days was a two to seven fold increase in plasma insulin, a 50% decrease in plasma glucose, a 40% decrease in plasma urea concentration, and a 30% fall in urinary excretion of urea N. After infusion plasma urea levels rapidly returned to those of the control period. The authors concluded that a major component of the protein-sparing effect, achieved by a surfeit feeding of carbohydrate, is mediated by insulin.

## Protein Intake and Dynamic Aspects of Whole Body Amino Acid and N Metabolism

In order to explore the mechanisms that may account for changes in body N balance and the efficiency of dietary N retention with alterations in protein and energy intakes, it is necessary to use isotope tracer techniques. Waterlow *et al.* (1978*b*) have reviewed the earlier studies in adults of Sprinson and Rittenberg (1949), Tschudy *et al.* (1959) and Kassenaar *et al.* (1960), involving $^{15}N$ tracers, but these studies, including our own (Steffee *et al.*, 1976), and those in children of Picou and Taylor-Roberts (1969) and Golden *et al.* (1977*a*,*b*) do not provide a clear picture of the impact of dietary change on body protein turnover in man. Therefore, it appeared to us that new and alternative isotope tracer methods would be necessary to define more precisely the changes in whole body N and amino acid dynamics that may occur in response to alterations in protein intake.

We have conducted a study in young men to examine changes in whole body leucine and lysine metabolism in response to alterations in dietary protein intake (Motil *et al.*, 1981*a*), using a primed constant infusion of 1-$^{13}C$-leucine and $\alpha$-$^{15}N$-lysine given simultaneously (e.g. Matthews *et al.*, 1980). Three protein intake levels were studied: 1·5 g, 0·6 g and 0·1 g protein kg/day. The 0·6 g protein/kg/day intake level was chosen to

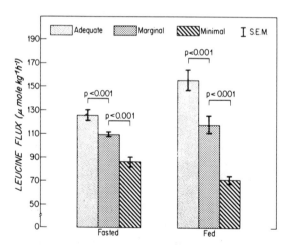

FIG. 3.   Whole body leucine flux in young men receiving three levels of protein intake: adequate (1·5 g/kg/day); marginal (0·6 g/kg/day); minimal (0·1 g/kg/day), where flux was determined after an overnight fast (fasted or post-absorptive) or while receiving small isocaloric isonitrogenous meals (fed). Drawn from Motil *et al.* (1981*a*).

represent a *near* maintenance level of intake (FAO/WHO, 1973) and to serve as the reference level for evaluation of the results at the other levels. At this intake our subjects were in negative N balance, whereas balance was markedly negative at the 0·1 g level and positive at the generous intake of 1·5 g/kg/day. As shown in Fig. 3, a change in dietary protein intake from a maintenance to a more generous or surfeit level (1·5 g/kg/day) caused an increase in leucine flux during both the post-absorptive and fed states. On the other hand, a restriction in protein intake to the inadequate level (0·1 g/kg/day) resulted in a reduction in flux. Furthermore, as summarized in Table 2, changes in the rates of oxidation and incorporation of leucine into body proteins occurred with these altered protein intakes; with increase in protein intake from a near requirement to surfeit level, the post-absorptive state is associated with an increased incorporation of leucine into body proteins. This accounted for 100 % of the change in flux since oxidation of leucine in this condition was not significantly altered. However, in the fed

TABLE 2

RATES OF WHOLE BODY LEUCINE OXIDATION, INCORPORATION INTO PROTEINS AND RELEASE FROM PROTEINS BY BREAKDOWN IN YOUNG MEN RECEIVING THREE LEVELS OF PROTEIN INTAKE, FOR THE FED AND POST-ABSORPTIVE STATES

| *Metabolic condition and parameter* | *Protein intake level (g/kg/day)* | | |
|---|---|---|---|
| | *1·5* | *0·6* | *0·1* |
| *Post-absorptive* | | | |
| Flux | 127·4 ± 3·7 | 109·7 ± 2·2 | 86·8 ± 5·3 |
| Oxidation | 17·6 ± 4·7 | 22·4 ± 2·7 | 12·9 ± 2·2 |
| Incorporation into protein | 113·1 ± 8·1 | 89·3 ± 2·4 | 76·1 ± 3·8 |
| Release from protein | 125·1 ± 3·7 | 107·4 ± 2·2 | 84·5 ± 5·3 |
| *Fed*[a] | | | |
| Flux | 157·3 ± 7·8 | 119·7 ± 6·1 | 72·7 ± 2·8 |
| Oxidation | 46·3 ± 3·8 | 21·6 ± 1·1 | 11·8 ± 0·7 |
| Incorporation into protein | 113·3 ± 6·7 | 102·4 ± 7·6 | 64·4 ± 3·8 |
| Release from protein | 67·2 ± 7·9 | 82·3 ± 6·0 | 65·7 ± 2·8 |

From Motil *et al.* (1981a). Groups of four subjects studied in post-absorptive and fed states at each level of protein intake. Values ($\mu$mol/kg/h) are means ± SEM.
[a] In the fed state subjects were offered, at hourly intervals, isocaloric and isonitrogenous meals, each equivalent to 1/12th of the daily protein and energy intake and beginning one hour before isotope infusion and three hours after their usual breakfast.

condition, leucine oxidation was substantially increased together with a higher rate of leucine incorporation into tissue proteins. This response is consistent with observations in rats showing higher rates of oxidation of indispensable amino acids when consumed in excess of needs for growth (e.g. Brookes *et al.*, 1972; Kang-Lee and Harper, 1977, 1978).

The effects of changes in protein intake within the submaintenance range are also summarized in Table 2. With the inadequate protein intake the reduced leucine flux during the post-absorptive phase is associated with a reduction in the rate of leucine incorporation into protein, which accounts for nearly 60 % of the change in flux, as well as with a lower rate of leucine oxidation. The latter response accounts for about 40 % of the decrease in leucine flux. During the fed condition, a major contribution to the change in leucine flux arises from a reduced rate of leucine incorporation into body protein, accounting for 81 % of the net decrease in leucine flux.

These findings suggest that more than one mechanism is responsible for the adaptive responses of whole body leucine metabolism and N balance to changes in dietary protein intakes. Also, if it can be reasonably assumed that 0·6 g protein/kg/day approximates the mean requirement for protein, these mechanisms appear to be integrated with the total N and amino acid needs of the host. For instance, at excess levels of protein a major change in leucine metabolism is the marked increase in oxidation in the fed state. On the other hand, within the sub-maintenance to maintenance levels of dietary protein intake the dramatic changes in leucine metabolism are associated with decreased rates of leucine incorporation into protein and decreased liberation of the amino acid into the metabolic pool via protein breakdown (protein turnover).

Lysine flux followed a pattern similar to that for leucine although the actual fluxes were lower for lysine than for leucine, especially at the generous and requirement levels of protein intake (Fig. 4). However, at the inadequate level of intake (0·1 g/kg/day) the fluxes were of similar magnitude. Similar discrepancies in leucine and lysine fluxes have been described in growing pigs (Simon *et al.*, 1978). While the mechanisms underlying these differences are uncertain, they may be related to differences in the sizes of the free amino acid pools in the tissues and particularly the different responses of the free pools of lysine and leucine to changes in dietary intake of these amino acids (e.g. Pion, 1973). This observation underscores the desirability of using multiple tracer probes to explore the responses of whole body amino acid metabolism to dietary factors.

From our recent experiments we conclude that the level of protein intake

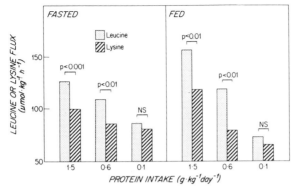

Fig. 4. Comparison of whole body fluxes for leucine and lysine in young men, studied while in the fed and fasted (post-absorptive) states, receiving three levels of dietary protein. Drawn from Motil *et al.* (1981*a*).

in adults does have a profound effect on the rate of total protein turnover. This is contrary to the conclusion drawn by Waterlow *et al.* (1978*b*), that in normal adults the level of protein intake has little or no effect on the rate of total protein turnover, but their conclusion was based on earlier and less extensive $^{15}N$ tracer studies.

## Response of Protein Turnover to Meals

Because several aspects of N, protein and amino acid metabolism show fluctuations during the day, the responses of whole body amino acid metabolism and protein turnover to the ingestion and composition of meals should be considered. Studies in rats have shown that fluctuations in the incorporation of labelled amino acids in a variety of tissues (e.g. Garlick *et al.*, 1973; Pocknee and Heaton, 1978; Buckley and Marquardt, 1980) and in the activities of liver enzymes (e.g. Watanabe *et al.*, 1968) are determined by the feeding schedule. Also, in humans, fluctuations in plasma amino acid levels are related to the level of protein in the diet (Fernstrom *et al.*, 1979), and to the composition and frequency of meals (Hussein *et al.*, 1971). We reported some years ago that the concentration of plasma tryptophan changed during the day and that the specific pattern of change depended upon the adequacy of tryptophan intake and frequency of meals (Young *et al.*, 1969). Furthermore, there is a rhythm in the urinary output of urea (Fig. 5) which is marked at generous protein intakes, but also occurs when these are inadequate.

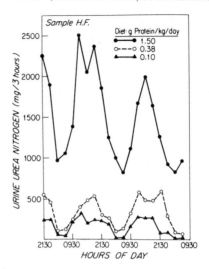

FIG. 5.    Fluctuation in urea-N excretion in a young adult male receiving three levels of dietary protein. Meals were given at 8.00 a.m., noon, 5.00 p.m. and 9.00 p.m. (unpublished data of Steffee, W. P. and Young, V. R.).

Our recent study (Motil *et al.*, 1981*a*) provides an opportunity to examine whole body leucine and lysine dynamics during the absorptive and post-absorptive phases of amino acid and energy metabolism. As indicated in Table 2, ingestion of small meals brought about a consistent reduction in the rate of body protein breakdown, because there was a lower rate of inflow of leucine (and lysine) into the metabolic pool from endogenous protein sources when meals were consumed. This effect was observed at all three levels of protein intake studied. These results imply that energy and protein may have separate roles in bringing about a reduction in body protein breakdown on consumption of meals. The dependence of changes in whole body amino acid dynamics on food ingestion is also revealed by the study of Golden and Waterlow (1977) which showed no diurnal variation in leucine flux and oxidation when subjects were fed continuously by intragastric tube throughout a 30-hour period.

The findings described above differ from those reported by Garlick *et al.*, (1980*a*). These authors explored the acute effect of feeding on whole body leucine dynamics in obese female subjects, and as summarized in Table 3, the rates of whole body synthesis and leucine oxidation during the night, when no food was given, were 67 % and 38 %, respectively, of the rates observed during the day when small meals were given hourly for 12 hours.

TABLE 3

RATES OF PROTEIN METABOLISM DERIVED FROM INFUSION OF $1-^{13}$C-LEUCINE IN OBESE FEMALES (96 kg, 52 YEARS)

|  | *Day* | *Night* | *Total* |
|---|---|---|---|
|  | *(g/12 h)* | | *(g/24 h)* |
| Dietary intake | 79 ± 2 | 0 | 79 ± 2 |
| Oxidation | 50 ± 8* | 19 ± 4 | 69 ± 7 |
| Synthesis | 129 ± 16* | 87 ± 25 | 216 ± 40 |
| Breakdown | 101 ± 16 | 107 ± 29 | 207 ± 40 |

From Garlick *et al.*, (1980*a*). Means ± SD
* Significantly different ($p < 0.001$) from night-time value.

In contrast, the rate of whole body protein breakdown remained constant throughout the entire 24-hour period and their findings lead to a distinctly different conclusion regarding the role of body protein breakdown in the daily economy of amino acid metabolism. Hence, it is worth considering possible reasons for the differences in the findings reported by Garlick *et al.* (1980*a*) as compared with ours.

First, differences in the experimental designs used by us and by Garlick *et al.* (1980*a*) should be emphasized. In our investigation separate subjects were studied in the post-absorptive state (at 8.00 a.m.) and fed state (at noon) and the results of whole body leucine dynamics for the two groups were compared. In contrast, in Garlick's experiment the same subjects were studied for the effects of meal feeding and this design offers a more strict control. However, it seems unlikely that this would be a sufficient explanation for the observed differences in results between the two experiments. A more likely explanation lies in the dietary inputs of both energy and protein which were substantially higher in our study than those given by Garlick *et al.* (1980*a*). Furthermore, the obese subjects had received an energy restricted diet for 2 days before the study. Because the N balance responses to changes in energy intake are not only dependent upon the absolute level of energy in the diet but on the amount of protein as well, as reviewed above, these differences in meal size and composition are probably responsible for the different observations made in the two experiments.

This view is further supported by our recent studies on the responses of amino acid metabolism to intravenous administration of glucose. As summarized in Table 4, whole body leucine flux and oxidation rates were essentially unaltered with a 150-minute infusion of glucose at a rate of 4 mg/kg/min, or about twice the basal rate of glucose production (Wolfe *et al.*, 1979). Although this rate of infusion of glucose caused a significant increase in plasma glucose and in circulating insulin levels it did not produce any change in the dynamic status of whole body leucine metabolism. However, in an earlier study in normal adult volunteers, when glucose infusion was given for a number of days in the presence of amino acids, whole body

TABLE 4

EFFECTS OF A 150-MINUTE INFUSION OF INTRAVENOUS GLUCOSE (4 mg/kg/min) ON WHOLE BODY LEUCINE DYNAMICS IN HEALTHY YOUNG MEN

|  | *Basal state* | *After glucose infusion* |
|---|---|---|
| Plasma glucose (mg/100 ml) | $89 \pm 1 \cdot 0^a$ | $152 \pm 4$ |
| Insulin ($\mu$U/ml) | $8 \cdot 2 \pm 1 \cdot 4^b$ | $38 \pm 2$ |
| *Leucine dynamics$^c$* ($\mu$mol/kg/h) |  |  |
| Flux | $99 \pm 4$ | $91 \pm 4$ |
| Oxidation | $20 \pm 2$ | $18 \cdot 6 \pm 2$ |
| Incorporation | $78 \pm 5$ | $70 \pm 4$ |

Unpublished data of Robert *et al.*
[a,b] Mean $\pm$ SEM for 27 and 9 subjects, respectively.
[c] Based on 6 subjects, studied with 1-$^{13}$C-leucine.

protein synthesis was found to be stimulated (Sim *et al.*, 1979). In addition, N balance studies in patients receiving parenteral solutions and in subjects undergoing weight reduction emphasize the importance of both amino acids and energy as determinants of N retention and body N balance (e.g. Elwyn *et al.*, 1978; Howard *et al.*, 1978). These diverse observations emphasize the importance and interactions of the amino acid (protein) and energy substrate supply in considering the responses of whole body protein turnover to the cyclical intakes of nutrients.

The *quantitative* significance of the level and source of the energy and protein (amino acid) components of meals on fluctuations in whole body amino acid and protein metabolism remains to be determined. For example, we have reported that meals providing adequate N but free of lysine still promote a reduction in whole body protein breakdown (Fig. 6). (Conway *et al.*, 1980). Therefore, this raises the question whether specific amino acids may be more important in the responses of protein breakdown to meals than total nitrogen intake *per se*. It also seems possible from the limited data available that the important dietary factor responsible for

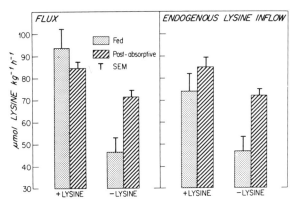

FIG. 6. Comparison of rates of lysine release from body proteins during the fasted (post-absorptive) and fed states in young men adapted to a diet containing adequate lysine or one devoid of this amino acid. Drawn from Conway *et al.* (1980).

bringing about changes in protein breakdown is the energy-yielding substrate of the meal. On the other hand, both protein and energy are probably significant in relation to the meal-dependent stimulation of whole body protein synthesis. Clearly, insufficient information is currently available and a more extensive investigation of this problem must be made before a definitive statement is possible about the responses of whole body amino acid and protein dynamics to acute changes in diet and eating patterns of human subjects.

## Energy Intake and Dynamic Aspects of Whole Body Amino Acid and N Metabolism

The foregoing has concentrated largely on the response of whole body protein turnover to dietary protein in healthy, well-nourished subjects. A further insight into the physiology of human protein metabolism can be achieved by considering a number of studies, particularly with obese subjects, in which energy intake and body energy balance have been the major focus of investigation. Although fasting in normal and obese subjects brings about a qualitatively similar decrease in the rate of urinary N output (e.g. Owen *et al.*, 1969) to that seen in healthy young adults given a protein-free but otherwise adequate diet (e.g. Young and Scrimshaw, 1968; Rand *et al.*, 1976), the mechanisms responsible for these adaptive changes may not necessarily be the same. For example, Table 5 shows that the changes in plasma amino patterns in these different nutritional conditions are quite distinct. Fasting is associated with a decrease in plasma alanine and an

TABLE 5

COMPARISON OF CHANGES IN SELECTED PLASMA AMINO ACID LEVELS WITH A PROTEIN-FREE DIET AND DURING A FAST IN NORMAL AND OBESE SUBJECTS

| Amino acid | Normal subjects | | Obese subjects |
| | Protein-free diet (1 week) | Fast (60 h) | Prolonged fast (5–6 wk)[c] |
| --- | --- | --- | --- |
| Threonine | 84 | 73 | 155 |
| Valine | 66 | 187 | 63 |
| Isoleucine | 84 | 233 | 105 |
| Leucine | 81 | 212 | 72 |
| Serine | 98 | 83 | 85 |
| Glutamate | 117 | 47 | — |
| Glycine | 127 | 80 | 137 |
| Alanine | 154 | 60 | 38 |

[a] Fujita *et al.* (1979), venous blood.
[b] Pozefsky *et al.* (1976), arterial blood.
[c] Felig *et al.* (1969), arterial blood.
Values expressed as percentage of initial concentration,

increase in leucine, at least early on in a fast, whereas the opposite changes occur with ingestion of a protein-free diet. Of course, these observations alone do not indicate differences in the response of dynamic aspects of whole body amino acid metabolism to fasting on the one hand and protein restriction on the other. However, in studies involving intravenous infusion of L-leucine at a rate of $75 \, \mu mol/min/m^2$ body surface for 3–4 hours, Sherwin (1978) estimated that the plasma delivery rate of leucine was about $33 \, \mu mol/min/m^2$ or 95 mmol/day during the post-absorptive state in obese and non-obese subjects, that this value was not affected by three days of fasting but was reduced during starvation for four weeks. This suggests reduced leucine turnover in obese subjects undergoing prolonged starvation, but not during the early period of adaptation to starvation. This evidence reveals some similarity with the responses of whole body leucine metabolism to altered protein intake in adults discussed above, but a more detailed survey of this problem reveals some interesting features.

First, we have explored the responses of whole body protein synthesis and breakdown rates with the aid of $^{15}$N-glycine in five moderately obese women before and at the end of a three week protein-sparing modified fast (PSMF) and then after a consecutive one week total fast (Winterer *et al.*, 1980). As summarized in Table 6, total body protein synthesis rates fell to 73 % ($p < 0.01$) of baseline values during the week-long total fast, but the rate of total body protein breakdown was unaffected by either one week of fasting or three weeks of PSMF. Neither parameter of whole body protein turnover was affected by the PSMF when comparisons were made with measurements taken during the baseline or control diet.

The significant decrease in total body protein synthesis during a one week fast, together with the unresponsiveness of whole body protein breakdown, agree with the observations by Garlick *et al.* (1980*b*), who found a reduction in protein synthesis when obese subjects received a low-energy (500 kcal) protein-free diet, but not when a similar diet contained protein (50 g). Similarly, using the $^{15}$N model of Picou and Taylor-Roberts (1969),

TABLE 6

EFFECTS OF REDUCED ENERGY INTAKE, WITH 'ADEQUATE' PROTEIN, AND A
SUCCEEDING BRIEF TOTAL FAST ON WHOLE BODY PROTEIN METABOLISM IN
OBESE SUBJECTS

|  | | *Diet* | |
| --- | --- | --- | --- |
|  | *Adequate* | *PSMF*[a] | *Fast*[b] |
| N balance (g/day) | $-0.4 \pm 1.2$ | $-0.4 \pm 1.5$ | $-5.8 \pm 0.6$ |
| *Whole body protein* | | | |
| Synthesis (g/day) | $154 \pm 22$ | $145 \pm 25$ | $112 \pm 30$ |
| Breakdown (g/day) | $150 \pm 22$ | $143 \pm 30$ | $143 \pm 31$ |
| *Muscle protein*[c] | | | |
| Breakdown (g/day) | $26 \pm 11$ | $19 \pm 12$ | $24 \pm 11$ |
| % of whole body | $16 \pm 5$ | $12 \pm 5$ | $16 \pm 4$ |

From Winterer *et al.* (1980). Means $\pm$ SD.
[a] PSMF lasted for three weeks. Protein turnover was measured during the final week.
[b] Fast lasted seven days, following earlier PSMF.
[c] Based on measurement of urinary $N^\tau$-methylhistidine excretion.

Pencharz et al. (1980) have reported that there is no significant effect on protein synthesis and breakdown rates with restriction in total energy intake in growing obese adolescents provided that protein intake was maintained at 1·5 g/kg/day ideal body weight for height.

Hence, all three studies provide a consistent picture and indicate that, in both modified and total fasting, endogenous lipid-derived fuels can potentially meet the energy needs for maintenance of protein synthesis. Only with an adequate intake of protein, however, can total body protein synthesis be maintained under these conditions of energy restriction. This provides a metabolic explanation for the protein-sparing effect of a hypocaloric protein diet during weight reduction (Marliss et al., 1978).

These investigations indicate that, providing there is an adequate endogenous energy source, rates of whole body protein synthesis can be maintained when the diet supplies essentially just protein at adequate intakes. The situation is different for healthy adults of normal weight, since we found that a reduction of the non-protein energy intake, provided intravenously but with the protein intake maintained, caused a significant fall in whole body protein synthesis and breakdown rates (Sim et al., 1979). Furthermore, the importance of an adequate dietary energy intake for protein synthesis has also been demonstrated by Golden et al. (1977a), who found a significant correlation between protein synthesis and ad libitum energy intake in growing children who were recovering from malnutrition. The significance of the level of exogenous energy intakes, where diets supply adequate protein, on whole body protein turnover clearly depends upon the physiological state of the host and whether there is an excess of stored energy in the adipose tissue depots.

Reduced energy intake in normal subjects profoundly affects N metabolism, as we have discussed, but there is also some evidence that intakes of energy in excess of needs cause changes in protein turnover. In a recent study we examined the possible basis for the increased N balance that occurs when energy intake is consumed in excess of needs (Motil et al., 1981b).

We chose a dietary protein intake level of 0·6 g/kg/day to explore responses of whole body leucine and lysine metabolism to excess intakes of dietary energy from the different energy sources. Maintenance energy intakes averaged 44 kcal/kg/day, which is comparable to an adequate energy intake for adult men (FAO/WHO, 1973). The excess intake, amounting to a 25 % increment above maintenance, was arbitrarily chosen as an increase in total daily food intake that would be acceptable to the subject.

Nitrogen balance improved with excess dietary energy intake. However, these significant changes in N retention were not associated with any major changes in whole body leucine or lysine fluxes, in either the fed or post-absorptive state (Table 7). The effect of excess energy intake on the contribution to amino acid flux due to inflow of leucine and lysine into the metabolic pool via body protein breakdown is also shown in Table 7. Again, we did not detect a significant effect of excess dietary energy intake, regardless of the energy source, on the inflow from endogenous sources of either leucine or lysine into the metabolic pool.

However, for both amino acids, the rate of entry into the metabolic pool via endogenous protein breakdown was significantly higher in the post-absorptive than in the fed state. Similarly, excess dietary energy resulted in a slight, but not statistically significant increase in leucine incorporation into protein in the fed state. When these small changes in response to excess energy are taken together by estimating *net leucine retention* (leucine outflow minus leucine inflow) this was increased, in the fed state, from $22 \pm 3$ to $27 \pm 3 \, \mu mol/kg/h$ when dietary energy intake was raised in excess of maintenance needs. Furthermore, the diets containing a relatively higher carbohydrate content were associated with a significantly greater change in net leucine retention, again as seen in the fed state.

By these criteria, increased dietary energy levels resulted in increased net retention of leucine. The relatively small changes in leucine oxidation and incorporation into body protein may, in part, be related to the marginally adequate level of protein intake used in this study because it is known that low levels of dietary protein intake limit the magnitude of the changes in amino acid metabolism and N retention in response to an excess energy intake. The issue can be analysed further by comparing the observed changes in parameters of leucine metabolism with those predicted from N balance data. As summarized in Table 8, daily N balance was improved by an average of about 22 mg N/kg/day with excess energy intake. Assuming body protein contains 3·7 mmol/g N (Waterlow *et al.*, 1978*b*) this change in N balance is equivalent to a leucine retention of 81 $\mu mol/kg/day$. In comparison with this, and assuming the daily retention of leucine is the average of the difference between leucine inflow into the metabolic pool and leucine outflow via protein synthesis as estimated for the post-absorptive and fed states, the mean difference in net leucine retention between the maintenance and excess energy intake diet approximates 71 $\mu mol/kg/day$. This general agreement between N balance data and those based on leucine kinetics is encouraging because the two estimates of leucine retention are based on entirely different measurements. More importantly, these

### TABLE 7
WHOLE BODY LEUCINE AND LYSINE DYNAMICS IN YOUNG MEN RECEIVING EXCESS ENERGY INTAKE FROM VARIOUS ENERGY SOURCES

| Level of energy intake | Metabolic state | No. of subjects | Leucine intake | Leucine flux | Leucine inflow | Leucine oxidation | Lysine intake | Lysine flux | Lysine inflow |
|---|---|---|---|---|---|---|---|---|---|
| Maintenance | Post-absorptive | 12 | 2·3 | 94·6 ±2·7 | 92·3 ±2·7 | 16·1 ± 1·9 | 1·9 | 83·1 ±3·4 | 81·2 ±3·6 |
| | Fed | 12 | 35·5 | 101·6 ±2·4 | 65·9 ±2·2 | 18·0 ± 2·4 | 23·9 | 90·5 ±2·0 | 66·6 ±2·0 |
| Excess | Post-absorptive | 12 | 2·3 | 94·3 ±3·6 | 91·9 ±3·5 | 15·7 ± 1·8 | 1·9 | 80·7 ±3·3 | 78·9 ±3·3 |
| | Fed | 12 | 35·5 | 100·5 ±3·1 | 64·9 ±2·9 | 12·4 ± 2·5 | 23·9 | 89·1 ±3·1 | 65·2 ±3·0 |

From Motil et al. (1981b). Values for flux, inflow and oxidation (expressed as $\mu$mol/kg/h) are means $\pm$ SEM.

TABLE 8

NITROGEN BALANCE OF YOUNG ADULT MEN RECEIVING A MAINTENANCE
AND EXCESS LEVEL OF DIETARY ENERGY FROM DIFFERENT SOURCES

| Source of excess energy | N balance for energy intake | | p |
|---|---|---|---|
| | Maintenance | Excess | |
| | (mg/kg/day) | | |
| Mixed | $-18 \pm 5$ | $+9 \pm 4$ | $<0.001$ |
| Carbohydrate | $-24 \pm 5$ | $+4 \pm 4$ | $<0.001$ |
| Fat | $-17 \pm 4$ | $-8 \pm 2$ | $<0.025$ |
| Combined data | $-20 \pm 3$ | $+2 \pm 2$ | $<0.001$ |

From Motil *et al.* (1981*b*).
Each value is mean $\pm$ SEM for eight subjects.

comparisons emphasize that a significant change in body N balance can be brought about by only a relatively small change in the disposition of leucine entering and leaving the metabolic pool.

In an earlier review of N balance data by Munro (1964) it was concluded that surfeit energy intakes enhance N balance primarily by affecting the utilization of amino acids between meals. However, our findings indicate that the time of meal ingestion and absorption is important in determining the effect of excess energy intake upon amino acid and N retention. Our observations do not rule out the possible importance of other periods of the 24-hour day with respect to the response of amino acid metabolism to energy intake because this problem has only received limited investigation in man so far. Clearly, the more extensive data available from animal models must be examined for their relevance to human nutrition and metabolism through further studies in man.

## Summary and Conclusions

Numerous observations confirm that body N balance is determined by both the level and source of energy and protein intake in human subjects in both healthy and various disease states. Only recently efforts have been made to explore, in depth, the possible changes in the dynamic status of whole body amino acid, N and protein metabolism to alterations in dietary protein and energy intakes. Although data are still limited, it is evident that in growing children and in mature adults rates of amino acid oxidation, whole body protein synthesis and breakdown change in response to acute and more prolonged alterations in energy and protein intakes. The quantitative contributions made by each of these dietary factors under differing nutritional states should be explored more extensively and this can be achieved using safe, non-invasive approaches that are based, in part, on stable isotope tracers (e.g. Waterlow *et al.*, 1978*b*; Young and Bier, 1980) However, the data also suggest that improvements are desirable in the models and analytical techniques used to achieve this goal. Finally, recent studies indicate that the responses of whole body amino acid dynamics to dietary change may be best described within the context of the protein and amino acid needs of the host. This suggests to us that a further clarification of the responses of protein turnover to dietary protein and energy will lead to more reliable and appropriate methods for assessing the quantitative relationships between diet, eating habits and the health of people.

## Acknowledgements

We thank our colleagues, especially Drs J. Burke, N. Scrimshaw, H. Munro, and Ms E. Murray, and C. Bilmazes for their contributions to our studies. The unpublished work of the authors was obtained with the financial support of NIH grants AM15856, AG01215, and RR88.

## Discussion

*Waterlow (London School of Hygiene and Tropical Medicine, UK):* How do you explain the difference between the leucine and lysine fluxes?

*Young:* As I understand it from the earlier discussion by Garlick (p. 308), it probably results from differences in the extent to which the plasma enrichment of the two amino acids reflects the enrichment of the specific amino acid in the free amino acid pools in body tissues. These problems may also depend upon the tissue in question as well as the amino acid under study. Garlick, would you agree?

*Garlick (London School of Hygiene and Tropical Medicine, UK):* As far as we know. It just happens that, by some quirk of amino acid transport, lysine transport relative to the production of unlabelled lysine within the cell is lower than leucine transport relative to its production within the cell. Why this should be we don't know, but each amino acid seems to have its own characteristic. It may also depend on the species, since the rat appears to be the opposite of the pig. In the rat the rate of transport (measured directly) relative to the rate of production of unlabelled lysine is higher than the equivalent value for leucine (Waterlow *et al.*, 1978*a*). This is consistent with a higher value for the plateau of intracellular specific activity relative to that of plasma for lysine (Waterlow and Stephen, 1968) than for leucine (Albertse, 1980), and a slightly higher rate of protein flux calculated from lysine (Waterlow and Stephen, 1967) than for leucine (Albertse, 1980).

*Waterlow:* Are you saying then that this is an artefact due to differences in the relation between intracellular and plasma specific activities?

*Young:* That is my interpretation. What is interesting is that these relationships appear to depend very much on the nutritional state of the host.

*Waterlow:* On the other hand, the difference between leucine and lysine fluxes, when each is expressed as amount of amino acid per kg per hour, could simply reflect the different proportions of these amino acids in body protein. For example, in your Fig. 4, in the fed state the leucine flux was about 160, and the lysine flux about 120 $\mu$mol/kg/h. These would give the same value for protein flux if the leucine content of whole body protein was 6·7%, as suggested by Reeds (p. 394) and the lysine content 5·5%. This brings us back to the problem which we discussed earlier.

I don't think the absolute values of flux with a particular amino acid are very important. The important thing is the comparison of values with a given amino acid under different conditions. It will, however, be a serious problem if the factors which influence the estimate of flux, such as the ratio of intracellular to plasma specific activity, alter significantly as conditions change.

*Garlick:* The very impressive data that Young presented showed that as dietary protein intake was reduced, the difference in flux between the fed and fasted states was reduced as well; Dr Clugston and I have gone one stage further and given a protein-free diet for 3 weeks (just 500 kcals of glucose daily). We found no difference between the fed and fasted state in any of the parameters we measured: protein synthesis, protein breakdown or leucine oxidation. This result was rather surprising because we would have expected glucose to make a difference to protein metabolism due to the requirement of gluconeogenesis for amino acids. In addition, the subjects, who had become protein-depleted, were then given a protein-containing meal, and instead of increasing protein synthesis, which is the usual response to a protein meal, there was no increase in synthesis at all, but a big decrease in protein breakdown. The response to a meal therefore seems to depend on the nutritional state over a period of time as well as on the actual dietary intake at that moment.

*Waterlow:* Young concluded in his paper that the level of protein intake has a profound effect on the rate of total protein turnover, and that this is contrary to the conclusion we had reached (Waterlow *et al.*, 1978a). In fact what we said at that time was that the short-term ingestion of protein does affect the rate of protein synthesis, but that the prevailing level of protein intake seems to have very little effect. On the first point—the short term effect of food—we are all in agreement, but on the second I am still not convinced by the data that Young has presented. In 1978 we were relying on results obtained with $^{15}$N-glycine in children by Golden *et al.* (1977) and in adults by Steffee *et al.* (1976). It seems to me that the results presented with leucine (Table 2) do not show a convincing effect of protein intake on synthesis rate, provided (a) that the measurements are made in the fed state, to avoid the interfering short-term effect of fasting; and (b) the subjects are not depleted, as people who have been receiving only 0·1 g protein/kg/day must surely be.

*Young:* I think the response depends on the nutritional state of the individual at that moment in time, on the level of protein and energy intake, and I think probably also on the source of both these components. As I showed in my paper (Table 4, p. 430), when an intravenous infusion of glucose was given at a rate equivalent to about twice the basal production rate of approximately 4 mg/kg/min, we found little change in either leucine oxidation or rates of incorporation into or release from protein. Without going into further details, our studies on the effects of excess energy intake of N balance (as illustrated in Tables 7 and 8 pp. 436, 437) clearly indicated that, because only a small change in the fate of leucine entering the pool is required to bring about a marked change in N balance, we were really stretching to the limit the precision of the isotope methodology that we applied. Hence the reason for my conclusion that we should work towards developing new and more precise models and methods.

*Reeds (Rowett Research Institute, Aberdeen, UK):* I agree about the effect of energy intake. As I indicated (p. 400), we have also been looking at responses to energy in growing animals, and again the response of protein synthesis is very small indeed—almost within the limits of precision of the method. Changes in N retention in response to changes in energy intake involve reductions in the rate of protein breakdown as well (Reeds *et al.*, 1980).

*Millward (London School of Hygiene and Tropical Medicine, UK):* In the lysine–leucine studies where your subjects were on a particular regime for seven days or so, did you measure the methylhistidine excretion?

*Young:* Not in this series of studies. In a previous series we did not observe a detectable change in the output of methyhistidine over a two week period of low-protein feeding in healthy young men.

*Millward:* Do you have any information on the endocrine status of the subjects?

*Young:* Not yet, but we are now beginning to look at the endocrine aspects in relation to the impact of glucose and amino acid supply on whole body amino acid kinetics.

*Harper (University of Wisconsin-Madison, USA):* Is there any problem with the techniques you are using, that a proportion of the amino acids from the gut, on reaching the liver, may be oxidized before they mix with the pool in the body?

*Young:* In this particular study the fed subjects were given, at hourly intervals, meals which were isonitrogenous and isocaloric and equivalent to 1/12th of the total daily intake. The meals began one hour before the administration of the priming doses of isotope, just prior to beginning the period of constant infusion of isotope. The extent to which there is incomplete mixing of amino acids entering the portal circulation with the intravenously administered isotope is uncertain and potentially may be a problem. However, during the course of the four to five hour infusion in the fed state, the enrichment of plasma leucine is constant and different from that observed during a four hour period of a fast.

*Harper:* Are there good measurements in studies lasting more than 24 hours of the rate of incorporation into gastrointestinal proteins and subsequently the rate at which they are degraded, contributing label that goes to the liver and then to muscle; this then recycles so that muscle continues to build up activity over a period of several days?

*Young:* There are some data (Fauconneau and Michel, 1970) obtained in rats, showing a return of $^{14}$C-labelled lysine, initially incorporated into gastrointestinal protein and later liberated through digestion, with the appearance of the label in skeletal muscle.

*Bier:* I think we should not over-simplify the problem just mentioned by Harper about absorption of the meal. The problem of calculating protein breakdown from amino acid flux is likely to be substantial, depending upon how much of the ingested amino acid is taken up by the liver the first time it passes through and how much flows through the portal–hepatic circuit. The fraction of an amino acid presented to the liver via the portal vein which eventually reaches the systemic pool where tracer dilution is sampled is different for each amino acid. I don't think we know all the answers to this yet, but calculations of endogenous amino acid inflow from protein breakdown which are done by simple subtraction of exogenous rate of amino acid ingestion from total flux are almost certainly an over-simplification.

*Garlick:* Does Harper mean that the label would be taken up into gut protein and then recycled later? If this occurred to any significant extent, there would be a rise in the plateau with infusions lasting more than 24 hours, and this does not appear to occur.

*Reeds:* Like Harper and Bier, when we first started this type of work, we were very concerned about the problem of whether amino acids absorbed from the diet are oxidized or otherwise removed by the liver before they mix with an infusion into the aorta. We set up dual infusions of $^{14}$C-leucine into the aorta of pigs and $^3$H-leucine into the hepatic portal vein (Reeds *et al.*, 1978). We also did the same thing with tyrosine (unpublished results). With leucine we got identical answers for flux with the two routes of infusion, and there was only a 10 % difference in the estimates with tyrosine. Apparently, therefore, the liver did not remove a significant amount of amino acid the first time round. This might, of course, be because we tend to feed much higher protein intakes to our animals than humans ever receive.

*Garrow (Clinical Research Centre, Harrow, UK):* Halliday has done some studies (see Paper 25, p. 295) with $^{15}$N-labelled yeast protein with 48 hour intragastric infusions in man; the level of labelling continued to go up in plasma α-amino-N and in urinary $NH_3$ and never reached a plateau. At first we thought this was because we had got the primer dose wrong, but even in people who are in a steady state the same thing happened. One explanation is that if you give whole labelled yeast protein into the gut it labels a lot of gut protein rather highly and this is effectively a second source of highly labelled material which prevents a plateau being reached.

# References

ALBERTSE, E. C. (1980). Protein metabolism in diabetes. Ph.D. thesis, University of London.

BROOKES, I. M., OWEN, F. N. and GARRIGUS, U. S. (1972). Influence of amino acid level in diet upon amino acid oxidation by the rat, *J. Nutr.*, **102**, 27–36.

BUCKLEY, W. T. and MARQUARDT, R. R. (1980). Diurnal variation in oxidation of L-[1-$^{14}$C]leucine and L-[1-$^{14}$C]lysine in rats, *J. Nutr.*, **110**, 974–81.

CALLOWAY, D. H. (1975). Nitrogen balance of men with marginal intakes of protein and energy, *J. Nutr.*, **105**, 914.

CALLOWAY, D. H. and MARGEN, S. (1971). Variation in endogenous nitrogen excretion and dietary nitrogen utilization as determinants of human protein requirement, *J. Nutr.*, **101**, 205–16.

CALLOWAY, D. H. and SPECTOR, H. (1954). Nitrogen balance as related to caloric and protein intake in active young men, *Am. J. Clin. Nutr.*, **2**, 405–12.

CONWAY, J. C., BIER, D. M., MOTIL, K. J., BURKE, J. F. and YOUNG, V. R. (1980). Whole body lysine flux in young adult men: Effects of reduced total protein and of lysine intake, *Am. J. Physiol.*, **239**, E192–200.

CRIM, M. C. and MUNRO, H. N. (1979). Protein-energy malnutrition and endocrine function, in *Endocrinology* (Eds L. J. DeGroot, C. F. Cahill, Jr., L. Martini, D. Nelson, W. D. O'Dell, J. Potts, Jr., E. Steinberger and A. I. Winegrad), Vol. 3, Grune and Stratton, New York, 1987–2000.

DEAN, R. T. (1980). Mechanisms and regulation of protein degradation of animal cells, in: *Biochemistry of Cellular Regulation* (Ed. M. Ashwell), Vol. II. Clinical and scientific aspects of the regulation of metabolism, CRC Press, Inc., Boca Raton, Florida, p. 101–22.

ELWYN, D. H., GUMP, F. R., ILES, M., LONG, C. L. and KINNEY, J. M. (1978). Protein and energy sparing of glucose added in hypocaloric amounts to peripheral infusions of amino acids, *Metabolism*, **27**, 325–31.

ELWYN, D. H., GUMP, F. R., MUNRO, H. N., ILES, M. and KINNEY, J. M. (1979). Changes in nitrogen balance of depleted patients with increasing infusions of glucose, *Am. J. Clin. Nutr.*, **32**, 1597–611.

FAO/WHO, 1973. *Energy and Protein Requirements.* Report of a Joint FAO/WHO Ad Hoc Expert Committee, WHO Tec. Rept. Ser. No. 522, World Health Organization, Geneva, Switzerland.

FAUCONNEAU, G. and MICHEL, M. C. (1970). The role of the gastrointestinal tract in the regulation of protein metabolism, in *Mammalian Protein Metabolism*, Vol. IV (Ed. H. N. Munro), Academic Press, New|York.

FELIG, P., OWNEN, O. E., WAHREN, J., CAHILL, J. F., JR. (1969). Amino acid metabolism during prolonged starvation, *J. Clin. Invest.*, **48**, 584–94.

FERNSTROM, J. D., WURTMAN, R. J., HAMMERSTRAM-WIKLUND, B., RAND, W. M., MUNRO, H. N. and DAVIDSON, C. S. (1979). Diurnal variations in plasma neutral amino acid concentrations among patients with cirrhosis: Effect of dietary protein, *Am. J. Clin. Nutr.*, **32**, 1932–33.

FUJITA, Y., YAMAMOTO, T., RIKIMARU, T. and INOUE, G. (1979). Effect of low protein diets on free amino acids in plasma of young men: Effect of wheat gluten diet, *J. Nutr. Sci. Vitaminol.*, **25**, 427–39.

FULLER, M. F., WEEKES, T. E. C., CODENHEAD, A. and BRUCE, J. B. (1977). The protein-sparing action of carbohydrate. 2. The role of insulin, *Brit. J. Nutr.*, **38**, 489–96.

GARLICK, P. J., CLUGSTON, G. A., SWICK, R. W. and WATERLOW, J. C. (1980a). Diurnal pattern of protein and energy metabolism in man, *Am. J. Clin. Nutr.*, **33**, 1983–6.

GARLICK, P. J., CLUGSTON, G. A. and WATERLOW, J. C. (1980b). Influence of low-energy diets on whole-body protein turnover in obese subjects, *Am. J. Physiol.*, **238**, E235–44.

GARLICK, P. J., MILLWARD, D. J., JAMES, W. P. T. (1973). Diurnal response of muscle and liver protein synthesis *in vivo* in meal-fed rats, *Biochem. J.*, **136**, 935–45.

GARZA, C., SCRIMSHAW, N. S. and YOUNG, V. R. (1976). Human protein requirements: The effect of variations in energy intake within the maintenance range, *Am. J. Clin. Nutr.*, **29**, 280–7.

GARZA, C., SCRIMSHAW, N. S. and YOUNG, V. R. (1977). Human protein requirements: Evaluation of the 1973 FAO/WHO safe level of protein intake for young men at high energy intakes, *Brit. J. Nutr.*, **37**, 403–20.

GARZA, C., SCRIMSHAW, N. S. and YOUNG, V. R. (1978). Human protein requirements: Interrelationships between energy intake and nitrogen balance in young men consuming the 1973 FAO/WHO safe level of egg protein, with added non-essential amino acids, *J. Nutr.*, **108**, 90–6.

GOLDEN, M. H. N. and WATERLOW, J. C. (1977). Total protein synthesis in elderly people: A comparison of results with ($^{15}$N)glycine and ($^{14}$C)leucine, *Clin. Sci. Mol. Med.*, **53**, 277–88.

GOLDEN, M. H. N., WATERLOW, J. C. and PICOU, D. (1977a). Protein turnover, synthesis and breakdown before and after recovery from protein-energy malnutrition, *Clin. Sci. Mol. Med.*, **53**, 473–7.

GOLDEN, M. H. N., WATERLOW, J. C. and PICOU, D. (1977b). The relationship between dietary intake, weight change, nitrogen balance, and protein turnover in man, *Am. J. Clin. Nutr.*, **30**, 1345–8.

HOWARD, L., DOBS, A., CHODOS, R., CHU, R. and LOLUDICE, T. (1978). A comparison of administering protein alone and protein plus glucose on nitrogen balance, *Am. J. Clin. Nutr.*, **31**, 226–9.

HUSSEIN, M. A., YOUNG, V. R., MURRAY, E. and SCRIMSHAW, N. S. (1971). Daily fluctuation of plasma amino acid levels in adult men: Effect of dietary tryptophan intake and distribution of meals, *J. Nutr.*, **101**, 61–70.

INOUE, G., FUJITA, Y., KISHI, K., YAMOMOTO, S. and NIIYAMA, Y. (1974). Nutritive values of egg protein and wheat gluten in young men, *Nutr. Rept. Intl.*, **10**, 201–7.

INOUE, G., FUJITA, Y. and NIIYAMA, Y. (1973). Studies on protein requirements of young men fed egg protein and rice protein with excess and maintenance energy intakes, *J. Nutr.*, **103**, 1673–87.

KANG-LEE, T. A. and HARPER, A. E. (1977). Effect of histidine intake and hepatic histidase activity on the metabolism of histidine *in vivo*, *J. Nutr.*, **107**, 1427–43.

KANG-LEE, T. A. and HARPER, A. E. (1978). Threonine metabolism *in vivo*: Effect of threonine intake and prior induction of threonine dehydratase in rats, *J. Nutr.*, **108**, 163.

KASSENAAR, A., DE GRAEFF, J. and KOUWENHOVEN, A. T. (1960). Glycine studies of protein synthesis during re-feeding in anorexia nervosa, *Metabolism*, **9**, 831–7.

KISHI, Y., MIYATANI, S. and INOUE, G. (1978). Requirement and utilization of egg protein by Japanese young men with marginal intakes of energy, *J. Nutr.*, **108**, 658–69.

MARLISS, E. B., MURRAY, F. T. and NAKHOODA, A. F. (1978). The metabolic response of hypocaloric protein diets in obese man, *J. Clin. Invest.*, **62**, 468–79.

MATTHEWS, D. E., MOTIL, K. J., ROHRBAUGH, D. K., BURKE, J. F., YOUNG, V. R. and BIER, D. M. (1980). Measurement of leucine metabolism in man from a primed continuous infusion of L-[1-$^{13}$C]leucine, *Am. J. Physiol.*, **238**, E473–9.

MOTIL, K. J., MATTHEWS, D. E., BIER, D. M., BURKE, J. F., MUNRO, H. N. and YOUNG, V. R. (1981a). Whole body leucine and lysine metabolism: response to dietary protein intake in young men, *Am. J. Physiol.*, in press.

MOTIL, K. J., BIER, D. M., MATTHEWS, D. E., BURKE, J. F. and YOUNG, V. R. (1981b). Whole body leucine and lysine metabolism studied with [1-$^{13}$C]-leucine and [α-$^{15}$N]lysine: Response in healthy young men given excess energy intake, *Metabolism*, in press.

MUNRO, H. N. (1951). Carbohydrate and fat as factors in protein utilization and metabolism, *Physiol. Rev.*, **31**, 449.

MUNRO, H. N. (1964). General aspects of the regulation of protein metabolism by diet and by hormones, in *Mammalian Protein Metabolism*, Vol. 1 (Eds H. N. Munro and J. B. Allison), Academic Press, New York, pp. 381–481.

OWEN, O. E., FELIG, P., MORGAN, A. P., WAHREN, J. and CAHILL, G. J., JR. (1969). Liver and kidney metabolism during prolonged starvation, *J. Clin. Invest.*, **48**, 574.

PENCHARZ, P. B., MOTIL, K. J., PARSONS, H. G. and DUFFY, B. J. (1980). The effect of an energy-restricted diet on the protein metabolism of obese adolescents: Nitrogen-balance and whole-body nitrogen turnover, *Clin. Sci.*, **59**, 13–18.

PICOU, D. and TAYLOR-ROBERTS, T. (1969). The measurement of total protein synthesis and catabolism and nitrogen turnover in infants in different nutritional states and receiving different amounts of dietary protein, *Clin. Sci.*, **36**, 283–96.

PION, R. (1973). The relationships between the levels of free amino acids in blood and muscle and the nutritive value of proteins, in *Proteins in Human Nutrition* (Eds J. W. G. Porter and B. A. Rolls), Academic Press, New York, pp. 329–42.

POCKNEE, R. C. and HEATON, F. W. (1978). Changes in organ growth with feeding patterns. The influence of feeding frequency on circadian rhythm of protein synthesis in the rat, *J. Nutr.*, **108**, 1266–73.

POZEFSKY, T., TANCREDI, R. G., MOXLEY, R. T., DUPRE, J. and TOBIN, J. D. (1976). Effect of brief starvation on muscle amino acid metabolism in non-obese man, *J. Clin. Invest.*, **57**, 444–9.

RAND, W. M., SCRIMSHAW, N. S. and YOUNG, V. R. (1976). Change of urinary N excretion in response to low-protein diets in adult humans, *Am. J. Clin. Nutr.*, **29**, 639–44.

REEDS, P. J., FULLER, M. F., CADENHEAD, A. and LOBLEY, G. E. (1980). The effects of dietary energy and protein on protein turnover and nitrogen balance in growing pigs, in *Protein Metabolism and Nutrition* (Eds H. T. Olsage and

446   *V. R. Young, J. J. Robert, Kathleen J. Motil, D. E. Matthews and D. M. Bier*

K. Rohr), Information Centre of Bundesforschungsanstalt für Landwirtschaft, Braunswieg, FDR, pp. 67–72.

REEDS, P. J., FULLER, M. F., LOBLEY, G. E., CADENHEAD, A. and MCDONALD, J. D. (1978). Protein synthesis and amino acid oxidation in growing pigs, *Proc. Nutr. Soc.*, **37**, 106A.

RICHARDSON, D. P., WAYLER, A. H., SCRIMSHAW, N. S. and YOUNG, V. R. (1979). Quantitative effect of an isoenergetic exchange of fat for carbohydrate on dietary protein utilization in healthy young men, *Am. J. Clin. Nutr.*, **32**, 2217–26.

SHERWIN, R. S. (1978). Effects of starvation on the turnover and metabolic response to leucine, *J. Clin. Invest.*, **61**, 1471–81.

SIM, A. J., WOLFE, B. M., YOUNG, V. R., CLARKE, D. and MOORE, F. D. (1979). Glucose promotes whole body protein synthesis from infused amino acids in fasting man, *Lancet*, **i**, 68–72.

SIMON, R., MUNCHMEYER, R., BERGNER, H., ZEBROWSKA, T. and BURACZESKA, L. (1978). Estimation of rate of protein synthesis by constant infusion of labelled amino acids in pigs *Brit. J. Nutr.*, **40**, 243–52.

SPRINSON, D. B. and RITTENBERG, D. (1949). The rate of utilization of ammonia for protein synthesis, *J. Biol. Chem.*, **180**, 707–14.

STEFFEE, W. P., GOLDSMITH, R. S., PENCHARZ, P. B., SCRIMSHAW, N. S. and YOUNG, V. R. (1976). Dietary protein intake and dynamic aspects of whole body nitrogen metabolism in adult humans, *Metabolism*, **25**, 281–97.

TSCHUDY, D. P., BACCHUS, H., WEISSMAN, S., WATKINS, D. M., EWBANKS, M. and WHITE, J. (1959). Studies on the kinetics of nitrogen balance using [15]N aspartic acid, *J. Clin. Invest.*, **38**, 892–901.

WATANABE, M., POTTER, V. R. and PITOT, H. C. (1968). Systematic ascillations in tyrosine transaminase and other metabolic functions in liver of normal and adrenalectomized rats on controlled feeding schedules, *J. Nutr.*, **95**, 207–27.

WATERLOW, J. C. (1968). Observations on the mechanisms of adaptation to low protein intakes, *Lancet*, **ii**, 1091–7.

WATERLOW, J. C., GARLICK, P. J. and MILLWARD, D. J. (1978a). *Protein Turnover in Mammalian Tissues and in the Whole Body*, Elsevier North Holland, Amsterdam, p. 143.

WATERLOW, J. C., GARLICK, P. J. and MILLWARD, D. J. (1978b). *Protein Turnover in Mammalian Tissues and in the Whole Body*, Elsevier North Holland, Amsterdam, p. 804.

WATERLOW, J. C. and STEPHEN, J. M. L. (1967). Measurement of total lysine turnover in the rat by intravenous infusion of L-[U-[14]C-]lysine, *Clin. Sci.*, **33**, 489–506.

WATERLOW, J. C. and STEPHEN, J. M. L. (1968). The effect of low protein diets on the turnover rates of serum, liver and muscle proteins in the rat, measured by continuous infusion of L-[[14]C]-lysine, *Clin. Sci.*, **35**, 287–305.

WINTERER, J., BISTRIAN, B. R., BILMAZES, C., BLACKBURN, G. L. and YOUNG, V. R. (1980). Whole body protein turnover, studied with [15]N glycine and muscle protein breakdown during protein-sparing diet and brief total fast, *Metabolism*, **29**, 575–81.

WOLFE, R. R., ALLSOP, J. R. and BURKE, J. F. (1979). Glucose metabolism in response to intravenous glucose infusion, *Metabolism*, **28**, 210–19.

YOUNG, V. R. and BIER, D. M. (1980). Stable isotopes ($^{13}$C and $^{15}$N) in the study of human protein and amino acid metabolism and requirements, in: *Nutritional Factors: Modulating Effects on Metabolic Processes* (Eds R. F. Beers and E. G. Bassett), Raven Press, New York, pp. 267–308.

YOUNG, V. R., HUSSEIN, M. A., MURRAY, E. and SCRIMSHAW, N. S. (1969). Tryptophan intake, spacing of meals, and diurnal fluctuations of plasma tryptophan in men, *Am. J. Clin. Nutr.*, **22**, 1563–7.

YOUNG, V. R. and SCRIMSHAW, N. S. (1968). Endogenous nitrogen metabolism and plasma free amino acids in young adults given a 'protein-free' diet, *Brit. J. Nutr.*, **22**, 9–20.

YOUNG, V. R. and SCRIMSHAW, N. S. (1978). Nutritional evaluation of proteins and protein requirements, in *Protein Resources and Technology* (Eds. M. Milner, N. S. Scrimshaw and D. I. C. Wang), pp. 136–73.

YOUNG, V. R., TAYLOR, Y. S. M., RAND, W. M. and SCRIMSHAW, N. S. (1973). Protein requirements of man: Efficiency of egg protein utilization at maintenance and submaintenance levels in young men, *J. Nutr.*, **103**, 1164–74.

# 36

# Protein Turnover and Metabolic Rate in Obesity

J. S. GARROW

*Clinical Research Centre, Harrow, UK*

There is no reason to believe that protein turnover in obese people is in any way systematically different from protein turnover in lean people. There are two reasons why obese subjects have been studied by various techniques: (i) obesity is a common complaint, and severe cases are usually willing to be admitted to hospital for metabolic investigation under controlled dietary conditions. Thus, regardless of the relevance of protein turnover to obesity, the obesity clinic is a rich hunting ground for volunteers for studies of protein turnover; and (ii) obesity is a disorder of energy balance, so any factor which effects energy expenditure is relevant to obesity. It is at least a plausible hypothesis that the rate of protein synthesis is an important factor in determining resting metabolic rate, and hence energy expenditure.

At this symposium, much attention has been given to the validity of techniques for measuring protein turnover. At the most elementary level it is obvious that if a protein molecule inside a cell is broken down and resynthesized, without using any precursor from outside that cell, that event cannot be detected by any tracer study. Any experimental conditions which increase the probability that this will happen will tend to cause tracer studies to underestimate the rate of protein turnover. Since the process of intracellular recycling (or any recycling which does not involve exchange of tracer with whatever metabolic pool we are monitoring) is unmeasurable, we can only guess what conditions would affect it. For example, a reduction in the protein content of the diet may increase the probability of recycling, as defined above.

To test the hypothesis mentioned above it is possible to consider those conditions in which metabolic rate changes, and to see if there are corresponding changes in protein turnover. The result of this approach is set out in Table 1.

Few publications offer a direct comparison of metabolic rate and rate of

## TABLE 1
CONDITIONS ASSOCIATED WITH ALTERATION IN RESTING METABOLIC RATE (RMR) IN MAN

| Condition | Change in RMR (%) | Change in protein synthesis (%) | Reference | Method for measuring protein synthesis |
|---|---|---|---|---|
| (a) *Increased resting metabolism* | | | | |
| Thyrotoxic | +10 | +39 | Nair et al. (1981) | $^{15}$N-glycine-NH$_4$ |
| Growth | ?+10 | +92 | Golden et al. (1977) | $^{15}$N-glycine-urea |
| Infection | ?+10–20 | +37 | Garlick et al. (1980d) | $^{15}$N-glycine-NH$_4$ |
| Overfeeding | +15 | ? | Sims et al. (1973) | none |
| Cold | +50 | ? | Wyndham et al. (1968) | none |
| Anxiety | +10 | ? | Blaza and Garrow (1980) | none |
| (b) *Decreased resting metabolism* | | | | |
| Underfeeding | −15 | −50 | Sender et al. (1975) | $^{14}$C-tyrosine |
| | −12 | −20 | Nair et al. (1981) | $^{15}$N-glycine-NH$_4$ |
| | ? | −10 | Garlick et al. (1980a) | several |
| Old age | ?−10 | −10–30 | Golden and Waterlow (1977) | $^{14}$C-leucine/ $^{15}$N-glycine |
| Sleep | ? | ? | Garlick et al. (1978) | $^{14}$C-leucine |
| | — | — | Adam and Oswald (1980) | none |

protein turnover in the same subjects under two conditions in which the metabolic rate differs. The results of Nair et al. (1981) indicate that, if a group of obese patients is kept on a diet supplying 800 kcal (3·4 MJ) and 100 g protein/day, both metabolic rate and protein flux decrease. If such patients are given 120 μg/day of tri-iodothyronine both metabolic rate and protein flux increase. The change in both variables, relative to the starting value, indicates that 100 % change in protein flux would be associated with 45 % change in resting metabolism, or that a change of 100 g/day in flux is associated with a change of 200 kcal (850 kJ) in energy expenditure. This does not prove, or even strongly suggest, that the change in protein synthesis is the sole cause of the change in metabolic rate. *In vitro* studies of tissue slices show that most of the thermogenesis caused by thyroid is attributable to the energy cost of sodium transport, but it is still not clear if this is true in the intact animal (Smith and Edelman, 1979).

Other conditions associated with a change in resting metabolic rate are listed in Table 1. During growth there is an increase in metabolic rate (Ashworth, 1969) and protein flux (Golden *et al.*, 1977): during the rapid phase of recovery of malnourished children the increase in flux is 92 % above that of recovered children, but no exact comparison of change in metabolic rate is possible, so the value of 10 % increase has been based on Ashworth's data.

Garlick *et al.* (1980*d*) found an increase of 37 % in protein synthesis following vaccination, and from the fever induced the increase in metabolic rate was probably about 10–20 %. During prolonged overfeeding (Sims *et al.*, 1973), exposure to cold (Wyndham *et al.*, 1968) and anxiety (Blaza and Garrow, 1980), there is an increase in metabolic rate, but so far as I know there have been no measurements of protein metabolism under these conditions.

Underfeeding certainly is associated with a decrease in metabolic rate (Benedict *et al.*, 1919) but the effects on protein turnover are less clear. Sender *et al.* (1975) found a 50 % decrease after three weeks of a protein-free low-energy diet, but Garlick *et al.* (1980*a*) found very little decrease if the protein content of the diet was maintained. This result is contrary to that of Nair *et al.* (1981), who found a decrease in both metabolic rate and protein flux after three weeks on a diet which supplied 800 kcal (3·4 MJ) and 100 g protein.

In old age there is a decrease in metabolic rate (Boothby *et al.*, 1936) and also in protein flux (Golden and Waterlow, 1977), but measurements of both variables have not been done on the same subjects, so it is difficult to give relative percentage decreases. The effect of sleep on metabolic rate and protein synthesis is disputed: from the constant infusion studies of Golden *et al.* (1977) and Garlick *et al.* (1978) it appears that the diurnal changes are related to feeding rather than to sleep. This conclusion has been disputed by Adam and Oswald (1980) who believe that protein synthesis is accelerated at night, when food intake is least; however, it is difficult to accept their evidence for this view (Garlick *et al.*, 1980*c*).

Inspection of Table 1 shows that although there is a trend for metabolic rate and protein flux to change in the same direction, there are few studies in which both changes were measured simultaneously, and no firm case can be made for a quantitative relationship between the two variables. Protein synthesis must incur some energy cost, which will be reflected in resting metabolic rate, but it is possible to say that the energy cost of protein synthesis cannot be as much as half the total resting metabolism, or a stronger association would emerge in Table 1. However, we may be looking in the wrong place if we seek correlations between steady state protein turnover and resting metabolic rate. The most striking change in protein turnover occurs during catch-up growth in malnourished infants (Golden *et al.*, 1977), and here the change in metabolic rate occurs as a greatly enhanced thermogenic response after a meal (Ashworth, 1969). If this burst of increased metabolism is not related to protein synthesis, what is the alternative explanation?

The relevance of all this to the management of obesity in man is two-fold. First, if it is true that the thermogenic response to food is related to increased protein turnover, and if obese people have a reduced thermogenic response to food (Pittet *et al.*, 1976), this implies that obese people may have some abnormality in the control of protein turnover. To prove this association we need better methods for measuring protein flux from hour to hour, so that changes can be related to concurrent changes in metabolic rate. If the association is confirmed it opens the possibility that the susceptibility of obese people to disease (James, 1976) is a reflection of this abnormality in protein metabolism. If the association is not confirmed, an alternative explanation is required for the diminished thermogenic response to food of obese people.

The second reason for wanting to understand the relationship of protein flux to metabolic rate concerns the maintenance of weight loss by obese people. Metabolic rate falls with weight loss, and this fall can be largely explained by a reduction in lean body mass (Doré, pers. comm.). If it is true that a reducing diet with adequate protein does not cause a decrease in protein turnover (Garlick *et al.*, 1980*a*) does this also prevent the decrease in metabolic rate? Here the findings of Garlick *et al.* (1980*a*) conflict with those of Nair *et al.* (1981), who found a decrease in both protein flux and metabolic rate, despite a high concentration of protein in the low-energy diet. The difference in result may reflect a difference in the measurement technique, or in the time at which the measurement was made. It is important that the differences should be resolved, because anything which prevents a fall in metabolic rate with weight loss in obese patients increases the chance that weight loss will be maintained. This is the major outstanding problem in the management of obesity.

## Discussion

*Young (Massachusetts Institute of Technology, USA):* There are apparently two schools of thought with regard to protein turnover in obesity: on the one hand, your results (Nair *et al.*, 1981) show a decrease in synthesis rate in obese patients on a diet which is low in energy but contains adequate protein, whereas there are three studies (Garlick *et al.*, 1980*a*; Pencharz *et al.*, 1980; Winterer *et al.*, 1980) which suggest that protein synthesis in such individuals is relatively well maintained when the diet is low in energy but adequate in protein, and is only decreased when patients are put on a regime low in energy and inadequate in protein. Is there any information on resting metabolic rate in these two weight-reducing conditions?

*Garrow:* As I have shown (p. 450), our subjects, who were fed 100 g milk protein and studied by the single shot glycine-$NH_4$ method, showed a decrease in metabolic rate and, in contrast to the results of Garlick *et al.* (1980*a*), a decrease in protein flux. I should be interested to see the metabolic rates of their subjects, if they were measured.

*Waterlow (London School of Hygiene and Tropical Medicine, UK):* In the 10 subjects who were studied over 24 h by continuous infusion of $^{14}$C-leucine (Garlick *et al.*, 1980*b*), there was no relationship between resting metabolic rate and rates of protein synthesis and breakdown, either in the fed or the fasted state.

*Golden (University of the West Indies, Kingston, Jamaica):* What do you think is the most appropriate reference measurement for comparing protein synthesis rates in obese patients? Would you not expect to find differences in protein turnover between obese and normal individuals when expressed per kg body weight?

*Garrow:* Yes, I think the results should be expressed in terms of lean body mass until we find a better reference basis. I do not know of any evidence that there is a difference in metabolic rate between fat and thin people expressed per kg lean body mass; that may simply mean that there are not enough data to show the difference, if it exists, because there is a very large range of values for resting metabolic rate among both obese and thin people and the distributions probably overlap.

*Waterlow:* We do not have measurements of lean body mass, but we extrapolate from creatinine output or height to get some estimate of what the real body is like inside all that fat.

*Garrow:* Both protein synthesis rate and resting metabolic rate go down with age and I am not sure whether this is simply a function of the fact that lean body mass also decreases with age. We have a fairly large series (140) on women in whom we have measured body composition and metabolic rate. We have done a multiple regression on these results, taking into account lean body mass, age, weight, etc., and it is obvious that there is a significant negative term related to age. i.e. in our experience for a given weight and lean body mass, metabolic rate still drops with age. Therefore it is not simply that as you get older your lean body mass becomes smaller. This would fit in with the fact that per unit of lean body mass protein turnover also decreases.

## References

ADAM, K. and OSWALD, I. (1980). Protein synthesis and breakdown after vaccination. *Brit. Med. J.*, **281**, 809.

ASHWORTH, A. (1969). Metabolic rates during recovery from protein calorie malnutrition: the need for a new concept of specific dynamic action. *Nature*, **223**, 407–9.

BENEDICT, F. G., MILES, W. R., ROTH, P. and SMITH, M. (1919). *Efficiency under prolonged restricted diet*. Publication 446, Carnegie Institution of Washington, Washington, DC.

BLAZA, S. E. and GARROW, J. S. (1980). The effect of anxiety on metabolic rate. *Proc. Nutr. Soc.*, **39**, 13A.

BOOTHBY, W. M., BERKSON, J. and DUNN, H. L. (1936). Studies of the energy metabolism of normal individuals: a standard for basal metabolism with a nomogram for clinical application. *Am. J. Physiol.*, **116**, 468–84.

GARLICK, P. J., CLUGSTON, G. A., SWICK, R. W., MEINERTZHAGEN, I. H. and WATERLOW, J. C. (1978). Diurnal variations in protein metabolism in man. *Proc. Nutr. Soc.*, **37**, 33A.

GARLICK, P. J., CLUGSTON, G. A. and WATERLOW, J. C. (1980*a*). Influence of low-energy diets on whole-body protein turnover in obese subjects. *Am. J. Physiology*, **238**, E235–44.

GARLICK, P. J., CLUGSTON, G. A., SWICK, R. W. and WATERLOW, J. C. (1980*b*). Diurnal pattern of protein energy metabolism in man. *Am. J. Clin. Nutr.*, **33**, 1983–6.

GARLICK, P. MCNURLAN, M., FERN, E., TOMKINS, A. and WATERLOW, J. C. (1980*c*). Protein synthesis and breakdown after vaccination. *Brit. Med. J.*, **281**, 1215–16.

GARLICK, P. J., MCNURLAN, M., FERN, E. B., TOMKINS, A. M. and WATERLOW, J. C. (1980*d*). Stimulation of protein synthesis and breakdown by vaccination. *Brit. Med. J.*, **281**, 263–5.

GOLDEN, M. H. N. and WATERLOW, J. C. (1977). Total protein synthesis in elderly people: a comparison of results with $^{15}$N-glycine and $^{14}$C-leucine. *Clinical Science and Molecular Medicine*, **53**, 277–88.

GOLDEN, M. H. N., WATERLOW, J. C. and PICOU, D. (1977). Protein turnover, synthesis and breakdown before and after recovery from protein-energy malnutrition. *Clinical Science and Molecular Medicine*, **53**, 473–7.

JAMES, W. P. T. (Compiler). (1976). *Research on Obesity*. A report of the DHSS/MRC Group, HMSO, London.

NAIR, K. S., HALLIDAY, D. and GARROW, J. S. (1981). Protein metabolism in obese patients on caloric restriction and on $T_3$ therapy. *Recent Advances in Obesity Research. Proc. 3rd Int. Congr. on Obesity*, Rome, 1980, in press.

PENCHARZ, P. B., MOTIL, K. J., PARSONS, H. G. and DUFFY, B. J. (1980). The effect of an energy-restricted diet on the protein metabolism of obese adolescents: nitrogen-balance and whole-body nitrogen turnover. *Clin. Sci.*, **59**, 13–18.

PITTET, PH., CHAPPUIS, PH., ACHESON, K., DE TECHTERMANN, F. and JÉQUIER, E. (1976). Thermic effect of glucose in obese subjects studied by direct and indirect calorimetry. *Brit. J. Nutr.*, **35**, 281–92.

SENDER, P. M., JAMES, W. P. T. and GARLICK, P. J. (1975). Protein metabolism in obesity. In: *Regulation of Energy Balance in Man*, Jequier, E. (Ed.), Editions Médecine et Hygiène, Genève, pp. 224–7.

SIMS, E. A. H., DANFORTH, E. JR., HORTON, E. S., BRAY, G. A., GLENNON, J. A. and SALANS, L. B. (1973). Endocrine and metabolic effects of experimental obesity in man. *Recent Progress in Hormone Research*, **29**, 457–76.

SMITH, T. J. and EDELMAN, I. S. (1979). The role of sodium transport in thyroid thermogenesis. *Fed. Proc.*, **38**, 2150–3.

WINTERER, J., BISTRIAN, B. R., BILMAZES, C., BLACKBURN, G. L. and YOUNG, Y. R. (1980). Whole body protein turnover, studied with [15]N-glycine and muscle protein breakdown during protein-sparing diet and brief total fast. *Metabolism*, **29**, 575–81.

WYNDHAM, C. H., WILLIAMS, C. G. and LOOTS, H. (1968). Reactions to cold. *J. Appl. Physiol.*, **24**, 282–7.

# 37

# Metabolism of Plasma Proteins

W. P. T. JAMES and W. A. COWARD

*MRC Dunn Clinical Nutrition Centre,*
*Addenbrookes Hospital, Cambridge, UK*

Plasma proteins, their nature, function and metabolism, have received far more attention than any other series of proteins in the body. Investigations which deal solely with the turnover of these proteins far outnumber the experiments on all other body proteins, so it is not surprising that we have a far better understanding of the significance and cause of altered circulating concentrations of individual plasma proteins. Most, if not all, circulating proteins should now be considered individually because each protein is likely, in due course, to be found to have a specific function with separate regulatory factors geared to the protein's particular role, for example, the function of complement in immunology, transferrin in metal transport or high density lipoprotein in cholesterol metabolism. It is evident that no short summary of plasma protein turnover can encompass many common features. Classic reviews have been written or compiled on this topic by Schultze and Heremans (1966), Putman (1975), Allison (1976), Rosenoer *et al.* (1977), and there are also published annual colloquia on protides of the biological fluids and those stemming from the small group of scientists who held biennial meetings on the physiological and pathological significance of plasma protein metabolism, e.g. Rothschild and Waldmann (1970), Ciba Symposium (1973) and Bianchi *et al.* (1976).

Freeman, one of the major contributors to the study of plasma proteins since World War II, classified proteins not in terms of their physical characteristics, specific biological function or their turnover kinetics, but as either biophilic or suicidal proteins (Freeman, 1967). The biophilic proteins survive for some time after performing their function, e.g. albumin, transferrin and low density lipoprotein, whereas the suicidal proteins, e.g. fibrinogen or complement are destroyed in the process of functioning. The physical characteristics of circulating proteins seem to be unrelated to their function except for albumin, which by virtue of its relatively small size has an important role in maintaining plasma oncotic pressure.

The major group of biophilic proteins are the transport proteins, many of which can transport substantial amounts of their specific molecules before they are themselves destroyed; the control of the synthesis and breakdown of these proteins may be affected by an increased requirement with, for example, an increase in transferrin synthesis in iron deficiency states (Morton *et al.*, 1976), and a reduction in albumin synthesis when plasma oncotic pressure rises (Oratz, 1970). This adjustment in the synthesis of a plasma protein may exert an important effect on body function, since an increased concentration of a carrier protein may affect hormone binding and thereby modify the metabolic response to the hormone. With suicidal proteins we might expect, from Freeman's analysis, that the control of turnover would be linked to protein catabolism induced as the protein performs its function, but in practice this is not always the case. Thus, fibrinogen turnover is a controlled process with about 30% of the intravascular pool being catabolized each day in normal man (McFarlane *et al.*, 1964), but with much higher turnover rates when fibrinogen is activated and organized into fibrin polymers in the process of blood clotting. Under these circumstances there seems to be a general rather than a specific feed-back system inducing increased fibrinogen synthesis, this 'stress' reaction leading to the additional synthesis of other proteins, such as haptoglobin, which can then exert a role in scavenging for free haemoglobin. Though there are important differences in the rates of turnover of individual proteins expressed as the proportion of the intravascular (IV) pool catabolized each day (Table 1), these are small when compared with the variations in concentration and absolute catabolic rate. The latter no doubt reflect a diversity of functions and quantitatively different requirements. On the other hand, similar fractional turnover rates for plasma proteins may be related to their co-existence in plasma and interstitial fluid and a common proximity to sites of degradation.

Many attempts have been made to find common factors which might affect the concentration of plasma proteins, but the only two general environmental stimuli which are known to have an appreciable effect on several circulating proteins at once are those involving stress, which leads to an increase in the concentration of the 'acute phase' proteins, and nutritional change which seems only to affect transport protein concentrations; in practice nutritional insufficiency rarely occurs in the absence of stress reactions.

## TABLE 1

PLASMA PROTEINS; THEIR TOTAL MASS, PLASMA CONCENTRATION AND CATABOLISM

| Protein | Mass (g) | Plasma concentration (g/l) | Daily catabolism (g) | (% IV pool) | Reference |
|---|---|---|---|---|---|
| Albumin | 280 | 44 | 9·7 | 9 | Rossing (1967) |
| Transferrin | 18 | 2·3 | 1·7 | 23 | Katz (1961) |
| Fibrinogen | 13 | 2·9 | 3·2 | 31 | McFarlane *et al.* (1964) |
| $\alpha_2$-Macroglobulin | 8·3 | 2·6 | 0·7 | 8 | Norberg *et al.* (1970) |
| $\gamma$-Globulins | 105 | 18 | 6·7 | 11 | Hoffenberg *et al.* (1966) |
| $\gamma$-Macroglobulins | 2 | 0·5 | 0·2 | 11 | Jensen (1969) |
| Low density lipoproteins | 6 | 1·2 | 1·0 | 28 | Calvert and James (1979) |

Results recalculated assuming body weight 70 kg and plasma volume 3 litres.

# Nutritional Effects on Plasma Proteins

The early demonstration that the total plasma protein concentration was reduced in malnutrition led to an exhaustive study of the relationship between malnutrition and the concentration of individual plasma proteins (see Waterlow *et al.*, 1960). It soon became apparent that human malnutrition was associated with recurrent infections and that this would complicate the interpretation of the data. When subjects are maintained first on a normal protein intake of 70 g/day and then provided with only 15 g protein daily, there is only a small and insignificant fall in plasma $\gamma$-globulin concentrations with no change in the catabolic rates of these proteins (Hoffenberg *et al.*, 1966). Tables 2 and 3 show recent data on the effect of reducing the protein intake under controlled conditions in obese and normal individuals. Clearly, albumin concentrations do fall eventually, but despite the large amounts of albumin which have to be synthesized each day, the rate of fall is much slower than that for other proteins with a much smaller mass.

    The simplicity of the concept that plasma protein concentrations are affected by nutrient intake masks what is a much more complex situation. Rate of protein synthesis is likely to be the component of the system dependent on diet but rates of catabolism and utilization, degree of

TABLE 2

CHANGES IN CONCENTRATION OF PLASMA PROTEINS WITH RESTRICTION OF PROTEIN AND
ENERGY INTAKE IN GROUPS OF OBESE PATIENTS (SHETTY, 1979)

| *Plasma proteins* | *Protein intake (g/day)* | *Normal energy* | *Restricted energy* | |
|---|---|---|---|---|
| Days on diet | | 10 | 12 | 24 |
| Plasma albumin | 80 | $42·3 \pm 2·9$ | $43·8 \pm 1·6$ | $37·8 \pm 0·5$ |
| (g/l) | 40 | $40·5 \pm 2·6$ | $40·5 \pm 3·3$ | $43·0 \pm 1·8$ |
| | 20 | $40·8 \pm 2·3$ | $36·5 \pm 1·5$ | $38·7 \pm 2·2$ |
| Transferrin | 80 | $1·9 \pm 0·1$ | $2·3 \pm 0·3$ | $1·9 \pm 0·1$ |
| (g/l) | 40 | $1·6 \pm 0·2$ | $1·8 \pm 0·1$ | $1·6 \pm 0·1$ |
| | 20 | $1·7 \pm 0·1$ | $1·4 \pm 0·1$ | $1·3 \pm 0·1^a$ |
| Thyroid binding | 80 | $29·8 \pm 1·4$ | $20·8 \pm 4·3^a$ | $19·3 \pm 1·9^b$ |
| pre-albumin | 40 | $29·0 \pm 4·0$ | $16·7 \pm 2·0^a$ | $15·0 \pm 1·0^a$ |
| (mg/dl) | 20 | $24·0 \pm 3·2$ | $14·7 \pm 3·4^a$ | $12·7 \pm 2·9^a$ |
| Retinol binding | 80 | $5·7 \pm 0·5$ | $5·4 \pm 0·3$ | $3·6 \pm 0·5^a$ |
| (mg/dl) | 40 | $5·4 \pm 0·5$ | $3·3 \pm 0·9^a$ | $3·4 \pm 0·9^a$ |
| | 20 | $3·7 \pm 0·4$ | $2·9 \pm 1·1$ | $3·3 \pm 0·6^a$ |

$^a$ $p < 0·05$
$^b$ $p < 0·005$
Means $\pm$ SEM; 4 patients/group.

TABLE 3

CHANGES IN ALBUMIN CONCENTRATION, MASS AND CATABOLIC
RATE IN NORMAL ADULTS WITH RESTRICTION OF PROTEIN INTAKE
(HOFFENBERG *et al.*, 1962)

| | *Pre- or post depletion* | *Low protein feeding* |
|---|---|---|
| Protein intake (g/day) | $130·0 \pm 5·8$ | $3·7 \pm 0·4$ |
| Plasma albumin (g/l) | $40·6 \pm 0·7$ | $36·3 \pm 1·3$ |
| Intravascular albumin mass (g) | $104·2 \pm 7·2$ | $93·7 \pm 6·5$ |
| Extravascular albumin mass (g) | $163·1 \pm 8·8$ | $140·3 \pm 17·8$ |
| Absolute catabolic rate (g/day) | $9·1 \pm 2·6$ | $5·2 \pm 0·5$ |

Means $\pm$ SEM for six men studied before and after three to four
weeks on a low protein isoenergetic diet.

hydration and the extent of intravascular–extravascular exchange all have effects on plasma protein concentration. There are, therefore, two possibilities for research likely to lead to a fuller understanding of plasma protein metabolism in malnutrition. Wide-ranging investigations of changes in plasma protein concentrations in defined states of malnutrition may lead to useful, but empirical, knowledge of cause–effect relationships; however, ultimately fundamental investigations on the interactions between plasma protein synthesis, distribution and catabolism will be required.

With these differing rates of response of the plasma proteins to dietary deficiency one might well wonder whether the changes in the plasma concentrations of transferrin, thyroid binding pre-albumin (TBPA), or retinol binding protein (RBP) might serve as indices of the nutritional state of an animal or human. Ingenbleek *et al.* (1975) found that in children with kwashiorkor the circulating concentrations of the last two proteins were only about 30 % of normal. However, these proteins doubled their concentration within a week of treatment at a stage when malnutrition was still severe, so it could be argued that these proteins are more useful as indices of dietary adequacy than of nutritional state as such. Serum transferrin did not fall as low as TBPA and RBP and also occupied an intermediate position between these proteins and serum albumin in the response of treatment. Of the four proteins listed in Table 2, only the concentration of albumin has been investigated thoroughly enough to be shown to be of prognostic significance (Whitehead *et al.*, 1973). The use of plasma transferrin as an index of malnutrition is complicated by the widespread prevalence of co-existing iron deficiency and infections, both of which will stimulate rather than depress transferrin synthesis (Ismadi *et al.*, 1971; Ingenbleek *et al.*, 1975; Delpeuch *et al.*, 1980). Vitamin A deficiency can also affect the release of RBP from the liver so that RBP accumulates in the liver as the plasma concentration falls. However, after treatment with retinyl palmitate plasma RBP increases within 10 h without any associated change in TBPA (Large *et al.*, 1980). Hepatic synthetic capacity also affects the concentration of RBP in liver since marasmic children appear to accumulate RBP to a greater extent than children with kwashiorkor. The type of the malnutrition is also reflected in the plasma holo–RBP concentrations achieved after dosing with retinyl palmitate. Children with kwashiorkor have the lowest, the marasmic children the highest and marasmic–kwashiorkor children intermediate values, a finding conforming with the recognized range in albumin concentrations.

In cross-sectional studies in Nigeria, TBPA was found to be substantially below normal ($16.5 \pm 0.8$ vs $23.8 \pm 0.9$ mg/100 ml) in stunted children with normal arm circumference and triceps skinfold thickness (Ogunshina and Hussain, 1980). This may be taken as further evidence that dietary patterns may be reflected by changes in the concentrations of the rapidly turning over proteins as well as in small effects on growth. Although monitoring these proteins will not be useful if one wishes to identify only those children at severe risk of malnutrition, they may be useful indices in determining whether ill patients are responding to nutritional therapy, e.g. intravenous feeding.

## Albumin Metabolism

The relative resistance to change in plasma albumin concentration in some forms of malnutrition, e.g. marasmus, may in part reflect the re-routing of amino acids from peripheral sites to the liver (Whitehead and Alleyne, 1972). Short-term resistance will depend on the buffering effect of the larger peripheral mass of protein in the well-nourished individual. In malnourished children there is a greater fall in albumin synthesis than in well-nourished children when the two groups are fed a low protein diet (James and Hay, 1968). Nevertheless, any inflow of amino acids from the periphery should also allow the generation of the smaller amounts of TBPA and RBP, despite their more rapid turnover rates than serum albumin. Thus, if we take the half-lives of these proteins as equivalent to 2 days (Socolow et al., 1965) and 12 h (Peterson, 1971) respectively, and take the plasma concentrations listed in Table 2, then their combined synthetic rate amounts to little more than $0.5$ g daily compared with $1.7$ g for transferrin (Katz, 1961) and 10 g for albumin. On this basis one would expect a greater change in albumin metabolism once protein intake was reduced despite the buffering effect of amino acids derived from the periphery. Whether there is preferential routing of amino acids for albumin synthesis is unclear, but this seems unlikely in view of the equivalent effects of protein deficiency on transferrin and albumin synthesis in experimental animals (Morgan and Peters, 1971). Albumin synthesis cannot be maintained for any length of time by re-directing amino acids derived from intracellular hepatic protein breakdown, and in vitro experiments, perfusion studies and measurements of albumin synthesis in vivo have all demonstrated the importance of extra-hepatic amino acids in determining the rate of albumin synthesis (Morgan and Peters, 1971; Rothschild et al., 1977). In man, direct measurements of the synthetic rate of albumin are few in number, e.g. Tavill et al. (1968), but there is clear evidence that the provision of amino acids by infusion will enhance synthesis (Skillman *et al.*, 1976).

The earlier work on synthesis rates depended on inferring the rate from measurements of the catabolic rate measured under steady state conditions. Nevertheless, it is clear that albumin synthesis is much more sensitive to protein intake than one would infer simply from the plasma albumin concentration (Hoffenberg *et al.*, 1966). The protective mechanisms involved in maintaining plasma albumin have been described elsewhere (James *et al.*, 1976), but very little work has been undertaken to delineate the mechanisms responsible for the early fall in the extravascular mass and the later slowing of the catabolic rate.

## *Extravascular Albumin Mass*

Albumin is not unique in having a very substantial portion of its mass located outside the vascular system. Indeed, an analysis of the distribution of several plasma proteins (Table 4) suggests that the use of the term 'plasma

TABLE 4

EXTRAVASCULAR PLASMA PROTEINS

| *Protein* | *Molecular weight (approx.)* | *Transcapillary escape rate ($\%/h$)* | *% Pool extravascular* | *Reference* |
|---|---|---|---|---|
| Albumin | 69 000 | 5·4 | 55 | Rossing *et al.* (1976) |
| Transferrin | 90 000 | — | — | — |
| Fibrinogen | 341 000 | — | 25 | McFarlane *et al.* (1964) |
| $\alpha_2$-Macroglobulin | 820 000 | — | 7 | Norberg *et al.* (1970) |
| Ig G | 175 000 | 3 | 50 | Rossing *et al.* (1976) |
| Ig M | 1 000 000 | 1–2 | 20 | Jensen (1969) |
| Total $\gamma$-glubulins | — | — | 49 | Hoffenberg *et al.* (1966) |
| Haptoglobin | 100 000 | — | $\approx 20$ | Bottiger and Malin (1968) |

Transcapillary escape rates are expressed as a percentage of the intravascular mass escaping per hour.

protein' is misleading since there are substantial amounts of many of the plasma proteins in the extravascular spaces. Many of the circulating proteins also exhibit their principal function in the extravascular space, e.g. the immunoglobulins, and too little emphasis has been given to investigating the importance of the extravascular circulation of these proteins. The earlier work was detailed by Schultze and Heremans (1966), who noted that the exchange of proteins with different tissues varied considerably; the mass flow and transit time through the liver, muscle and skin differ by a factor of ten (Reeve and Chen, 1970). In man, the skin, muscle, gastrointestinal tract and spleen account for a very substantial amount of the extravascular albumin (Rothschild et al., 1955); in muscle much of it is located within the connective tissue between muscle fibres (Reeve and Chen, 1970). In addition, albumin is now recognized to occur intracellularly, particularly in muscle, and can also become incorporated into the matrix of bone (Owen and Triffitt, 1976).

Much of the evidence on the control of extravascular albumin has come indirectly from studies with radiolabelled albumin which show that changes in extravascular albumin are comparatively slow, considering that the transcapillary escape of albumin is approximately ten times the rate of catabolism. A fall in extravascular albumin mass can, however, be demonstrated in nephrotic rats (Katz et al., 1970) and in animals given a low protein diet (Coward and Sawyer, 1977). This fall appears to begin within the first 1–2 days of low protein feeding (James and Hay, 1968; Coward and Sawyer, 1977). Analyses of the factors controlling the extravascular circulation of proteins such as those of Renkin (1964) and Taylor et al. (1973) show that these changes could be associated with either an increase in the lymphatic return of albumin or a fall in capillary permeability to this protein. Lymph flow rates and transcapillary escape rates for albumin have not been measured simultaneously when animals switch from a high to a low protein diet, but it is recognized that small reductions in plasma oncotic pressure may increase transcapillary loss of water and lymph flow rate (Coward and Fiorotto, 1979). It is also possible that a general hypocirculatory state could reduce considerably the total capillary surface area available for albumin transfer from the plasma into the interstitium.

These interrelationships between plasma and interstitial albumin and water have important consequences in preventing oedema (Fadnes et al., 1978). Direct measurements of the colloid osmotic pressure of interstitial fluid show a precipitous drop on low protein feeding which initially parallels the fall in plasma colloid osmotic pressure (Table 5) and thereby maintains a negative interstitial fluid pressure. Eventually, however, the fall in

TABLE 5

PLASMA AND INTERSTITIAL PRESSURES ON LOW PROTEIN FEEDING (FIOROTTO AND COWARD, 1979)

| *Pressure* | *Normal* | *Weeks on low protein diet* | | |
|---|---|---|---|---|
| | | *2* | *5* | *18–20* |
| Plasma colloid osmotic pressure ($\pi_p$) | $270 \pm 5$ | $203 \pm 6$ | $192 \pm 3$ | $119 \pm 5$ |
| Interstitial osmotic pressure ($\pi_i$) | $138 \pm 6$ | $70 \pm 4$ | $64 \pm 7$ | $13 \pm 3$ |
| Interstitial fluid pressure ($P_i$) | $-15 \pm 2$ | $-26 \pm 1$ | $-33 \pm 4$ | $-1 \pm 1$ |
| Sum of forces opposing filtration ($\pi_p - \pi_i + P_i$) | $119 \pm 11$ | $107 \pm 7$ | $95 \pm 7$ | $101 \pm 1$ |

From studies conducted on rats fed on either a 21 or 0·55 % protein diet and interstitial pressures measured in implanted Guyton capsules.
Pressures are given in mm water, means ± SEM.

interstitial osmotic pressure cannot match that of the plasma and interstitial fluid pressure rises and precipitates oedema. At this stage the outflow of water has exceeded the capacity of the lymphatic system to return fluid to the blood stream.

This end-stage, classically seen in children with kwashiorkor or the nephrotic syndrome, is the outcome of extreme changes which outweigh homoeostatic mechanisms. In other conditions, however, these mechanisms are themselves affected. Thus, Parving *et al.* (1979) have recently re-emphasized the observation that in myxoedema the formation of oedema is associated with an increased rather than a reduced extravascular albumin mass. The transcapillary escape of albumin is increased by 50 % and lymphatic return is slowed; the extravascular albumin mass averaged 3337 $\mu$mol in the hypothyroid state compared with a value of 1925 $\mu$mol after treating the myxoedema with thyroxine. This increase in extravascular albumin may be associated with an increased synthesis of mucopolysaccharide–protein complexes in the connective tissues. Direct measurements of subcutaneous [131]I-albumin dispersion from the leg in myxoedema show that the clearance rate is only half normal, whereas in hypoproteinaemic states there is the expected increased return of albumin from the subcutaneous tissues (Langgård, 1963).

These changes in peripheral circulation of albumin emphasize the importance of the physico-chemical characteristics of the interstitial space, but altered mucopolysaccharides in the interstitium in myxoedema are unlikely to account for the surprising increase in the transcapillary escape rate of albumin unless the basement membrane of the capillaries is also altered by changes in mucopolysaccharide composition. Rossing *et al.* (1976) have undertaken a series of studies which show the way in which changes in peripheral permeability and in blood pressure can alter the transcapillary escape rate of albumin in man. Changes in transcapillary escape are not confined to albumin but the escape rate will affect rather than determine the extravascular mass of proteins. Although the transcapillary escape of large molecules probably does occur by vesicular transport, an analysis of the extravascular mass of protein in relation to their molecular size suggests that large molecules are excluded from the interstitial spaces. This may depend more on molecular sieving in the extravascular connective tissue than on a failure of vesicular transport into the tissues (Reeve, 1977).

## Plasma Protein Catabolism

Most of the kinetic studies on plasma protein metabolism have focussed on measurements of the catabolic rate. Figure 1 is a composite diagram derived from Schultze and Heremans (1966) and Waldmann (1977) demonstrating the extraordinary variety of relationships between changes in the plasma concentration of a protein and its rate of catabolism. Albumin is unique in showing a substantial reduction in its breakdown rate as plasma albumin concentrations fall. Transferrin breakdown rates do not

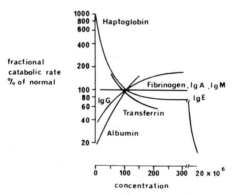

Fig. 1.    The relationship between the circulating concentrations of a plasma protein and its rate of catabolism.

appear to respond to changes in the circulating mass of carrier proteins and this may well explain the more rapid decline in plasma transferrin as dietary protein intake falls. The capacity for adaptation in albumin catabolism is, however, substantial with rates of only 1·5 g/day in hypoproteinaemia, compared with a normal of 10 g daily (Wilkinson and Mendenhall, 1963). The possible mechanisms accounting for albumin catabolism have been sought for years, but we are still little further than McFarlane's view of ten years ago that catabolism occurs in a distinct pool close to, but not identical with, the plasma pool (McFarlane and Koj, 1970). The suggestion that there should be a close link between the transcapillary escape rate, aging of the albumin molecule and its catabolism (Rossing *et al.*, 1976) does not match the data in myxoedema where increased transcapillary escape rates are associated with reduced catabolic rates (Parving *et al.*, 1979). Those tissues, such as the liver and gastrointestinal tract, which have very rapid transcapillary escape rates of albumin, so that intracellular lysosomal enzymes could operate in a manner appropriate for a pool closely linked with the vascular bed, have been exhaustively investigated as sites of normal catabolism (Hoffenberg, 1970; Waldmann, 1977). The pathological loss of albumin from the gut is an important precipitating factor in the hypoalbuminaemia of many children succumbing to kwashiorkor (Coward and Lunn, 1981) and in intestinal disease associated with gastrointestinal protein-losing enteropathy. Extensive loss of albumin from the extravascular pool also occurs in burns. These processes give no clue, however, to the normal mode of albumin catabolism.

The mechanisms controlling the catabolism of other plasma proteins are not defined with any greater precision. The continuing emphasis on asialation as the key step determining catabolism of plasma glycoproteins (Gregoriadis, 1976) accords with the much faster rates of catabolism of the glycoproteins once one or more sialic acid groups are removed, but the physiological role of neuraminidase-like enzymes in blood or tissues remains uncertain and is unlikely to affect all glycoproteins (Wong *et al.*, 1974). The relationships between catabolism and circulating plasma levels depicted in Fig. 1 suggest that the synthesis rate of these proteins is the principal method for controlling their circulating concentrations.

## Conclusions

The wealth of evidence which has now accumulated on the metabolic role and turnover of plasma proteins suggests that the initial emphasis given to research on serum albumin was appropriate for a molecule which has the greatest synthetic rate of all the circulating proteins. Nevertheless, it is becoming apparent that, despite the diversity of roles which albumin can perform (Rosenoer *et al.*, 1977), the need for the other plasma proteins may be greater since the rare genetic conditions where these proteins are absent are associated with a greater metabolic disturbance and impaired health than that observed in analbuminaemia. The specificity of roles of the plasma proteins, their small pools and rapid turnover emphasize the importance of plasma proteins in the normal regulation of body function and in the defence against environmental stress. The increase in the synthesis of many of these proteins in response to infections and trauma is one of the principal changes induced by environmental influences, but nutritional factors have more sustained effects which are reflected in changes in the metabolism of a wide variety of proteins from serum albumin to the mineral and vitamin transport proteins and the lipoproteins. The interest in plasma protein metabolism is likely to grow with greater emphasis on the environmental factors controlling their synthesis. If their apparent role in chronic diseases is confirmed, then further kinetic studies on a variety of different plasma proteins which have received little attention so far seem warranted.

## Discussion

*Munro* (*Tufts University, Boston, USA*): Many plasma proteins may have a distinctive mechanism and control for catabolism, different from that for albumin because they are glycosylated. Serum transferrin concentrations may also be affected by factors other than protein intake; for example, an increase has been demonstrated in iron deficiency (Morton and Tavill, 1977).

*James:* The glycosylation of proteins may not be necessarily the key to their catabolic regulation, as already mentioned in relation to transferrin metabolism which does indeed respond to iron deficiency and stress.

*Millward* (*London School of Hygiene and Tropical Medicine, UK*): Are the transcapillary exchange rates relevant to the control of albumin catabolism? The intracellular degradation of albumin by peripheral tissues may be an important mechanism accounting for the transfer of amino acids into cells.

*James:* Calculations of the transcapillary escape rate across different vascular beds suggest that the rapidly exchanging tissues are in the splanchnic bed. Extravascular albumin in muscle and skin exchanges very slowly and is unlikely to be an appreciable source of albumin for peripheral intracellular degradation since albumin catabolism has been shown to be rapidly responsive and closely related to the intravascular compartment. It would be tempting, however, to suggest that lysosomes in the splanchnic tissues were the effective site of albumin degradation because of the rapid albumin exchange in this area. The liver does not, however, seem to be a major catabolic site.

*Alleyne* (*University of the West Indies, Kingston, Jamaica*): I am not convinced that an altered transcapillary escape rate of albumin in naturally occurring or even experimentally induced hypertension necessarily means that hydrostatic pressure alone accounts for the changes in albumin escape rates. Altered basement membrane permeability is also recognized in these circumstances.

*James:* There are no systematic studies on capillary permeability and protein escape rates in hypertension, so that an alteration in membrane characteristics cannot be excluded.

*Waterlow* (*London School of Hygiene and Tropical Medicine, UK*): There is no evidence of a constant albumin catabolic rate; fractional catabolic rates change with albumin mass and this is a normal regulatory process. Other proteins may exhibit changes in catabolic rate which have a different relationship to changes in concentration (Freeman, 1965).

*James:* There is no disagreement here; I am impressed by the small range in values for the fractional catabolic rate of plasma proteins when expressed as a percentage of the vascular mass. This consistency occurs despite large differences between proteins in their total whole body masses.

# References

ALLISON, A. C. (ed.) (1976). *Structure and Function of Plasma Proteins*, Vols. I and II, Plenum Press, New York.

BIANCHI, R., MARIANI, G. and McFARLANE, A. S. (Eds) (1976). *Plasma Protein Turnover*, Macmillan Press Ltd, London.

BOTTIGER, L. E. and MALIN, L. (1968). Turnover of [131]I- and [125]I-labelled haptoglobin in man. *Acta Med. Scand.*, **184**, 187–90.

CALVERT, G. D. and JAMES, H. M. (1979). Low-density lipoprotein turnover studies in man. Evaluation of the integrated equations method, use of a whole-body radioactivity counter and the problem of partial denaturation. *Clin. Sci.*, **56**, 71–6.

Ciba Symposium (1973). Protein turnover. Elsevier, Excerpta Medica, North Holland, Amsterdam.

COWARD, W. A. and FIOROTTO, M. (1979). The pathogenesis of oedema in kwashiorkor—the role of plasma proteins. *Proc. Nutr. Soc.*, **38**, 51–9.

COWARD, W. A. and LUNN, P. G. (1981). The biochemistry and physiology of kwashiorkor and marasmus. *Brit. Med. Bull.*, **37**, 19–24.

COWARD, W. A. and SAWYER, M. B. (1977). Whole-body albumin mass and distribution in rats fed on low-protein diets. *Brit. J. Nutr.*, **37**, 127–34.

DELPEUCH, F., CORNU, A. and CHEVALIER, P. (1980). The effect of iron-deficiency anaemia on two indices of nutritional status, prealbumin and transferrin. *Brit. J. Nutr.*, **43**, 375–9.

FADNES, H. O., REED, R. K. and AUKLAND, K. (1978). Mechanisms regulating interstitial fluid volume. *Lymphology*, **11**, 165–9.

FIOROTTO, M. and COWARD, W. A. (1979). Pathogenesis of oedema in protein energy malnutrition: the significance of plasma colloid osmotic pressure. *Brit. J. Nutr.*, **42**, 21–31.

FREEMAN, T. (1965). Gamma globulin metabolism in normal humans and in patients. *Series Haematologica*, **4**, 76–86.

FREEMAN, T. (1967). The function of plasma proteins. In *Protides of the Biological Fluids*, Vol. 15, H. Peeters (Ed.) Elsevier Publishing Co., Amsterdam, pp. 1–14.

GREGORIADIS, G. (1976). The role of sialic acid in the catabolism of plasma glycoproteins. In *Structure and Function of Plasma Proteins*, Vol. II, Allison, A. C. (Ed.), Plenum Press, New York, pp. 145–62.

HOFFENBERG, R. (1970). Control of albumin degradation in vivo and in the perfused liver. In *Plasma Protein Metabolism*, Rothschild, M. A. and Waldmann, T. (Eds), Academic Press, New York, pp. 239–55.

HOFFENBERG, R., BLACK, E. and BROCK, J. F. (1966). Albumin and γ-globulin tracer studies in protein depletion states. *J. Clin. Invest.*, **45**, 143–51.

HOFFENBERG, R., SAUNDERS, S., LINDER, G. C., BLACK, E. and BROCK, J. F. (1962). I[131]-albumin metabolism in human adults after experimental protein depletion and repletion. In *Protein Metabolism. An International Symposium, Leyden*, Springer Verlag, Berlin.

INGENBLEEK, Y., VAN DEN SCHRIECK, H.-G., DE NAYER, P. and DE VISSCHER, M. (1975). Albumin, transferrin and the thyroxine-binding pre-albumin/retinol-binding proteins (TBPA–RBP) complex in assessment of malnutrition. *Clin. Chim. Acta*, **63**, 61-7.

ISMADI, S. D., SUSHEELA, T. P., NARASINGA RAO, B. S. (1971). Usefulness of plasma ceruloplasmin and transferrin levels in the assessment of protein calorie malnutrition. *Indian J. Med. Res.*, **59**, 1581–7.

JAMES, W. P. T. and HAY, A. M. (1968). Albumin metabolism: effect of the nutritional state and the dietary protein intake. *J. Clin. Invest.*, **47**, 1958–72.

JAMES, W. P. T., SENDER, P. M. and WATERLOW, J. C. (1976). Nutritional aspects of plasma protein metabolism: the relevance of protein turnover rates during malnutrition and its remission in man. In *Plasma Protein Turnover*, Bianchi, R., Mariani, G. and McFarlane, A. S. (Eds), Macmillan Press Ltd, London, pp. 251–63.

JENSEN, K. B. (1969). Metabolism of human γ-macroglobulin (1gM) in normal man. *Scand. J. Clin. Lab. Invest.*, **24**, 205–14.

KATZ, J. H. (1961). Iron and protein kinetics studied by means of doubly labelled human crystalline transferrin. *J. Clin. Invest.*, **40**, 2143–52.

KATZ, J., BONORRIS, G., GOLDEN, S. and SELLERS, A. G. (1970). Extravascular albumin mass and exchange in rat tissues. *Clin. Sci.*, **39**, 705–24.

LANGGÅRD, H. (1963). The subcutaneous absorption of albumin in edematous states. *Acta Med. Scand.*, **174**, 645–50.

LARGE, S., NEAL, G., GLOVER, J., THANANGKUL, O. and OLSON, R. E. (1980). The early changes in retinol-binding protein and prealbumin concentrations in plasma of protein-energy malnourished children after treatment with retinol and an improved diet. *Brit. J. Nutr.*, **43**, 393–402.

MCFARLANE, A. S. and KOJ, A. (1970). Short-term measurement of catabolic rates using iodine-labelled plasma proteins. *J. Clin. Invest.*, **49**, 1903–11.

MCFARLANE, A. S., TODD, D. and CROMWELL, S. (1964). Fibrinogen catabolism in humans. *Clin. Sci.*, **26**, 415–20.

MORGAN, E. H. and PETERS, T. JR. (1971). The biosynthesis of rat serum albumin. V. Effect of protein depletion and re-feeding on albumin and transferrin synthesis. *J. Biol. Chem.*, **246**, 3500–7.

MORTON, A., HAMILTON, S. M., RAMSDEN, D. B. and TAVILL, A. S. (1976). Studies on regulatory factors in transferrin metabolism in man and the experimental rat. In *Plasma Protein Turnover*, Bianchi, R., Mariani, G. and McFarlane, A. S. (Eds), Macmillan Press Ltd, London, pp. 165–77.

MORTON, A. G. and TAVILL, A. S. (1977). The role of iron in the regulation of hepatic transferrin synthesis. *Brit. J. Haematol.*, **36**, 383–94.

NORBERG, R., BIRKE, G., HEDFOSS, E. and PLANTIN, L.-O. (1970). Metabolism of $\alpha_2$-macroglobulin. Regulation and distribution. In *Plasma Protein Metabolism*, Rothschild, M. A. and Waldmann, T. (Eds), Academic Press, New York, pp. 427–36.

OGUNSHINA, S. O. and HUSSAIN, M. A. (1980). Plasma thyroxine binding prealbumin as an index of mild protein-energy malnutrition in Nigerian children. *Amer. J. Clin. Nutr.*, **33**, 794–800.

ORATZ, M. (1970). Oncotic pressure and albumin synthesis. In *Plasma Protein Metabolism*, Rothschild, M. A. and Waldmann, T. (Eds), Academic Press, New York, pp. 223–38.

OWEN, M. and TRIFFITT, J. T. (1976). Extravascular albumin in bone tissue. *J. Physiol.*, **257**, 293–307.

PARVING, H.-H., HAUSEN, J. M., NIELSEN, S. L., ROSSING, N., MUNCK, O. and LASSEN, N. A. (1979). Mechanisms of edema formation in myxedema—increased protein extravasation and relatively slow lymphatic drainage. *New Eng. J. Med.*, **301**, 460–5.

PETERSON, P. A. (1971). Demonstration in serum of two physiological forms of human retinol binding protein. *European J. Clin. Invest.*, **1**, 437–44.

PUTMAN, F. W. (Ed) (1975). *The Plasma Proteins. Structure, Function and Genetic Control*, Vols I and II, Academic Press, New York.

REEVE, E. B. (1977). Interstitial albumin. In *Albumin Structure, Function and Uses*, Rosenoer, V. M., Oratz, M. and Rothschild, M. A. (Eds), Pergamon Press, Oxford, UK, pp. 283–303.

REEVE, E. B. and CHEN, A. Y. (1970). Protein distribution. Regulation of interstitial albumin. In *Plasma Protein Metabolism*, Rothschild, M. A. and Waldmann, T. (Eds), Academic Press, New York, pp. 89–109.

RENKIN, E. M. (1964). Transport of large molecules across capillary walls. *The Physiologist*, 7, 13–28.

ROSENOER, V. M., ORATZ, M. and ROTHSCHILD, M. A. (Eds) (1977). *Albumin Structure, Function and Uses*, Pergamon Press, Oxford, UK.

ROSSING, N. (1967). The normal metabolism of $I^{131}$-labelled albumin in man. *Clin. Sci.*, 33, 593–602.

ROSSING, N., PARVING, H.-H. and LASSEN, N. A. (1976). Albumin transcapillary escape rate as an approach to microvascular physiology in health and disease. In *Plasma Protein Turnover*, Bianchi, R., Mariani, G. and McFarlane, A. S. (Eds), Macmillan Press Ltd, London, pp. 357–70.

ROTHSCHILD, M. A., BAUMAN, A., YALOW, R. S. and BERSON, S. A. (1955). Tissue distribution of $I^{131}$-labelled human serum albumin following intravenous administration. *J. Clin. Invest.*, 34, 1354–8.

ROTHSCHILD, M. A., ORATZ, M. and SCHREIBER, S. S. (1977). Albumin synthesis. In *Albumin Structure, Function and Uses*, Rosenoer, V. M., Oratz, M. and Rothschild, M. A. (Eds), Pergamon Press, Oxford, UK, pp. 227–53.

ROTHSCHILD, M. A. and WALDMANN, T. (Eds) (1970). *Plasma Protein Metabolism. Regulation of Synthesis, Distribution and Degradation*, Academic Press, New York.

SCHULTZE, H. E. and HEREMANS, J. F. (1966). *Molecular Biology of Human Proteins with Special Reference to Plasma Proteins*, Vol. I: Nature and metabolism of extracellular proteins, Elsevier Publishing Co., Amsterdam.

SHETTY, P. S. (1979). *Studies on Protein and Energy Restriction and Dietary Thermogenesis in Obesity and Chronic Undernutrition*. Ph.D. Thesis, Cambridge University, UK.

SKILLMAN, J. J., ROSENOER, V. M., SMITH, P. C. and FANG, M. S. (1976). Improved albumin synthesis in postoperative patients by amino acid infusion. *New Engl. J. Med.*, 295, 1037–40.

SOCOLOW, E. L., WOEBER, K. A. and PURDY, R. H. (1965). Preparation of $I^{131}$-labelled human serum pre-albumin and its metabolism in normal and sick patients. *J. Clin. Invest.*, 44, 1600–9.

TAVILL, A. S., CRAIGIE, A. and ROSENOER, V. M. (1968). The measurement of the synthetic rate of albumin in man. *Clin. Sci.*, 34, 1–28.

TAYLOR, A. E., GIBSON, D. H., GRANGER, H. J. and GUYTON, A. C. (1973). The interaction between intracapillary and tissue forces in the overall regulation of intestitial fluid volume. *Lymphology*, 6, 192–208.

WALDMANN, T. A. (1977). Albumin catabolism. In *Albumin Structure, Function and Uses*. Rosenoer, V. M., Oratz, M. and Rothschild, M. A. (Eds.), Pergamon Press, Oxford, pp. 255–73.

WATERLOW, J. C., CRAVIOTO, J. and STEPHEN, J. M. L. (1960). Protein malnutrition in man. *Advances in Protein Chemistry*, 15, 131–238.

WHITEHEAD, R. G. and ALLEYNE, G. A. O. (1972). Pathophysiological factors of importance in protein-calorie malnutrition. *Brit. Med. Bull.*, **28**, 72–8.

WHITEHEAD, R. G., COWARD, W. A. and LUNN, P. G. (1973). Serum albumin concentration and the onset of kwashiorkor. *Lancet*, **i**, 63–6.

WILKINSON, P. and MENDENHALL, C. L. (1963). Serum albumin turnover in normal subjects and patients with cirrhosis measured by $^{131}$I-labelled human albumin. *Clin. Sci.*, **25**, 281–92.

WONG, K-L., CHARLWOOD, P. A., HATTON, M. W. C. and REGOECZI, E. (1974). Studies of the metabolism of asialotransferrins: evidence that transferrin does not undergo desilylation *in vivo. Clin. Sci. Mol. Med.*, **46**, 763–74.

# 38

# Protein Turnover in Individual Tissues with Special Emphasis on Muscle

D. J. MILLWARD, J. G. BROWN and B. ODEDRA

*Clinical Nutrition and Metabolism Unit,*
*London School of Hygiene and Tropical Medicine,*
*London, UK*

## Introduction

This paper briefly reviews current understanding of how rates of protein synthesis and degradation in individual tissues are regulated. At the outset it is important to state that there are almost certainly major differences in regulation between different tissues. This point has been made many times (e.g. Millward and Garlick, 1972; Millward, 1980; Millward *et al.*, 1980*a*; Waterlow *et al.*, 1978). The rate of protein turnover in several tissues is presented, together with a brief survey of their responsiveness to malnutrition. The major part of the paper is concerned with the regulation of protein turnover in skeletal muscle, both because the process in this tissue is particularly sensitive to a wide range of insults and stimuli, and because the regulation of protein balance in muscle is of central importance to the organism as a whole.

## The Extent of Protein Turnover in Individual Tissues

From the earliest recognition of the dynamic state of tissue proteins, higher rates of protein turnover have been expected to occur in liver and visceral tissues than in the rest of the body, and this was confirmed by Waterlow and Stephen (1966, 1968). Since then most work has been concerned with making these measurements more precise, particularly in such tissues as gastrointestinal tract and liver, which pose particular problems for measurement. Some recent measurements made in the young rat are shown in Table 1. In order to compare RNA concentrations (as an indication of

TABLE 1

PROTEIN SYNTHESIS IN RAT TISSUES

| Tissue | Fractional synthesis rate (%/day) | Capacity (g RNA/g protein × 10³) | RNA activity (g protein synthesized/ g RNA/day) |
|---|---|---|---|
| Jejunal mucosa[a] | 123 | 69 | 17·9 |
| Liver[a] | 86 | 49 | 17·5 |
| Jejunal serosa[a] | 52 | 29 | 18·2 |
| Heart[b] | 18·4 | 15 | 12·5 |
| Soleus[b] | 19·4 | 15·9 | 12·4 |
| Plantaris[b] | 14·0 | 10·0 | 13·8 |
| Gastrocnemius[a] | 16·9 | 10·1 | 16·7 |
| Gastrocnemius[b] | 11·5 | 7·3 | 15·8 |

[a] McNurlan (1980); large dose of phenylalanine
[b] Odedra and Millward (unpublished); tyrosine infusion
Two values for the gastrocnemius are included to show that the two methods give similar rates of protein synthesis per unit RNA.

the capacity for protein synthesis) with the rate of protein synthesis it is convenient to express RNA concentration as the RNA/protein ratio (Millward et al., 1973). This enables the RNA activity to be calculated from the rate divided by RNA/protein, giving g protein synthesis/g RNA/day. These measurements enable at least one important question about the regulation of protein synthesis to be answered. The highest rates of protein synthesis are present in tissues with the highest concentration of RNA, a fact long known (see Munro, 1964). However, the rate of protein synthesis per unit RNA appears not to be significantly different in any of the tissues. Thus, if ribosomal RNA is the same proportion of total RNA in these tissues, then the translation rate—the rate of protein synthesis per ribosome—is also similar in all these tissues.

## The Distribution of Protein Synthesis between Tissues

When the fractional rates of protein synthesis shown in Table 1 are multiplied by the amounts of protein in the various tissues, the distribution of whole body protein synthesis can be determined. Such values are important in trying to interpret changes in whole body protein turnover in man and animals. As shown in Table 2, in a young rat, the liver and gastrointestinal tract together, and muscle each contribute about a quarter of the total. The rate of protein synthesis in skin is particularly difficult to measure but preliminary measurements in our department (Preedy and Garlick, unpublished) indicate that it might account for more than 25 % of the total. This fact, together with the observation that skin protein is mobilized in malnutrition (Mendes and Waterlow, 1958), makes it an important (and much neglected) tissue in protein homoeostasis.

TABLE 2

DISTRIBUTION OF PROTEIN SYNTHESIS IN THE RAT

|  | *Fractional rate* ($\%/day$) | *Protein content of tissue* ($g/100\,g$ body weight) | *Protein synthesis* ($\%$ of whole body synthesis) |
|---|---|---|---|
| Whole body | $34^a$ | $16^c$ | 100 |
| Liver | $105^a$ | $0.7^b$ | 14 |
| Gastrointestinal tract | $93^a$ | $0.61^b$ | 11 |
| Skin | $32^b$ | $4.87^d$ | 29 |
| Muscle | $17^a$ | $7.2^b$ | 23 |
| Remainder | 50 | 2.6 | 24 |

[a] McNurlan (1980); large dose of leucine
[b] Preedy and Garlick (unpublished)
[c] Millward (unpublished); McNurlan (1980)
[d] Waterlow and Stephen (1966)

These values are for young rats (100 g) and change with age. In particular the rate of protein synthesis in muscle falls with age (e.g. Millward *et al.*, 1975; Waterlow *et al.*, 1978), but the fall in the rate is compensated for by the increased proportion of muscle mass in lean tissue mass. Thus, the muscle's share of total protein synthesis in the rat increases from about 14 % at weaning to 21 % at maturity (Millward *et al.*, 1981). In man the distribution of protein synthesis is probably different since the body composition is different, with muscle accounting for a larger fraction of lean tissue (see Munro, 1969). According to measurements by Halliday and McKeran (1975) and more recent measurements (Rennie, Halliday, Edwards, Clugston, Matthews and Millward, unpublished), the rate of protein synthesis in adult muscle is between 2 and 3 %/day, which accounts for about one half of the protein turnover in the whole body.

## Sensitivity to Malnutrition of Protein Turnover in Various Tissues

Table 3 shows the response of jejunal mucosa, liver, heart and skeletal muscle to starvation and a protein-free diet. The rate of protein synthesis fell in all tissues with both treatments. There are several important implications of these changes. First, while protein synthesis at the level of translation is reduced in all tissues, judging by the reduced RNA activity, in muscle the capacity for protein synthesis, i.e. the RNA concentration, is also markedly reduced. Thus, unlike the liver, in which protein and RNA are

TABLE 3

EFFECT OF MALNUTRITION ON PROTEIN SYNTHESIS IN RAT TISSUES EXPRESSED AS A PERCENTAGE OF CONTROL VALUES

| Tissue | Starvation | | | Protein deficiency | | |
|---|---|---|---|---|---|---|
| | Fractional synthesis rate | RNA activity | RNA/ protein | Fractional synthesis rate | RNA activity | RNA/ protein |
| Jejunal mucosa | 74 | 77 | 95 | 77 | 68 | 114 |
| Liver | 84 | 83 | 100 | 80 | 70 | 115 |
| Heart | 61 | 73 | 82 | 50 | 74 | 57 |
| Gastrocnemius | 34 | 50 | 68 | 24 | 62 | 38 |

Data of McNurlan (1980); large dose of phenylalanine
Hundred-gram male rats were starved for 2 days or fed a protein-free diet for 9 days

lost at an equal rate (Millward *et al.*, 1974), in muscle ribosomes are selectively lost in malnutrition. It is this combination of reduced capacity and reduced ribosome activity which results in such large proportional suppression of muscle protein synthesis. Secondly, although the percentage fall in synthesis is small in the gastrointestinal tract and liver compared with muscle, because the actual rate is so fast such a fall can have a marked effect on the protein content of the tissue. For example, in jejunum the fall on the protein-deficient diet involved a reduction of 23 %/day in protein synthesis, which, with no change in the degradation rate, would induce a halving of the mucosal protein mass every 2·5 days. Since the protein mass had probably stabilized in the gut at this time (9 days of the diet) this implies that degradation had probably fallen. In the liver the situation is more complicated because a considerable proportion of protein synthesis includes the exported plasma proteins, and part of the fall in synthesis involves a selective fall in albumin synthesis (see Garlick, 1980). Nevertheless, the point can still be made that a small proportional fall in protein synthesis in the liver can have marked effects on the organ size even without changes in degradation.

However, it is known that degradation rates do increase in liver during fasting or protein deficiency (see Mortimore and Schworer, 1980), when the acute loss of protein is occurring (e.g. Millward *et al.*, 1974). Such losses cannot be maintained for long and when liver mass has stabilized after 9 days of a protein-deficient diet the degradation rate must return to normal or even be depressed (Garlick *et al.*, 1975). In the case of muscle, because turnover is relatively slow, growth accounts for a considerable proportion of total protein synthesis, so that growth suppression alone will involve a fall of one-third in the rate of synthesis. However, the fall in synthesis rate is much greater than this in both the dietary situations shown in Table 3. This means that protein can be mobilized from muscle as a result of a reduction in synthesis without a necessary increase in degradation. In human muscle our recent measurements indicate a fall of over $50\%$ in the rate of protein synthesis after an overnight fast, which accounts for the total fall in the whole body rate (Rennie, Halliday, Edwards, Clugston, Matthews and Millward, unpublished).

The way in which the degradation rate changes can be determined by comparing the measured rate of synthesis with the rate of net change in protein mass. When this is done two types of response are observed (see Millward and Waterlow, 1978; Waterlow *et al.*, 1978). The first occurs in response to a protein-deficient diet in rats. In addition to the marked fall in the rate of protein synthesis (Table 3), the degradation rate also falls. Since synthesis falls more than degradation, net catabolism can and does occur. Similar changes probably occur in man during fasting since the rate of methylhistidine excretion falls (Young *et al.*, 1973). The second response occurs after several days' fasting in rats. Degradation is increased so that, coupled with the fall in synthesis, mobilization of muscle protein is markedly accelerated (Millward and Waterlow, 1978). Although in terms of net effects the only difference between the two responses is the rate at which muscle protein is lost, the mechanism of the responses can be considered separately because the degradation rate changes in different directions in each case.

It is our working hypothesis that the most important factor in inducing these changes in muscle is the hormonal balance in the animal. Thus, the rest of this paper examines the way in which three hormones, insulin, triiodothyronine and corticosterone, the active glucocorticoid in the rat, appear to regulate protein turnover in muscle.


## The Regulation of RNA Activity in Skeletal Muscle

Our measurements of RNA activity *in vivo* have been made with the constant infusion of $^{14}C$-tyrosine, which gives similar results to measurements with the large dose method (see Table 1). Results obtained in rats with hormonal treatments are shown in Table 4 (and will be referred to by the number assigned in that table).

The first point, which has been known for some time (Hay and Waterlow, 1967; Jefferson et al., 1977), is that RNA activity is reduced in the diabetic rat (1) and insulin will stimulate RNA activity. Insulin treatment during a 6-h infusion (100 mU/h + glucose + amino acids) (2), or by continuous infusion over 7 days by an Alza minipump implanted in the peritoneal cavity (3 and 4) will restore RNA activity to normal values (Odedra and Millward, unpublished). However, a high plasma insulin concentration is not always sufficient for activation of muscle RNA activity.

Corticosterone will suppress RNA activity even when the levels of insulin are supranormal (6). This suggests that the suppression of RNA activity induced by fasting could result either from increased corticosterone or from reduced insulin. In fact, a 2-day fast in adrenalectomized rats (9) reduces RNA activity as effectively as in intact rats (Odedra and Millward, unpublished). Thus, the suppression of RNA activity in fasting or in protein deficiency is not necessarily corticosterone-dependent.

Although the glucocorticoid does appear to have a specific effect on initiation (Rannels and Jefferson, 1980), it has always been difficult to judge the extent of its action because of the hyperinsulinaemia following treatment with the hormone *in vivo*. We have administered minimal levels of insulin by implanted minipump (0·6U/day or 1·2U/day) and shown that even when plasma insulin levels are in the low–normal range, RNA activities in corticosterone-treated rats (7) do not fall as much as in diabetic rats (1). However, in the diabetic rat the corticosterone levels may be elevated and will be completely unopposed by insulin. Thus, we cannot determine whether low insulin is a more potent suppressor of RNA activity than elevated glucocorticoids.

The next question which can be asked is whether insulin is obligatory for optimum RNA activity. We have some preliminary evidence that it is not (Brown et al., 1981). In hypophysectomized rats the reduced RNA activity

## TABLE 4

RNA ACTIVITY IN MUSCLE FROM RATS RECEIVING VARIOUS HORMONAL TREATMENTS

| Hormone examined | No. in text | Condition of rats | Treatment | RNA activity (g protein synthesized/day/g RNA) | Plasma insulin (mU/ml) |
|---|---|---|---|---|---|
| Insulin[a] | 1 | Diabetic | Untreated | 6·0 ± 1·0 | 4 |
| | 2 | Diabetic | Insulin infusion (6 h) | 15·0 ± 3·0 | 114 |
| | 3 | Diabetic | 0·6 U/day insulin (7 days)[c] | 12·1 ± 1·4 | 23 |
| | 4 | Diabetic | 1·2 U/day insulin (7 days)[c] | 13·7 ± 0·8 | 47 |
| Corticosterone[a] | 5 | Adrenalectomized | Untreated | 15·8 ± 3·2 | 10 |
| | 6 | Adrenalectomized | 5 mg/day corticosterone (6 days) | 8·4 ± 1·4 | 144 |
| | 7 | Insulin-maintained diabetic[c] | 5 mg/day corticosterone (6 days) | 9·4 ± 1·5 | 14 |
| | 8 | Adrenalectomized | Untreated | 13·8 ± 1·0 | 8 |
| | 9 | Adrenalectomized | Fasted (2 days) | 8·6 ± 1·4 | <1 |
| Triiodothyronine (T$_3$)[b] | 10 | Normal | Untreated | 13·9 ± 4·6 | — |
| | 11 | Thyroidectomized | Untreated (5 days post op.) | 17·7 ± 3·1 | — |
| | 12 | Thyroidectomized | Untreated (16 days post op.) | 15·0 ± 4·9 | 18 |
| | 13 | Thyroidectomized | T$_3$ 6 days (16 days post op.) | 19·6 ± 2·0 | — |
| | 14 | Hypophysectomized | Untreated | 6·8 ± 2·1 | <2·0 |
| | 15 | Hypophysectomized | T$_3$ 7 days | 13·4 ± 1·2 | <2·0 |

[a] Results of Odedra and Millward (unpublished)
[b] Results of Brown *et al.* (1981)
[c] Insulin administered by osmotic minipump

(14) can be increased by $T_3$ treatment (15). Furthermore, this occurs with no appreciable increase in the very low plasma insulin level. Thus, it appears that $T_3$ can independently stimulate RNA activity in muscle. However, $T_3$ is certainly not obligatory since, in the absence of $T_3$ following thyroidectomy, RNA activity is maintained (11–13).

Summarizing these findings, it would appear that as far as the regulation of muscle RNA activity is concerned:

1.    Insulin is an independent activator but may not be obligatory when $T_3$ is elevated.
2.    $T_3$ is not an obligatory activator when insulin is present but may be an independent stimulator.
3.    Corticosterone is an independent suppressor, will override insulin, but is not obligatory for the starvation response.

The reduction of RNA activity in fasting or protein deficiency is, therefore, not surprising given these results, since insulin is depressed, corticosterone is elevated (Coward et al., 1977) and $T_3$ is depressed (Cox et al., 1981; Cox and Millward, unpublished and see Table 6).

These measurements of RNA activity do not indicate the mechanism of the changes; however, the extent of the changes has implications. Although initiation is usually considered to be the major site for regulation (Jefferson et al., 1980; Kay, 1980) we have argued that, since the changes in RNA activity are much greater than the reported changes in ribosomal aggregation, it is likely that marked changes in the rate of elongation occur in these various states (see Brown et al., 1981).

## The Regulation of RNA Content

The RNA concentration in muscle changes acutely in response to feeding and fasting (Millward et al., 1973, 1974), falling to very low levels after extended periods of protein deficiency (Millward et al., 1975), Hormonal factors affecting muscle RNA are shown in Table 5. One difficulty in evaluating the involvement of hormones in the regulation of RNA content of muscle is the separation of direct from indirect effects. The majority of treatments which depress RNA activity also involve a loss of RNA. It is usually argued that when reduced RNA activity involves a reduced rate of initiation, polysomes become disaggregated and this increases the susceptibility of ribosomes to degradation. For example, in diabetic rats and in glucocorticoid-treated rats (Table 4; 1, 6 and 7) the reduced RNA activity is accompanied by a loss of RNA (Table 5; 1, 5, 6 and 7) and we do not know from these results whether this loss of RNA is a direct response to lack of insulin or to corticosterone treatment, or an indirect response to the effect of those hormones on translation. Although insulin has been shown to stimulate RNA synthesis in incubated muscles (see Manchester, 1970), the evidence for it playing an important role *in vivo* is not very strong. For example, the presence of insulin is not able to prevent the loss of RNA which occurs following thyroidectomy (Brown *et al.*, 1981; Table 5; 10–12). Also the marked hyperinsulinaemia is not able to prevent the loss of RNA following glucocorticoid treatment (Odedra and Millward, 1981, Table 5; 4–7).

The loss of RNA in response to glucocorticoid treatment is more likely to reflect a direct effect of the hormone since glucocorticoids act by binding (with their receptor proteins) to the nucleus, suppressing DNA synthesis (e.g. Goldberg and Goldspink, 1975) and possibly altering the pattern of gene expression. Thus, a direct inhibition of RNA synthesis would not be surprising but it has yet to be demonstrated. However, one observation which suggests that corticosterone is not directly involved in the loss of RNA during fasting is that such a loss still occurs when the fasting rats have been adrenalectomized and consequently produce no corticosterone (Table 5; 8 and 9). It is obvious, therefore, that measurements of RNA synthesis *in vivo* in relation to insulin and glucocorticoid hormone concentrations, are needed.

We believe that $T_3$ undoubtedly plays a key role in regulating muscle RNA. Following thyroidectomy, RNA loss from muscle can be detected as early as 5 days afterwards (Table 5; 10–12). Furthermore, this occurs without a fall in RNA activity as shown in Table 4 (10–12), so a primary effect on ribosome production can be expected. this was indicated by measurements made of RNA synthesis in thyroidectomized rats. RNA concentration fell and RNA synthesis was also depressed (Grimble and Millward, unpublished). Since $T_3$ levels fall in starvation and protein deficiency (Cox *et al.*, 1981; Cox and Millward, unpublished) the loss of RNA in these states may reflect decreased synthesis owing to direct hormonal action on RNA synthesis as well as increased degradation as a secondary response to a fall in RNA activity, reduced initiation and consequent ribosomal dissaggregation. One difference between $T_3$ and other homones is the fact that the response to the hormone changes as the dose increases. When $T_3$ is administered (by Alza minipump), low doses given to thyroidectomized rats increase RNA levels, but very high doses depress them (Table 5; 14–19).

TABLE 5

RNA CONCENTRATIONS IN MUSCLE FROM RATS RECEIVING VARIOUS HORMONAL TREATMENTS

| Hormone examined | No. in text | Condition of rats | Treatment | RNA/DNA (means ± 1SD) |
|---|---|---|---|---|
| Insulin | 1 | Diabetic | Untreated | 0·80 ± 0·2 |
| | 2 | Diabetic | 0·6 U/day insulin (7 days) | 1·64 ± 0·16 |
| | 3 | Diabetic | 1·2 U/day insulin (7 days) | 2·00 ± 0·12 |
| Corticosterone[a] | 4 | Adrenalectomized | Untreated, well fed | 1·73 ± 0·18 |
| | 5 | Adrenalectomized | 5 mg/day corticosterone | 1·43 ± 0·16 |
| | 6 | Adrenalectomized | 10 mg/day corticosterone (6 days) | 1·07 ± 0·03 |
| | 7 | Insulin-maintained diabetic | 5 mg/day corticosterone (6 days) | 0·78 ± 0·20 |
| | 8 | Adrenalectomized | Untreated, fed | 1·81 ± 0·32 |
| | 9 | Adrenalectomized | Untreated, fasted (2 days) | 1·33 ± 0·22 |

Triiodothyronine $(T_3)^b$

| | | | |
|---|---|---|---|
| 10 | Normal | Untreated, fed | $2\cdot71 \pm 0\cdot59$ |
| 11 | Thyroidectomized | Untreated, fed (5 days post op.) | $2\cdot02 \pm 0\cdot26$ |
| 12 | Thyroidectomized | Untreated, fed (16 days post op) | $1\cdot18 \pm 0\cdot22$ |
| 13 | Thyroidectomized | $T_3$ treatment for 16 days (16 days post op.) | $2\cdot29 \pm 0\cdot08$ |
| 14 | Thyroidectomized | Untreated, fed (7 days post op.) | $1\cdot56 \pm 0\cdot11$ |
| 15 | Thyroidectomized | 300 g/day $T_3$,[c] fed (7 days post op.) | $1\cdot81 \pm 0\cdot03$ |
| 16 | Thyroidectomized | 750 g/day $T_3$,[c] fed (7 days post op.) | $1\cdot92 \pm 0\cdot09$ |
| 17 | Normal | Untreated, fed | $1\cdot87 \pm 0\cdot06$ |
| 18 | Thyroidectomized | 2·0 g/day $T_3$,[c] fed (7 days post op.) | $2\cdot06 \pm 0\cdot16$ |
| 19 | Thyroidectomized | 20·0 g/day $T_3$,[c] fed (7 days post op.) | $1\cdot58 \pm 0\cdot17$ |

[a] Data of Odedra and Millward (unpublished)
[b] Brown et al. (1981) and unpublished results
[c] Administered by implanted osmotic minipump

In summary, the role of these hormones in the regulation of RNA concentration can be described as follows:

1.  While insulin maintains muscle RNA, and corticosterone induces losses, it is not clear whether these are direct effects on RNA synthesis or secondary to the changes in ribosomal aggregation induced by these hormones.
2.  $T_3$ regulates RNA concentrations independently of any effect on translation through changing the rate of RNA synthesis.

## Regulation of Protein Degradation in Muscle

There are two changes in muscle protein degradation to be explained; the increase during fasting and the fall during protein deficiency. The low insulin and elevated corticosterone are both candidates for inducing the increase in fasting. Thus, in diabetes protein degradation is increased *in vivo* (Albertse, 1980) and insulin lack has been implicated in the increase in degradation which occurs when muscles are incubated or perfused *in vitro* (e.g. Jefferson *et al.*, 1977). Of course the unopposed action of corticosterone may also be responsible for the increased degradation in diabetes.

An elevated corticosterone concentration certainly appears to stimulate degradation as indicated by an increased $N^\tau$-methylhistidine excretion, at least at very high doses (Tomas *et al.*, 1979). Although we have questioned the assumption that excreted $N^\tau$-methylhistidine originates from muscle (Millward *et al.*, 1980b) our measurements of the concentration of free $N^\tau$-methylhistidine in muscle indicate elevated levels in steroid-treated rats and humans (Rennie *et al.*, 1980). This is good evidence for increased degradation in muscle. However, our attempts to demonstrate an increase in degradation in rats treated with corticosterone at 5 mg/day, sufficient to increase $N^\tau$-methylhistidine excretion (Tomas *et al.*, 1979), were not very successful, since the changes in degradation were small compared with the marked falls in synthesis (see Fig. 1). Because we were concerned that the effect of corticosterone on degradation might be counteracted by the hyperinsulinaemia, we evaluated the effect of the hormone in diabetic rats maintained on low physiological levels of insulin by the Alza minipump (Odedra and Millward, unpublished).

An increase in degradation was apparent in the gastrocnemius but not in the plantaris (see Fig. 1). However, it is significant that the increase in

Protein Turnover in Individual Tissues with Special Emphasis on Muscle 487

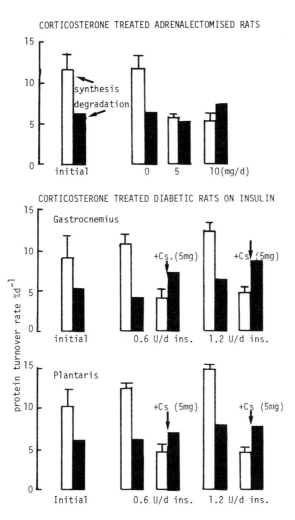

FIG. 1. Effect of corticosterone treatment on protein synthesis and degradation in rat skeletal muscle. The hormone was given to either adrenalectomized rats (top) or diabetic rats maintained on insulin dispensed by an implanted Alza osmotic minipump. Values are shown for rates of protein turnover at the start of the treatment (initial groups) and after treatment with or without the hormone (Odedra and Millward, unpublished).

degradation was not as great as that observed in starvation (Millward et al., 1976). Thus, the response induced in starvation may require the combination of low insulin and elevated corticosterone to produce the marked increase in degradation which is eventually observed.

We believe that the reduced $T_3$ levels may be largely responsible for the fall in degradation in protein deficiency and in fasting adult man. Free $T_3$ levels are reduced in protein-deficient and energy-restricted rats (Table 6).

TABLE 6

THYROID STATUS IN MALNOURISHED RATS

|  | Total $T_3$ (ng/ml) | Free $T_3$ (pg/ml) |
|---|---|---|
| (1)  Fasting[a] | | |
| fed | $1·83 \pm 0·2$ | — |
| starved 1 day | $1·27 \pm 0·2$ | — |
| 2 days | $0·97 \pm 0·08$ | — |
| 3 days | $0·67 \pm 0·08$ | — |
| 4 days | $0·36 \pm 0·15$ | — |
| (2)  Protein-energy malnutrition[b] | | |
| 20% casein ad lib. | 1·32 | 5·7 |
| 10% casein ad lib. | 2·09 | 5·78 |
| 5% casein ad lib. | 2·24 | 4·27 |
| 20% casein restricted | 1·28 | 4·05 |

[a] Cox and Millward (unpublished)
[b] Cox et al. (1981)

Furthermore, degradation falls following thyroidectomy and is restored on $T_3$ treatment (Brown et al., 1981). We recently examined the dose response of protein synthesis and degradation to $T_3$ levels dispensed by minipumps over 7 days (Brown and Millward, unpublished). The results clearly show that in the gastrocnemius and plantaris muscles the rate of degradation depends on the dose rate of $T_3$ (Fig. 2). Furthermore, the fall in degradation occurs at low levels of $T_3$, such as are found in the malnourished rat; at very high doses of $T_3$, degradation is increased so that growth is suppressed. The changes in degradation occur in concert with changes in lysosomal proteinases (de Martino and Goldberg, 1978; Millward et al., 1980a). Thus, in the same way as $T_3$ regulates the capacity for protein synthesis (i.e. RNA) in a dose-dependent way, the hormone similarly regulates the capacity for protein degradation.

In conclusion, it would appear that increased degradation in muscle, as

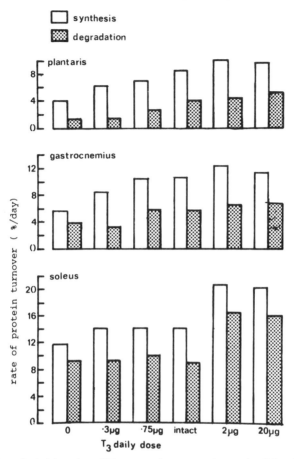

FIG. 2.    Effect of administration of $T_3$ on protein turnover in muscle of thyroidectomized rats. The hormone was given by implanted Alza minipump for 7 days (Brown and Millward, unpublished).

observed in fasting or severe food restriction, is associated with a particular combination of increased glucocorticoids and decreased insulin. Perhaps because these combinations of hormonal changes are also observed in protein deficiency, though to a less marked extent (Coward *et al.*, 1975), and without an increase in degradation, we should not rule out the involvement of other agents in the fasting response. At the moment, however, there are no obvious candidates. The fall in degradation during protein deficiency may result from the reduced $T_3$ concentrations (Cox *et al.*, 1981).

The role of thyroid hormones in setting a particular rate is particularly important, with several physiological implications. The most obvious is that during adaptation to fasting, with a reduction in the need for gluconeogenesis from muscle protein as the brain switches to the oxidation of ketones, the fall in $T_3$ will not only conserve energy by reducing the metabolic rate; it will also, by reducing protein degradation in muscle, conserve the lean body mass which would otherwise be rapidly lost owing to the fall in protein synthesis.

## Acknowledgements

This work was generously supported by the Muscular Dystrophy Group of Great Britain and the Medical Research Council.

## Discussion

*Felig (Yale University School of Medicine, USA):* I should like to take up the point you raised about corticosterone being in excess both in the diabetic and in the starved animal. We have not observed an increase in urinary excretion of glucocorticoids or in plasma cortisol in starvation in human subjects. Also, as I understand it, if one looks at the insulin-deprived animal or the human patient with diabetes and replaces glucocorticoids in physiological amounts, one finds the full expression of insulin deficiency. Obviously, the Houssay preparation shows that if you remove ACTH you can protect against diabetes. To what extent is there evidence that an increase in glucocorticoids is necessary to produce nitrogen wasting in starvation or diabetes, in terms of changes in synthesis or degradation rates?

*Millward:* As far as man is concerned, starvation seems to be very much like protein deficiency in the rat, that is, there is very little evidence for any change in degradation. In fact, it falls; so this would be quite consistent with the fact that in fasting man there is no increase in glucocorticoids. The changes I was referring to are those one sees in the diabetic rat where the changes in corticosterone are not particularly great, but there is a small increase. The point I was making was that in the complete absence of insulin this might be a potent stimulator of degradation of muscle protein. In the diabetic human, I do not know how hard the data on protein breakdown rates in muscle are.

*Felig:* In thyrotoxicosis there may be not only changes in muscle but also changes in rates of gluconeogenesis in the liver. We have demonstrated that in the thyrotoxic human, there is a very substantial increase in the amount of alanine that has been taken up by the liver, and it seems to be related more to an increase in extraction ratio than to delivery in terms of either arterial levels or even the increased blood flow (Wahren *et al.*, 1981). Thus, you may be dealing with unavailability of amino acids for recapture because of changes in gluconeogenesis.

*Munro (Tufts University, Boston, USA):* In our studies with thyroidecto-mized rats we find that the stage of thyroidectomy is important. Young rats recently thyroidectomized, e.g. after 10 days, will respond very well to a maintenance dose of thyroxine, whereas after 3 months, when the weight has reached a plateau (which many people take to indicate that thyroidectomy has taken its toll), they are quite refractory until very large doses are given (Burini *et al.*, 1981). Just as in the case of corticosteroids, the exact route and the type and level of dose and the condition of the animal can be significant factors in the outcome of the experiment.

# References

ALBERTSE, E. (1980). *Protein Metabolism in Diabetes*. Ph.D. Thesis, University of London.

BROWN, J. G., BATES, P. C., HOLLIDAY, M. A. and MILLWARD, D. J. (1981). Thyroid hormones and muscle protein turnover: the effect of thyroid hormone deficiency and replacement in thyroidectomised and hypophysectomised rats. *Biochem. J.*, **194**, 771–82.

BURINI, R., SANTIDRIAN, S., MOREYRA, M., BROWN, P., MUNRO, H. N. and YOUNG, V. R. (1981). Interaction of thyroid status and diet on muscle protein breakdown in the rat, as measured by $N^\tau$-methylhistidine excretion. *Metabolism*, in press.

COWARD, W. A., WHITEHEAD, R. G. and LUNN, P. G. (1977). Reasons why hypoalbuminaemia may or may not appear in protein-energy malnutrition. *Brit. J. Nutr.*, **38**, 115–26.

COX, M. D., DALAL, S. S. and HEARD, C. R. C. (1981). The importance of free $T_3$ measurements in assessing thyroid status in protein-deficient rats. *Proc. Nutr. Soc.*, **40**, 39A.

DE MARTINO, G. N. and GOLDBERG, A. L. (1978). Thyroid hormones control lysosomal enzymes. *Proc. Natl. Acad. Sci. USA*, **75**, 1369–73.

GARLICK, P. J. (1980). Protein turnover in the whole animal and in specific tissues. In *Comprehensive Biochemistry*, Vol. 19B(1), Florkin, M., Neuberger, A. and van Deenan L. L. M. (Eds), Elsevier North Holland, Amsterdam, p. 77.

GARLICK, P. J., MILLWARD, D. J., JAMES, W. P. T. and WATERLOW, J. C. (1975). Effect of protein deprivation and starvation on the rate of protein synthesis in tissues of the rat. *Biochim. Biophys. Acta*, **414**, 71–84.

GOLDBERG, A. L. and GOLDSPINK, D. (1975). Influence of food deprivation and adrenal steroids on DNA synthesis in various mammalian tissues. *Am. J. Physiol.*, **228**, 310–17.

HALLIDAY, D. and McKERAN, R. O. (1975). Measurement of muscle protein synthesis rate from serial muscle biopsies and total body protein turnover in man by continuous infusion of L(alpha-$^{15}$N)lysine. *Clin. Sci. Mol. Med.*, **49**, 581–90.

HAY, A. M. and WATERLOW, J. C. (1967). The effect of alloxan diabetes on muscle and liver protein synthesis in the rat measured by constant infusion of L($C_{14}$)lysine. *J. Physiol.*, **191**, 111.

JEFFERSON, L. S., BOYD, T. A., FLAIM, K. E. and PEAVY, D. E. (1980). Regulation of protein synthesis in perfused preparations of rat heart, skeletal muscle and liver. *Biochem. Soc. Trans.*, **8**, 282–3.

JEFFERSON, L. S., LI, J. B. and RANNELS, S. R. (1977). Regulation by insulin of amino acid release and protein turnover in the perfused rat hemi-corpus. *J. Biol. Chem.*, **252**, 1476–83.

KAY, J. E. (1980). Protein synthesis during activation of lymphocytes by mutagens. *Biochem. Soc. Trans.*, **8**, 288–9.

MANCHESTER, K. L. (1970). Sites of hormonal regulation of protein metabolism. In *Mammalian Protein Metabolism*, Vol. IV, Munro, H. N. (Ed.), Academic Press, New York and London, pp. 229–98.

McNURLAN, M. (1980). *Protein Synthesis in the Liver and Small Intestine of the Rat.* Ph.D. Thesis, University of London.

MENDES, C. B. and WATERLOW, J. C. (1958). The effect of a low-protein diet, and of refeeding, on the composition of liver and muscle in the weanling rat, *Brit. J. Nutr.*, **12**, 74–88.

MILLWARD, D. J. (1980). Protein degradation in muscle and liver. In *Comprehenisve Biochemistry*, Vol. 19B (1), Florkin, M., Neuberger, A. and van Deenan, L. L. M. (Eds), Elsevier North Holland, Amsterdam, pp. 153–232.

MILLWARD, D. J., BATES, P. C., BROWN, J. G., ROSOCHACKI, S. R. and RENNIE, M. J. (1980*a*). Protein degradation and the regulation of protein balance in muscle. In *Protein Degradation in Health and Disease*, Ciba Symposium 75, Excerpta Medica, Amsterdam, pp. 307–29.

MILLWARD, D. J., BATES, P. C., GRIMBLE, G. K., BROWN, J. G., NATHAN, M. and RENNIE, M. J. (1980*b*). The contribution of non-skeletal muscle tissues to urinary 3-methyl-histidine excretion in the rat. *Biochem. J.*, **190**, 225–8.

MILLWARD, D. J., BATES, P. C. and ROSOCHACKI, S. R. (1981). The extent and nature of protein degradation in the tissues during development. *Reproduction, Nutrition, Développement*, **21**(2), 265–77.

MILLWARD, D. J. and GARLICK, P. J. (1972). The pattern of protein turnover in the whole animal and the effect of dietary variations. *Proc. Nutr. Soc.*, **31**, 257–64.

MILLWARD, D. J., GARLICK, P. J., JAMES, W. P. T., NNANYELUGO, D. O. and RYATT, J. S. (1973). Relationship between protein synthesis and RNA content in skeletal muscle. *Nature*, **241**, 204–5.

MILLWARD, D. J., GARLICK, P. J., NNANYELUGO, D. O. and WATERLOW, J. C. (1976). The relative importance of muscle protein synthesis and breakdown in the regulation of muscle mass. *Biochem. J.*, **156**, 185–8.

MILLWARD, D. J., GARLICK, P. J., STEWART, R. J. C., NNANYELUGO, D. O. and WATERLOW, J. C. (1975). Skeletal muscle growth and protein turnover. *Biochem. J.*, **150**, 235–43.

MILLWARD, D. J., NNANYELUGO, D. O., JAMES, W. P. T. and GARLICK, P. J. (1974). Protein metabolism in skeletal muscle: the effect of feeding and fasting on muscle RNA, free amino acids, and plasma insulin concentrations. *Brit. J. Nutr.*, **32**, 127–42.

MILLWARD, D. J. and WATERLOW, J. C. (1978). Effect of nutrition on protein turnover in skeletal muscle. *Fed, Proc.*, **37**, 2283–90.

MORTIMORE, G. E. and SCHWORER, C. M. (1980). Application of liver perfusion as an *in vitro* model in studies of intracellular protein degradation. In *Protein Degradation in Health and Disease*, Ciba Symposium 75, Excerpta Medica, Amsterdam, pp. 281–98.

MUNRO, H. N. (1964). Historical introduction: the origin and growth of our present concepts of protein metabolism. In *Mammalian Protein Metabolism*, Vol. 1, Munro, H. N. (Ed.), Academic Press, New York and London.

MUNRO, H. N. (1969). Evolution of protein metabolism in mammals. In *Mammalian Protein Metabolism*, Vol. III, Munro, H. N. (Ed.), Academic Press, New York and London.

ODEDRA, B. R. and MILLWARD, D. J. (1981). The effect of corticosterone administration on skeletal muscle protein turnover *in vivo* in rats, *Biochem. J.* (Submitted for publication.)

RANNELS, S. R. and JEFFERSON, L. S. (1980). Effects of glucocorticoids on muscle protein turnover in perfused rat hemicorpus. *Am. J. Physiol.*, **238**, 564–72.

RENNIE, M. J., ROSOCHACKI, S., QUARTEY-PAPAFIO, P. and MILLWARD, D. J. (1980). Intracellular free 3-methyl-histidine concentration as an index of protein degradation. *Biochem. Soc. Trans.*, **8**, 355.

TOMAS, F. M., MUNRO, H. N. and YOUNG, V. R. (1979). Effect of glucocorticoid administration on the rate of muscle protein breakdown *in vivo* in rats as measured by urinary excretion of $N^r$-methyl-histidine. *Biochem. J.*, **178**, 139–46.

WAHREN, J., WENNLUND, A., NILSSON, H. and FELIG, P. (1981). Influence of hyperthyroidism on splanchnic exchange of glucose and gluconeogenic precursors. *J. Clin. Invest.*, **67**, 1056–63.

WATERLOW, J. C., GARLICK, P. J. and MILLWARD, D. J. (1978). *Protein Turnover in Mammalian Tissues and in the Whole Body*, Elsevier North Holland, Amsterdam.

WATERLOW, J. C. and STEPHEN, J. M. L. (1966). Adaptation of the rat to a low protein diet: the effect of a reduced protein intake on the pattern of incorporation of $L(^{14}C)$lysine. *Brit. J. Nutr.*, **20**, 461–84.

WATERLOW, J. C. and STEPHEN, J. M. L. (1968). The effect of a low protein diet on the turnover rate of serum liver and muscle protein in the rat measured by the constant infusion of $L(^{14}C)$lysine. *Clin. Sci.*, **35**, 287–305.

YOUNG, V. R., HAVERBERG, L. N., BILMAZES, C. and MUNRO, H. N. (1973).
Potential use of 3-methyl-histidine excretion as an index of the progressive
reduction in muscle protein catabolism during starvation. *Metabolism*, **23**,
1429–36.
YOUNG, V. R. and MUNRO, H. N. (1980). Muscle protein turnover in human beings
in health and in disease. In *Degradative Processes in Heart and Skeletal
Muscle*, Wildenthal, K. (Ed.), Elsevier North Holland, Amsterdam, pp. 271–94.

# 39

## Use of $N^\tau$-Methylhistidine Excretion as an *in vivo* Measure of Myofibrillar Protein Breakdown

H. N. MUNRO

*USDA Human Nutrition Research Center on Aging,*
*Tufts University, Boston, Massachusetts, USA*

and

V. R. YOUNG

*Department of Nutrition and Food Science,*
*Massachusetts Institute of Technology,*
*Cambridge, Massachusetts, USA*

## Introduction

Skeletal muscle constitutes about 45 % of body weight in the young adult mammal, and therefore represents a major component of body protein. Consequently, changes in uptake or release of amino acids by the musculature of the whole body can cause substantial changes in protein metabolism. For example, some 25 % of the whole body protein turnover of an adult man (about 200–300 g/day) is accounted for as muscle protein (Uauy *et al.*, 1978). This represents about 60 g protein, or 10 g amino acid-N, and is thus considerable by comparison with nitrogen balance of the same subject which usually lies within $\pm 2$ g N of equilibrium.

Accordingly, it would be valuable to have a method for studying factors affecting the rate of breakdown of skeletal muscle protein in intact human subjects in health and disease. This role could be served by an amino acid residue present in muscle protein that is not reutilized but is excreted in the urine. Such an amino acid would have to be produced by post-translational modification of muscle protein, released on breakdown of muscle protein, not reutilized for protein synthesis or metabolized in the body, and undergo quantitative and rapid excretion in the urine. Such criteria have been validated in the rat and man for $N^\tau$-methylhistidine, an amino acid formed by methylation of certain histidine residues in the myofibrillar proteins actin and myosin (Young and Munro, 1978). An additional desirable feature would be the demonstration that muscle is the major source of $N^\tau$-methylhistidine, a requirement that is more difficult to meet. This brief paper deals with the basis of the use of $N^\tau$-methylhistidine, and with some metabolic applications. This topic has been discussed by us in more detail elsewhere (Young and Munro, 1978).

## Evidence for the Concept

The concept is illustrated diagrammatically in Fig. 1. There is good evidence that $N^\tau$-methylhistidine satisfies the criteria for a suitable marker of muscle protein catabolism of being non-reutilized. Unlike histidine, $N^\tau$-methylhistidine does not charge muscle tRNA, and thus cannot be reutilized for synthesis of muscle protein (Young *et al.*, 1972). Second, when $^{14}CH_3$-labelled $N^\tau$-methylhistidine was administered orally or parenterally to rats, recoveries of $^{14}C$ in urine, faeces and expired air indicated complete excretion of the administered radioactivity in the urine (Table 1). The radioactivity could be accounted for by two compounds, namely the unchanged $N^\tau$-methylhistidine and its N-acetyl derivative.

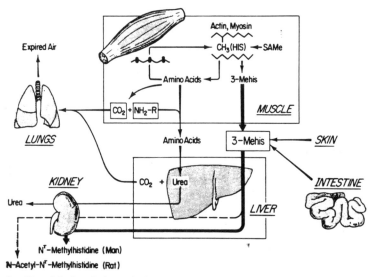

FIG. 1. Pictorial description of the proposed relationship between muscle protein breakdown and urinary output of $N^\tau$-methylhistidine excretion in the rat and human subject.

TABLE 1

FATE OF AN ORAL OR PARENTERAL DOSE OF $(^{14}CH_3)$-$N^{\tau}$-METHYL-
HISTIDINE IN THE RAT AND HUMAN (YOUNG AND MUNRO, 1978)

| Route of excretion | Rat | | Man |
|---|---|---|---|
| | Oral | Parenteral | Parenteral |
| Urine | | | |
| 0–24 h | 92 | 93 | 76 |
| 24–48 h | 4 | 7 | 17 |
| 48–120 h | 3 | 1 | 7 |
| Total | 99 | 101 | 101 |
| Faeces (total) | 2 | 2 | — |
| Expired air (total) | 0·1 | 0·1 | 0·1 |
| Recovery (% dose) | 101 | 103 | 101 |

In a similar fashion, human subjects receiving injections of $^{14}C$-labelled $N^{\tau}$-methylhistidine show essentially complete excretion of the administered radioactivity in the urine (Table 1), only 5 % of radioactivity being excreted in the form of N-acetyl-$N^{\tau}$-methylhistidine. The same recoveries have been recorded for adult rabbits (Harris et al., 1977), cattle (Harris and Milne, 1978) and the adult sheep (Harris and Milne, 1980). However, because of extensive trapping of $N^{\tau}$-methylhistidine as a dipeptide (balenine) in the muscle of the growing sheep, output of $N^{\tau}$-methylhistidine is less than production from muscle protein breakdown (Harris and Milne, 1980). A similar conclusion has been drawn for $N^{\tau}$-methylhistidine metabolism in the pig (Milne and Harris, 1978). Thus, the presence of a pool of balenine has to be evaluated as a factor in $N^{\tau}$-methylhistidine output, especially in growing animals which retain the $N^{\tau}$-methylhistidine in their tissues as the peptide.

The other major criterion is that urinary $N^{\tau}$-methylhistidine should be mainly a product of muscle protein turnover and not of other tissues. We analysed the major organs and tissues of the rat and showed that skeletal muscle is the main repository for $N^{\tau}$-methylhistidine (Haverberg et al., 1975b). Funabiki and his colleagues have analysed the skin of the rat, a tissue not included in our survey, and have concluded that this tissue accounts for 10 % of the total body pool of $N^{\tau}$-methylhistidine (Nishizawa et al., 1977a). They computed that turnover of protein in the skin and in the intestine may contribute about 17 % of the urinary $N^{\tau}$-methylhistidine of the rat.

Recently, Millward et al. (1980) have concluded that skeletal muscle, skin and gastro-intestinal tract contribute only 25, 7 and 10%, respectively, of the $N^\tau$-methylhistidine excreted by the adult rat and, in consequence, that more than half of the amount excreted comes from other unidentified organs. They based their estimates for turnover of $N^\tau$-methylhistidine of individual tissues on rates of synthesis of $N^\tau$-methylhistidine from $^{14}CH_3$-labelled methionine, using measurements of the specific activity of the tissue S-adenosylmethionine pools to correct for precursor specific activity. It is important to recognize that their data were obtained following a 6-h period of infusion of the labelled amino acid, whereas it is known that skeletal muscle varies in protein synthetic capacity throughout the day in relation to meal consumption pattern (Millward and Garlick, 1972). Their low estimate of the contribution of skeletal muscle to the daily urinary output of $N^\tau$-methylhistidine is based on an estimated daily turnover rate of 1% of the protein containing this amino acid compared with their estimate of 4% for the mixed proteins of skeletal muscle. Since the major source of $N^\tau$-methylhistidine in muscle is actin (Haverberg et al., 1974), this implies that actin has a rate of turnover four times longer than that of whole muscle

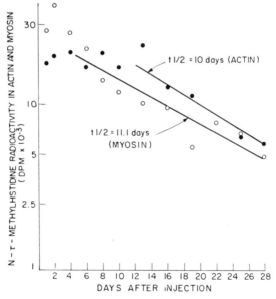

FIG. 2(a).   Change in specific activity with time of labelled $N^\tau$-methylhistidine isolated from actin and myosin in hind leg muscles of rats following a single injection of [$^3$H-methyl]-methionine. Unpublished data of Haverberg, Munro and Young. Each value based on analysis of pooled samples obtained from 18–22 rats.

protein. It would also have to be much longer than that of myosin, the other major protein of the myofibril, since the intact myofibril has a fractional rate of turnover of 3·4 % in adult rats according to these same authors (Waterlow *et al.*, 1978).

Accordingly, we have examined the turnover rate of ($^3$H-methyl)-$N^r$-methylhistidine and ($^3$H-methyl)-methionine in hind limb skeletal muscle protein of growing rats that received the latter amino acid and were killed at various times during the ensuing 28 days. Figure 2(a) shows that the specific activities of $N^r$-methylhistidine in myosin and in actin decline with half-lives of about 11 and 10 days, respectively, while $^{14}CH_3$-methionine labelling of these proteins shows rates of turnover of 13 and 18 days, respectively (Fig. 2(b)). Since the two proteins of the myofibril thus show rather similar turnover rates, it may be concluded that neither differs much from the

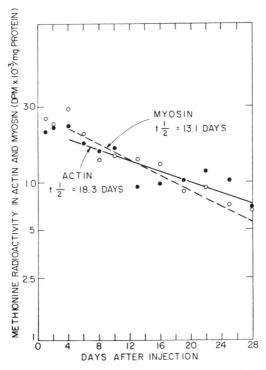

FIG. 2(b). Change in specific activity of methionine isolated from actin and myosin in hind leg muscles of rats following a single injection of [$^3$H-methyl]-methionine. Unpublished data of Haverberg, Munro and Young. Each value based on a pooled sample prepared from 18–22 rats.

overall myofibrillar turnover rate of $3 \cdot 4 \%$/day for adult rats (Waterlow *et al.*, 1978), a value more than three times greater than the estimate for actin made by Millward *et al.* (1980). In support of our data for actin turnover, it should be noted that Lobley and Lovie (1979) have made a careful study of the synthesis of actin and myosin isolated in various ways from rabbit muscle and have concluded that the synthesis rates of these two myofibrillar proteins are identical when isolated directly from the myofibril. They point out that other published procedures for actin isolation (e.g. Swick and Song, 1974) can yield values for its turnover somewhat longer than those for myosin in the same muscle, but they regard actin isolated by such procedures as being less representative of actin as it exists in the myofibril. Even so, none of these estimates for actin turnover suggests a rate as low as that obtained by Millward *et al.* (1980). Clearly, alternative approaches to measuring the relative contributions of different tissues of urinary $N^r$-methylhistidine output are needed, such as precise estimates of arterio-venous differences of the plasma concentrations of this amino acid across various major organs of the body.

## Effects of Diet and Hormones

Changes in the excretion pattern of $N^r$-methylhistidine have been used to study the effects of diet and hormones on muscle protein metabolism in health and disease. Our observations (Young *et al.*, 1973) showed that obese subjects on a prolonged fast undergo a reduction in $N^r$-methylhistidine output, an observation confirmed by Marliss *et al.* (1978), using hypocaloric diets with varying amounts of protein. Studies on young rats receiving diets deficient in protein (Haverberg *et al.*, 1975a) showed that the protein deprivation also results in a rapid reduction in $N^r$-methylhistidine output which is restored by re-introducing protein into the diet, an effect which has been confirmed in man by Nishizawa *et al.*, (1977b) and by us on malnourished children (Young and Munro, 1978). The increased $N^r$-methylhistidine output of the starving rat, and the capacity of refeeding to lower the excretion, has been demonstrated by Ogata *et al.*, (1978). This is confirmed by the lowering in $N^r$-methylhistidine output when carbohydrate is fed to fasting rats for one or more days (Wannemacher and Dinterman, 1980).

Our studies on rats show that corticosteroids and thyroxine affect the urinary output of $N^r$-methylhistidine, presumably through changes in muscle protein breakdown rate. It is well recognized that corticosteroid treatment of rats has a net catabolic effect on skeletal muscle along with enlargement of the liver (Goodlad and Munro, 1959). In a recent study (Tomas *et al.*, 1979), in which we gave growing adrenalectomized rats different replacement levels of corticosteroids over a 7-day period, we found that hormone doses less than 1 mg/100 g body weight daily produced no change in growth, or in liver and muscle weight, whereas 5 or 10-mg doses resulted in prompt cessation of growth, with loss of weight of leg muscles and gain in liver weight and in addition a marked rise in output of Nʳ-methylhistidine. We have shown (Fig. 3) that this response to high doses of

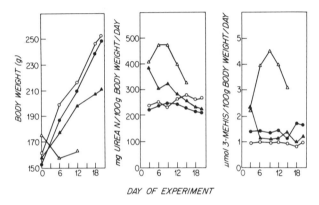

FIG. 3. Body weight changes and urinary urea-N and Nʳ-methylhistidine output in intact rats receiving vehicle injection (●——●) and adrenalectomized rats receiving either vehicle (○——○) or 10 mg of corticosterone/100 g body weight daily, subcutaneously (△——△) or intraperitoneally (▲——▲). Injections were given over a 20-day period. All animals were given an adequate diet and pair-fed with the average food intake of the adrenalectomized rats receiving vehicle injection. For body weight each point represents the mean value for four rats, whereas for urinary output each entry represents a pooled sample from the four rats. Adrenalectomized rats receiving 10 mg of the steroid subcutaneously died before planned completion of experiment.

corticosteroid is not obtained by administering the steroid intraperitoneally, but only by subcutaneous injection, indicating that liver inactivation of the hormone can determine the response to high doses (Santidrian *et al.*, 1981).

It has also been shown that thyroid hormones affect muscle protein turnover. Goldberg *et al.* (1980) found that thyroid deprivation in the rat resulted in diminished muscle protein synthesis and breakdown which can be corrected by treatment with physiological doses of thyroid hormone, while larger doses result in even more extensive muscle protein catabolism uncompensated by increased protein synthesis. We have shown that thyroidectomy of the rat reduces the weight of the liver and increases the weight of the leg muscles relative to body weight, responses that were reversed by increasing doses of thyroxine (Burini *et al.*, 1981). Output of $N^r$-methylhistidine was diminished by thyroidectomy and increased again with thyroxine treatment. This confirms that there is a control of muscle protein breakdown owing to physiological levels of this hormone.

## Conclusions

Partial validation of the use of urinary $N^r$-methylhistidine, as a quantitative index of the rate of myofibrillar protein breakdown in the intact animal has been obtained for the rat and human subject. However, definitive studies of the quantitative contribution made by various tissues are required before the precise proportion of the total daily urinary output of the amino acid owing to myofibrillar protein breakdown in skeletal muscle can be stated. Application of this *in vivo* index has proved useful for exploring the role of hormones in the regulation of muscle protein breakdown. Care is necessary in the interpretation of changes in urinary $N^r$-methylhistidine excretion in relation to muscle protein turnover. However, this method provides, at present, the only viable, non-invasive approach for exploring quantitative aspects of muscle protein breakdown in intact human subjects.

## Discussion

*Millward* (*London School of Hygiene and Tropical Medicine, UK*): I should like to respond to what Munro has said with two tables. Table 2 shows our basic data to which Munro referred. The key measurement is the turnover of methylhistidine in muscle, which is relatively slow, that is, 1·1 %/day. This would only account for about a quarter of the methylhistidine excreted. in the urine. Now there are obvious problems about the accuracy of the measurement of methylhistidine turnover in muscle. First, what was measured was the synthesis rate not the degradation rate, but because it was measured in non-growing rats we assumed that synthesis is the same as degradation; if anything, the synthesis rate would be expected to be slightly higher than the degradation rate because these measurements were made early in the morning in rats which had been feeding all night. Therefore, degradation might actually be slower than the

## TABLE 2

MEASUREMENT OF METHYLHISTIDINE EX-
CRETION BY ADULT FEMALE RATS AND TURN-
OVER IN THE TISSUES FOLLOWING INFUSION OF
$^{14}$C-METHYLMETHIONINE (MILLWARD *et al.*,
1980)

| Tissue | Turnover ($\%/day$) | Excretion ($\%/day$) |
|---|---|---|
| Whole body | 3·25 | 100 |
| GIT | 9·6 | 10 |
| Skin | 2·6 | 7 |
| Muscle | 1·1 | 25 |
| Remainder | 14·4 | 58 |

rate given here. Secondly, we made these measurements by infusing methyl-labelled methionine and measuring the specific activity of the precursor, S-adenosyl methionine (SAM) and of the product, methylhistidine, isolated from the protein. Now the precursor SAM may be compartmented so that our measured value for its specific activity might not be that of the real precursor; however, we have no information on this. The third possible problem is that this is a measurement of the rate of methylation; we assume that methylation occurs synchronously with peptide bond formation and that at least during the 6-h period of the infusion there was no difference between the rate of methylation and the rate of protein synthesis. Of course, over the entire day these two rates have to be the same. Nevertheless, although we can see no reason why these numbers should be inaccurate, we thought it important to try and get at the turnover of methylhistidine in muscle protein by another way. We therefore measured the relative turnover of methylhistidine compared with that of average muscle protein by injecting rats with $^{14}$C-labelled histidine and an hour later isolating protein-bound methylhistidine and histidine and measuring their relative specific activities. We would expect that the precursor pool that is methylated in actin would not be different from that for the histidine which is incorporated into muscle proteins in general.

We carried out this experiment in adult non-growing female rats, and to avoid the possibility of there being slight differences in synthesis in the fed and fasted states we made the measurements both in rats that were fed and in rats that were fasted for 24 h. Table 3 shows that the ratio of the specific activity of histidine to that of methylhistidine was 2·45 in the fed and 2·41 in

TABLE 3

TURNOVER OF METHYLHISTIDINE IN SKELETAL MUSCLE FROM NON-GROWING ADULT FEMALE RATS

| | | |
|---|---|---|
| 1. Overall turnover rate of mixed muscle protein (by infusion of $^{14}C$-tyrosine) (%/day) | | 4·18 |
| 2. Ratio of specific activity of histidine to that of methylhistidine: | | |
| | fed | 2·45 ± 0·5 |
| | fasted 24 h | 2·41 ± 0·7 |
| 3. Calculated methylhistidine turnover rate (%/day) | | 1·71 |
| Percentage of excretion rate | | 45 |

the fasted rats. In other words, there was no difference between the two states, but more importantly, methylhistidine seems to be replaced at a slower rate than the average for histidine in mixed muscle proteins. Now the overall synthesis rate measured in these rats by infusion of $^{14}C$-tyrosine was 4·18 %/day, so if we assume that the incorporation of label into the histidine is equivalent to a fractional rate of 4·18 %/day, then the methylhistidine replacement rate comes out to be 1·71 %/day. If we multiply this by the methylhistidine pool in skeletal muscle, then it would appear that 45 % of the excreted methylhistidine in these rats came from skeletal muscle—so the figure is a little higher than in our first set of experiments (Table 2), but it still indicates that skeletal muscle accounts for less than half the excreted methylhistidine. We are carrying out further experiments because we are still not satisfied that this is a definitive answer.

*Neuberger* (*Charing Cross Hospital Medical School, London, UK*): You made the point that your measurement of methylhistidine turnover in muscle was really a measurement of the methylation rate, not the actual rate of peptide bond formation. Is it right to assume that the rate of peptide bond formation and the methylation rates are in fact the same? Are the same positions methylated in actin, and is it possible to get some proteins that are not methylated?

*Millward:* As far as I know, a specific histidine residue in actin is methylated and all actin molecules that are functional contain this particular methylated residue. In fact, there are mechanisms within the cell for dealing with proteins that contain errors in their amino acid structure or in the extent to which they are modified during translation. If any of these errors or structural abnormalities lead to changes in the function of the protein, then these proteins are often degraded by proteases which seem to exist purely for that purpose.

*Young:* The amino acid sequence of actin is known and the position of the methylated histidine residue is also known (see Young and Munro, 1978). The amino acid occurs in the globular head of the myosin heavy chain and it constitutes the 73rd residue of the actin polypeptide chain.

*Cohen (University of Wisconsin-Madison, USA):* To what extent are there variations in the concentration of methylhistidine in different muscles?

*Young:* There are some differences in the methylhistidine concentration when expressed as a percentage of myofibrillar or mixed proteins in different muscles. It should be remembered that there are differences in the myosin molecules present in different muscle types and only the myosin heavy-chain species from white or fast-twitch muscle fibres contains methylhistidine; that from red muscle or cardiac muscle does not, but actin is methylated in all muscle types.

*Rennie (University College Hospital Medical School, London, UK):* I was interested in the radioisotopic data showing the complete recovery of injected methylhistidine in man. When I have made measurements of the recovery of $N^\tau$-methylhistidine given as meat (250 g/day) in food to people on otherwise meat-free diets, I find a lot of variation in the methylhistidine excretion. I wonder whether this reflects differences between people in bacterial digestion of methylhistidine so that varying amounts escape into the gut? I have some information showing that gut bacteria are very active *in vitro* in catabolizing $N^\tau$-methylhistidine.

*Munro:* This may be a possibility but it has not been investigated extensively. We have recorded a study on two subjects who received either a meat-free diet or one containing meat; the latter increased urinary output of $N^\tau$-methylhistidine, so we can conclude that exclusion of dietary sources of $N^\tau$-methylhistidine is important in applying this procedure to the study of muscle metabolism (Young and Munro, 1978).

*Walser (Johns Hopkins University, Baltimore, USA):* Do variations in protein intake of meat-free diets affect methylhistidine excretion in adults or children?

*Young:* Our experience with healthy adults has shown that variations in protein intake provided by a meat-free diet do not bring about changes over a 2-week period in the excretion of methylhistidine. Studies by Holmgren (1974) with young children given two levels of protein, one low and one generous, showed, as I recall, that methylhistidine excretion was increased at the higher protein intake.

*Munro:* This reflects the difference between growth and adulthood. We have done some work in Guatemala on re-habilitation which showed an increased breakdown when the increment in growth occurred (Young and Munro, 1978).

*Neuberger:* Methylhistidine is in a similar position as a marker of muscle protein as hydroxyproline peptides are in relation to collagen metabolism.

*Munro:* The hydroxyproline peptides are very often released from pre-collagen rather than from the breakdown of collagen. There is a contribution from collagen which is not yet stabilized. Secondly, hydroxyproline has a metabolic pathway leading to various intermediary metabolites, whereas methylhistidine is a final product.

*Neuberger:* But if one wanted to study the metabolism of collagen, which is quite an important part of the total body protein, measurement of hydroxyproline peptides in the urine might give one at least a new approach provided it can be quantitatively related to collagen turnover.

*Millward:* This is very pertinent to the concept of wastage, because these losses of collagen during its maturation are exactly the same as those which we assume to occur as part of the increased degradation during muscle growth which we call wastage. However, as Munro says, the problem with quantitating collagen turnover from hydroxypyroline excretion is that most of it (about 6/7) is oxidized before excretion (see Waterlow *et al.*, 1978, Chapter 15).

## References

BURINI, R., SANTIDRIAN, S., MOREYRA, M., BROWN, P., MUNRO, H. N. and YOUNG, V. R. (1981). Interaction of thyroid status and diet on muscle protein breakdown in the rat, as measured by $N^r$-methylhistidine excretion. *Metabolism*, in press.

GOLDBERG, A. L., TISCHLER, M., DE MARTINO, G. and GRIFFIN, G. (1980). Hormonal regulation of protein degradation and synthesis in skeletal muscle. *Fed. Proc.*, **39**, 31–6.

GOODLAD, G. A. J. and MUNRO, H. N. (1959). Diet and the action of cortisone on protein metabolism. *Biochem. J.*, **73**, 343–8.

HARRIS, C. I. and MILNE, G. (1978). Urinary excretion of 3-methylhistidine in cattle as a measure of muscle protein degradation. *Proc. Nutr. Soc.*, **38**, 11A.

HARRIS, C. I. and MILNE, G. (1980). The urinary excretion of $N^r$-methyl histidine in sheep: an invalid index of muscle protein breakdown. *Br. J. Nutr.*, **44**, 129–40.

HARRIS, C. I., MILNE, G., LOBLEY, G. E. and NICHOLAS, G. A. (1977). 3-Methylhistidine as a measure of skeletal-muscle protein catabolism in the adult New Zealand white rabbit. *Biochem. Soc. Trans.*, **5**, 706–8.

HAVERBERG, L. N., MUNRO, H. N. and YOUNG, V. R. (1974). Isolation and quantitation of $N^r$-methylhistidine in actin and myosin of rat skeletal muscle: use of pyridine elution of protein hydrolysates on ion exchange resin. *Biochim. Biophys. Acta*, **371**, 226–37.

HAVERBERG, L. N., DECKELBAUM, L., BILMAZES, C., MUNRO, H. N. and YOUNG, V. R. (1975a). Myofibrillar protein turnover and urinary $N^r$-methylhistidine output: response to dietary supply of protein and energy. *Biochem. J.*, **152**, 503–10.

HAVERBERG, L. N., OMSTEDT, P. T., MUNRO, H. N. and YOUNG, V. R. (1975b). $N^r$-methylhistidine content of mixed proteins in various rat tissues. *Biochim. Biophys. Acta*, **405**, 67–71.

HOLMGREN, G. (1974). Effect of low, normal and high dietary protein intakes on urinary amino acid excretion and plasma aminogram in children. *Nutrition and Metabolism*, **16**, 223–37.

LOBLEY, G. E. and LOVIE, J. M. (1979). The synthesis of myosin, actin and the major protein fractions in rabbit and skeletal muscle. *Biochem. J.*, **182**, 867–74.

LONG, C. L., HAVERBERG, L. N., YOUNG, V. R., KINNEY, J. M., MUNRO, H. N. and GEIGER, J. W. (1975). Metabolism of 3-methylhistidine in man. *Metabolism*, **24**, 929–35.

MARLISS, E. B., MURRAY, F. T. and NAKHOODA, A. F. (1978). The metabolic response to hypocaloric protein diets in obese man. *J. Clin. Invest.*, **62**, 468–79.

MILLWARD, D. J., BATES, P. C., GRIMBLE, G. K., BROWN, J. G., NATHAN, M. and RENNIE, M. J. (1980). Quantitative importance of non-skeletal-muscle sources of $N^r$-methylhistidine in urine. *Biochem. J.*, **190**, 225–8.

MILLWARD, D. J. and GARLICK, P. J. (1972). The pattern of protein turnover in the whole animal and the effect of dietary variations. *Proc. Nutr. Soc.*, **31**, 257–63.

MILNE, G. and HARRIS, C. I. (1978). The inadequacy of urinary 3-methylhistidine excretion as an index of muscle protein degradations in the pig. *Proc. Nutr. Soc.*, **37**, 18A.

NISHIZAWA, N., NOGUCHI, T., HAREYAMA, S. and FUNABIKI, R. (1977a). Fractional flux rates of $N^r$-methylhistidine in skin and gastrointestine: the contribution of these tissues to urinary excretion of $N^r$-methylhistidine in the rat. *Br. J. Nutr.*, **38**, 149–51.

NISHIZAWA, N., SHIMBO, M., HAREYAMA, S. and FUNABIKI, R. (1977b). Fractional catabolic rates of myosin and actin estimated by urinary excretion of $N^r$-methylhistidine: the effect of dietary protein level on catabolic rates under conditions of restricted food intake. *Br. J. Nutr.*, **37**, 345–53.

OGATA, E. S., FOUNG, S. K. H. and HOLLIDAY, M. A. (1978). The effects of starvation and refeeding on muscle protein synthesis and catabolism in the young rat. *J. Nutr.*, **108**, 759–65.

SANTIDRIAN, S., MOREYRA, M., MUNRO, H. N. and YOUNG, V. R. (1981). Effect of corticosterone and its route of administration on muscle protein breakdown, measured *in vivo* by urinary excretion of $N^r$-methylhistidine in rats: response to different levels of dietary protein and energy. *Metabolism*, in press.

SWICK, R. W. and SONG, H. (1974). Turnover rates of various muscle proteins. *J. Anim. Sci.*, **38**, 1150–7.

TOMAS, F. M., MUNRO, H. N. and YOUNG, V. R. (1979). Effect of glucocorticoid administration on the rate of muscle protein breakdown *in vivo* in rats, as measured by urinary excretion of N$^\tau$-methylhistidine. *Biochem. J.*, **178**, 139–46.

UAUY, R., WINTERER, J. C., BILMAZES, C., HAVERBERG, L. N., SCRIMSHAW, N. S., MUNRO, H. N. and YOUNG, V. R. (1978). Changing pattern of protein metabolism in aging humans: relationships among rates of whole body protein breakdown, N$^\tau$-methylhistidine excretion and obligatory urinary nitrogen losses. *J. Gerontol.*, **33**, 663–71.

WANNEMACHER, R. W. and DINTERMAN, R. E. (1980). Diurnal response in endogenous amino acid oxidation of meal-fed rats. *Biochem. J.*, **190**, 663–71.

WATERLOW, J. C., GARLICK, P. J. and MILLWARD, D. J. (1978). *Protein Turnover in Mammalian Tissues and in the Whole Body*, Elsevier North Holland, Amsterdam.

YOUNG, V. R., ALEXIS, S. D., BALIGA, B. S., MUNRO, H. N. and MUECKE, W. (1972). Metabolism of administered 3-methylhistidine: lack of muscle tRNA charging and quantitative excretion as 3-methylhistidine and its N-acetyl derivative, *J. Biol. Chem.*, **247**, 3592–600.

YOUNG, V. R., HAVERBERG, L. N., BILMAZES, C. and MUNRO, H. N. (1973). Potential use of 3-methylhistidine excretion as an index of progressive reduction in muscle protein catabolism during starvation. *Metabolism*, **22**, 1929–36.

YOUNG, V. R. and MUNRO, H. N. (1978). N$^\tau$-methylhistidine (3-methylhistidine) and muscle protein turnover: an overview. *Fed. Proc.*, **37**, 2291–300.

# 40

# Protein Metabolism During Exercise

M. J. RENNIE, R. H. T. EDWARDS,
*Department of Human Metabolism,*
*University College Hospital Medical School, London, UK*

D. HALLIDAY,
*Clinical Research Centre, Harrow, UK*

C. T. M. DAVIES,
*Department of Physiology,*
*University of Nottingham Medical School,*
*Nottingham, UK*

D. E. MATTHEWS
*Department of Internal Medicine,*
*Washington University School of Medicine,*
*St. Louis, Missouri, USA*

and

D. J. MILLWARD
*Clinical Nutrition and Metabolism Unit,*
*London School of Hygiene and Tropical Medicine,*
*London, UK*

Surprisingly little work has been carried out on the effects of exercise on protein metabolism. Gontzea *et al.* (1975) showed that multiple daily bouts of intense exercise induces negative nitrogen balance in healthy normal subjects. Studies on athletes competing in long distance events suggested increased amino acid oxidation, and a fall in the size of the $\alpha$-amino-N pool (Refsum and Strömme, 1974; Decombaz *et al.*, 1979), but in common with many other workers the authors seemed to be concerned mainly with the significance of the energy supplied by protein compared to the total energy requirement of exercise. On this basis the energy supplied by protein is only a small proportion at best, a conclusion also reached after many classical studies (Fick and Wislicenus, 1866; Cathcart, 1925). However, gluconeogenesis from amino acids may be very significant in terms of supplying glucose as an irreplaceable fuel for the nervous system; hypoglycaemia has been recognized as a limiting factor in long-term exercise for many years (Christensen and Hansen, 1939). The studies of Felig and Wahren and their collaborators (Ahlborg *et al.*, 1974; Wahren *et al.*, 1976) on the splanchnic and muscle exchange of amino acids have been outstanding in the area, but the picture they provide of increased branched chain amino acid uptake and alanine and glutamine release by muscle, leading to increased alanine and glutamine uptake by the splanchnic tissues, provides us with few answers to questions about the mechanisms of adaptation of the metabolic machinery to repeated exercise or possible changes in protein requirements.

It is now well known (Holloszy and Booth, 1976) that repeated moderate exercise results in increases in muscle mitochondrial respiratory capacity, blood supply and oxygen extraction. There may also be changes in the contractile properties of muscle fibres, possibly even changes of fibre type (Saltin *et al.*, 1977). All these changes would lead to increased performance of endurance exercise owing to an increase in maximal aerobic capacity ($\dot{V}_{O_2 max}$) and stamina, and they must come about through alterations in the metabolic architecture of muscle, and perhaps of other tissues as well. It is apparent, therefore, that the alterations in composition must be owing to alterations in protein synthesis and degradation about which we know almost nothing.

Rather than looking at relatively long-term effects we have decided that the acute effect of exercise may give us clues to the long-term adaptive changes and we have studied the effects of 2–4-h long bouts of exercise on amino acid oxidation and protein flux, and on $N^{\tau}$-methylhistidine excretion.

## Effects of Exercise on Catabolism

We have studied subjects who were fed and also fasted, with and without extra ingested glucose. We found that exercise always caused a significant increase in nitrogen production (Rennie *et al.*, 1980; Rennie *et al.*, 1981).

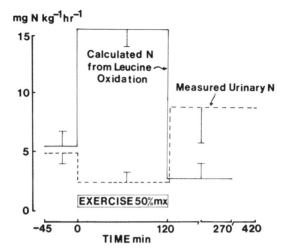

FIG. 1. Changes in N excretion during and after 2 h exercise at 50 % $\dot{V}_{O_2 max}$ in the fasted state (a) measured and (b) calculated from $1\text{-}^{13}C$-leucine oxidation, assuming that the same proportion of the leucine flux and the total N flux are oxidized. Means for 4 subjects $\pm$ SD.

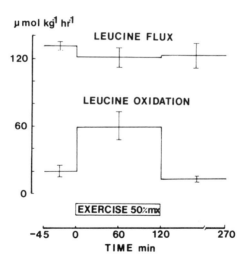

FIG. 2. Leucine flux and leucine oxidation, calculated from labelling of plasma leucine and breath $CO_2$, in 4 men before, during and after exercise for 2 h at 50 % $\dot{V}_{O_2 max}$ in the fasted state.

However, because of decreased urine production during exercise, there was a considerable lag between the conduct of the exercise and the appearance of increased rates of nitrogen excretion (Fig. 1). The use of 1-$^{13}$C-leucine, either ingested with liquid food at regular intervals, or continuously infused, helped to resolve the problem of when the increased amino acid catabolism actually took place. Exercise apparently always caused an increase in oxidation of labelled leucine, shown by an increase in the $^{13}CO_2$ production rate which was confined to the period of exercise (Fig. 2). The effect of additional glucose on amino acid oxidation was impossible to assess because of interference from $^{13}CO_2$ produced from the glucose syrup, which was itself relatively highly labelled. Nevertheless, nitrogen excretion was diminished, suggesting a decrease in the catabolism of amino acids, and plasma concentrations of branched chain keto acids fell during exercise (see later).

## Effects of Exercise on Protein Turnover

We have also made two sorts of measurements of amino acid turnover, using the $^{15}$N-ammonia end product method (Golden and Waterlow, 1977) or constant infusion of 1-$^{13}$C-leucine (Matthews et al., 1980). The $^{15}$N-glycine ammonia end product method was used in subjects who were fed periodically throughout 2 days of investigation, on the second of which they exercised for 4 h. The leucine infusion studies were performed on subjects who were either fasted or who ingested small amounts of glucose every 15 min throughout the period of investigation. The flux measurements in the $^{15}$N-glycine study showed that exercise caused an apparent increase in protein flux by about 50%. After exercise flux was still elevated above the resting level. When the increased amino acid oxidation (on the basis of nitrogen excretion) was assigned entirely to the exercise period, the unknown components of protein turnover, i.e. synthesis and breakdown, could be calculated (Fig. 3). The results suggested that during exercise protein synthesis was depressed, and that protein breakdown was elevated; after exercise protein breakdown fell slightly but was still elevated above the resting value, whereas protein synthesis rose above breakdown. Thus, exercise appeared to have caused a negative nitrogen balance in the whole body during exercise and a net positive balance following exercise. A possible mechanism was, therefore, apparently available for adaptations of protein metabolism as a result of exercise, with adaptive changes taking place owing to increased synthesis in the post-exercise period. The changes

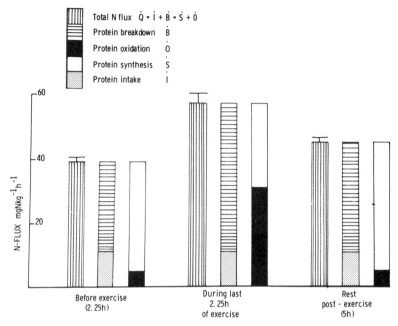

FIG. 3.   Total N flux and its components (dietary intake, whole body protein breakdown, whole body protein synthesis and amino acid oxidation) before, during and after 3·75 h exercise at 50–60% $\dot{V}_{O_2 max}$ in the fed state. Means for 6 studies on 4 subjects ± SD.

in plasma hormones (insulin, cortisol and glucagon) were entirely consistent with such a scheme.

The results of experiments conducted with leucine infusion in subjects completely fasted or fasted except for receiving glucose confirmed an increase in amino acid oxidation during exercise, but in these experiments exercise caused no change in the overall protein flux (Fig. 2). It appeared, therefore, that although protein synthesis was depressed, as found in the [15]N-glycine study, protein breakdown (which in the absence of dietary intake would be equal to the whole of the flux) was not increased as a result of exercise. In the period following exercise protein oxidation was depressed below the pre-exercise value, but because the total flux remained similar, the proportion of the flux owing to protein synthesis appeared to increase after exercise. The difference between the results for protein breakdown between the two studies may be a consequence of the differences in dietary state or the different methods used. A particular focus of attention is the possibility that exercise affects the source of urinary ammonia, so that the apparent increase in flux observed with $^{15}$N-glycine during exercise may be owing to some other mechanism.

## *Effects of Exercise on Muscle Protein Degradation*

We were particularly interested in possible changes in myofibrillar protein turnover as a result of exercise and expected that, if anything, exercise might increase muscle protein breakdown. However, we were surprised to find that in fed man (on a meat-free diet) and during fasting $N^\tau$-methylhistidine excretion fell during exercise. In the fasted state, when extra glucose was given, it had no effect on the absolute levels of $N^\tau$-methylhistidine excretion.

TABLE 1

PLASMA AND MUSCLE CONTENT OF $N^\tau$-METHYLHISTIDINE, AND RATES OF EXCRETION OF $N^\tau$-METHYLHISTIDINE AND CREATININE DURING AND AFTER EXERCISE

|  | *Rest* | *Exercise* | *Post-exercise* |
|---|---|---|---|
| Muscle $N^\tau$-methylhistidine (nmol/ml i.c. water) | $17 \cdot 0 \pm 2 \cdot 4$ | $12 \cdot 0 \pm 3 \cdot 6$ | $16 \cdot 0$ |
| Plasma $N^\tau$-methylhistidine (nmol/ml) | $12 \cdot 0 \pm 3 \cdot 0$ | $12 \cdot 5 \pm 2 \cdot 8$ | $11 \cdot 8 \pm 3 \cdot 1$ |
| $N^\tau$-methylhistidine excretion (nmol/kg/h) | $141 \pm 18$ | $32 \pm 30 \cdot 1$ | $151 \pm 28$ |
| Creatinine excretion ($\mu$mol/kg/h) | $7 \cdot 6 \pm 1 \cdot 2$ | $3 \cdot 5 \pm 0 \cdot 66$ | $8 \cdot 9 \pm 1 \cdot 6$ |
| $N^\tau$-methylhistidine/creatinine $\times 10^3$ | $18 \cdot 4 \pm 1 \cdot 3$ | $9 \cdot 17 \pm 3 \cdot 66$ | $17 \cdot 0 \pm 1 \cdot 9$ |

Muscle $N^\tau$-methylhistidine estimated in needle biopsy samples of |quadriceps. Rates of excretion are means for 4 men studied on 8 occasions over a complete day at rest before exercise, during $3 \cdot 75$ h exercise at $50 \%$ $\dot{V}_{O_2 max}$, and during 5 h post-exercise.
Means $\pm$ S.E.M.

Even when the fall in excretion was corrected for by expressing $N^\tau$-methylhistidine as a ratio with creatinine, the $N^\tau$-methylhistidine/creatinine ratio was depressed by up to $50 \%$ during exercise (Table 1). The $N^\tau$-methylhistidine/creatinine ratio should be directly proportional to fractional degradation of myofibrillar protein and the results, therefore, suggested that exercise decreased muscle protein degradation. This impression was strengthened when we discovered that neither plasma $N^r$-methylhistidine nor intracellular free $N^r$-methylhistidine in muscle, obtained by muscle biopsy before and after exercise, had increased as a result of the exercise. In fact, the intracellular free $N^r$-methylhistidine concentration appeared to fall as a result of exercise.

We still know almost nothing of the long-term effects of moderate exercise on muscle protein turnover itself, but we have some indication from cross-sectional studies on untrained subjects and on highly trained

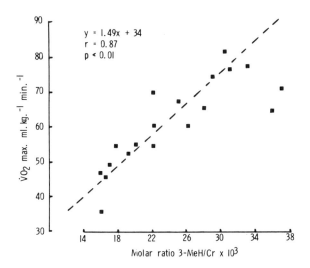

FIG. 4.   Relationship between maximal oxygen consumption during exercise ($\dot{V}_{O_2 max}$) and the $N^r$-methylhistidine/creatinine molar ratio estimated at rest for 19 subjects, including 11 ultra-marathon runners.

runners, who compete over distances between 25 and 100 miles, that $N^r$-methylhistidine excretion (on a meat-free diet) is related to the subject's maximal oxygen consumption during exercise (Davies *et al.*, 1981) (Fig. 4). These data are difficult to interpret since there are a number of possible explanations, e.g. that repeated exercise increases protein turnover and therefore $N^r$-methylhistidine output, or that well-trained subjects have a natural or acquired preponderance of slow twitch muscle fibres which may have a faster turnover rate for $N^r$-methylhistidine containing myofibrillar proteins.

# Where do the Changes in Whole Body Protein Synthesis and Degradation Occur?

Although the information available (except for data on $N^\tau$-methylhistidine) gives no indication where changes in protein turnover are occurring, we feel that during exercise depression of protein synthesis and possible elevation of breakdown is likely to occur in the viscera. Since blood flow to the splanchnic region is decreased during exercise, the decreased amino acid and oxygen supply might conceivably have this effect (Mortimore and Schworer, 1980; Taegtmeyer and Lesch, 1980). Some support for this comes from measurements made by Felig and Wahren and co-workers (Wahren *et al.*, 1976) who showed that during long-term exercise there was a net loss of amino acids, particularly of the branched chain amino acids, by the splanchnic regions and net uptake and possibly catabolism by leg muscle. Further evidence to support the idea comes from the finding that rats exercised to exhaustion by treadmill running actually show a fall in liver protein mass (Dohm *et al.*, 1980*b*).

In work on rats, treadmill exercise or electrical stimulation of perfused muscle appeared to cause a depression of muscle protein synthesis rate (Rennie *et al.*, 1980; Bates *et al.*, 1980; Dohm *et al.*, 1980*a,b*). In recent studies with more appropriate techniques than we previously used, we have not been able to find evidence of an effect of 40 min moderate treadmill exercise on synthesis rates of rat mixed muscle protein (Bates, Rennie and Millward, unpublished results), although this does not rule out effects on specific organelles or effects of more intense or longer exercise. In any event, any elevation of muscle protein synthesis rate appears to be confined to the post-exercise period (Rennie *et al.*, 1980; Bates *et al.*, 1980; Rogers *et al.*, 1979; Dohm *et al.*, 1980*a,b*). On *a priori* grounds, any substantial increase in protein synthesis is likely to depend on increases of ribosomal capacity rather than on translation rate and must, therefore, await arrival of new ribosomes.

# Effects of Exercise on Metabolism of Branched Chain Amino and Keto Acids

The branched chain amino acids (BCAA) have been suggested to have a controlling role in protein synthesis and degradation (Fulks *et al.*, 1975). We were, therefore, naturally interested in the effect of exercise on plasma amino acid (including BCAA) concentrations and also on the concentrations of the branched chain α-keto acids (BCKA). We found that during exercise plasma BCAA rose slightly within the first 0·5–1 h, but that following this, even in the fed state, plasma BCAA fell. The concentrations of the plasma BCKA fell early in exercise but then rose after an hour. The rise in plasma BCKA was maximal at the end of exercise, but was sustained throughout the post-exercise period—in some cases for as long as samples were taken, some 7 h after exercise. These results, taken together with the changes in $^{13}CO_2$ derived from leucine, suggest that the BCAA are subjected to increasing transamination and decarboxylation during exercise. A model which implies that concentrations of BCAA control protein synthesis would be consistent with the observed changes in increased BCAA oxidation and decreased protein synthesis during exercise, but not necessarily with an increase in protein synthesis following exercise. However, if the modulating variable were the BCKA concentration, then the model would fit rather well with the observed changes.

## Implications of the Effect of Exercise on Amino Acid Metabolism and Whole Body Protein Turnover

The finding that amino acid oxidation is increased by exercise suggests that chronic exercise would result in an increased dietary protein requirement. This of course fits in well with the preconceptions of many coaches and athletes and the dietary practice of many manual labourers, but the proposition may well be very difficult to prove. There may indeed be an adaptation to repeated exercise which minimizes the effects of exercise on amino acid catabolism. Gontzea *et al.* (1975) have reported such an adaptation with a fall in the nitrogen excretion induced by exercise after one week of intense exercise for 2 h daily. In any case it is likely that a small increased requirement for dietary protein would be easily met in fulfilling the extra energy requirement of physical work, since most food staples eaten for their energy content would carry with them sufficient protein.

Another likely implication of the results is the possibility that regular exercise might have an effect in controlling obesity, not through the direct energy expenditure incurred during exercise, but through the extra metabolic cost of the post-exercise increases in protein turnover which, integrated over a substantial period, may be considerable.

## Acknowledgements

This work was supported by the Muscular Dystrophy Group of Great Britain, Muscular Dystrophy Association of America, Medical Research Council, Action Research, and The Wellcome Trust.

## Discussion

*Waterlow* (*London School of Hygiene and Tropical Medicine, UK*): For clarification, in Fig. 4, where $\dot{V}_{O_2 max}$ is related to methylhistidine/creatinine ratio, was the methylhistidine measured in samples taken during exercise?

*Rennie:* No, those measurements were carried out at rest.

*Waterlow:* Secondly, you said the evidence shows no increased breakdown of muscle protein; where does the increased alanine come from, and the increased ammonia, which Felig and others (Felig and Wahren, 1971) have shown quite clearly are produced in exercised muscle?

*Rennie:* It has been suggested many times that the nitrogen of alanine might come from increased transfer from the branched chain amino acids and the carbon might come from increased pyruvate flux through muscle. We have shown that there is increased leucine oxidation when we use a $^{13}$C-leucine tracer, and also that the plasma concentrations of the branched chain α-keto acids all rise towards the end of the exercise period. As far as ammonia production is concerned, a sizeable amount of ammonia is produced during exercise only when the amount of ATP starts to become limiting. The deamination of AMP to IMP with the release of ammonia may be a mechanism which, by removing a product of the myokinase reaction, allows it to preserve the ATP/ADP ratio at a level sufficiently high to allow the myofibrillar apparatus to continue working. I doubt whether the purine nucleotide cycle contributes much ammonia at the level of exercise (50 % $\dot{V}_{O_2 max}$) which our subjects were undergoing. There might be glutamine production which we do not know about, but ammonia production appears to go up with lactate production, and here that was relatively small.

*Felig* (*Yale University School of Medicine, USA*): We still have difficulty in explaining amino acid balance. Our data on prolonged exercise agree with yours, in that we find a net output of branched chain amino acids from the liver in very prolonged exercise. Furthermore, there is a net extraction of these amino acids by the exercising leg. The problem we still have, on the basis of AV differences, is to explain the source of the nitrogen for the increase in alanine output that is occurring even with short-term exercise. The data I showed earlier (p. 51) were all from subjects exercised up to 40 min at varying intensity, and in those subjects alanine output was stimulated. I think this problem remains unresolved.

*Young (Massachusetts Institute of Technology, USA):* We have begun collaborative studies to explore the effects of exercise and training on some aspects of whole body protein metabolism. Our preliminary data indicate a consistent reduction in leucine flux, contrasting with your observation in subjects studied during fasting or in the post-absorptive state. We do, however, note a marked increase in leucine oxidation. The level of physical activity that our subjects underwent for a 2-h period was approximately 65 % of their $\dot{V}_{0_2 max}$. Have you explored the effects of varying intensities of exercise?

*Rennie:* In our studies, over the period of exercise there is a slight rise in the specific activity of plasma leucine, so there might be a fall in leucine flux. We have yet to see whether increasing the intensity of exercise affects leucine flux. If it is true that the natural abundance of $^{13}$C is greater in carbohydrate than in fat and that the harder the work the more carbohydrate is burnt, then one might expect to see an increased $^{13}CO_2$ production from pre-existing stores, which would complicate the issue. However, Millward has just carried out an experiment to examine this at 30, 50 and 75 % of $\dot{V}_{0_2 max}$ and there was no increase in $^{13}CO_2$ production at higher levels of activity.

*Stein (University of Pennsylvania School of Medicine, Philadelphia, USA):* We wondered whether the increased amounts of protein taken by athletes led to an increase in the basal protein synthesis rate. We measured the resting nitrogen flux and synthesis rates in two groups of males matched for age, weight and height. One group consisted of sedentary office workers who engaged in no voluntary physical activity. The other group were members of a light heavyweight boat crew who were in training for the 1980 Olympics. Protein synthesis rates were determined with $^{15}$N-glycine using a primed continuous oral administration method. Surprisingly, there was no difference in the whole body protein synthesis rate between the two groups (345 and 335 g protein/day).

*Felig:* We have recently carried out some studies to look at the effect of training on the sensitivity of tissues, particularly muscle, to insulin, and we are using net glucose uptake as the index. We find that tissue sensitivity to insulin is increased by training and this correlates directly with changes in $\dot{V}_{O_2 max}$.

We have also shown that changes in glucose exchange across the liver may be mediated by changes in blood glucose *per se*, so long as insulin is present in minimal amounts. The other point to keep in mind is that in exercise, while arterial or venous insulin concentrations may be falling, the delivery of insulin to specific tissues, such as muscle, is increasing because of a greater blood flow, so that there may be no deficiency of insulin in the target tissues.

*Rennie:* The changes in cortisol, insulin and glucagon during our study, when subjects were fed, are entirely consistent with the idea that gut is the source of the protein that was being broken down. The decrease in insulin and the increase in cortisol and glucagon would fit quite nicely with that.

*Alleyne (University of the West Indies, Kingston, Jamaica):* Some of the data which you presented may be complicated by changes in renal haemodynamics and acid-base status. For example, the rise in plasma urea and the fall in urinary urea N would be adequately explained by a drop in glomerular filtration rate and many of the other changes in the plasma may be related to this fall. Do you have any data on whole blood pH? It is well known that blood ammonia will rise and hepatic production of urea may fall as arterial blood pH falls.

*Rennie:* I fully accept the problems of kidney blood flow. The work of Castenfors (1967) tends to support the idea that there is a decreased renal blood flow during exercise. As far as pH is concerned, we have not measured blood pH. The amount of lactate in the blood was only about 2·2 mM, so there was no lactic acidosis to speak of. In the swimming rat you studied, exercise might have caused lactic acidosis, but even then there was only a 15 % increase in ammonia production in tissues taken from rats after swimming. Therefore I do not think that would have been important in our studies.

*Munro (Tufts University, Boston, USA):* Does your work have any implications for requirements for branched chain amino acids in relation to the degree of exercise? The studies I did 45 years ago (Cuthbertson *et al.*, 1937) showed that the effects on nitrogen balance of exercise, without compensatory increments in the intake of energy, were more severely adverse when the exercise was carried out in the absorptive period after a meal than in the fasting state.

*Rennie:* It seems logical that, if oxidation of branched chain amino acids is increased during exercise, that would increase the requirements in subjects taking large amounts of exercise. However, if they have simply increased their dietary energy to take account of the extra work they were doing, they would certainly get enough protein.

I am, of course, aware of the work of Sir David Cuthbertson, Professor Munro and McGirr in the 1930s, showing a larger effect of exercise on protein catabolism after a meal. Before reading of those early studies we have been rather surprised to find more protein catabolism in the study we did in the fed state than in the one conducted in the fasting state, although of course the methods used were different, as was the work time. It is possible that when people have actually adapted their muscles by training there might not necessarily be big differences in protein turnover.

*Garrow (Clinical Research Centre, Harrow, UK):* It is surprising the extent to which the renal excretion of $N^r$-methylhistidine decreased during exercise (Table 1); the amount of $N^r$-methylhistidine/ml intracellular water in the muscle biopsy was also decreased. Could it be that during exercise catabolism of muscle had increased with an increased release of $N^r$-methylhistidine, but that this was disguised by an increased water content of exercising muscle? Secondly, Munro showed us that when a dose of $N^r$-methylhistidine is injected intravenously it takes a day or two to come out. Does the evidence actually contravert the idea that there was increased catabolism?

*Rennie:* We measured the wet and dry weight and used intra- and extracellular sodium to try to calculate the partition. As far as we could tell there was no oedema of the muscle. I am at rather a loss to explain why we got this fall in urinary $N^r$-methylhistidine, except to say it suggests a fall in actin degradation rate. It does not seem likely that the subjects lost $N^r$-methylhistidine in sweat, because it appears to be present only in very low abundance in sweat relative to urine (Gitlitz *et al.*, 1974).

*Neuberger (Charing Cross Hospital Medical School, London, UK):* You take your muscle from one point. It may not be representative of the muscle mass as a whole.

*Rennie:* Yes, that is a good point.

# References

AHLBORG, G., FELIG, P., HAGENFELDT, L., HENDLER, R. and WAHREN, J. (1974). Substrate turnover during prolonged exercise in man. *J. Clin. Invest*, **53**, 1080–90.

BATES, P. C., DE COSTER, T., GRIMBLE, G. K., HOLLOSZY, J. O., MILLWARD, D. J. and RENNIE, M. J. (1980). Exercise and muscle protein turnover in the rat. *J. Physiol.*, **303**, 41P.

BUSE, M. G. and REID, S. S. (1975). Leucine: a possible regulator of protein turnover in muscle. *J. Clin. Invest.*, **58**, 1251–61.

CASTENFORS, J. (1967). Renal function during exercise. *Acta Phys. Scand.*, **70**, Suppl. 293.

CATHCART, E. P. (1925). The influence of muscle work on protein metabolism. *Physiol. Rev.*, **5**, 225–43.

CHRISTENSEN, E. H. and HANSEN, O. (1939). Hypoglykamie, Arbeitsfähigkeit und Ehrährung. *Skand Arch. Physiol.* **81**, 172–9.

CUTHBERTSON, D. P., McGIRR, J. L. and MUNRO, H. N. (1937). A study of the effect of overfeeding on protein metabolism of man. *Biochem. J.*, **31**, 2293–305.

DAVIES, C. T. M., EDWARDS, R. H. T., NATHAN, M., RENNIE, M. J. and THOMPSON, M. W. (1981). A relationship between $\dot{V}_{O_2 max}$ and muscle protein turnover in man. *J. Physiol.*, **313**, 47P.

DECOMBAZ, J., REINHARDT, P., ANANTHARAM, K., VAN GLUTZ, G. and POORTMANS, J. R. (1979). Biochemical changes in a 100 km run: free amino acids, urea and creatinine, *Eur. J. Appl. Physiol.*, **41**, 61–72.

DOHM, G. L., KASPAREK, G. J., TAPSCOTT, E. B., BARAKAT, H. A. and BEECHER, G. R. (1980*a*). Protein synthesis in liver and muscle during recovery from exercise. *Fed. Proc.*, **39**, 290. 130A.

DOHM, G. L., KASPAREK, G. J., TAPSCOTT, E. B. and BEECHER, G. R. (1980*b*). Effect of exercise on synthesis and degradation of muscle protein. *Biochem J.*, **188**, 255–62.

FELIG, P. and WAHREN, J. (1971). Amino acid metabolism in exercising man. *J. Clin. Invest.*, **50**, 2703–14.

FICK, A. and WISLICENUS, J. (1866). On the origin of muscular power. *London, Edinburgh and Dublin Philosophical Magazine*, **31**, 485–503.

FULKS, R. M., LI, J. B. and GOLDBERG, A. L. (1975). Effects of insulin, glucose, and amino acids on protein turnover in rat diaphragm. *J. Biol. Chem.*, **250**, 290–8.

GITLITZ, R. H., SONDERMAN, F. W. and HOHNADEL, D. C. (1974). Ion-exchange chromatography of amino acids in sweat collected from healthy subjects during sauna bathing. *Clin. Chem.*, **20**, 1305–12.

GOLDEN, M. H. N. and WATERLOW, J. C. (1977). Total protein synthesis in elderly people: a comparison of results with ($^{15}$N) glycine and ($^{14}$C) leucine. *Clin. Sci. Mol. Med.*, **53**, 277–88.

GONTZEA, I., SUTZESCU, R. and DUMITRACHE, S. (1975). The influence of adaptation to physical effort on nitrogen balance in man. *Nut. Reports Intl*, **11**, 231–6.

HOLLOSZY, J. O. and BOOTH, F. W. (1976). Biochemical adaptations to endurance exercise in muscle. *Ann. Rev. Physiol.*, **38**, 273–91.

MATTHEWS, D. E., MOTIL, K. E., ROHRBAUGH, D. K., BURKE, J. F., YOUNG, V. R. and BIER, D. M. (1980). Measurement of leucine metabolism in man from a continuous infusion of L-(1-$^{13}$C) Leucine. *Am. J. Physiol.*, **238**, E473–E479.

MORTIMORE, G. E. and SCHWORER, C. M. (1980). Application of liver perfusion as an *in vitro* model in studies of intracellular protein degradation. In *Protein Degradation in Health and Disease*, Ciba Foundation Symposium 75, Excerpta Medica, Amsterdam, p. 281.

RENNIE, M. J., EDWARDS, R. H. T., DAVIES, C. T. M., KRYWAWYCH, S., HALLIDAY, D., WATERLOW, J. C. and MILLWARD, D. J. (1980). Protein and amino acid turnover during exercise. *Biochem. Soc. Trans.*, **8**, 499–501.

RENNIE, M. J., HALLIDAY, D., DAVIES, C. T. M., EDWARDS, R. H. T., KRYWAWYCH, S., MILLWARD, D. J. and MATTHEWS, D. E. (1981). Exercise induced increase in leucine oxidation in man and the effect of glucose. In *Metabolism and Clinical Implications of Branched Chain Amino and Keto Acids*, Walser, M. and Williamson, J. R. (Eds.), Elsevier North Holland, Amsterdam.

REFSUM, H. W. and STRÖMME, S. B. (1974). Urea and creatinine production and excretion in urine during and after prolonged heavy exercise. *Scand. J. Clin. Lab. Invest.*, **33**, 247–54.

ROGERS, P. A., JONES, G. H. and FAULKNER, J. A. (1979). Protein synthesis in skeletal muscle following acute exhaustive exercise. *Muscle & Nerve*, **2**, 250–6.

SALTIN, B., HENRIKKSEN, J., NYGAARD, E. and ANDERSEN, P. (1977). Fiber types and metabolic potentialities of skeletal muscles in sedentary man and endurance runners. *Ann. N.Y. Acad. Sci.*, **301**, 3–29.

TAEGTMEYER, H. and LESCH, M. (1980). Altered protein and amino acid metabolism in myocardial hypoxia and ischaemia. In *Degradative Processes in Heart and Skeletal Muscle*, Wildenthal, K. (Ed.), Elsevier North Holland, Amsterdam.

WAHREN, J., FELIG, P. and HAGENFELDT, L. (1976). Effect of protein on splanchnic and leg metabolism in normal man and in patients with diabetes mellitus. *J. Clin. Invest.*, **57**, 987–94.

# 41

# Turnover in Pathological States

M. B. CLAGUE

*Royal Victoria Infirmary,
Newcastle upon Tyne, UK*

## Introduction

Although increased losses of urinary nitrogen in man during disease (Vogel, 1854) and following trauma (Malcolm, 1893) had been documented for a long time, the implication of this loss for protein metabolism was not fully realized until half a century ago. From simple observations and careful analyses on injured individuals, Cuthbertson and his colleagues were able to characterize this nitrogen loss. They found that the metabolic response to injury appeared to be subdivided into two phases, the second or 'flow phase' being associated with increased metabolism and a net loss of nitrogen from the body (Cuthbertson, 1942). This loss appeared to be related to the severity of injury (Cuthbertson, 1932), the age of the patient (Cuthbertson, 1932), the pre-injury nutritional status of the individual (Calloway *et al.*, 1955) and nutritional intake following trauma (Cuthbertson, 1936). Administration of extracts of endocrine glands (Cuthbertson *et al.*, 1941) or hormonal responses (Johnston, 1967) could also modify the nitrogen loss, which was thought to arise from increased muscle protein catabolism (Cuthbertson, 1930).

Proteins are continually being synthesized and degraded so that a negative nitrogen balance can be achieved by reducing the rate of protein synthesis, by increasing breakdown, or by a combination of both these mechanisms. By whatever means the negative nitrogen balance is attained in disease or following trauma, whether accidental or surgical, results of protein turnover studies must provide an explanation for the above observations.

## Protein Turnover Following Trauma

Three groups of research workers, quite independently and using two different techniques, have demonstrated reduced whole body protein synthesis with no change in protein breakdown in man following elective surgery (O'Keefe *et al.*, 1974*a*; Crane *et al.*, 1977; Kien *et al.*, 1978*a*) (see Table 1). However, there was a concomitant reduction in nutritional intake post-operatively in all groups of patients.

TABLE 1

PROTEIN SYNTHESIS (S) AND BREAKDOWN (B) FOLLOWING ELECTIVE SURGERY AS MEASURED BY THREE GROUPS OF WORKERS

| | $I^a$ | | $II^b$ | | $III^c$ | |
|---|---|---|---|---|---|---|
| | S | B | S | B | S | B |
| | (g/day) | | (g/kg/day) | | (g/kg/day) | |
| Pre-operative | 276 | 299 | $3·83 \pm 0·73$ | $3·88 \pm 0·66$ | $3·9 \pm 0·9$ | $3·4 \pm 0·7$ |
| Post-operative | 242 | 302 | $2·94 \pm 0·83$ | $3·66 \pm 0·65$ | $3·3 \pm 1·1$ | $3·3 \pm 1·1$ |
| Difference | −34 | +3 | $−0·88 \pm 0·90$ | $−0·17 \pm 0·59$ | −0·6 | −0·1 |
| | $P < 0·001$ | NS | $P = 0·01$ | NS | $P < 0·05$ | NS |

Means ± SD
[a] O'Keefe *et al.* (1974*a*). Patients fasted post-operatively.
[b] Crane *et al.* (1977). Patients with reduced intake post-operatively.
[c] Kien *et al.* (1978*a*). Patients with reduced intake post-operatively.

Results of limited studies on injured animals have supported the concept of increased catabolism (Levenson and Watkin, 1959). Results of three studies in man (following burn injury in sepsis, and after major skeletal trauma) have also shown increased whole body protein breakdown (Long *et al.*, 1977; Kien *et al.*, 1978*b*; Birkhahn *et al.*, 1980) (see Table 2). Interestingly, protein synthesis also increased, but in these studies nutritional intake was either maintained or increased in the diseased or injured state. Kien and his colleagues have demonstrated reduced synthesis under one set of circumstances (post-operatively) and increased breakdown and synthesis under another (burn injury), which suggests that in some way both mechanisms must play a role.

This apparent disparity in the results described could be attributed to the specificity of the response to different disease states, or to the varying nutritional status and degree of trauma in each group of individuals studied, the two sets of results representing opposite ends of the same spectrum of response. The conflict could be resolved if patients could be

TABLE 2

PROTEIN SYNTHESIS (S) AND BREAKDOWN (B) IN SEPSIS OR FOLLOWING ACCIDENTAL
INJURY AS MEASURED BY THREE GROUPS OF WORKERS

| % burn surface area | $I^a$ S (g/kg/day) | B | | $II^b$ S (g/kg/day) | B | | $III^c$ S (g/kg/day) | B |
|---|---|---|---|---|---|---|---|---|
| 0 | 3·4 | 3·0 | Controls | 3·695 | 4·379 | Controls | 2·33 ± 0·11 | 2·67 ± 0·14 |
| 0·5–25 | 4·0 | 3·2 | | | | | | |
| 25–60 | 5·1 | 3·8 | Sepsis | 4·479 | 5·298 | Trauma | 3·49 ± 0·21 | 4·78 ± 0·36 |
| 60 | 7·1 | 6·3 | | | | | | |
| | | | Difference | +0·784 | +0·919 | Difference | +1·16 | +2·11 |
| | | | | | | | P.< 0·001 | P < 0·001 |

Means ± SEM
[a] Kien *et al.* (1978*b*). Intake increased following stress.
[b] Long *et al.* (1977). Intake maintained following stress.
[c] Birkhahn *et al.* (1980). Intake maintained following stress.

studied at different, but controlled, levels of nutritional intake, while undergoing varying but measurable surgical stress. The wide range of protein turnover in the normal population also makes it imperative that control studies are carried out in the same individual before surgery.

## Assessing the Severity of Injury

During a constant infusion of L-1-$^{14}$C-leucine some of the infused labelled amino acid becomes incorporated into plasma proteins. The amount so incorporated can be readily determined and was found to be fairly constant in pre-operative asymptomatic individuals even on varying dietary intake (Clague *et al.*, 1981). However, the value rose in the presence of active disease, benign or malignant, and post-operatively. Furthermore, the rise correlated with staging of colorectal neoplasia and was maximal in the early post-operative period and after more major surgical procedures. A universally applicable technique for assessing the degree of trauma or advancement of disease in surgical patients during turnover studies was therefore available.

## Effect on Protein Turnover of Post-operative Nutritional Intake

Protein turnover was determined in seven patients undergoing elective surgery. An initial study was undertaken pre-operatively whilst the patients

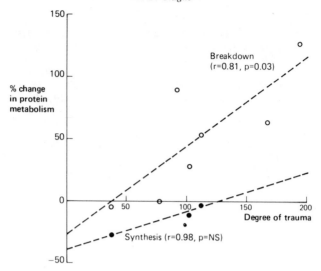

FIG. 1.   The percentage change in protein breakdown (○) and synthesis (●) post-operatively with the degree of trauma in patients fed pre-operatively but fasted post-operatively.

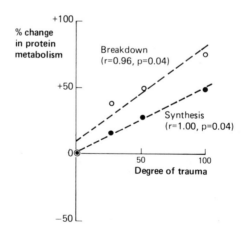

FIG. 2.   The percentage change in protein breakdown (○) and synthesis (●) post-operatively with the degree of trauma in patients in whom nutritional intake was maintained at the same level throughout.

were being fed, whereas the post-operative study was carried out with patients fasted. The results of the latter were calculated as the percentage change from the pre-operative value for each individual and plotted against the degree of trauma, calculated as above, and shown in Fig. 1.

Protein turnover was then determined in three patients with a similar protocol except that nutrition was maintained at the same level throughout. The results are shown in Fig. 2. A fourth point has been added—no change in synthesis or breakdown in the non-traumatized situation. The degree of trauma was limited in this series of studies as severely injured patients will not eat or cannot eat because of associated ileus.

## Concept of Protein Turnover and the Metabolic Response to Trauma

The results described here can be combined with the results of others to advance a concept of protein turnover in the metabolic response to trauma under differing circumstances and, as will be seen, explain away the apparent disparity in the results obtained by previous workers. The concept is illustrated in Fig. 3.

Figure 3(b) illustrates the response when nutritional intake is kept the same. This is derived from the results in the second series of patients (Fig. 2) and supported by the results in trauma (see Table 2). Both protein synthesis and breakdown rise along with the degree of trauma; breakdown rises more rapidly than synthesis, thus accounting for the increased negative nitrogen balance with increasing severity of injury (Cuthbertson, 1932).

Figure 3(a) illustrates the response when the nutritional intake is reduced post-operatively. This is derived basically from the results in the first group of patients (Fig. 1). Compared with the previous situation (Fig. 3(b)), the increased protein loss in a patient who is fasted post-operatively is brought about by a marked suppression in protein synthesis, although breakdown may also fall slightly. The same response to fasting has been demonstrated in normal individuals, with a marked reduction in synthesis and only a small fall in breakdown (Garlick *et al.*, 1978).

Examination of the results shown in Fig. 1 shows that there is a point where breakdown remains unaltered but synthesis is reduced (i.e. degree of trauma of 40). The results for patients studied by previous workers following elective surgery, with reduced nutritional intake post-operatively (see Table 1), could be scattered about this point, the mean results suggesting that the principal component in the response to trauma is one of

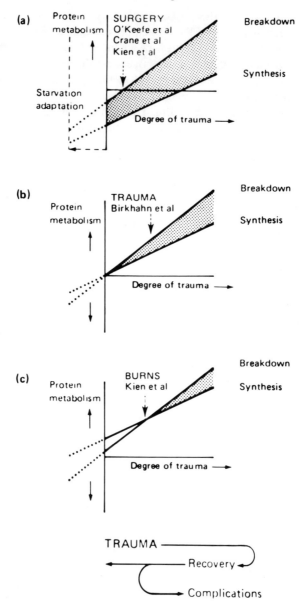

FIG. 3. Diagramatic representation of the concept of protein turnover and the metabolic response to trauma. (a) Fasted state, (b) identical nutritional intake, (c) increased nutritional intake.

reduced synthesis. Close examination of the published data from these workers, in fact, shows that protein breakdown is increased in about half their patients following surgery. It is not possible in retrospect to assess the degree of trauma in these patients but a good correlation is obtained when changes in protein breakdown in individual patients post-operatively are plotted against their changes in synthesis. Such a correlation, shown in Fig. 4, would be expected if individual patients were undergoing differing

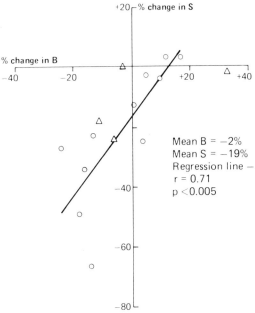

FIG. 4. Graph to show the correlation between changes in protein breakdown (B) and synthesis (S) in two groups of post-operative fasted patients studied previously. ◯, data of Crane *et al.*, 1977; △, O'Keefe *et al.*, 1974*a*.

degrees of trauma, scattered around the above suggested point, at the time of their post-operative study protein synthesis only increasing in those patients with markedly increased breakdown rates. The gradient and intercept might differ from ours as the majority of those patients previously studied did receive some nutritional intake post-operatively. That the reduced synthesis in this group of patients is largely due to the restricted intake is also supported by the finding that synthesis is restored towards normal on intravenous infusion of amino acids post-operatively (O'Keefe *et al.*, 1974*b*).

At this point one can speculate regarding the effects of prolonged reduced intake and aging on the response, since nitrogen losses are reduced under both circumstances (Cuthbertson, 1932; Calloway *et al.*, 1955). Reduced synthesis and breakdown have been demonstrated in the elderly (Golden and Waterlow, 1977) and those on a reduced intake for several weeks (Sender *et al.*, 1975). This would result in a shift of the zero degree of trauma to the left (see Fig. 3(a)). Trauma will then cause a move to the right, depending on the degree of trauma, but as the patient started further to the left than normal he will only reach a position where protein breakdown and synthesis are not as divergent as if he had started in the normal place, and so he will excrete less nitrogen than the normally fasted or younger individual undergoing the same stress.

Figure 3(c) depicts the anticipated response when patients are given an increased intake post-operatively. The nitrogen loss is reduced as synthesis is increased towards the elevated breakdown rate. Breakdown is largely related to the severity of injury, whereas synthesis responds to substrate availability. This is supported by the results following burn injury (see Table 2; Kien *et al.*, 1978*b*).

The concept advocated here of changes in protein turnover in the metabolic response to trauma is common to most forms of stress (e.g. sepsis, accidental injury or elective surgery), differences arising only from the varying degree of trauma and nutritional status, age and possibly sex of the patients. Small differences will probably exist between conditions arising out of the differing amino acid profiles of tissues affected and the specific effects of certain conditions (e.g. thermal losses and energy requirements following burn injury), but our present techniques are not sufficiently refined to define them.

## Specific Responses

The general metabolic response is altered or modified in certain specific conditions in man.

### Endocrine Disorders

Protein synthesis and breakdown are controlled at tissue and cellular level by several hormones (Waterlow *et al.*, 1978), so that it is not surprising to discover that endocrine disorders can give rise to disordered protein metabolism. Myxoedema or hypothyroidism has been shown to be associated with reduced synthesis which was restored with supplementation by tri-iodothyronine (Crispell *et al.*, 1956). Protein synthesis was elevated in acromegaly but returned to normal following hypophysectomy (Haak *et al.*, 1962). Diabetes mellitus caused reduction in both protein turnover and synthesis, which were restored to normal by insulin treatment (Waterlow *et al.*, 1977), although a more recent study has failed to confirm this (personal communication). These findings are in agreement with the known action of the hormones *in vitro*.

## Neoplasia

Neoplastic conditions are often associated with marked cachexia, particularly in the advanced stages of the disease, so that one would expect to find disparity between the rates of synthesis and breakdown with advancement of disease. This, however, we were unable to demonstrate, synthesis and breakdown increasing in parallel (see Fig. 5; Carmichael *et al.*, 1980), but rates of synthesis and breakdown were reduced in anorectic patients.

The explanation for the lack of crossover and divergence of the lines for synthesis and breakdown could lie in the fact that the result is a summation of host and tumour metabolism. The enlarging tumour, with its nitrogen trapping, cancels out the nitrogen loss in the host tissues if the tumour/host conglomerate is considered as a single entity.

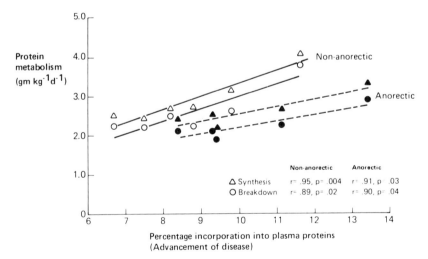

FIG. 5. The relationship between rates of protein synthesis and breakdown and advancement of disease in anorectic (▲, ●) and non-anorectic (△, ○) patients with colorectal neoplasia.

## Conclusion

Techniques for measuring whole body protein turnover in man in pathological conditions have been of value for research purposes. Further development and refinement of the available techniques could lead to even greater understanding of protein metabolism in disease in the intact human. A technique is already available for simultaneous measurement with a stable isotope of specific tissue and whole body protein metabolism (Halliday and McKeran, 1975). Such methodology could shed light on the problems of tissue specific activity and responses of different tissues (e.g. muscle, rectal tumours) in pathological states. At present the latter can only be assessed to a limited degree by employing separate techniques.

Ultimately it is to be hoped that a safe, reliable and rapid test would become available that would enable the clinician to monitor protein metabolism in the critically ill and assess its response to treatment.

## Discussion

*Young (Massachusetts Institute of Technology, USA):* I think you implied that, irrespective of the adequacy of the protein intake, the amount of radioactivity incorporated into plasma proteins was constant before operation. Surely it is not reasonable to suppose that there was no effect of the level of protein intake on plasma protein synthesis?

*Garrow (Clinical Research Centre, Harrow, UK):* In very early experiments at the TMRU in 1954, when I injected $S^{35}$-methionine into malnourished children, I was surprised to find that not only was the label incorporated, but more of it was incorporated into plasma proteins in malnourished than in recovered children (Garrow, 1957). Allison's protein-depleted dogs behaved in exactly the same way; when they were injected with a pulse label of $S^{35}$-methionine, the more depleted the dog the more label went into plasma protein for reasons that now become evident (Garrow, 1959). I think it is very interesting that Clague can use the incorporation of a label into plasma protein as an index of trauma, which obviously correlates with the seriousness of the operation or the extent of the tumour. Whether the trauma is having the same effect as protein depletion or whether it is in fact protein depletion, it is clear that in these conditions plasma protein has preference for synthesis over other tissues.

*Clague:* The incorporation into plasma proteins was higher on the first post-operative day than on the second. The patients were fasted continually throughout that period, and if the effect had been due to protein depletion I would have thought there would be a higher value on the second day than on the first.

*Cohen (University of Wisconsin-Madison, USA):* I am not clear whether the incorporation of activity was specifically into the albumin fraction.

*Clague:* I have tried over 2 years to find out where the radioactivity is going and have not yet found the answer. I thought at first it was going into the albumin, but later I found a lot in the fibrinogen fraction.

*Cohen:* As you know, albumin has a great affinity for free fatty acids and binds them very firmly. I wonder whether the α-keto analogue of leucine would also bind to albumin and whether you might have radioactivity bound to proteins rather than incorporated.

*Clague:* I tried mixing cold leucine and cold keto-leucine and then diffused out the diffusible fraction, but I still got the same answer. So it may be bound, but if so, it does not come off.

*Garrow:* I think we can tell Cohen that the radioactivity is in fact in albumin. Initially, when we observed in malnourished children a greater incorporation of radioactivity into plasma protein, we thought it must be because the children were infected and were producing γ-globulins. However, fractionation showed that it was in the albumin and not anything to do with binding of lipid, because it also happened with methionine. The experiments in which Waterlow (1959) injected pulse labels of amino acids into rats demonstrated that in a protein-depleted rat more of the activity goes into the viscera and less into the carcass than in control rats, and the plasma more closely reflects visceral activity than carcass activity.

*Waterlow (London School of Hygiene and Tropical Medicine, UK):* Thank you for drawing attention to those experiments. I think there is some evidence from injured rats that something similar happens; the synthesis rate or fractional synthesis rate of total mixed liver protein is said to be increased in rats which have been burnt or suffered from fractured femurs (Levenson *et al.*, 1976).

*Garlick (London School of Hygiene and Tropical Medicine, UK):* Pain and I recently injected some mice which had tumours with labelled phenylalanine and measured rates of protein synthesis in liver and muscle (Pain and Garlick, 1980). Not only was the muscle synthesis rate decreased by 44% as in malnutrition, but the liver synthesis rate was increased by 40%. That would give more incorporation into plasma proteins, assuming that their production was not selectively falling in liver, because the isotope that was not being incorporated into muscle would then be incorporated into liver and plasma proteins.

*Clague:* The effect I find appears to be independent of nutritional intake. If an individual is fed, synthesis of plasma protein and also of muscle will be increased. My simplistic view is that there are two components at work: one is something that is directing synthesis more towards the liver than towards the periphery; the second is the absolute value which depends on the nutritional intake.

*Garlick:* That is what I was trying to say: that nutritional factors tend to cause all tissues to change in the same direction. For example, starvation and protein deprivation in the rat result in a fall in protein synthesis in both muscle and liver (Garlick *et al.*, 1975; McNurlan and Garlick, 1979). In the mice with tumours, there was a fall in synthesis in muscle but a rise in liver, which could result in a more pronounced redistribution of isotope in favour of liver synthesized proteins.

*Waterlow:* I do not altogether agree with what Garlick has said, because, as Garrow pointed out, in a depleted rat there is a change in the distribution of protein synthesis.

*Stein (University of Pennsylvania School of Medicine, Philadelphia, USA):* I think that in rats one can go to a much more extreme situation: one can feed the rat, pre-impose trauma and then starve it. If you starve the rat pre- and post-trauma, protein synthesis in muscle is reduced by 50% pre-trauma and is virtually negligible post-trauma. The amount of isotope incorporated into muscle protein was not detectable in the system we used. On the other hand if the animal is fed adequately, there is an increase in synthesis.

*Munro (Tufts University, Boston, USA):* Do you think it is possible that the increment in activity incorporated into plasma protein represents stimulation by corticosteroids?

*Clague:* I think that may be so, and that the effect is probably related to the hormonal pattern. I do not know whether it is corticosteroids that are the cause; I have not measured any hormone levels in these patients. Does anyone know what are the hormonal changes in patients with tumours?

*Munro:* In animal tumours there is an increase in adrenocortical size and probably increased activity also (Goodlad, 1964).

*Picou* (*Mount Hope Medical Complex Task Force, Port of Spain, Trinidad*): I am very glad that the surgeons are entering this field; they should be encouraged. I should like to make a reservation about these studies: even if the incorporation of radioactivity into the plasma protein represented the synthesis of plasma protein *per se*, which it may not, it would not necessarily reflect what is happening in muscle or total body protein. There is no alternative but to do this type of study, which I hope is becoming easier, on the sort of model that we developed and used in Oxford (Crane *et al.*, 1977). In your study the patients were not in negative nitrogen balance. Our patients underwent rather severe elective operations where large negative nitrogen balances were to be expected. We wanted to show that if you fed them post-operatively, the negative nitrogen balance would be reduced if not eliminated, and although it was not easy to persuade patients to eat much in these circumstances, we were able to suppress the negative balance.

## References

BIRKHAHN, R. H., LONG, C. L., FITKIN, D., GEIGER, J. W. and BLAKEMORE, W. S. (1980). Effects of major skeletal trauma on whole body protein turnover in man measured by 1-(1-$^{14}$C) leucine, *Surgery*, **88**, 294–9.

CALLOWAY, D. H., GROSSMAN, M. I., BOWMAN, J. and CALHOUN, W. K. (1955). The effect of previous level of protein feeding on wound healing and the metabolic response to injury, *Surgery*, **37**, 935–46.

CARMICHAEL, M. J., CLAGUE, M. B., KIER, M. J. and JOHNSTON, I. D. A. (1980). Whole body protein turnover, synthesis and breakdown in patients with colorectal carcinoma, *Brit. J. Surg.*, **67**, 736–9.

CLAGUE, M. B., CARMICHAEL, M. J., KIER, M. J., WRIGHT, P. D. and JOHNSTON, I. D. A. (1981). Increased incorporation of an infused labelled amino acid into plasma proteins as a means of assessing the severity of injury in surgical patients, *Ann. Surg.* (in press).

CRANE, C. W., PICOU, D., SMITH, R. and WATERLOW, J. C. (1977). Protein turnover in patients before and after elective orthopaedic operations, *Brit. J. Surg.*, **64**, 129–33.

CRISPELL, K. R., PARSON, W. and HOLLIFIELD, G. (1956). A study of the rate of protein synthesis before and during administration of l-tri-iodothyronine to patients with myxoedema and healthy volunteers using N$^{15}$ glycine, *J. Clin. Invest.*, **35**, 164–9.

CUTHBERTSON, D. P. (1930). The disturbance of metabolism produced by bony and non-bony injury, with notes on certain abnormal conditions of bone, *Biochem. J.*, **24**, 1244–63.

CUTHBERTSON, D. P. (1932). Observations on the disturbance of metabolism produced by injury to the limbs, *Quart. J. Med.*, **1**, 233–46.

CUTHBERTSON, D. P. (1936). Further observations on the disturbance of metabolism caused by injury, with particular reference to the dietary requirements of fracture cases, *Brit. J. Surg.*, **23**, 505–20.

CUTHBERTSON, D. P. (1942). Post-shock metabolic response, *Lancet*, **i**, 433–7.

CUTHBERTSON, D. P., SHAW, G. B. and YOUNG, F. G. (1941). The anterior pituitary gland and protein metabolism. II. The influence of anterior pituitary extract on the metabolic response of the rat to injury, *J. Endocrinol.*, **2**, 468–74.

GARLICK, P. J., MILLWARD, D. J., JAMES, W. P. T. and WATERLOW, J. C. (1975). The effect of protein depletion and starvation on the rate of protein synthesis in tissues of the rat, *Biochim. Biophys. Acta*, **414**, 71–84.

GARLICK, P. J., CLUGSTON, G. A., SWICK, R. W., MEINERTZHAGEN, I. H. and WATERLOW, J. C. (1978). Diurnal variations in protein metabolism in man, *Proc. Nutr. Soc.*, **37**, 33A.

GARROW, J. S. (1957). $S^{35}$ methionine uptake in protein-depleted Jamaican children, in *Amino-acid Malnutrition. XIIIth Annual Protein Conference* (Ed. W. Cole) Rutgers University Press, New Brunswick, p. 14.

GARROW, J. S. (1959). The effect of protein depletion on the distribution of protein synthesis in the dog, *J. Clin. Invest.*, **38**, 1241–50.

GOLDEN, M. H. N. and WATERLOW, J. C. (1977). Total protein synthesis in elderly people: a comparison of results with ($^{15}$N) glycine and ($^{14}$C) leucine, *Clin. Sci. Mol. Med.*, **53**, 277–88.

GOODLAD, G. A. J. (1964). Protein metabolism and tumor growth, in *Mammalian Protein Metabolism*, Vol. II (Eds. H. N. Munro and J. B. Allison), Academic Press, New York, pp. 415–44.

HAAK, A., KASSENAAR, A. A. H. and QUERIDO, A. (1962). In *Protein Metabolism: Influence of Growth Hormone, Anabolic Steroids and Nutrition in Health and Disease* (Ed. F. Gross), Springer-Verlag, Berlin, p. 150.

HALLIDAY, E. and McKERAN, R. O. (1975). Measurement of muscle protein synthetic rate from serial muscle biopsies and total body protein turnover in man by continuous intravenous infusion of 1-($\alpha$-$^{15}$N) lysine, *Clin. Sci. Mol. Med.*, **49**, 581–90.

JOHNSTON, I. D. A. (1967). The role of the endocrine glands in the metabolic response to operation, *Brit. J. Surg.*, **54**, 438–41.

KIEN, C. L., YOUNG, V. R., ROHRBAUGH, D. K. and BURKE, J. E. (1978a). Increased rates of whole body protein synthesis and breakdown in children recovering from burns, *Ann. Surg.*, **187**, 383–91.

KIEN, C. L., YOUNG, V. R., ROHRBAUGH, D. K. and BURKE, J. F. (1978b). Whole body protein synthesis and breakdown rates in children before and after reconstructive surgery of the skin, *Metab.*, **27**, 27–34.

LEVENSON, S. M. and WATKIN, D. M. (1959). Protein requirements in injury and certain acute and chronic diseases, *Fed. Proc.*, **18**, 1155–90.

LEVENSON, S. M., CROWLEY, L. V. and SEIFTER, E. (1976). Effects of injury on wound healing and wound infection, in *Metabolism and the Response to Injury* (Eds. A. W. Wilkinson and D. P. Cuthbertson) Pitman Medical, Tunbridge Wells.

LONG, C. L., JEEVANANDAM, M., KIM, B. M. and KINNEY, J. M. (1977). Whole body protein synthesis and catabolism in septic man, *Am. J. Clin. Nutr.*, **30**, 1349–52.

MALCOLM, J. D. (1893). *The Physiology of Death from Traumatic Fever*, Churchill, London.

McNURLAN, M. A. and GARLICK, P. J. (1979). Rates of protein synthesis in rat liver and small intestine in protein deprivation and diabetes, *Proc. Nutr. Soc.*, **38**, 133A.

O'KEEFE, S. J. P., SENDER, P. M. and JAMES, W. P. T. (1974a). Catabolic loss of body nitrogen in response to surgery, *Lancet*, **ii**, 1035–8.

O'KEEFE, S. J. D., SENDER, P. M., CLARK, G. C. and JAMES, W. P. T. (1974b). The effect of varying the postoperative regime on the dynamics of protein metabolism following operative trauma, in *Proceedings of the Congres International de Nutrition Parenterale*, Montpellier, France, pp. 333–9.

PAIN, V. M. and GARLICK, P. J. (1980). The effect of an Ehrlich ascites tumour on the rate of protein synthesis in muscle and liver of the host, *Biochem. Soc. Trans.*, **8**, 354.

SENDER, P. M., JAMES, W. P. T. and GARLICK, P. J. (1975). Protein metabolism in obesity, in *Regulation of Energy Balance in Man* (Ed. E. Jequier), Editions Medicine et Hygiene, Geneva, pp. 224–7.

VOGEL (1854). Quoted by D. P. Cuthbertson (1979) in *Symposium on Nutrition and Surgery*, British Nutrition Society, Glasgow.

WATERLOW, J. C. (1959). Effect of protein depletion on the distribution of protein synthesis, *Nature*, **184**, 1875–6.

WATERLOW, J. C., GARLICK, P. J., ELL, S. and REEDS, P. J. (1977). Some results obtained by measuring protein turnover with ($^{15}$N) glycine as tracer and urinary ammonia as end-product, *Clin. Sci. Mol. Med.*, **52**, 17P.

WATERLOW, J. C., GARLICK, P. J. and MILLWARD, D. J. (1978). *Protein Turnover in Mammalian Tissues and in the Whole Body*, Elsevier North-Holland, Amsterdam, pp. 625–746.

# 42

# Protein Turnover in Infection: Studies with $^{15}$N

J. C. WATERLOW and A. M. TOMKINS

*Department of Human Nutrition,*
*London School of Hygiene and Tropical Medicine,*
*London, UK*

Several studies have been published in which $^{15}$N-labelled amino acids have been used to measure protein turnover in injury and infection. Crane *et al.* (1977), using $^{15}$N-glycine with urea as end product, found a small decrease in synthesis with no change in breakdown in patients after elective orthopaedic operations. The injury, as judged by the degree of negative nitrogen balance, was probably not very great. Bilmazes *et al.* (1978) by the same method found an increase in whole body protein breakdown in children with burns, with a parallel increase in the excretion of $N^r$-methylhistidine. These patients were on extremely high protein intakes. Long *et al.* (1977) found an increase of 20% in both synthesis and breakdown rates in three septic patients compared with two normal subjects. The rates were determined by compartmental analysis of $^{15}$N excretion in both urea and ammonia after a single dose of $^{15}$N-alanine while the subjects were on a nitrogen-free diet. In general, these results seem to fit reasonably well with those obtained by Clague (see p. 530) with $^{14}$C-leucine, the extent and duration of the changes in synthesis and breakdown depending both on the severity of the injury and on the dietary intake.

We have been pursuing this problem by the method which, as Garlick (p. 357) has pointed out, is probably the most suitable for general clinical use, because it is both simple and quick. With $^{15}$N-glycine given in a single dose and urinary ammonia as end product, the test is completed in 9 h.

Table 1 shows the effect of vaccination, which produces a mild fever, in healthy volunteers. If the results 10 days after vaccination are taken as 'normal', the vaccination produced an increase of 37% in the rate of synthesis and 55% in the rate of breakdown. Synthesis is greater than breakdown because these measurements were made during the day, in the fed state.

TABLE 1

EFFECT OF VACCINATION ON PROTEIN TURNOVER

| Time | Synthesis | Breakdown |
|------|-----------|-----------|
|      | (g protein/9 h) | |
| Day of vaccination | 81 | 62 |
| 3 days later | 66[a] | 47 |
| 10 days later | 59[a] | 40[b] |

Difference from mean on day of vaccination:
[a] $P < 0.05$ [b] $P < 0.01$.

Table 2 shows results obtained in infected and malnourished children in a rural hospital in Northern Nigeria. The malnourished children who were not infected (group E) had rates of synthesis and breakdown which were rather lower than the value of about 1·5 g protein/kg/9 h found in similarly malnourished children in Jamaica (Golden *et al.*, 1977). In all the infected groups the turnover rates were greatly increased. Comparison of groups C and D shows the effect of malnutrition in counteracting this increase, again in agreement with Clague's findings.

The high rates of protein turnover were accompanied by increases in creatinine and methylhistidine output, particularly in the children with measles (group A). Creatine in the urine could not be measured, but it is probable that much of what was measured as creatinine was in fact excreted as creatine.

One of the problems which arises in studies of this kind is that the infection may produce an acidosis with increased output of ammonia. This would invalidate the estimates of flux if there is a change in the precursors of ammonia. In the vaccination experiment there was no difference in ammonia output between the vaccination and post-vaccination tests. In the Nigerian study, however, there is a suggestion of increased ammonia output in the infected children. We have therefore undertaken some preliminary studies in healthy volunteers of the effects of acidosis produced by calcium chloride.

Table 3 shows that on average the acidosis produced a decrease in the estimate of flux of the order of 15–20 %. In this small series the change is not statistically significant. Moreover, the direction of the change is opposite to that which one would expect from the argument that the high turnover rates in the infected children are an artefact caused by acidosis. Admittedly the experiments on volunteers were acute. So far one experiment has been

TABLE 2

EFFECTS OF INFECTION AND MALNUTRITION ON PROTEIN TURNOVER IN CHILDREN IN NORTHERN NIGERIA

| Clinical group | No. of children | Synthesis (g protein/kg/9h) | Breakdown (g protein/kg/9h) | $NH_3$ ($mg.N/kg/9h$) | Excretion Creatinine ($\mu mol/kg/9h$) | Methylhistidine ($\mu mol/kg/9h$) |
|---|---|---|---|---|---|---|
| A. Acute measles | 6 | 3·57 | 4·66 | 18·1 | 245 | 5·25 |
| B. Early convalescent measles[a] | 6 | 2·97 | 3·37 | 14·7 | 94 | 2·5 |
| C. Mild PEM with infection | 6 | 4·19 | 4·55 | 21·6 | 62 | 2·0 |
| D. Severe PEM with infection | 5 | 2·47 | 2·40 | 14·6 | 42 | 1·6 |
| E. Severe PEM, no infection | 4 | 1·04 | 0·89 | 9·6 | 54 | 0·9 |

[a] Same cases as Group A.

TABLE 3
EFFECT OF CALCIUM CHLORIDE ACIDOSIS ON FLUX MEASURED
WITH $^{15}$N-GLYCINE AND URINARY $NH_4^+$ AS END PRODUCT
(MEAN OF 5 SUBJECTS)

|  | Control | Acidotic |
|---|---|---|
| $NH_4^+$ excretion (mmol/h) | 1·72 | 3·24 |
| Flux (g N/h) | 2·27 | 1·92 |
| % of dose of $^{15}$N excreted in $NH_4^+$ in 9 h | 1·16 | 2·38 |

performed in which a metabolic acidosis was produced over a period of a week. This produced an apparent increase in flux of 20%—a change which is trivial compared with the very large differences between the different groups of Nigerian children.

These findings are preliminary, but from the results so far we conclude that the simple method of measuring protein turnover with ammonia as end product is capable of giving informative results.

# References

BILMAZES, C., KIEN, C. L., ROHRBAUGH, D. K., UAUY, R., BURKE, J. F., MUNRO, H. N. and YOUNG, V. R. (1978). Assessment of the quantitative contribution by skeletal muscle to elevated rates of whole body protein breakdown in burned children as measured by 3-methylhistidine output, *Metabolism*, **27**, 671–6.

CRANE, C. W., PICOU, D., SMITH, R. and WATERLOW, J. C. (1977). Protein turnover in patients before and after elective orthopaedic operations, *Brit. J. Surg.*, **64**, 129–33.

GOLDEN, M. H. N., WATERLOW, J. C. and PICOU, D. (1977). Protein turnover, synthesis and breakdown before and after recovery from protein-energy malnutrition, *Clin. Sci. Mol. Med.*, **53**, 473–7.

LONG, C. L., JEEVANANDAM, M., KIM, B. M. and KINNEY, J. M. (1977). Whole body protein synthesis and catabolism in septic man, *Amer. J. Clin. Nutr.*, **30**, 1340–4.

# Final Remarks

A. NEUBERGER

*Charing Cross Hospital Medical School,
London, UK*

When you asked me to sum up, I had the firm intention of making detailed notes and giving you a survey of what had been said. But I do not think now that this would be the most useful thing I could do. If you will allow me, I shall give instead a general survey of where I believe we stand and how this meeting may have advanced our general knowledge.

Nowadays, a biologist who is interested in what is happening in man, or eukaryotic systems generally, may be a biochemist or an enzymologist who works with subcellular particles, such as mitochondria, identifying enzymes, their activators or inhibitors. He thus acquires a lot of knowledge which is accurate, which can be easily quantified, and which involves a minimum of arbitrary assumptions. However, when he tries to relate this knowledge to what is happening in the whole organism, he is in a very difficult position. He might, for example, assume that the enzyme which he measures *in vitro* has the same activity *in vivo*, and this assumption may be wrong. The enzymologist studies an enzyme with a certain conformation, but when this enzyme is membrane-bound it may have a different conformation, and may, for instance, have greater or lesser affinity for substrate. Therefore, without doing any work on the whole animal, he finds it almost impossible to relate his findings with any certainty to what happens in the whole organism. On the other hand, the whole animal approach of the physiologist, which was popular with biochemists 30–40 years ago, has become less favoured among the biological community. However I am sure that this whole body approach is important, and we probably cannot wait until some indeterminate time when the observations of the enzymologist, the molecular biologist, or other people working with well-defined subcellular particles or molecules can be integrated into a complete understanding of what is happening in the whole organism.

By having this meeting, we are giving support to the whcle animal approach that has been used by the great majority of people who are present here, combining it, if appropriate, with investigations on isolated systems. One special problem which concerns us is the nitrogen metabolism in the whole organism, how it is regulated, and how it varies under conditions such as disease. The use of radio-isotopes and stable isotopes is still the basis of most methods used, and the improvements in such techniques, which have been described, for measuring isotopes in very small quantities are impressive.

We have also heard a lot about isolated systems. For instance, the information we have on plasma proteins is likely to have given us accurate values, especially those obtained by the use of reliable iodination techniques in which oxidation and denaturation are eliminated. Not all the problems have been solved, and James has indicated some of the uncertainties in this field. We also know a lot about the metabolism of haemoglobin largely arising from the work done by Shemin and Rittenberg, which showed that this particular protein has a fixed life span. There has been some discussion whether fixed life span kinetics may apply to other structures in the body. Waterlow and his colleagues have given good evidence that it does not apply to the structural proteins of skeletal muscle, contrary to an assumption which had been put forward earlier. However, I do not think one should completely exclude this possibility for other systems. It is quite possible that some structural components of mitochondria or peroxysomes, for example, may show fixed life span characteristics.

Apart from the use of isotopes, we have other techniques to measure the turnover of proteins. Thus, it has been clearly shown that the muscle components actin and myosin contain methylhistidine, that methylation is post-translational, that this modified amino acid is not re-incorporated into proteins but is excreted. Thus, the excretion of methylhistidine can be used as a measure of the catabolism of muscle proteins. Similar methods could in principle be used to measure the turnover of other proteins containing amino acids which are modified after translation, are not re-utilized, and are specifically associated with such proteins, e.g. hydroxyproline and hydroxylysine. Peptides containing such modified amino acids appear in the urine, but these amino acids are also degraded by defined metabolic pathways. It is thus at present uncertain whether the urinary excretion of peptides containing such amino acids can be quantitatively related to the catabolism of proteins such as collagen, elastin, and the component of complement, Clq.

Work with isotopes and other methods has clearly shown that an organ such as the liver contains proteins with widely differing turnover rates. Thus, we have information on enzymes occurring in this organ with a very short life, such as tryptophan pyrrolase, $\delta$-aminolae vulinic acid synthetase, or ornithine decarboxylase, which have half-lives of two hours or less. On the other hand, there appear to be some fractions of collagen in the liver which appear to have an average life span of several weeks. A similar situation is likely to apply in almost all tissues in the body.

Proteins which turn over very slowly are the components of connective tissue, collagen and elastin. However, even here there are marked differences, and the study of the rates of metabolism of different collagen fractions may have great physiological significance, an example being the turnover of collagen associated with the involution of the uterus.

The present symposium has shown that the knowledge which has been acquired by the various groups over the last 10 years has been very important. It may be worthwhile, however, to examine carefully the assumptions which underlie such work, and which perhaps are not stated quite as often as one might wish. For instance, in assessing the turnover of nitrogenous compounds in the whole organism, we have generally neglected the peptides which are removed during the maturation of those proteins which are excreted from the cell. Most of these proteins have either an N-terminal or C-terminal part of the chain, which are split off and are hydrolysed to free amino acids, these becoming part of the amino acid pool. We cannot be certain as yet whether this phenomenon is confined to excreted proteins, or whether structural proteins also suffer some proteolytic change before they become incorporated into structural components. It is also quite possible that a certain amount of degradation is unavoidably associated with the biosynthesis of some or most proteins. With collagen, for example, an increase in synthesis seems to be almost always associated with the excretion from the cell of peptides containing hydroxyproline residues. In other words, it is quite possible that a considerable amount of degradation is associated with protein synthesis, and thus most of the methods which we employ at present would tend to underestimate the number of peptide bonds formed by an animal or a cell in a given period of time. This becomes particularly important if we want to get an assessment of the energy cost of protein synthesis.

Another aspect which has worried all the workers who have been active in this field is the assumption that there is complete equilibration between the various fractions forming the free amino acid pool, part of which delivers the building stones for protein synthesis. Quite often the assumption has been made that the specific radioactivity of the amino acids present in plasma gives a quantitative indication of the specific activity of the amino acid used for the biosynthesis of a particular protein. In skin or muscle, which have a relatively poor rate of blood flow, and where there may also be a large permeability barrier between plasma and cell, equilibration between serum and the tissue concerned is slow, and thus the specific activity of the intracellular amino acid is used. We may also not be fully justified in basing our calculations on the specific radio-activity of the amino acid present inside the cell. It is becoming increasingly clear that the cytosol is not homogeneous, but is most probably structured, and thus there may not be complete equilibration between different regions of the cytosol, apart from possible lack of equilibration between the cytosol and structural elements of the cell. Some research workers have tried to overcome these difficulties by basing their calculations on the specific activities of amino acid residues associated with tRNA. This to a large extent eliminates many of the problems, but the technique is slow and laborious.

These general comments are not meant to be destructive. The main purpose of making them is to encourage workers to eliminate as far as possible the uncertainties stated, and to look for alternative methods of measuring protein synthesis and degradation. I am sure in time it will also be possible to relate observations made on the whole animal with, on the one hand, knowledge which is being acquired to an increasing extent on the regulation of protein synthesis in isolated subcellular systems, and on the other hand with the large body of information which is being acquired on proteolytic enzymes in tissues. In any case, the observations which are now being made are of considerable value, in particular if we compare the effects of different physiological and pathological conditions on rates of protein synthesis. The absolute values obtained may not be completely correct, but the relative values are likely to yield a lot of useful information. The present meeting has provided the opportunity to assess the progress which has been made, and has shown that most workers in the field are conscious of the difficulty encountered with the remaining problems. Further work is likely to give us much more detailed and reliable knowledge than we possess at the present time.

# List of Participants

Professor G. A. O. Alleyne
Dr N. Anderson
Dr H. Betton
Dr D. M. Bier
M. B. Clague, Esq. FRCS
Professor P. P. Cohen
Professor H. Coore
Dr P. Felig
Dr L. Fowden FRS
Dr P. J. Garlick
Dr J. S. Garrow
Dr M. H. N. Golden
Dr D. Halliday
Professor A. E. Harper
Dr A. Jackson
F. Jahoor, Esq.
Dr W. P. T. James
Dr D. S. Kerr

Dr P. D. Klein
Dr P. Lund
Dr D. E. Matthews
Dr D. J. Millward
Professor H. N. Munro
Professor A. Neuberger FRS
Professor W. L. Nyhan
Professor D. Picou
Dr O. E. Pratt
Dr P. J. Reeds
Dr M. J. Rennie
Dr T. P. Stein
Dr J. M. L. Stephen
Professor R. W. Swick
Professor M. Walser
Professor J. C. Waterlow
Professor V. R. Young

# Index